Social and Economic Impact of Earth Sciences

Vineet K. Gahalaut · M. Rajeevan
Editors

Social and Economic Impact of Earth Sciences

Springer

Editors
Vineet K. Gahalaut
CSIR-National Geophysical Research Institute
Hyderabad, India

M. Rajeevan
Ministry of Earth Sciences
New Delhi, India

ISBN 978-981-19-6928-7 ISBN 978-981-19-6929-4 (eBook)
https://doi.org/10.1007/978-981-19-6929-4

This Springer imprint is published by the registered company Springer Nature Singapore Pte Ltd.
The registered company address is: 152 Beach Road, #21-01/04 Gateway East, Singapore 189721, Singapore

Foreword

Earth system science considers interactions between the earth's spheres, atmosphere, hydrosphere, cryosphere, lithosphere, and biosphere, as well as the impact of human societies on these components. It provides a physical basis for understanding the world in which we live and upon which humankind seeks to achieve sustainability.

Earth system science services in terms of weather, climate, ocean, and seismological services provide us with significant economic and societal benefits. Measures taken by the developed countries in natural hazard mitigation save more than six times the investment made on mitigation and research grants. In developing countries, like India, this ratio could be even higher. The best example in Indian scenario is tropical cyclones. The adverse impact due to tropical cyclones in terms of deaths and mitigation cost has reduced significantly during the recent years. This book has been designed to review and evaluate the economic and societal impacts of research and mitigation measures taken up by the Ministry of Earth Sciences (MoES), India. In this book, the societal and economic impacts of forecasts and warnings for weather, monsoon, air pollution and fog, ocean state monitoring for fisheries and other ecological services, technology development for desalination of seawater, deep ocean research and technology, beach restoration efforts, tsunami warning, landslide and earthquake monitoring, and seismic hazard microzonation have been assessed.

Weather/climate forecasts and warnings are used extensively by various user sectors like agriculture, water resources, power, disaster management, tourism, energy, health, and defense. Over the recent years, the accuracy of weather forecasts and warnings in India has improved due to systematic investments and research efforts made by the institutions in India and abroad. Substantial investments made in designing appropriate modeling strategy, strengthening of observational network using latest technologies, improved data assimilation methods, and systematic research on physics and human resources development have led to this significant improvement in weather and climate forecasts. This improvement has ultimately culminated in saving precious lives from natural disasters like tropical cyclones and heat waves and bringing appreciable economic benefits to farmers and fishermen in terms of increasing their income.

As majority of the landslides are caused by precipitation, accurate weather forecasting coupled with the identification of vulnerable slopes has resulted in very effective landslide hazard assessment and monitoring. After the giant 2004 Sumatra Andaman earthquake, a tsunami warning center has been established in India, which monitors tsunami and surges in the Indian Ocean. Seismic hazard microzonation has been proved to be very effective in mitigating the seismic risk due to future large earthquakes through incorporating safer building design. Earthquake monitoring in India started sometime after the 1897 Shillong plateau earthquake. Currently, there are more than 150 earthquake monitoring observatories in India. Real-time monitoring of tsunamigenic earthquakes and accurate estimation of earthquake parameters in a short time, assessment of potential regions for earthquake and tsunami hazards have helped the government and other stakeholders to quickly take up the rescue, relief, and rehabilitation measures and plan for constructing safer structures and shelters.

In this book, there are 21 chapters contributed by subject experts on earth system science services provided by the MoES. The authors lucidly discuss the details of the improvements in skill of forecasts and warnings, and the gap areas where we need to now focus. Oceans provide us with many resources like water, energy, food, drugs, and minerals. A few chapters are devoted in discussing non-living ocean resources and technology development for their peaceful exploration and ultimately exploitation. The indigenous technology developed for desalination plants in the Lakshadweep islands is the best example of how science and technology can be effectively used for the benefits of people in islands. The Deep Ocean Mission, launched by the Government of India, will further explore the development of technologies for exploration of ocean resources and marine biodiversity.

I am very optimistic that this book will be extensively used by policymakers, researchers, and students and will also help in showcasing the social and economic benefits to the society. I congratulate the editors Drs. Vineet K. Gahalaut and M. Rajeevan for editing this book and all the authors for their contributions.

Chandrima Shaha
President
Indian National Science Academy
New Delhi, India

Preface

Research in earth system science is mainly focused on atmosphere, hydrosphere, and lithosphere. It focuses on the natural dynamic processes involved in the system and their interaction, economic and societal benefits, and mitigation of hazards and risk due to the extreme events arising from the dynamic processes. Thus, earth system science brings together researchers across several disciplines. The Ministry of Earth Science (MoES), under the Government of India, probably is the only ministry in the world which provides this unique opportunity to scientists to focus on various aspects related to social and economic benefits of earth sciences to general public. It is mandated to provide services for weather, climate, ocean and coastal state, hydrology, seismology, and natural hazards, to explore and harness marine living and non-living resources in a sustainable manner for the country and for the three poles of the earth (the Arctic, the Antarctic, and Himalaya). The MoES has ten institutes which are involved in research related to earth system sciences. Although social and economic benefits of research carried out by the MoES are quite visible and are available in public domain, we decided to bring out a volume dedicated only to this aspect. We carefully selected different topics to be covered in this volume.

Understanding of weather and climate and the forecasts have a direct influence on agriculture and economy of the country and it is undoubtedly the most important topic of research in the MoES. Understanding, forecasting, and assessment of severe weather events (for example, cyclones, heavy rainfall, heat, and cold waves), help in mitigating risk due to these events. Accordingly, there are nine chapters in the volume devoted to these topics. There is a growing concern on air pollution and fog on the economy and public health. This topic has been covered quite adequately in the volume. Sustainable use of ocean resources for economic growth, improved livelihoods, and jobs while preserving the health of ocean ecosystem is a new paradigm. In the recent years, lots of efforts have gone into these aspects. The MoES has given special emphasis on satellite-based ecological services, exploration of deep sea minerals, developing technology for ocean exploration, desalination of ocean water, and beach restoration.

After the 2004 Sumatra Andaman earthquake and its ensuing tsunami, which caused disaster in the coastal regions around the Bay of Bengal, a major initiative was taken and a tsunami warning center was established. Landslides are frequent in the Sahyadri (Western Ghats) and the Himalayan mountainous region. Majority of them are triggered by heavy rainfall. Earthquake monitoring in the past few years has improved due to an increase in the number of seismological stations which also has helped in identifying earthquake-prone regions. Understanding and assessing seismic hazards is the first step towards seismic risk mitigation. In recent years, the seismic vulnerability of urban regions has increased due to growth in population, urbanization, and poor built-up structures. Accordingly, seismic hazard microzonation of a few Indian cities has been taken up for future safer structures and retrofitting of existing structures.

We hope that this volume will help general public and stakeholders in better understanding the role of earth sciences in socio-economic development and also in mitigating the natural hazards.

Hyderabad, India Vineet K. Gahalaut
New Delhi, India M. Rajeevan

Acknowledgements

We thank Dr. A. K. Singhvi who invited and encouraged us to bring out this volume, highlighting the social and economic impact of earth sciences and the work done through research by the Ministry of Earth Sciences and its institutes. We also thank the Indian National Science Academy (INSA) for agreeing to publish it through Springer Nature.

We thank several reviewers (namely, Drs. Shailesh Nayak, M. Ravichandran, R. R. Kelkar, K. J. Ramesh, S. S. C. Shenoi, M. Ravikumar, Gopal Iyengar, S. L. Jain, Shyam Lal, K. T. Damodaran, B. Nagender Nath, M. R. Ramesh Kumar, G. Srinivasan, S. C. Bhan, D. R. Kothawale, Abhijit Sarkar, Abhilash, A. K. Mitra, D. S. Pai, Parthasarathi Mukhopadhyay, Manik Bali, S. Prabhakar, S. Iniyan, A. K. Sahai, S. K. Roy Bhowmik, G. Sankar, A. K. Mishra, Rahul Sharma, M. Radhakrishna, Chandan Ghosh, Sumer Chopra, Rajeev Bhatla, A. D. Rao, M. Ravindran, Ananthanarayanan, and a few anonymous reviewers) who spent time in promptly reviewing these articles and helped the authors and editors to improve the articles and their presentations.

Finally, we thank all the authors for contributing their research work to this volume.

Vineet K. Gahalaut
M. Rajeevan

Introduction

Earth system science is the science of local as well as global importance. The scientific understanding of the earth system not only helps to improve the prediction of climate, weather, and natural hazards but also leads to sustainable use of living and non-living resources. During the last 15 years, there have been major improvements in weather, climate and monsoon forecasting, prediction of hazards such as tsunami and cyclones, sustainable use of fishery resources, and exploring mineral and energy resources for future. The knowledge about the earth system and its application to address societal issues has brought immense socio-economic benefits to the country. The development of these societal services was the result of investments in long-term measurements of earth system variables, focused process studies and development of numerical weather and earth system models, conducting ocean expeditions to discover new phenomena and resources, development of information systems and web-and location-based services along with setting up state-of-the-art computing system, and development of human resources. This book describes these studies which lead to improving the quality of life of people.

Weather prediction is a primary requirement for agriculture, irrigation, shipping, aviation, sports, and many other sectors. Numerical weather prediction modeling has evolved in the country and is able to simulate atmosphere–land–ocean–cryosphere subsystems and facilitate forecasting weather and climate phenomena in short-range, extended range, and seasonal time scales. INSAT satellite data including temperature and humidity profiles, providing information on intensification, propagation, and decay of weather systems, are assimilated in weather prediction models, and have become a mainstay of the weather prediction.

Monsoon forecasting is very vital for India as well as for the world. There is a large temporal variability in rainfall which has a significant impact on Indian agriculture and population. In this book, the seasonal forecast of monsoon has been provided by using both statistical and dynamical methods. It has been found that numerical models provide reliable information, especially for below-normal monsoon conditions. The changing climate has made an impact on the amount and pattern of seasonal rainfall. The projected changes during near-term, middle of the century, and long-term have been estimated by using IPCC Coupled Model Inter-comparison Project-5 (CMIP5) for intermediate and high emission scenarios. These projections are quite useful for planning for the management of water resources in future.

The potential risk from natural hazards such as cyclones, extreme rainfall, floods, heat and cold waves, earthquakes, landslides, tsunamis, coastal erosion, forest fires, etc., is very high, and most places in India are vulnerable to one or other hazard. There has been a significant improvement in understanding hazard generating processes, their distribution, timing, severity, and their impact on terrain, settlements, and society.

The improvements in observations in hydro-meteorological parameters and their assimilation in high-resolution global (12 km) and regional models (3 km) have led to improved forecast for cyclones, heavy rainfall, and heat waves, up to 10 days in advance. The increase in lead time and reliability of forecast has helped in minimizing the loss of lives. Forecasting severe weather events, such as thunderstorms, hailstorms and heavy rainfalls, and heat and cold waves, are important to alleviate their impact on people. They are being now predicted with reasonable accuracy with 24–48 h lead time by using high-resolution (12.5 km) global models. This information helps to respond to administrators to take timely and appropriate actions. The impact of climate change and variability on frequency and intensity of natural hazards has been addressed in the book. The frequency of heat waves has increased in North, West, and East India, while the number of cold waves has decreased in these regions.

Air pollution affects human health and crop productivity. However, the estimation of economic cost of 640 billion USD on account of premature mortalities seems to be an exaggeration. At the same time, the economic loss of 1–1.5 billion USD due to the effect of surface ozone on crop looks to be miniscule compared to the total value of crops. The validation of these models is desirable. The exposure to air pollution and its effect on health depends on many aspects, and it may be useful to conduct local studies to derive appropriate models applicable to India.

Predictive capabilities for earthquakes, landslides, and avalanches are yet to be developed. The monitoring of seismic activity is critical for proving earthquake information in near-real time. The national seismological network has been providing earthquake magnitude and location with its hazard potential to all stakeholders and has helped to plan relief and rescue operations. About 500 towns and cities are vulnerable to earthquakes in India. One of the major requirements for these urban areas is to develop seismic microzonation maps. These maps provide seismological and geophysical characteristics of subsurface formations. Such maps for few cities have been prepared. These maps are to be used by city planners for constructing safe buildings and infrastructure and facilitate seismic risk mitigation. A plan for preparing such maps for all cities and towns should be prepared.

Landslide-prone areas cover almost 12% of India's area. Landslides can be triggered by heavy rainfall or by an earthquake. A case study has been carried out to determine the probability of landslide initiation due to heavy rainfall. NWP models were employed for short-range forecast of rainfall in the Nilagiri area with reasonable success. Such forecasts can help for warning people in advance about impeding danger. Such studies need to be validated in other vulnerable regions and can be part of disaster management strategy.

Coastal erosion is prevalent on all along the Indian coast. Coastal protection measures such as the construction of sea wall and groins have limited success. A novel approach of comprising the construction of a submerged artificial reef and beach nourishment has been found to be very effective in restoring beach in Puducherry. A similar approach should be adopted for other areas facing severe coastal erosion.

Tsunami occurred in India on 26 December 2004 was one of the worst disasters of this century. The setting up of the tsunami warning system in October 2007 has provided advisories not only to India but also the countries in the Indian ocean. Apart from providing advisories, major efforts were directed to building community awareness and preparedness. During the last 14 years, no single false warning was issued, which helped to build trust in tsunami advisories in coastal communities and avoiding evacuation.

Oceans are storehouse of minerals. Major exploration work has been undertaken for poly-metallic nodules (Cu, Ni, Co) and hydrothermal systems on mid-oceanic ridges (Cu, Zn and Au, Ag, etc.). Though mining of these mineral resources is not likely to take place in the near future, the knowledge about occurrence and distribution is critical for economic development of the country in a long term. India is committed to meet its energy requirement of 50% from non-fossil fuel resources. Ocean waves, winds, tides, and thermal energy are potential sources. Technologies for harnessing wave energy through floating wave-powered devices, offshore wind energy farms, ocean currents and ocean thermal energy are under development. Experiments on cage culture have been demonstrated successfully near Rameshwaram. Technological challenges in harvesting, post-harvesting, and product development need to be undertaken for a commercial venture.

The development of deep-sea technologies, autonomous vehicles, remotely operable vehicles, drilling and mining equipment, and manned submersibles is critical for exploration and exploitation of living and non-living resources. The investment made in the development of these technologies is vital for ushering in an era of the blue economy.

The coastal development and resultant increase in population in coastal regions will increase the demand of freshwater. The low-temperature thermal desalination (LTTD) technology has been successfully utilized to provide drinking water in the Lakshadweep islands and helped to reduce health-related issues in population. Similar approach has also been used in providing freshwater in thermal power stations located on the coast. This technology has tremendous scope to scale it up and provide water to coastal cities.

The distribution and health of various ecosystems, their components, functions, and dynamics should be understood for conservation, preservation, and management of vital and critical ecosystems. These ecosystems support livelihood of coastal population through tourism, recreational activities, fishery, etc. The satellite-based Algal Bloom Information Service (ABIS), Coral Bleaching Alert System (CBAS) for monitoring phytoplankton bloom and coral bleaching, respectively, provide information about the health of marine and coral reef ecosystems. A conceptual framework has been prepared to initiate Jellyfish Aggregation Advisory Service (JAAS), useful for atomic and thermal power stations that draw large amount of sea water.

The forecasting for ocean state and potential fishery zone helps navigation and efficient fishing, respectively. The accurate, reliable, and timely delivery of these services has brought economic and societal benefits to coastal communities. The fishing community, port, and maritime authorities, coast guards and navy have been extensively using these services. The recent use of NAVigation with Indian Constellation (NAVIC) and Gagan Enabled Mariner's Instrument for Navigation and Information (GEMINI) systems facilitated the dissemination of these advisories to those who operate in deep seas.

The sustainable development of a country or a region requires not only understanding of the earth system processes and resources but also communicating their impact to society. I am sure that this book will fill this gap. I would like to thank all authors for their contributions and Dr. M. Rajeevan and Dr. Vineet K. Gahalaut for editing this publication.

Shailesh Nayak
National Institute of Advanced Studies
Bengaluru, India

Contents

About the Editors

Vineet K. Gahalaut, Ph.D., is Chief Scientist at CSIR-National Geophysical Research Institute, Hyderabad, India. He did his masters and Ph.D. from the University of Roorkee (now IIT Roorkee) in geophysics in the year 1989 and 1995, respectively. He has also served as Director of National Center for Seismology, Ministry of Earth Sciences, New Delhi (from 2015–2019). He is involved in research related to earthquake processes, tectonic geodesy, seismology, and geodynamics. His research work mainly focusses on plate convergence, strain accumulation, earthquake occurrence processes in the plate interior and plate boundary regions of the Indian plateau. He has more than 150 research publications in peer reviewed journals and is Fellow of the Indian Academy of Sciences, Bengaluru, and Indian National Science Academy, Delhi, India.

M. Rajeevan, Ph.D., is Distinguished Scientist at the Ministry of Earth Sciences (MoES), the Government of India. He did his masters and Ph.D. in physics from Madurai Kamaraj University in 1983 and from the University of Pune in 1997, respectively. He has served the Tata Institute of Fundamental Research (TIFR), India Meteorological Department (IMD), National Atmospheric Research Laboratory (NARL), and Indian Institute of Tropical Meteorology, before he became the Secretary at the MoES in December 2015. His expertise includes monsoon variability and monsoon prediction, climate change and extreme weather events. He made significant contributions in developing many application tools and prediction models for societal applications like monsoon seasonal prediction models, gridded climate data sets and climate application products for regional climate services. He has more than 150 research papers to his credit published in several peer reviewed journals. He is Fellow of all the three science academies of India and also an academician of the International Academy of Astronautics. He is an expert member in the Research Board of World Meteorological Organization (WMO).

Short to Medium Range Weather Forewarning System in India

M. Mohapatra, Naresh Kumar, Monica Sharma, Sunitha S. Devi, Soma Senroy, K. Sathi Devi, and K. K. Singh

Abstract India experiences several types of natural disasters including cyclones, floods, droughts, earthquakes, landslides, heat waves, cold waves, thunder squalls, and tornadoes. Most of these natural disasters (about 80%) are hydro-meteorological in nature. The risk management of these natural disasters includes several steps based on their severity and importance, such as (i) hazard analysis, (ii) vulnerability analysis, (iii) preparedness and planning, (iv) early warning, and (v) prevention and mitigation. The early warning component includes (i) skill in monitoring and prediction of natural hazards, (ii) effective warning products generation and dissemination, (iii) coordination with emergency response units, and (iv) public awareness and perception about the credibility of the official predictions and warnings. There has been a significant improvement in multi-hazard monitoring and warning system in recent years due to various initiatives of the Ministry of Earth Sciences (MoES)/India Meteorological Department (IMD) and the policy framework of the Government of India. The resolution of global models has increased 10 times from 120 to 12 km and that of regional models 3 times from 27 to 3 km in these 10 years. The lead period of the models has also increased from 48 h to 10 days forecast in the interval of 3 h. Improved warnings with increased accuracy and high lead period against hazards like cyclones, heavy rainfall, and heat wave have helped disaster managers and general public to minimize loss of lives and property. Especially the deaths due to cyclones and heat waves have been limited to <100 in recent years. There is still scope for improvement of forecast and early warning at district and station levels in terms of improving the mesoscale hazard detection and monitoring, improving the spatial and temporal scale of forecasts, sectoral applications of early warning and warning communication to last mile and disaster managers through state-of-the-art technology. All these aspects have been discussed in this chapter with special emphasis on prediction skills and socio-economic impacts, existing gap areas, and future scope.

Keywords Weather forecasting · Early warning · Socio-economic benefit

M. Mohapatra (✉) · N. Kumar · M. Sharma · S. S. Devi · S. Senroy · K. S. Devi · K. K. Singh
India Meteorological Department, Ministry of Earth Sciences, Government of India, Lodi Road, New Delhi 110003, India
e-mail: mohapatraimd@gmail.com

V. K. Gahalaut and M. Rajeevan (eds.), *Social and Economic Impact of Earth Sciences*,
https://doi.org/10.1007/978-981-19-6929-4_1

1 Introduction

It has always been challenging to predict weather, especially the extreme weather events over the South Asian region including India due to mesoscale convective processes in the tropics, varied physiography of the region, and their complex nonlinear interaction with large-scale atmospheric circulation. The South Asian region is affected by different hazards including cyclone, depression, heavy rainfall, thunderstorm/squall/hailstorm, storm surge, coastal inundation, etc. Most of these natural disasters (about 80%) are hydro-meteorological in nature. There are 13 coastal states/Union Territories (UT) encompassing 96 coastal districts which are affected by cyclones. It includes 72 districts touching the coast and 24 districts, not touching the coast, but very close to the coast (within 100 km) (Mohapatra et al. 2012a; Mohapatra 2015). Four states (Andhra Pradesh, Odisha, Tamil Nadu, and West Bengal) and one UT (Pondicherry) on the East Coast and one state (Gujarat) on the West Coast are more vulnerable to cyclones and associated storm surges. The states falling within the periphery of "India Flood Prone Areas" are West Bengal, Orissa, Andhra Pradesh, Kerala, Assam, Bihar, Gujarat, Uttar Pradesh, Haryana, and Punjab (Tripathi 2015). Heat waves are more frequent over the Indo-Gangetic plains of India (Uttar Pradesh, Bihar, Haryana, Delhi, Rajasthan, Madhya Pradesh, Chhattisgarh, Odisha, Andhra Pradesh, and Telangana) according to Naresh Kumar et al. 2017 and Pai et al. (2013). Northern and eastern states such as Rajasthan, Punjab, Haryana, New Delhi, Jammu and Kashmir, Himachal Pradesh, Madhya Pradesh, Bihar, and Tripura are affected by the cold waves (Naresh Kumar et al. 2017). Thunderstorms are more frequent and intense over east and northeastern India and severe fog occurs over Indo-Gangetic plains in the winter months (Roy et al. 2021, 2019).

The risk management of these natural disasters depends on several factors including (i) hazard analysis, (ii) vulnerability analysis, (iii) preparedness and planning, (iv) early warning, and (v) prevention and mitigation (NDMA 2019). The detection and prediction of these hazards with sufficient lead period and reasonable accuracy is very essential for their management. The IMD, MoES is the Nodal National Meteorological agency mandated for issuing seamless operational weather forecasts and warnings for various meteorological hazards across the country. IMD also provides forecast guidance to the South Asian countries and members of WMO/ESCAP panel countries as the WMO-designated Regional Specialized Meteorological Center (RSMC). The regional guidance on severe weather events includes cyclone, heavy rain, strong wind, sea waves, and storm surge (RSMC New Delhi 2020).

The warnings and advisories by IMD for national purpose are issued for different temporal ranges and spatial domains (IMD 2012). The temporal domain of forecasts and warnings ranges from nowcasting which is valid only for a couple of hours to long range which has the validity of months to a whole season. The components of a forecasting and warning system include (i) detection, monitoring, and warning through global, regional, national, and local observations, (ii) numerical weather prediction, (iii) forecasts on different timescales (nowcasting to several days), (iv) timely dissemination of authoritative warning information, and (v) risk analysis and impact assessment.

Mitigation and response actions of recipients depend on the content and clarity of the warning, the credibility of issuing organization, and the preparedness of receiving authorities. Disaster management needs collaborative efforts of warning organizations and communities including other government organizations, local officials, emergency managers, media, voluntary, non-government organizations, etc. Short-to-medium-range forecasting plays a dominate role for management of weather-related hazards.

There has been a significant improvement in multi-hazard monitoring and warning system in recent years due to various initiatives of the Ministry of Earth Sciences (MoES)/India Meteorological Department (IMD) and the policy framework of the Government of India. Improved warnings with increased accuracy and high lead period against hazards like cyclones, heavy rainfall, and heat wave have helped disaster managers and general public to minimize loss of lives and property. Especially, the deaths due to cyclones and heat wave have been limited to <100 in recent years. There is still scope for improvement of forecast and early warning at district and station levels in terms of improving the mesoscale hazard detection and monitoring, improving the spatial and temporal scale of forecasts, sectoral applications of early warning and warning communication to last mile and disaster managers through state-of-the-art technology. All these aspects have been discussed in this chapter with special emphasis on prediction skills and socio-economic impacts, existing gap areas, and future scope.

2 Weather Forecasting and Warning Services of IMD

IMD weather analysis and forecasting is the succession of the following tasks:

I. To monitor the actual condition of the atmosphere at all-time scales (weather analysis);
II. To obtain the pertinent information from numerical models and observational systems including satellite and radar and to assess the future evolution of the atmosphere in order to determine the most likely scenario;
III. To deduce the consequences of the expected synoptic situation in terms of weather elements (weather elements forecasting) and to evaluate the risk of the occurrence of hazardous phenomena (risk assessment);
IV. To prepare the meteorology-related information (weather warnings) to be directed toward the various internal or external users.

2.1 Monitoring Process

The entire process of extreme weather monitoring and forecasting is shown in a schematic diagram (Fig. 1). The observational network for monitoring consists of land-based surface and upper-air stations, observations from Doppler Weather Radars (DWRs), and data from geo-stationary and polar-orbiting satellites. In addition, observations from ships and buoys are of immense importance for the analysis and forecasting of extreme weather, especially ocean-related hazards like cyclones, monsoonal heavy rainfall, etc. It is important to correctly determine the location and intensity of the system, as initial error in location and intensity can lead to an increase in error in forecast of location and intensity (Mohanty et al. 2010; Osuri et al. 2012). Hence, there is a need for dense observational network over the sea and along the coast. IMD does have an excellent network for observing surface climatic change which is up to the mark of the standards of the World Meteorological Organization (WMO). There are 559 departmental manned surface observatories of IMD at present. There are 62 pilot balloon observatories. High-quality upper-air observational data network using the GPS technology has now been extended to 43 locations. There has been the augmentation of surface observational network over the Indian landmass by increasing the number of Automatic weather stations (AWS) and by introducing Automatic Rain-Gauges (ARG) leading to the availability of meteorological observations from all the districts of the country. The meteorological data thus collected all over these stations are used on real-time basis for monitoring the weather.

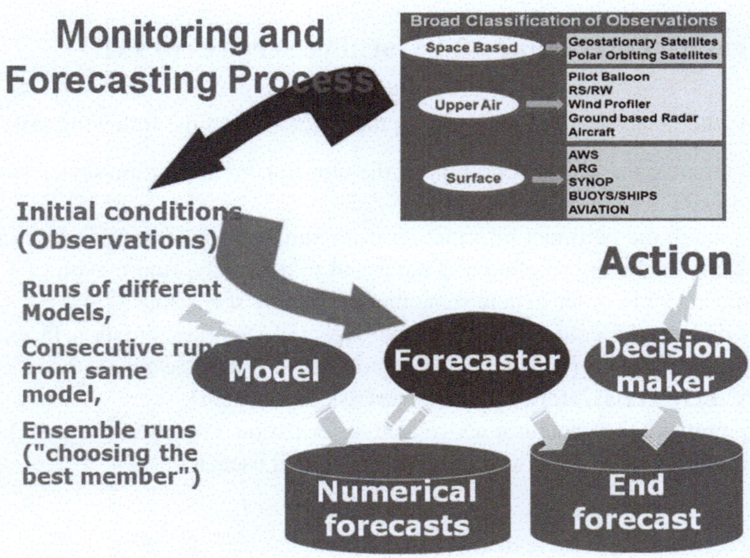

Fig. 1 Weather monitoring and forecasting process of IMD

Under the National Data Buoy Program (NDBP) at the National Institute of Ocean Technology (NIOT), Chennai, 12 moored data buoys are deployed currently in the NIO which are equipped with sensors to measure air pressure and temperature, wind speed and direction, sea surface temperature, etc. These buoys have resulted in better monitoring and reduction of location and intensity error of cyclonic disturbances in association with ship and satellite observations. As per the guidelines issued to the Indian voluntary observing fleet (IVOF), synoptic observations are made at the main standard times: 0000, 0600, 1200, 1800, 0300, 0900, 1500, and 2100 UTC by the ships. The sea surface wind as estimated by scatterometer-based satellites (ASCAT, Windsat, and SCATSAT) is very useful in monitoring weather systems over the seas (Osuri et al. 2012).

There are 27 numbers of C/S-band DWR in India. A DWR renders velocity and spectral width data apart from meteorological, hydrological, and aviation products (Raghavan 1997, 2013). The geostationary satellite, INSAT 3D and 3DR, provides imageries in visible (VIS), infrared (IR), and water vapor (WV) channels. In addition to the above, the products like outgoing long wave radiation, quantitative precipitation estimates, sea surface temperatures (SST), cloud motion vectors, water vapor derived wind vector, and isotherm analysis on enhanced infrared images are also analyzed on operational mode for monitoring. The microwave imageries are also used for monitoring and guidance, especially the internal structure of weather systems. Apart from the Indian satellite, products from other international satellites are also used for monitoring (IMD 2021). Numerous studies (Dvorak 1984; Kalsi 2002; Velden et al. 2006) have built up a repository on satellite applications in weather monitoring in the NIO.

2.2 *Forecasting System*

A variety of observational data have been used in India till the 1960s to forecast the extreme events. Satellite era, since the 1960s, added another feature. There has been rapid development in objective techniques since the 1970s and especially in recent years for forecasting extreme weather events. To summarize, currently following methods are used by IMD for short-to-medium-range forecasting.

- Statistical technique (Analogue, Persistence, Climatology, Climatology, and persistence (CLIPER).
- Synoptic technique—Empirical technique.
- Satellite technique—Empirical technique.
- Radar technique—Empirical technique.
- Numerical weather prediction (NWP) models.
- Dynamical statistical models.

There are three types of NWP models, viz., individual deterministic models, Multi-Model Ensemble (MME), and single-model Ensemble Prediction System (EPS) for different ranges of forecast as mentioned below (RSMC New Delhi 2021):

- Global forecasting system (GFS) global model (horizontal resolution of 12 km and forecast up to 10 days).
- Unified model (horizontal resolution of 12 km and forecast up to 10 days).
- Global Ensemble Forecasting System (GEFS) global probabilistic model (horizontal resolution of 12 km and forecast up to 10 days).
- Unified model ensemble prediction system (horizontal resolution of 12 km and forecast up to 10 days).
- Weather research forecast (WRF) Mesoscale model (horizontal resolution of 3 and 9 km and forecast up to 3 days.
- Unified Mesoscale regional model(horizontal resolution of 4 km and forecast up to 3 days.
- Hurricane WRF (HWRF) for cyclone prediction (horizontal resolution of 2 km and forecast up to 5 days).
- MME-based cyclone track prediction (forecast up to 5 days).
- There is a suit of dynamical statistical models for prediction including genesis of tropical cyclone (Kotal and Bhattacharya 2013), track of cyclone based on multi-model ensemble (MME) technique (Kotal and Roy Bhowmik 2011), intensity of cyclone (Kotal et al. 2008), storm surge (rise of sea water above the astronomical tide due to cyclone), and coastal inundation (Murty et al. 2017).
- All the above models are operational at MoES using High-Power Computing Systems (HPCS). The models are run twice/four times in a day (00, 06, 12, and 18 UTC). IMD also uses NWP products prepared by other operational NWP Centers such as GFS-USA, Japan Meteorological Agency (JMA), ECMWF, etc. The MME technique for cyclone track forecast (Kotal and Roy Bhowmik 2011) is based on a statistical linear regression approach using five operational NWP models (ECMWF, GFS(IMD), GFS(NCEP), UKMO, and JMA). The single model-based ensemble forecast products from ECMWF (50+1 Members), NCEP (20+1 Members), UKMO (23+1 Members), Meteorological Services, Canada (MSC) (20+1 Members), and JMA (20+1 Members) are available near real time for NIO region. These products includes deterministic and ensemble track forecasts of cyclones, strike probability maps. The super-ensemble has also been developed based on thevabove ensembles. In the ensemble prediction system (GEFS and UMEPS) run in MoES, various probabilistic forecast products are generated for severe weather events, probabilistic occurrence of different thresholds of rainfall (heavy rain and deficient rain), temperature(heat wave and cold wave), wind (squally wind and gale winds), etc., at different locations/regions.

2.3 Decision-Making Process

In the synoptic method, prevailing environmental conditions, NWP model analysis, and forecasts are considered. On the one hand, the synoptic, statistical, and satellite/radar guidance aids in predicting short-range forecasts (up to 12/24 h); on the other hand, NWP guidance is mainly used for long-range forecasts of up to 24–120 h forecasts. The analysis, prediction, and decision-making process is made by blending scientifically based conceptual models, dynamical and statistical models, meteorological datasets, technology, and expertise (IMD 2012, 2013, 2021). The decision support system (DSS), digitally is used to plot and analyze weather parameters, satellite, Radar and NWP model products, etc. This is further supported by modern graphical and GIS applications for producing high-quality analysis and forecast products. It includes:

- Analysis of all synoptic, satellite, Radar and NWP model products for genesis, intensity, movement, and growth/dissipation of weather systems.
- Preparation of past weather and forecast up to 120 h.
- Depiction of uncertainty in forecast based on probabilistic assessment.

An SOP (IMD 2012, 2013, 2021) is followed for analyzing various monitoring and forecast guidance available from different sources to arrive at decisions on forecast and formulation of forecast and warning bulletins.

2.4 *Warning System of IMD*

The goal of the warning system is to maximize actions for safety. To carry out the warning process, there is a well-defined organization and SOP is in place (IMD 2012, 2013, 2021). The warning criteria are defined for each parameter. Linkages with Media and Disaster authorities are well defined. Scheduled time of issue and frequency of bulletins along with the content (text and graphics) are also designed as per the requirement of users. The warnings and advisories by IMD for national purpose are issued for different temporal ranges (from a few hours (nowcast) to 5 days (short-to-medium-range forecast) and different spatial domains (IMD 2012). The spatial domains range from venue, city, district, and state/meteorological subdivisions, each having different temporal domains. However, short- and medium-range forecast, especially during the southwest monsoon season (June–September) and cyclonic disturbances period continues to remain in the highest demand by the different users including farming community and other stakeholders and therefore accurate prediction in this range occupies greater significance. Followings are the various types of forecasts and warnings issued by IMD.

2.4.1 Cyclone Warning System

Types of bulletins and warnings issued in the interest of Mariners in cyclone-specific situations are:

- Sea area bulletins (for shipping on high seas and for ships plying in coastal waters (up to 75 kms off the coast line),
- Bulletins for the Indian Navy,

- Port warnings,
- Fisheries warnings,
- Four-stage warning for disaster managers,
- Bulletin for All India Radio (AIR),
- Warnings for Designated/Registered users, and
- Bulletins for Press.

The cyclone warnings are issued to central and state government officials in four stages. The **First Stage** warning known as "**PRE CYCLONE WATCH**" issued at least 72 h in advance contains the early warning about the development of a cyclonic disturbance in the north Indian Ocean. This early warning is addressed to the Cabinet Secretary and other senior officers of the Government of India including the Chief Secretaries of concerned maritime states. The **Second Stage** warning known as "**CYCLONE ALERT**" is issued at least 48 h prior to the onset of adverse weather condition over the coastal areas. Location and intensity of the storm, likely direction of its movement, intensification, advice to fishermen, general public, media and disaster managers, etc., are shared in the Cyclone Alert bulletin. This is issued by the concerned Area Cyclone Warning Centers (ACWCs)/Cyclone Warning Centers (CWCs) and Cyclone Warning Centre (CWD), New Delhi, at national level. The **Third Stage** warning, known as "**CYCLONE WARNING**", is issued at least 24 h in advance of the expected commencement of adverse weather over the coastal areas. Landfall point is forecast more precisely at this stage. The ACWCs/CWCs/and CWD at 3 hourly intervals release the latest position of cyclone and its intensity, likely point and time of landfall, associated heavy rainfall, strong wind and storm surge along with their impact and advice to general public, media, fishermen, and disaster managers. The **Fourth Stage** of warning known as "**POST LANDFALL OUTLOOK**" is issued by the concerned ACWCs/CWCs/and CWD at headquarter at least 12 h in advance of the expected time of landfall. It gives likely the direction of movement of the cyclone after its landfall and adverse weather likely to be experienced in the interior areas. However, the above procedure is applicable to the cyclones developing over the open sea area. In case the genesis takes place near the coast and thus restricts the lead time before the landfall, the above procedure may not be applicable. The cyclone warning may be issued directly bypassing the pre-cyclone watch and cyclone alert (RSMC New Delhi 2021; IMD 2021).

At national level, the bulletins are issued from the stage of depression onwards. During depression/deep depression stages, it is issued 00, 03, 06, 09, 12, 15, 18, and 21 UTC (three hourly intervals) based on previous observations. The cyclone warning is provided on a real-time basis to the Control Room of the Ministry of Home Affairs, Government of India, besides other Ministries and Departments of the Central Government, and TV, electronic and print media, and concerned state governments. Since the post-monsoon season of 2006, different color codes are used at different stages of the cyclone warning bulletins. An example of graphical presentation of track forecast along with the cone of uncertainty and actual observed track in case of cyclone Hudhud is presented in Fig. 2.

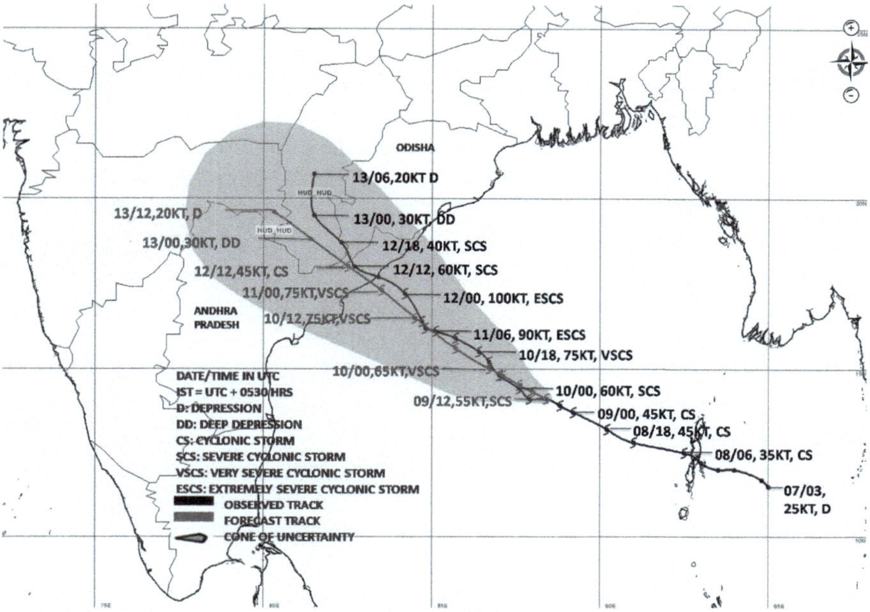

Fig. 2 Observed and forecast track of ESCS Hudhud along with cone of uncertainty

2.4.2 Thunderstorm/Dust Storm/Lightning Warning

Severe thunder storms are localized phenomena with a life period of about a few hours. These can be detected on real time with a dense network of DWR in the country. The currently available array of numerical weather prediction systems is not capable to provide accurate location-specific forecast of these thunderstorms and associated weather with reasonable lead time. Worldwide nowcasting, the very short-range forecasting up to 3–6 h is applied to predict and warn against these severe thunderstorms with the help of DWRs supported by dense observational network, satellite observations, and suitable numerical modeling and expert system. In India, this nowcasting is provided for 1084 cities and for all the 739 districts. The district level nowcast against the thunderstorm and associated adverse weather is issued by IMD through its state-level meteorological centers to the disaster managers and general public. As soon as IMD issues any severe weather warning, it is flashed to general public through mass media and registered mobile users. IMD through its state-level offices is issuing SMS-based warning to All India Radio stations, Doordarshan for immediate broadcast/telecast to general public (Roy et al. 2019, 2021).

2.4.3 Heavy Rainfall and Flood Warning

IMD issues heavy rainfall warning at meteorological sub-divisional level through its National Weather Forecasting Center (NWFC) located at New Delhi. The terminology in use at present to indicate and forecast the rainfall intensity (24 h accumulated rainfall ending at 0830 h. IST of each day) in the short- and medium-range weather forecasts are given in Table 1.

The color codes (red, orange, yellow, and green corresponding to warning, alert, watch, and no warning) are used for warning advisories in association with rainfall and snowfall. In order to decide upon the color to be assigned to a given weather forecast situation under the 5-day forecast scheme, IMD follows the matrix (Fig. 3), giving thrust on the probability of occurrence of the event as well as its impact assessment. The probability of occurrence for D1–D5 (days) is derived based on the ensemble probabilistic forecasts provided by NCMRWF, IITM, and various global centers (Mohapatra et al. 2021a, b). For an impact assessment, the forecaster goes ahead with the conceptual model of the weather associated with the expected evolution of the synoptic situation based on a proper interpretation of the numerical model outputs along with the topography and land use pattern of the region under consideration. It is carried out through both objective consensus derived from the models and the subjective consensus derived through tele/video conferencing of forecasters within the country every day utilizing knowledge, experience, and expertise.

Flood forecasting and warning in India is entrusted with the Central Water Commission (CWC) of the Ministry of Water Resources (MoWR). The Hydromet Division of IMD at New Delhi caters to the need of hydromet inputs, particularly of water resources development and water-related disaster (like floods and drought) monitoring/management (IMD 2021).

A WMO Project for the South Asia Flash Flood Guidance System (SAFFGS) has been adapted in collaboration with WMO/South Asia Countries and Hydro

Table 1 Categorization of rainfall & snowfall

Descriptive term	Rainfall (mm)	Descriptive term	Snow depth (in cm)
Heavy rain	64.5–115.5	Heavy snowfall	64.5–115.5
Very heavy rain	115.6–204.4	Very heavy snowfall	115.6–204.4
Extremely heavy rain	>204.5	Extremely heavy snowfall	>204.5

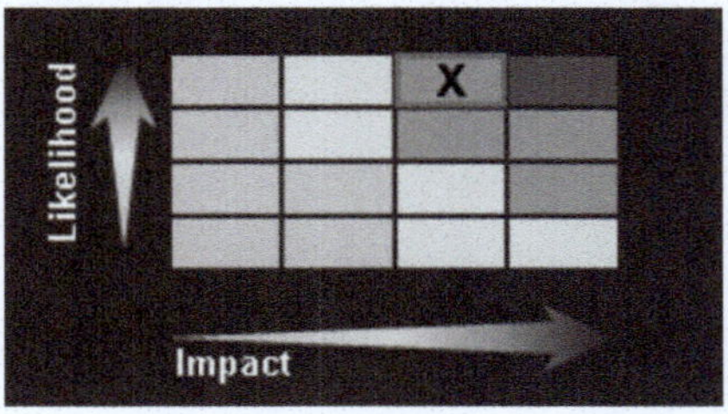

Fig. 3 The matrix used by IMD to provide impact-based forecast and warning against weather hazards

Research Centre (HRC), USA. It aims to provide flash flood guidance to Nepal, Bhutan, Bangladesh, Myanmar, Sri Lanka, and Maldives (IMD 2021).

2.4.4 Heat and Cold Waves

In the country, appreciable rise in maximum temperatures as well as heat waves are found to be more in the months of April, May, and June. The Ministry of Earth Sciences has implemented a state-of-the-art coupled climate model for seasonal forecasting. The model has been used to prepare monsoon forecasts since 2012. The model was used the first time to prepare a temperature outlook for the 2016 hot weather season (April–June) and also the cold weather season (December–February) since 2016–17. It is interpreted to provide the outlook on sub-divisional levels for expected heat wave and cold wave conditions in the country during the season. It is followed by the extended range outlook on summer maximum and winter minimum temperatures issued every Thursday valid for the next two weeks on meteorological sub-divisional level. In short-to-medium range, heat wave and cold wave warnings are issued daily valid for up to 5 days (IMD 2021). The color code system indicating the severity of an expected heat wave and cold wave used by IMD is given in Table 2.

As an adaptive measure, IMD, in collaboration with local health departments has started heat action plan in many parts of the country to forewarn about the heat waves and also advising action to be taken during such occasions. Heat action plan became operational since 2013. The Heat Action Plan is a comprehensive early warning system and preparedness plan for extreme heat events. The plan presents immediate as well as longer term actions to increase preparedness, information-sharing, and

Table 2 IMD criteria for impact based heat and cold wave warning

(a) Cold wave		
Green (No action)	Normal night	Minimum temperatures are near normal
Yellow Alert (Be updated)	Cold Alert	Cold wave conditions at isolated pockets persists on 2 days
Orange Alert (Be prepared)	Severe Cold Alert for night	(i) Severe cold wave conditions persists for 2 days (ii) Through not severe, but cold wave persists for ≥4 days
Red Alert (Take Action)	Extreme cold Alert for night	(i) Severe cold wave persists for more than 2 days (ii)Total number of cold/severe cold wave nights > 6 days.

(b) Heat wave		
Green (No action)	Normal Day	Maximum temperatures are near normal
Yellow Alert (Be updated)	Heat Alert	Heat wave conditions at isolated pockets persists on 2 days
Orange Alert (Be prepared)	Severe Heat Alert for day	(i) Severe heat wave conditions persists for 2 days (ii) Through not severe, but heat wave persists for ≥4 days
Red Alert (Take Action)	Extreme Heat Alert for day	(i) Severe heat wave persists for more than 2 days. (ii) Total number of heat/severe heat wave days > 6 days.

response coordination to reduce the health impacts of extreme heat on vulnerable populations. NDMA and IMD are working with 23 states prone to high temperatures leading to heat wave conditions to develop heat action plans. IMD has started the Forecast Demonstration Project (FDP) on heat waves for the hot weather season under which a detailed daily report including realized data of heat waves, weather systems leading to the occurrence of heat waves, diagnosis on the basis of Numerical Model outputs and forecast and warnings for 5 days is prepared. This bulletin is disseminated to all concerned including health departments.

During the cold season, associated with the passage of the Western Disturbances, north and northwest India and adjoining central India are frequently affected by adverse weather elements like fog, cold wave to severe cold wave, and cold day to severe cold day conditions. By the end of November, IMD issues Press-Release regarding Seasonal Outlook for the temperatures during December–February which brings out the temperature scenario with respect to mean temperature, mean minimum and maximum temperatures during the cold weather season. During the cold weather season, IMD also issues Press-Releases whenever any place or region is likely to experience cold/severe cold waves. In addition to these, a Forecast Demonstration Project (FDP) for winter weather systems has been initiated in 2016, and it has brought together several institutes other than IMD also to enhance the monitoring and forecast of weather elements during cold weather season. Accordingly, an FDP bulletin is prepared and issued during November–February on daily basis. From November 2020 onwards, IMD started issuing a special bulletin related to winter weather systems (All India Multi-hazard Winter Warning Bulletin) which provide the details of color-coded warning for 5 days for the adverse weather elements, along with present weather scenario related to cold wave, cold day, etc.

2.4.5 Fog

Frequent occurrence of fog in different parts of northern India is common during the winter months of December and January. Low visibility conditions due to fog disrupt normal public life. Visibility conditions heavily affect both surface and air transport. A number of flights are either diverted or canceled every year during the winter season due to low visibility conditions, experienced at different airports in north India. Thus, fog and visibility forecasts over the plains of north India become very important during winter months. There has been drastic improvement in the fog monitoring and forecasting systems in winter of recent years. The visibility forecast during fog obtained from NWP models provides a reasonably good indication of the spatial extent of fog in advance of one day (Singh et al. 2018). The fog intensity is also predicted fairly well. It has also been possible because of the successful and effective implementation of three new projects, viz., (i) installation of the latest observational instruments, (ii) Winter fog experiment (WIFEX), and (iii) forecasting demonstration project (FDP) in recent years. Details of fog monitoring and forecasting are discussed in IMD (2021).

2.5 *Forecast Accuracy*

2.5.1 Heavy Rainfall

There is considerable improvement in heavy rainfall skill in recent years (Fig. 4i) for Day 1 (D1), Day 2 (D2), and Day 3 (D3). Improvement in D3 skill scores is at higher rate as compared to D1 and D2. For D1 heavy rainfall warning, False Alarm Rate (FAR) has reduced from 42 to 32% and Probability of Detection (PoD) increased from 45 to 70% during 2017–19 as compareed to 2002–16. Comparing year-wise, the PoD in case of heavy rainfall at meteorological subdivision levels has increased from 50% during 2014 to 77% during 2020 for 24 h lead period, from 48 to 70% for 48 h lead period, and from 37 to 66% for 72 h lead period. The heavy rainfall forecast issued in 2020 five days ahead has the accuracy of about 59% against the forecast accuracy of 50% in 2014 only 24 h ahead. Thus there is a gain of four days in lead period of forecast of heavy rainfall in 2020 as compared to 2014. The missing rate in case of heavy rainfall has decreased from 50% during 2014 to 23% during 2020 for 24 h lead period, from 52 to 30% for 48 h lead period and from 63 to 34% for 72 h lead period.

2.5.2 Cyclone

Noteworthy improvement was made in track and intensity forecast of the tropical cyclones (24 h forecast error in track prediction reduced from 141 km to 69 km and Landfall error from 99 to 27 km during 2006–2019). Increase in lead time along with an increase in accuracy of the forecast of genesis, track, and intensity (Mohapatra et al. 2012b, c, 2013a, b, c, 2014, 2017; Mohapatra and Sharma 2019) and adverse weather (Mohapatra 2015) in recent years have helped in better management of disaster due to cyclone (Fig. 4i). Comparing the track forecast errors of 2016–2020 against that of 2011–2015 as shown in Fig. 4ii, the track forecast error has decreased from 97, 145, and 183 km during 2011–15 to 77, 117, and 159 km during 2016–2020 for forecast issued 24, 48, and 72 h ahead. Similarly, the skills of cyclone track forecast have improved from 49, 63, and 69% during 2011–15 to 64, 76, and 78% during 2016–2020 for forecast issued 24, 48, and 72 h ahead. The annual average landfall point forecast errors for the year 2020 have been 18, 70, and 43 km for 24, 48, and 72 h lead periods against the long period average of the past 5 years during 2015–19 of 47, 70, and 110 km. The landfall point forecast error of 56, 94, and 106 km during 2011–15 has decreased to 32, 62, and 92 km during 2016–20 for forecast issued 24, 48, and 72 h ahead of landfall of cyclone. As regards improvement in intensity forecast over the past 10 years (not shown) there has been a decrease in errors. The intensity (wind) forecast errors have decreased from about 12, 17, and 18 knots during 2011–2015 to 8, 11, and 14 knots during 2016–2020 for the forecast issued 24, 48, and 72 h ahead.

2.5.3 Thunderstorm

Significant improvement in thunderstorms 24 h and for nowcast is also observed in recent years (Fig. 4iii). In 24 h forecast at meteorological sub-division levels, PoD was about 65% in 2019 as compared to about 30% in 2016. It has further increased to 80% in 2020 (not shown). The PoD of location-specific thunderstorms 3 h in advance issued at station levels has increased from 61% (2014) to 88% (2020).

2.5.4 Heat and Cold Wave

Forecast skill for heat wave and cold wave warning skills of IMD has also improved considerably in recent years. India as a whole and region-wise skill of heat wave and cold wave warning is shown in Fig. 4iv and v, respectively. Heat wave PoD was 92% for D1 and 42% for D5 in 2019. In region-wise, the PoD for northwest India was

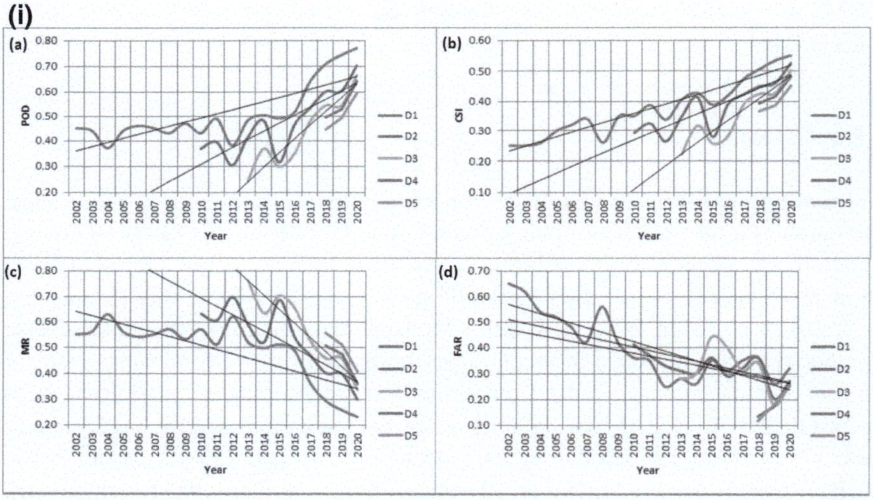

Fig. 4 **i**: Southwest Monsoon heavy rainfall scores, namely (a) PoD, (b) CSI, (c) MR, and (d) FAR for Day 1 (D1) (2002–20), Day 2 (D2) (2010–20), and Day 3 (D3) (2013–20). **ii**: Comparative average (a) Track forecast error, (b) track forecast skill, and (c) landfall point forecast errors of IMD during 2016–20 and 2011–15. **iii**: (a) 24 h of thunderstorm forecast skill scores during 2016–19 and (b) 3-hourly nowcast verification forecast skill scores during 2013–19 for ratio scores PoD, FAR, ETS, and CSI. PoD: Probability of Detection, FAR: False Alarm Rate, ETS: Equitable Threat Score, CSI: Critical Success Index, MR: Missing Ratio. **iv**: (a) Heat Wave (April, May, and June) 2019 skill scores for (a) different regions of India for Day 1 (D1) and (b) India as a whole for Day 1(D1), Day 2 (D2), Day 3(D3), Day 4 (D4) and Day 5 (D5). **v**: Cold Wave forecasting skill scores during 2019–20 (December 2019, January 2020 & February 2020) (a) for Day 1(D1) (24 h) for different regions of India and (b) for Day 1(D1), Day 2 (D2), Day 3(D3), Day 4 (D4), and Day 5 (D5) (24, 48, 72, 96, and 120 h, respectively) for India as a whole. (vi): The PoD and CSI of fog forecast issued by IMD for D1, D2, and D3. PoD: Probability of detection, HSS: Heidke's Skill Score, CSI: Critical Success Index, MR: Missing Ratio, and FAR: False Alarm Rate

(ii)

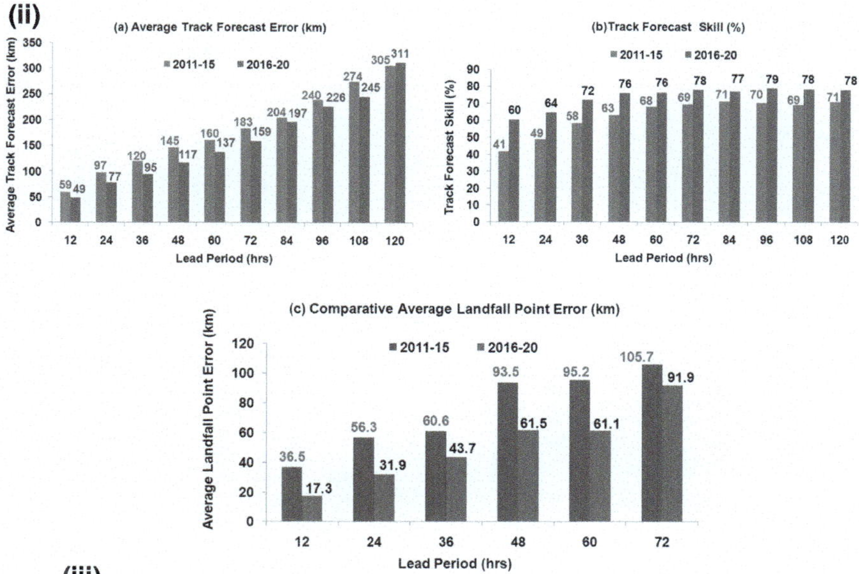

(iii)

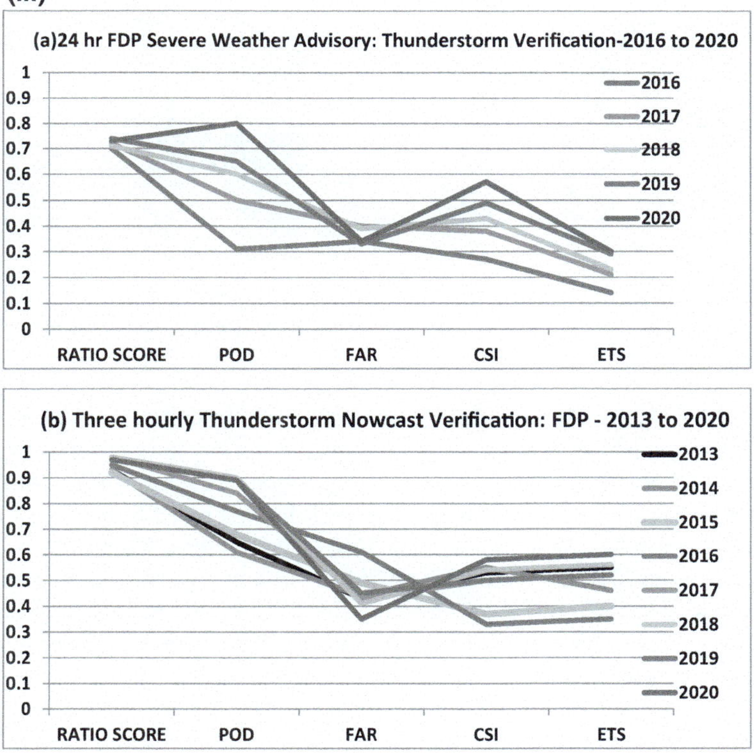

Fig. 4 (continued)

(iv)

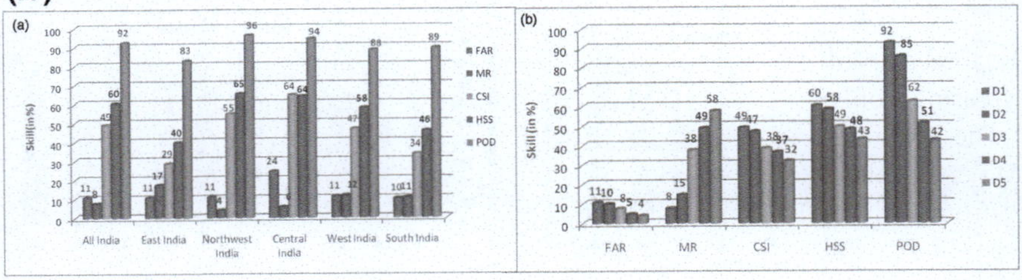

(v)

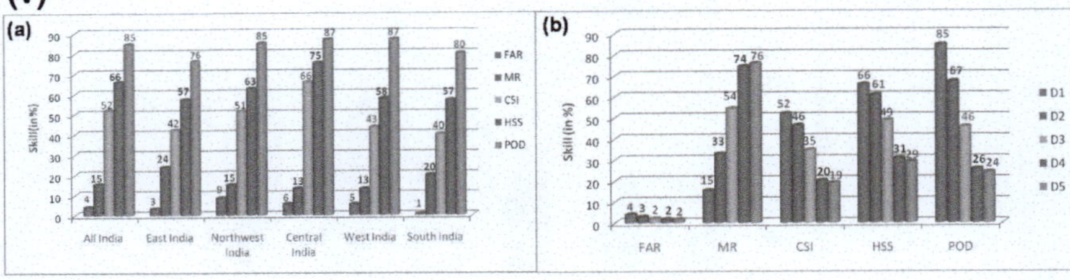

(vi)

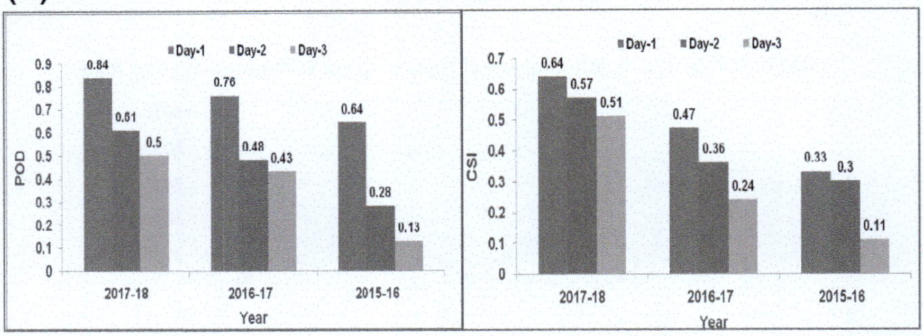

Fig. 4 (continued)

96 and 94% for central India in 2019. Cold wave warning skill is also high in recent years, India as a whole; it was 85% in D1 and 24% in D5. However, skill for D4 & D5 needs further improvement warning dissemination mechanism. The PoD in case of heat wave at meteorological subdivision levels has improved from 67% (2014) to 100% (2020) for 24 h lead period, from 50% (2014) to 95% (2020) for 48 h lead period, and from 27% (2014) to 90% (2020) for 72 h lead period (not shown). The forecast issued in 2020 five days ahead has the accuracy of about 62% against the forecast accuracy of 68% in 2014 only 24 h ahead. Thus there is a gain of 4 days in lead period of heat wave forecast in 2020 as compared to 2014.

2.5.5 Fog

The forecast skill for fog monitoring has also improved in recent years with PoD increasing from 64% (2015–16) to 84% (2017–18) for Day-1 as shown in Fig. 4(vi).

The warnings are disseminated to various users through telephone, fax, e-mail, SMS, All India Radio, FM and community radios, television, and other print and electronic media. These warnings/advisories are uploaded in the website of IMD (www.imd.gov.in). Also, the warning bulletins are disseminated by email and SMS to state and national disaster management authorities. In case of emergency, police wireless and telecommunication lines of railways and aviation authorities are also used. The Cyclone Advisories bulletin for WMO/ESCAP Panel countries and international airports is disseminated through global telecommunication system (GTS) and e-mail.

Recently, there is an initiative through NDMA for Common Alert protocol (CAP) for dissemination of thunderstorm warning. Once implemented, it will be very useful for quick dissemination to the last mile. Further, it is recommended that the synergized observational and forecast and warning products in text cum visual graphics format should be generated by IMD with a direct CAP-feed facility so as to reach the last mile with no loss of time. Similarly, the proposed last mile connectivity through satellite-based initiative of ISRO, viz., Gagan Messaging System and Navik in collaboration with MoES and state and central government will be very helpful to mitigate the disasters.

3 Achievements in Recent years

In the last decade, there have been considerable improvements in the quality of short-to-medium-range weather forecasting services. This was mainly due to augmentation of observational network, modeling, research efforts, and specialized training of scientists. Important achievements in short-to-medium-range weather forecasting services during the last decade are as follows:

- The Gramin Krishi Mausam Seva (GKMS) is rendered twice a week in collaboration with State Agricultural Universities (SAUs) and institutions of the Indian Council of Agricultural Research (ICAR). District-level weather forecasts for the next 5-days, weekly cumulative rainfall forecast, and crop-specific advisories are provided to farmers (Mahadevaiah et al. 2012). Presently, around 4 crore farmers are directly benefitted from this service being provided through SMS as compared to 0.5 crore farmers in April 2014.
- Introduction of scientific methodology and technologies in cyclone forecasting including (i) extension of lead period of cyclogenesis forecast from 1 day in 2008 to 3 days in 2014 and 5 days in 2018 (ii) cyclone track and intensity forecast from 24 h in 2008 to 72 h since 2009 and to 120 h since 2013 (Mohapatra et al. 2013a, b, c).

- Introduction of (i) extended range forecast (for next 2 weeks) for cyclogenesis in 2018, (ii) fishermen warning for the entire north Indian Ocean since 2018 in both graphic and textual form which was previously issued along the Indian coast only, (iii) district-wise impact-based forecast and warning in color-coded form, and (iv) verification of forecast from 2008 for building the confidence of disaster managers and public (Mohapatra and Sharma 2019).
- Scientific and technological interventions led to significant improvement in track forecast, intensity, landfall, structure, and associated adverse weather including heavy rain, gale wind, and storm surge forecast leading to extension of lead period of warnings up to 5 days, improvement in track forecast accuracy by 60–70% up to 72 h and landfall forecast accuracy by 70–80% up to 48 h by 2019 compared to 2009 (Mohapatra and Sharma 2019; Mohapatra 2015).
- Cyclone hazard proneness mapping is utilized in planning of cyclone shelters and other structures in cyclone-prone districts and for planning mitigation measures (Mohapatra et al. 2012a, b, c, 2021a, b; Mohapatra 2015).
- Increase in the validity period of rainfall and other weather parameters forecasts from 2 days in 2006 to 7 days in 2015 (Mohapatra et al. 2009) and spatial resolution from met. Sub-division to district-level and block-wise forecasts for selected blocks.
- Increase in city forecasts from around 30 cities in 2006 to 450 cities in 2020.
- Introduction of nowcasts (next 6 h) of thunderstorms for 898 cities by 2020 based on Doppler Weather Radars and highway and tourism forecasts (Roy et al. 2021).
- Quantitative precipitation forecast (QPF) to CWC for flood warnings increased from 125 to 153 river sub-basins. Validity of QPF has been increased from 5 to 7 days in 2015.
- Customized weather forecasts for Shri Amarnathji Yatra, Bhakra Beas Management Board (BBMB), Indian Coast Guard, and defense sectors.
- Services for optimum operation of activities like irrigation, shipping, and oil explorations were augmented. Specialized services were introduced for Power and health sectors.
- The above achievement in early warning system could be possible due to (i) Improvement in observational network (ocean, land, and atmosphere) and quality of data (DWR, Ship, Buoys, AWS, High wind speed recorders, etc.), (ii) Satellite images and derived products (Kalpana, INSAT 3D & 3A, SCATSAT) and other international satellites, (iii) Fast communication and data Exchange system, (iv) Superior computational capabilities, super computer facilities, (v) Improved Numerical modelling capabilities (modeling), (vi) Skilled Human Resource Capabilities, (vii) Improved tools and techniques of forecasting including Decision Support System (DSS), (viii) Excellent support and Inter-ministerial collaborations within different sister organizations of MoES and also with other R&D institutes, (ix) International collaborations, (x) Research and Development, (xi) Timely warning product (text & graphics) generation and dissemination with increased lead period utilizing latest information and communication technology, (xii) Effective triggering mechanism for disaster managers and Public, and (xiii)

reaching last mile through all available means of communication (Mohapatra et al. 2013a, b, c; Mohapatra and Sharma 2019).

4 Socio-Economic Benefit

The improvement in weather forecasting and early warning has resulted in socio-economic benefits in many socio-economic sectors. Here, the socio-economic impact on three specific sectors like agriculture, disaster management, and power sectors are presented and discussed.

4.1 Agriculture Sector

As per the assessment made by the National Council of Applied Economic Research (NCAER) (2015), 95% of surveyed farmers experienced improved reliability of service in recent years. The incremental profit due to AAS was assessed to be 10–25% of their net income. The Annual Economic Profit of 24% of farmers cultivating four principal crops (wheat, paddy, sugarcane, and cotton), after using AAS in 2010, was assessed at Rs. 38,463 crores which raised to Rs. 42,000 crores in 2015.

According to the National Center for Applied Economic Research (NCAER) (2020) study based on the survey of farmers adapting nine different crop practices using agrometeorological advisory services, 98% of farmers made modifications to at least one of nine critical practices based on weather advisories. Average annual income to surveyed farmers was Rs 2.43 Lakh with modification in 1–4 practices, Rs 2.48 Lakh with modification in 5–8 practices, Rs 3.02 Lakh in all the 9 practices, and Rs 1.98 Lakh with no modification in crop practices based on weather advisories issued by IMD. About 80% of farmers receiving information on weather reported to have reduced losses. The estimated additional annual income of Rs. 12,500 per agricultural household belonging to the below poverty line (BPL) category in rainfed areas. The total income gain is estimated at Rs. 13,331 crore per annum in rainfed districts. The use of twice a week of weather advisories by the farmers increased to 59% in 2019 from 7% before 2015.

4.2 Disaster Management Sector

The FRONT LINE Magazine in its November 1999 issue described the Scientific failures towards the management of the Odisha super cyclone, October 1999. It said that the scientific systems whose responsibility it was to predict the contours of the cyclone did a far from perfect job. To be able to do a better job next time around, an integrated approach to cyclone studies is needed. The same FRONT

LINE Magazine in its November 2013 issues published an article entitled "ACING THE STORM". It said, the India Meteorological Department, with improved models and observation systems and greater forecast skills, predicts accurately not only the intensity of cyclone Phailin but also its landfall. During Phailin, the genesis forecast was issued 3 days in advance. The track, intensity, and landfall forecast was issued 5 days in advance. The increase in confidence of disaster managers and general public led to effective management of cyclone and minimum loss of lives (Mohanty et al. 2015). Only 22 lives were lost in Phailin against 9885 in Odisha Super Cyclone (1999). The economic loss has been $4500 million due to Odisha Super Cyclone (1999) as against $696 million due to VSCS Phailin (2013). The ex gratia paid by the government during Odisha Super Cyclone (1999) was Rs. 593 crore as against Rs. 1.24 crore during VSCS Phailin (2013). (Assuming that ex gratia @ Rs. 6 lakhs was paid by the government in both the years). The area of evacuation has decreased from 500 km during the 1999 Odisha Super Cyclone to 180 km during Phailin. Thus, the cost of evacuation has reduced being Rs. 180 crore during Phailin (2013) against Rs. 500 crore during the Odisha Super Cyclone (1999) (Assuming the cost of evacuation as Rs. 1 crore per km for both the years). There has been a decrease in the area of evacuation by 300 km in 20 years and hence evacuation cost by 60% and a decrease in ex gratia paid by government to survivors by about 99% as compared to 1999. Due to the improvement in cyclone warning services, there has been an increase in the confidence of disaster managers and public to manage cyclones. Based on an independent survey by NCAER in Andhra Pradesh and West Bengal in 2015, more than 95% population believe and appreciate cyclone warnings by IMD. Thus, society benefitted in terms of security of life and property in India and neighboring countries. It has led to better implementation of evacuation plan and hence a decrease in death toll and management of natural resources and infrastructure.

The success story was repeated thereafter leading to <100 deaths due to any cyclones crossing the coast of India (Amphan 2020 (76), FANI 2019 (64), TITLI 2018 (78), HUDHUD 2014 (46), and PHAILIN 2013(21) compared to thousands due to similar cyclones in past like the Odisha Super Cyclone (1999) (10,000). It also led to minimum loss of live stocks, minimum government expenditure towards evacuation due to the decrease in area of evacuation through better landfall point forecast, and minimum government expenditure towards payment of exgratia for loss of lives. About Rs. 590 crores in ex gratia payments and Rs. 32 crores in evacuation are saved for each landfalling cyclone double the entire cost of modernization program of IMD from 2008 to 12 (Rs. 437 crores).

The comparative figures in Table 3 indicate the socio-economic benefits accrued due to the improvement in cyclone forecasting and early warning services. Figure 5a shows the number of lives lost due to cyclones crossing coasts of different countries bordering the north Indian Ocean. It indicates that the improved early warning system of cyclone developed by IMD/MoES helped immensely 13 other countries of the WMO/ESCAP Panel in the region also. There have been 1,40,000 deaths in Myanmar due to cyclone "Nargis" in 2008 against <100 deaths due to any cyclone in recent years (Sagar 2018 (53, Somalia), Mekunu 2018 (26, Oman), Luban 2018 (14, Yemen), Chapala 2015 (5, Yemen), Megh 2015(18, Yemen), Bulbul 2019 (6, Bangladesh),

Table 3 Comparative analysis of monitoring and forecasting of cyclone Phailin (2013) and Odisha Super Cyclone (1999)

Odisha Super Cyclone (1999)	Extremely severe cyclonic storm, Phailin (2013)
• No objective forecast • Lead period was less (24 h) • Accuracy was moderate • Poor Warning communication and triggering mechanism • Poor response and evacuation (44, 000 people)	• Objective track, intensity, and landfall forecast—5-day lead • Accurate impact-based warning (Rain, storm surge) • Effective communication and triggering mechanism • Effective response and evacuation (1 Million people)
• Loss of lives: 9887	• Loss of lives: 21
• Ex gratia paid by Govt. @ Rs. 0.6 Million: Rs. 5932.2 Million	• Ex gratia paid by Govt. @ Rs. 0.6 Million: Rs. 12.6 Million
• Area of evacuation: 500 km • Cost of evacuation (Rs.10 Million/km): 5000 Million	• Area of evacuation: 200 km • Cost of evacuation (Rs 10 Million/km): 2000 Million

Assumption: similar amounts would have been spent for evacuation and payment of ex gratia in 1999 as in 2013

and Amphan 2019 (18, Bangladesh) against lakhs in past (Nargis 2008 (1.4 lakhs, Myanmar)).

IMD earned appreciations globally including appreciations from the Honorable Prime Minister and the President of India, the Parliamentary Standing Committee, the Indian Science Congress, the United Nations, WMO, and WMO ESCAP Panel member countries.

In the past few years, despite heat wave becoming a major challenge, the endeavors of the National disaster management authority (NDMA) in leading the coordinated efforts of the Central Government Ministries, the Departments of State Governments, the District Administrations, the India Meteorological Department, and civil society in a planned way resulted in a significant reduction in mortality due to heat wave from 2,050 deaths in 2015 to 4 in 2020 (NDMA 2021) as shown in Fig. 5b.

4.3 Power Sector

Accurate anticipation of extreme weather conditions, i.e., snowstorms, dust storms, high winds, thunderstorms, heavy rain, cyclones, etc., facilitates advance operation planning, secure system operation, and early restoration of the affected areas. Timely and effective forecasting of weather is one of the key issues in enabling integration/higher penetration of renewable generation in the Indian Power System in the times to come. In future, meteorologists may be called upon to integrate solar

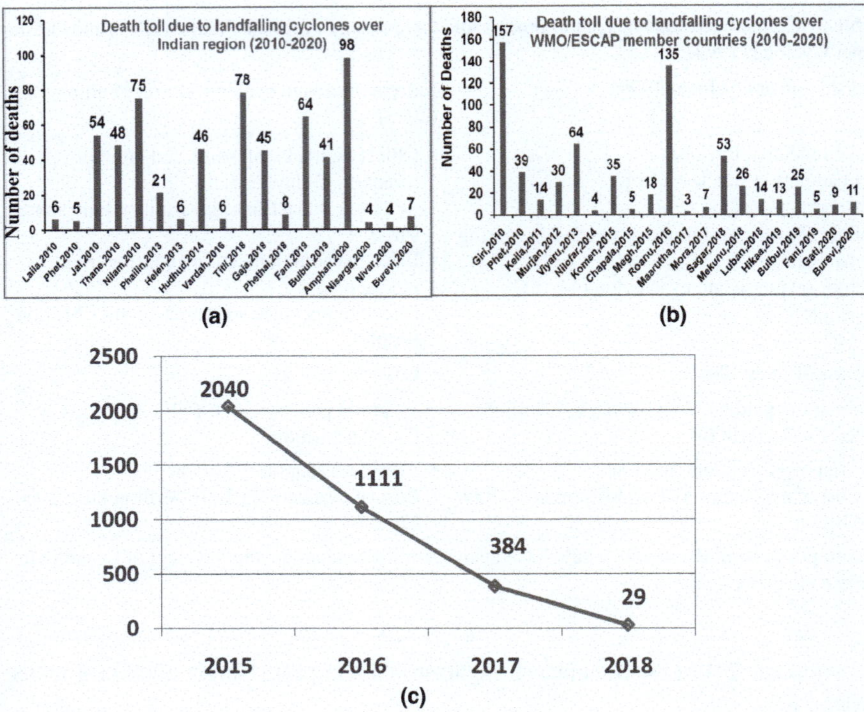

Fig. 5 **a** Number of deaths due to cyclones developing over the north Indian Ocean and crossing different coasts during 2010–2020. **b** Number of deaths due to heat waves over India during 2015–2020 (*Source* NDMA, Govt. of India)

irradiance/wind forecast into load for economic optimization of generation. Meteorologists can use their statistical background and forecast expertise to better bracket the intra-day and day-ahead renewable energy forecast and load forecast. They are uniquely equipped to integrate the science of meteorology into informed operational decisions in power system The Weather Portal provides both near real-time weather information and forecast of different weather parameters across the country. Accurate anticipation of weather variations would pave the way for efficient, secure, and reliable operation of the Indian power grid according to Power System Operation Corporation (POSOCO) Ltd., a Government of India Enterprise (POSOCO 2019).

All the Regional Load Despatch Centers (RLDCs), National Load Despatch Centers (NLDC), and constituents have started using the products/features available on the weather Portal on regular basis. The forecast available at the portal helps constituents to take proactive steps such as demand estimation, decisions on STOA (Short Term Open Access), etc. The near real-time information available from Radar and Satellite help system operators to take real-time decisions related to Grid operation. Meteogram, wind, and rain forecast for 27–29 May 2017 helped in better load assessment of UP control area by UP State Load Despatch Center (SLDC). As anticipated, UP demand went down from 19,000 MW to 17,000 MW due to the change

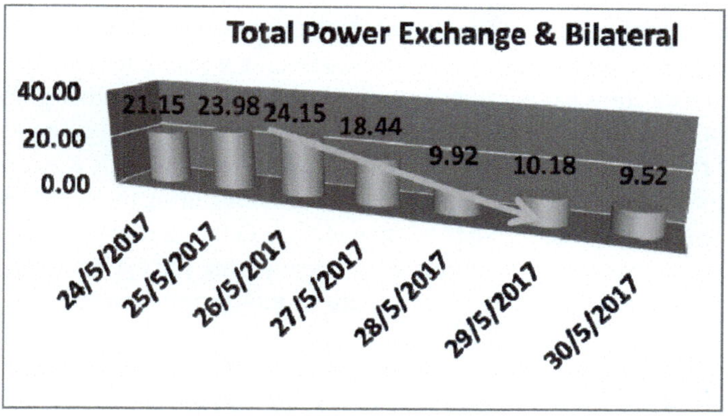

Fig. 6 Total power exchange and bilateral during 24–30 May 2017 (*Source* POSOCO, Govt. of India)

in weather conditions. Accordingly, STOA and purchase from power exchange of the order of 2000 MW was reduced, i.e., backing down of approximately 13 MU of costly thermal generation (Fig. 6).

Delhi's Discom BRPL has managed to take proactive action to manage its drawl from the grid and purchases from power exchange/bilateral contracts. The discom has made savings by utilizing weather portal. BRPL has used meteogram to predict the increasing temperature and therefore increase in demand, based on which requests were made to SLDC Delhi to bring on bar additional units. This also helped in following merit order despatch, by avoiding costly power exchange purchases during peak summer. In addition to meteogram, live radar imagery helped sending in timely revision for backing down of generating stations in anticipation of thunderstorms/rain, which saved under drawl of power thereby also ensuring grid security.

The Northern Region and the Eastern Region of the power system in India face foggy conditions during the night and early morning hours of winter months. The fog coupled with pollution leads to the tripping of EHV lines. To mitigate the adverse effect of tripping on power system, Grid operators have taken successful action based on near real-time color composite images of fog from a satellite image.

There was a depression over the northwest Bay of Bengal on 9 December 2017. It was known beforehand from the weather website that heavy rain at isolated places was very likely over West Bengal and Odisha. Anticipating lower demand and high frequency, Eastern Region Load Despatch Center (ERLDC) suggested NLDC to issue regulation down the instruction in advance which helped in maintaining frequency profile.

Earlier Bihar was not capable of predicting demand profile accurately, causing a large difference between schedule and actual exchange. With the introduction of weather portal the demand prediction of Bihar improved remarkably. Here is a comparison of schedule and actual exchange on 8 October 2016 and 2017.

Weather portal for Southern Region helped the system operator and the system reliability team to take precautionary measures in issuing warning notes to the southern states to prepare for a bad weather and a probable cyclone. Weather portal gave a timely warning on the depression formed over the southwest Bay of Bengal near the southeast Sri Lankan coast on the morning of 29 November 2017. It was identified that this depression would develop into a cyclonic storm during the next 12 h. System reliability and study group of SRLDC prepared an alert report explaining the probable cyclone and various precautions to be taken during the cyclone period by the system operator. The following aspects were of utmost importance for the system operation.

The improvements in forecast accuracy also resulted in significant gains to various sectors such as power sector saved around 500 crores each from cyclone warnings during Phailin (2013) and Hudhud (2014). There has been significant utilization of weather and climate information by the power sector in India. There has been a bilateral MoU with POSOCO, Ministry of Power, Government of India, and IMD for this purpose. A common website has been developed by the two organizations. It is being used by all power generation and distribution companies in India. It has resulted in minimizing the power losses and saving the economy. Taking an example, the Meteogram, wind, and rain forecast for 27–29 May 2017 helped in better load assessment of UP control area by UP SLDC. As anticipated, UP demand went down from 19,000 MW to 17,000 MW due to the change in weather conditions. Accordingly, STOA and purchase from Power Exchange of the order of 2000 MW were reduced, i.e., backing down of approximately 13 MU of costly thermal generation as shown in Fig. 6. Power sector has shown its ability to face the situation, like super cyclone Phailin with proper planning ahead to meet the contingency in operation during the incident, and in the aftermath taking fast restorative action to minimize disruption of electric supply using different facets of Smart Grid operation, and quick transportation of alternate means of transmission as well as decentralized generation (Mukhopadhyay et al. 2014). Due to the availability of all this relevant information the instances of flashovers/dc line faults which could be captured only by PMUs (and not by the SCADA), the operators became alert and could reduce the power order on the HVDC Talcher–Kolar Bi-pole in time along with shutdown/backing down of critical generating units and transmission lines.

5 Gap Areas

Weather prediction models are used for generating forecasts of weather and climate parameters. However, the present-day forecast models are known to have systematic errors and random biases. Inaccurate treatment of physical processes in the models (Physical parameterization schemes) is one of the largest sources of errors in weather and climate prediction models. Therefore, it is important to improve physical parameterization schemes in the models. This is an important, but difficult task. Development of new physical parameterization schemes or improvement of existing schemes will need many observational data of physical processes. These data are used for testing a physical hypothesis, validating physical processes in the models, and also for tuning empirical constants used in the physical parameterization schemes. Therefore, observational campaigns, physical process studies, and model development should go together.

At present, there is a skillful short-range prediction of weather up to 3–5 days over the Indian region. Since short-range forecasts depend on initial conditions, improvement in initial conditions can make a visible improvement in prediction skill. Therefore, observational network (conventional, satellite, and Radar) needs to be strengthened and the data assimilation methods to ingest the observations need to improve. More stress is needed to ensure good quality of data and their timely reception for data assimilation. The current skill level of 3–5 days can be improved to 7–8 days through an ensemble approach already adapted by IMD in forecasting in addition to the deterministic forecasting system at a very high horizontal resolution of about 5 km for global model and 1 km for regional mesoscale model.

Significant research has been carried out in the past to enhance the understanding on Indian monsoons, however, challenges and uncertainties still remain with respect to the accuracy of monsoon predictions and precipitation forecast skill improvements in short to medium range. The users also need forecasts for much smaller spatial and temporal scales, for example, at block/panchayat/station level. In many instances of extreme rainfall having potential of causing floods, flooding events are missed causing loss to economy and life as the meteorological forecasts are not readily usable by various stakeholders. Therefore, there is an urgent need to improve and customize meteorological forecasts specifically for floods. For formulating the flood forecast in the real time, the observed meteorological and flow data are used in the calibrated and validated real-time flood forecasting model to forecast the flood flow and corresponding water levels for different lead periods varying from few hours to few days depending on the size of catchment and purpose of the forecast.

Accurate forecasts of the intensity of Tropical Cyclones (TCs) still remain a gap area which sets the basis for research in improving the model. There is also scope to further improve prediction skill in predicting landfall and associated rainfall, storm surge, and winds. The recent research analysis suggests that models do not perform well in predicting associated heavy rainfall during the landfall of tropical cyclones.

Large research gaps include the understanding and prediction of thunderstorms and associated adverse weather. Denser Radar network not only would help detecting mesoscale convective system but also would help constraining the model parameters for better representation of convection. Assimilation of mesonetwork observational data would be necessary to generate mesoscale analyses/reanalyzes representing the regional heterogeneity. Predictability of convection initiation and scale interactions associated with the superposition of microscale, mesoscale, and synoptic-scale processes needs to be studied. Already, MoES initiated high-resolution forecast up to 48 h for lightning, squall, hailstorm, etc., since 2019. This remains as one of the most important gap areas of forecasting.

Though there has been significant improvement, there is still scope to improve in predicting prolonged heat and cold wave spells and improve forecast accuracy of flash flood events over the urban and complex topography. Another detrimental weather phenomenon during Indian winter months is the dense fog formation over the Gangetic Valley which has very high damage potential (physical as well as economical). Objective tools for short- and medium-range forecasting of timing of onset, intensity, and duration of fog well in advance falls into one of the largest gap areas. A comprehensive understanding of micro-meteorological and chemical processes responsible for the haze or fog formation over the Ganges Basin and fog impact on health and ecosystems is not clearly addressed yet, though there has been significant improvement by MoES through a special experiment, Winter fog experiment (WIFEX) in Delhi airport, air quality modeling, etc. High-resolution mesoscale model, coupled with interactive aerosol model with chemical details of species is also the need of the hour and is being addressed by MoES.

The urban development plans also should be supported by adequate weather-related support systems for information, management, and mitigation of urban hazards such as heat wave, flash floods, heat islands, air pollution dispersion, etc.

6 Conclusions and Future Plans

IMD continuously expands and strengthens its activities in relation to observing strategies, forecasting techniques, disseminating methods, and research relating to different aspects of extreme weather events to ensure the most critical meteorological support to disaster managers and decision-makers.

Demands from the public/private/government sectors are increasing for more accurate prediction of weather at various temporal and spatial scales. Improved and reliable forecast of weather and climate requires high-resolution dynamical models. Thus, intensive monitoring of various weather systems through different platform-based observing systems provide not only the necessary information about current weather systems but also their effective assimilation in numerical models provides important guidance for accurate forecasts. Hence, there is scope to improve further the weather services by providing high-resolution customized impact-based forecast (IBF) and risk-based warning products for different existing and emerging socio-economic sectors. IMD has already initiated IBF for various severe weather. There is a need to:

- strengthen the observational network (conventional, satellite, and Radar) on an urgent basis and to ensure good quality of data and their timely reception for data assimilation into numerical prediction models;
- improve quantitative precipitation forecasts (QPF) for river basins and major cities;
- develop a meteorological support system for flood warning and forecasting;
- improve the intensity forecast of tropical cyclones (TCs);
- improve in predicting anomalous/unprecedented severe weather episodes, viz., prolonged heat and cold wave spells, thunderstorms spells, dense fog, etc., and

improved forecast accuracy of these episodes in particular for flash flood events over the urban and complex topography;

- further enhance the understanding of Indian monsoons and to improve the forecast skill of monsoon predictions and precipitation forecast at different spaces and time scales;
- develop an advanced weather prediction system, for block/Panchayat/village level forecasts, skillful for next 3–5 days and develop advisories for sectors like Agriculture, Disaster Management, Water resources, power, tourism and pilgrimage, smart cities, renewable energy sector, and transport;
- carry out cutting-edge research studies on severe weather events, monsoons, meteorological droughts, desertification, and land use changes using observations and modeling;
- develop urban meteorological services with a dense mesonetwork, observational and modeling strategy to cater to the needs of growing cities in the country; and
- augment HPC and data archival/storage facilities for MoES institutes to achieve the above objectives, particularly on modeling and observational data storage/archival.

There is also scope for improvement of (i) sectorial applications of early warning, (ii) warning communication to last mile and disaster managers through state-of-the-art technology, (iii) developing synergized standard operation procedure among the early warning agencies and user agencies and, (iv) upgrading and enhancing the link between early warning service provider and disaster managers.

References

Dvorak VF (1984) Tropical cyclone intensity analysis using satellite data. Technical Report (NOAA TR NESDIS 11). National Oceanic and Atmospheric Administration, National Environmental Satellite, Data, and Information Service, pp 47

IMD (2012) Standard operation procedure: weather forecasting and warning. Cyclone Warning Division, IMD, New Delhi

IMD (2013) Cyclone warning services: standard operation procedure. Cyclone Warning Division, IMD, New Delhi

IMD (2021) Standard operation procedure: weather forecasting and Warning, IMD, New Delhi

Kalsi SR (2002) Use of satellite imagery for tropical cyclone intensity analysis and forecasting. Met. Monograph, Cyclone Warning Division No. 1/2002, pp 100

Kotal SD, Bhattacharya SK (2013) Tropical cyclone Genesis Potential Parameter (GPP) and it's application over the north Indian Sea. Mausam 64:149–170

Kotal SD, Roy Bhowmik SK (2011) A multi-model ensemble technique for cyclone track prediction over north Indian Sea. Geofizika 28:275–291

Kotal SD, Roy Bhowmik SK, Das AK (2008) A statistical cyclone intensity prediction model for Bay of Bengal. J Earth Syst Sci 117:157–168

Mahadevaiah M, Reddy DR, Sashikala G, Sreenivas G, Reddy PK, Redddy BB, Nagarani K, Rathore LS, Singh KK, Chattopadhyay N (2012) A framework to develop content for improving agromet advisories. In: The 8th Asian federation for information technology in agriculture (AFITA 2012), Taipei, September 3–6, 2012

Mohanty UC, Osuri KK, Routray A, Mohapatra M, Pattanayak S (2010) Simulation of Bay of Bengal tropical cyclones with WRF model: impact of initial and boundary condition. Mar Geodesy 33:294–314

Mohanty UC, Osuri KK, Tallapragada V, Marks FD, Pattanayak S, Mohapatra M, Rathore LS, Gopalakrishnan SG, Niyogi D (2015) A great escape from the Bay of Bengal "Super Sapphire-Phailin" tropical cyclone: a case of improved weather forecast and societal response for disaster mitigation. Earth Interactions 19. https://doi.org/10.1175/EI-D-14-0032.1

Mohapatra M (2015) Cyclone hazard proneness of districts of India. J Earth Syst Sci 124:515–526. https://doi.org/10.1007/s12040-015-0556-y

Mohapatra M, Sharma M (2019) Cyclone warning services in India during recent years: a review. Mausam 70(4):635–666

Mohapatra M, Hatwar HR, Bandyopadhyay BK, Subrahmanyam V (2009) Evaluation of heavy rainfall warning over India during summer monsoon season. Mausam 60(4):475–490

Mohapatra M, Mandal GS, Bandyopadhyay BK, Tyagi A, Mohanty UC (2012a) Classification of cyclone hazard prone districts of India. Nat Hazards 63:1601–1620

Mohapatra M, Bandyopadhyay BK, Tyagi A (2012b) Best track parameters of tropical cyclones over the North Indian Ocean: a review. Nat Hazards 63:1285–1317

Mohapatra M, Nayak DP, Bandyopadhyay BK (2012c) Evaluation of cone of uncertainty in tropical cyclone track forecast over north Indian Ocean issued by India Meteorological Department. Trop Cyclone Res Rev 2:331–339

Mohapatra M, Sikka DR, Bandyopadhyay BK, Tyagi A (2013a) Outcomes and challenges of forecast demonstration project (FDP) on landfalling cyclones over the Bay of Bengal. Mausam 64:1–12

Mohapatra M, Nayak DP, Sharma RP, Bandyopadhyay BK (2013b) Evaluation of official tropical cyclone track forecast over north Indian Ocean issued by India Meteorological Department. J Earth Syst Sci 122:589–601

Mohapatra M, Bandyopadhyay BK, Nayak DP (2013c) Evaluation of operational tropical cyclone intensity forecasts over north Indian Ocean issued by India Meteorological Department. Nat Hazards. https://doi.org/10.1007/s11069-013-0624-z

Mohapatra M, Bandyopadhyay BK, Tyagi A (2014) Construction and quality of best tracks parameters for study of climate change impact on tropical cyclones over the North Indian Ocean during satellite era, monitoring and prediction of tropical cyclones in the Indian ocean and climate change (Eds. Mohanty et al). Springer, Netherlands and Capital Publishing Company, New Delhi

Mohapatra M, Srivastava AK, Balachandran S, Geetha B (2017) Inter-annual variation and trends in tropical cyclones and monsoon depressions over the north Indian Ocean. In: Rajeevan MN, Nayak S (eds) Observed climate variability and change over the Indian region. Published by Springer Geology, Springer Geology, pp 89–106. https://doi.org/10.1007/978-981-10-2531-0_6

Mohapatra M, Sharma M, Devi SS, Kumar SVJ, Sabade BS (2021a) Frequency of genesis and landfall of different categories of tropical cyclones over the north Indian Ocean. Mausam 72:1–26

Mohapatra M, Kumar N, Mishra K et al (2021b) Evaluation of heavy rainfall warnings of India National Weather Forecasting Service for monsoon season (2002–2018). J Earth Syst Sci 130. https://doi.org/10.1007/s12040-020-01549-z

Mukhopadhyay S, Soonee SK, Agrawal VK, Narasimhan SR, Saxena SC (2014) Impact of super-cyclone Phailin on power system operation—defense mechanism and lesson learned. In: IEEE PES general meeting|conference & exposition, National Harbor, MD, USA, 2014, pp 1–5. https://doi.org/10.1109/PESGM.2014.6939338

Murty PLN, Padmanabham J, Srinivasa Kumar T, Kiran Kumar N, Ravi Chandra V, Shenoi SSC, Mohapatra M (2017) Real-time storm surge and inundation forecast for very severe cyclonic storm 'Hudhud'. Ocean Eng 131:25–35

NCAER (2015) Report on "Economic benefits of dynamic weather and ocean information and advisory services in india and cost and pricing of customized products and services of ESSO-NCMRWF & ESSO-INCOIS". National Council of Applied Economic Research (NCAER), New Delhi, August 2015

NCAER (2020) Report on "Estimating the economic benefits of investment in monsoon mission and high performance computing facilities". National Council of Applied Economic Research (NCAER), New Delhi, July 2020

NDMA (2019) National Disaster Management Plan. Published by National Disaster Management Authority, Ministry of Home Affairs, Government of India, pp 1–347

NDMA (2021) Early Planning for early action—heat wave, Aapda Samvaad, February 2021

Naresh Kumar AK, Jaswal MM, Kore PA (2017) Spatial and temporal variation in daily temperature indices in summer and winter seasons over India (1969–2012). Theoret Appl Climatol 129(3–4):1227–1239

Osuri KK, Mohanty UC, Routray A, Mohapatra M (2012) The impact of satellite-derived wind data assimilation on track, intensity and structure of tropical cyclones over the North Indian Ocean. Int J Rem Sens 33:1627–1652. https://doi.org/10.1080/01431161.2011.596849

Pai DS, Nair SA, Ramanathan AN (2013) Long term climatology and trends of heat waves over India during the recent 50 years (1961–2010)

POSOCO (2019) Weather information portal for indian power system reference document. Power System Operation Corporation Ltd. (A Government of India Enterprise) B-9, Qutub Institutional Area, Katwaria Sarai, New Delhi

Raghavan S (1997) Radar observations of tropical cyclone. Mausam 48:169–188

Raghavan S (2013) Observational aspects including weather radar for tropical cyclone monitoring. Mausam 64:89–96

RSMC New Delhi (2020) Report on cyclonic disturbances over north Indian Ocean during 2019. Published by Cyclone Warning Division, IMD, New Delhi

RSMC New Delhi (2021) Report on cyclonic disturbances over north Indian Ocean during 2020. Published by Cyclone Warning Division, IMD, New Delhi

Roy SS et al (2021) A new paradigm for short range forecasting of severe weather over the Indian region. Meteorol Atmos Phys. https://doi.org/10.1007/s00703-021-00788-z

Roy SS, Mohapatra M, Tyagi A, Roy Bhowmik SK (2019) A review of Nowcasting of convective weather over the Indian region. Mausam 70:465–484

Singh A, George JP, Iyengar GR (2018) Prediction of fog/visibility over India using NWP Model. J Earth Syst Sci 127:26. https://doi.org/10.1007/s12040-018-0927-2

Tripathi P (2015) Interdisc J Contemp Res 2:91–98

Velden CS, Harper B, Wells F, Beven JL II, Zehr R, Olander T, McCrone P (2006) The Dvorak tropical cyclone intensity estimation technique: a satellite-based method that has endured for over 30 years. Bull Amer Met Soc 87:1195–1210

Operational Seasonal Forecasting of the Southwest Monsoon Rainfall

O. P. Sreejith, Divya Surendran, Arti Bandgar, and D. S. Pai

Abstract In this paper, various methods used by the India Meteorological Department (IMD) to issue Long-Range Forecast (LRF) of Southwest monsoon rainfall and its current status are discussed. IMD started issuing long-range forecast for rainfall in 1886 using conditions like snow cover. Since then, several forecast techniques were developed for improving the accuracy of seasonal forecasts. Presently, the Statistical Ensemble Forecast System (SEFS) based on empirical method and Monsoon Mission Climate Forecast System (MMCFS) based on dynamic seasonal prediction system are used to issue monsoon rainfall forecast for the country. A comparison of the present forecasting method based on SEFS reveals improvement in the average absolute error by 6.57% during the recent 13 years period, as compared to the previous 13 years. The absolute error in the MMCFS simulations during the period 2007–2019 was slightly higher, 7.8% indicating that the performance of SEFS is better for all India rainfall seasonal forecasts for the recent years. The seasonal forecasts of all India rainfall for the southwest monsoon season based on the MMCFS model have shown comparatively good skill in predicting below-normal rainfall during the El Nino events (e.g., 2014 and 2015). Thus, advancement in research and development is needed to further improve the seasonal forecasting methods.

1 Introduction

Reliable climate information enables countries in planning adaptation and mitigation strategies under adverse climate events and is also useful for various sectoral applications. In recent years, many countries implemented the Global Framework of Climate Services (GFCS) in order to strengthen the production and delivery of science-based climate prediction and services. The agriculture sector is one of the important sectors which require skillful seasonal forecasts. The agricultural practices in India have traditionally been tied strictly to the annual cycle of rainfall which is dominated by two monsoons; southwest and northeast monsoons. The southwest

O. P. Sreejith (✉) · D. Surendran · A. Bandgar · D. S. Pai
Office of Climate Research and Services, India Meteorological Department, Pune, India
e-mail: op.sreejith@imd.gov.in

V. K. Gahalaut and M. Rajeevan (eds.), *Social and Economic Impact of Earth Sciences*,
https://doi.org/10.1007/978-981-19-6929-4_2

monsoon season extending from June to September is the principal rainy season for most parts of the country. Most parts of the country receive more than 75% of its annual rainfall during this season. In Gujarat, Saurashtra, and Kutch adjoining Rajasthan and Madhya Pradesh, the percentage is more than 90%. The summer monsoon accounts for almost 78% of the gross cropped area. About a third of the cropped area is still rainfed. The primary feature of the annual cycle of the southwest monsoon is its stability and regularity with all India southwest monsoon season rainfall (ISMR) being within ±30% of its long period average (LPA) during almost all years and within ±10% of LPA in about 70% of the years. On the regional scale, the variability of the rainfall can be much more than this. The very regularity of the monsoon makes agriculture susceptible to changes in the annual cycle of rainfall. Total monsoon rainfall during the season has a statistically significant positive relationship with the crop yield over the country. In general, a weak monsoon year with significantly low rainfall can cause a low crop yield. On the other hand, a strong monsoon is favorable for abundant crop yield, although sometimes too much rainfall may cause devastating floods. Parthasarathy et al. (1988) and later Gadgil (1996) and Webster et al. (1998) have shown an in-phase variation of rice production in India with the all India summer monsoon rainfall.

The relationship of ISMR with all India major crop production during the concurrent Kharif season and during the subsequent Rabi season for the period 1966–2018, respectively, is presented in Fig. 1a and b. The crop production time series were first detrended to remove the technology trend from the series and anomalies were computed using normal computed for the base period of 1971–2010. It is seen that there is in-phase relationship between the rainfall and the detrended crop production anomalies of the both Kharif and Rabi seasons. This indicates that above-normal monsoon rainfall in addition to providing favorable conditions for Kharif crops helps in maintaining improved soil moisture during the subsequent winter season which is vital for the crops during the Rabi season. However, the relationship between ISMR and Kharif crop production is stronger (CC = 0.70) than that between ISMR and Rabi crop production (CC = 0.38).

The seasonal forecasting of summer monsoon rainfall over India has been one of the first targets of endeavors at tropical climate predictions and it has a history of more than 130 years. Although the duration of monsoon over various parts of India varies from about 2 months to 6 months, long-range forecasts are generally issued mainly for monthly and seasonal (4 months of June–September) scale. The year-to-year variation in the Indian summer monsoon rainfall (ISMR) is primarily attributed to its association with the slowly varying boundary forcing such as sea surface temperature, snow cover, soil moisture, etc. This is the predictable part of the interannual variability. The unpredictable part of the variability is due to the fast-atmospheric variations associated with the day-to-day weather.

Two main approaches are used for the seasonal forecasting of the ISMR. The first approach is based on the empirical (statistical) method. The statistical approach uses the historical relationship between the ISMR and predictors derived from global atmosphere–ocean parameters (mainly derived from slowly varying boundary

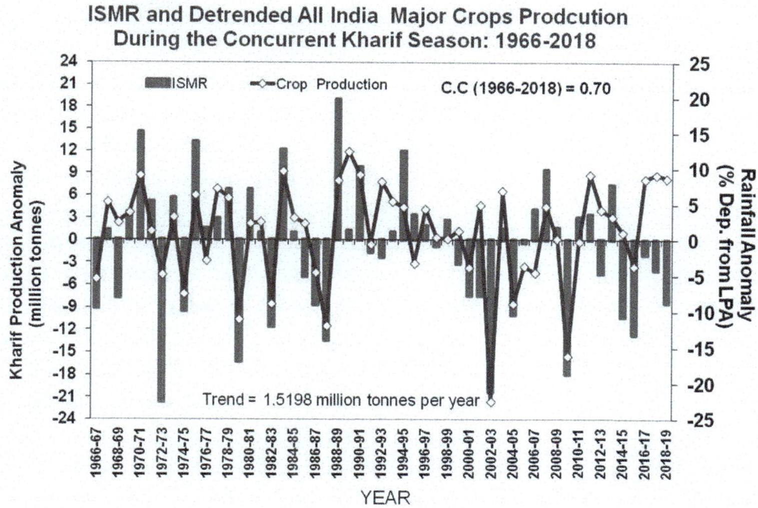

(a)

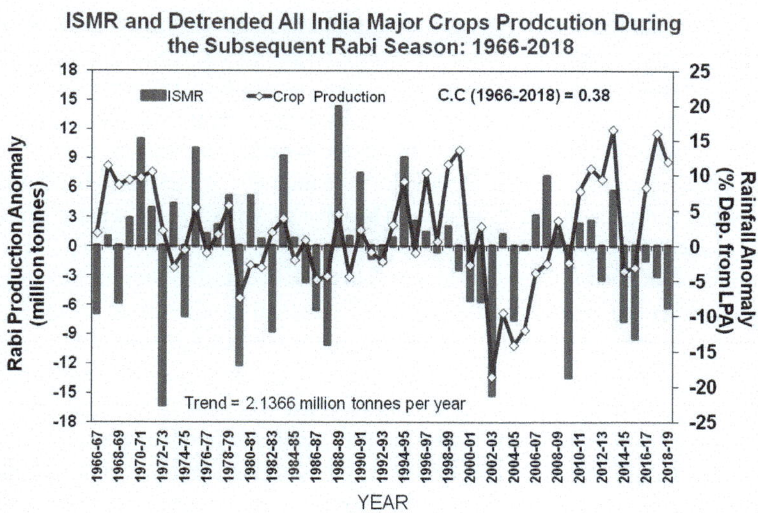

(b)

Fig. 1 **a** Relationship between the year-to-year variation of all India southwest monsoon season (June–September) rainfall anomaly and detrended all India major crops production anomaly during the Kharif season for the period 1966–2018 (*Data Source* https://eands.dacnet.nic.in/Advance_Estimates.htm). **b** Relationship between the year-to-year variation of all India southwest monsoon season (June–September) rainfall anomaly and detrended all India major crops production anomaly during the Rabi season for the period 1966–2018 (*Data Source* https://eands.dacnet.nic.in/Advance_Estimates.htm)

forcing). The second approach towards long-range forecasting is based on the dynamical method, which uses the General Circulation Models (GCM) of the atmosphere and oceans to simulate the summer monsoon circulation and associated rainfall. The India Meteorological Department (IMD) is the nodal government agency responsible for providing operational weather and climate services required for the country. For issuing operational seasonal forecast for the southwest monsoon rainfall, IMD uses the first approach based on state-of-the-art statistical models developed through in-house research and development work (Pai et al. 2011, 2017). For example, operational forecasts for the ISMR during the period 1988–2002 were prepared using 16-parameter power regression and parametric models (Gowariker et al. 1989, 1991). IMD introduced the new state-of-the-art statistical models in 2003–2007 (Rajeevan et al. 2004, 2007) following the review of an old forecasting system in 2002.

In the dynamical method, Coupled General Circulation Model (GCM) is used for seasonal forecasts. The basic conceptual hypothesis of seasonal predictability is based on the influence of boundary forcing at the Earth's surface as demonstrated by studies like Charney and Shukla (1981), Shukla and Wallace (1983), Livezey et al. (1996), etc. However, dynamical models have large systematic errors in the simulation of seasonal mean anomalies associated with changes in boundary conditions and hence predictability of Indian summer monsoon rainfall has been relatively low. Improvements in the coupled ocean general circulation models in the recent years have led to moderate skill in the prediction of monsoon rainfall (Rajeevan et al. 2012). From 2004 onwards, IMD has been also using a dynamical modeling approach to prepare experimental forecasts for the monsoon rainfall. At present, the dynamical coupled forecasting system (CFS) developed under the Monsoon Mission (MM) project of the Ministry of Earth Sciences (MoES) or MMCFS is used for preparing various seasonal forecasts.

The main objective of this review paper is to present details about the development techniques and performance of the various statistical models used in the present operational seasonal system for the southwest monsoon rainfall over India. Section 2 describes the brief history of the development of operational seasonal forecasting in India since 1886 when the first operational forecast was issued. Section 3 discusses details of the models used in the statistical forecasting system for preparing various operational seasonal forecasts. The performance of the operational seasonal forecast of southwest monsoon rainfall over the country as a whole during the period 1988–2019 is discussed in Sect. 4. Section 5 describes the experimental seasonal forecasting system based on the MMCFS and, finally, Sect. 6 presents the conclusions.

2 Brief History of Operational Seasonal Forecasting in India

The first operational seasonal forecasting of Indian summer monsoon rainfall for the region covering whole India and Burma was issued on June 4, 1886, by Sir H. F. Blanford who was the Imperial Meteorological Reporter in India and founder head of IMD in 1875. Forecasts during the initial years were subjective and qualitative. Blandford (1884) used the relationship between winter and spring snowfalls over the Himalayas and subsequent ISMR to prepare successful tentative forecasts from 1882 to 1885 and then to issue the first operational forecast for 1886. Over the years, the operational seasonal forecasting system in India underwent many changes in its approach and scope (Normand 1953; Jagannathan 1960; Thapliyal and Kulshrestha 1992; Hastenrath 1995; Krishna Kumar et al. 1995; Rajeevan 2001; Gadgil et al. 2005; Pai et al. 2011). Forecasting based on objective techniques such as correlation and regression analysis was first introduced by Sir Gilbert Walker. Between 1924 and 1987, operational forecasts were issued for Northwest India and Peninsular India using regression models initially developed by Walker and updated from time to time (Walker 1923, 1924). Forecast for the geographical regions was discontinued during 1988–1998. During 1988–2002, operational forecast for the season rainfall over the country as a whole was based on the 16-parameter power regression and parametric models (Gowariker et al. 1989, 1991). In view of increasing user demands, the operational forecasts for three geographical regions of the country (Northwest India, Peninsular India, and Northeast India) were reintroduced in 1999. The areas of these geographical regions were, however, different from that of Walker's geographical regions with the same names.

Since 2003 IMD adopted a new strategy for seasonal forecasting for the monsoon rainfall as suggested by Rajeevan et al. (2004). Accordingly, the forecasts were issued in two stages. The first stage forecast issued in April consisted of forecast for seasonal rainfall over the country as a whole and the second stage forecast issued in the end of June consisted of the update for the April forecast along with seasonal rainfall forecast for the geographical sub-regions of the country. These forecasts were issued based on the 8- and 10-parameter power regression and probabilistic discriminant analysis techniques. However, in recent years, the second stage forecast is issued in the end of May or early June. In 2004, an experimental dynamical forecasting system was implemented at IMD, Pune in collaboration with the Indian Institute of Science (IISc), Bangalore. The first dynamical prediction system was based on the seasonal forecast model (SFM) (Kanamitsu et al. 2002; Roads 2000) of the Experimental Climate Prediction Center (ECPC), USA, which was an atmospheric circulation model. In 2004, the country was reclassified into four sub-geographical regions, namely Northwest India, Central India, South Peninsula, and Northeast India and is continued till today. In 2007, a new statistical ensemble forecasting system (SEFS) was introduced for seasonal rainfall forecasting over the country as a whole (Rajeevan et al. 2007).

IMD started to issue additional forecast guidance with the experimental forecasts for the monsoon rainfall generated by MMCFS developed by the Indian Institute of Tropical Meteorology (IITM), Pune under the monsoon mission (MM) project of the Ministry of Earth Sciences (MoES) since 2012. The MMCFS was implemented in IMD, Pune, in 2017 for operational services. Various operational seasonal and monthly forecasts currently issued by IMD are listed in Table 1. As seen, in addition to the forecasts for the monsoon season, IMD also prepares seasonal forecasts for winter

Table 1 Various operational forecasts issued by IMD for southwest monsoon rainfall

Sr. no.	Forecast for	Region for which forecast issued	Issued in	Method/model
1	SW Monsoon season (June–September) rainfall	Country as a whole	April	Statistical, dynamical
2	SW Monsoon season (June–September) rainfall	Country as a whole	June	Statistical, dynamical
3	Southwest Monsoon onset	Kerala	May	Statistical
4	SW Monsoon season (June–September) rainfall	Four broad geographical regions: Northwest India, Northeast India, Central India, and South Peninsula	June	Statistical, dynamical
5	SW Monsoon monthly rainfall for July and August	Country as a whole	June	Statistical, dynamical
6	SW Monsoon second half of the season (August–September) rainfall	Country as a whole	July	Statistical, dynamical
7	September rainfall	Country as a whole	August	Statistical, dynamical

(January–March) precipitation (issued in January), northeast monsoon (October–December) rainfall over Southern Peninsula (issued in October), and subdivision wise temperature forecasts for the country for the hot and cold weather seasons. The temperature forecasts are prepared using MMCFS only and the rainfall/precipitation forecasts are prepared using both the MMCFS and state-of-the-art statistical models.

3 The Present Operational Seasonal Forecasting System Based on Statistical Models

At present, IMD issues seasonal forecasts for SW Monsoon rainfall over the country as a whole and four broad geographical regions (Northwest India, Central India, Northeast India, and South Peninsula) with useful skill.

There are three major changes in the SEFS used from 2007 for the operational seasonal forecasting of rainfall over the country as a whole from the 8/10-parameter power regression models used rom 2003 to 2006. These are (a) use of a new and smaller number predictors, (b) use of both linear and nonlinear statistical techniques, and (c) generation of ensemble forecasts similar to that in the dynamical forecasting system.

The set of eight predictors (Table 2) presently used in the SEFS has a statistically stable and significant relationship and strong physical linkage with the Indian southwest monsoon rainfall. For the April forecast, the first five predictors listed in Table 2 are used. For the updated forecast issued in May/June, the last six predictors are used that include three predictors used for April forecast. A schematic diagram of the statistical ensemble forecasting system is shown in Fig. 2. According to this forecasting system, the forecast for the seasonal rainfall over the country as a whole was computed as the ensemble forecasts prepared from the best few out of all possible models constructed using two techniques. The first set of models was constructed using multiple regression (MR) technique and the second set of models was constructed using projection pursuit regression (PPR) technique (Friedman and Stuetzle 1981; Chan and Shi 1999), which is a nonlinear regression technique. The independent forecasts for each model were prepared based on the moving training method with a fixed window period of 23 years. The model selection is done based on the performance of these models during some fixed independent test period. Performance of the April and June SEFS for the independent test period of 1981–2019 computed using the new ensemble method is shown in Fig. 3 and Fig. 4, respectively. The standard errors of the April and June SEFS for the period 1981–2019 are 65.7 mm (7.5% of long period average (LPA)) and 62.2 mm (7.1% of LPA), respectively. The CC between observed and forecast rainfall of the April and June SEFS for the period 1981–2019 are 0.58 and 0.61, respectively.

In addition to the quantitative forecast, the ensemble forecasting system has also been used to generate a five-category probabilistic rainfall forecast based on the forecast error distribution of the ensemble forecasting system. The five rainfall categories defined based on the observed data for the period 1901–2005 are deficient (<90% of LPA), below-normal (90–96% of LPA), normal (96–104% of LPA), above-normal (104–110% of LPA), and excess (>110% of LPA). The climatological probabilities of these five categories are 16%, 17%, 33%, 16%, and 17%, respectively. The long-period average of the seasonal rainfall over the country as a whole computed for the base period of 1961–2010 is 880.6 mm. The five-category probability forecast is prepared using normal probability distribution with the ensemble average of the forecast from the ensemble forecasting system as the mean and the standard error of the independent test period as the standard deviation.

From 1999 to 2003, IMD was issuing long-range forecasts for seasonal rainfall over the three broad geographical regions of India, viz., Northwest India, Northeast India, and Peninsular India using three individual power regression models based on different sets of predictors. In 2004, the country was reclassified into four geographical sub-regions, viz., Northwest India, Central India, Northeast India, and South Peninsular India (Fig. 5). These forecasts are prepared using separate principal component regression (PCR) models each based on a different set of predictors and with a model error of $\pm 8\%$.

The PCR model uses the moving training period method for preparing independent forecasts with a constant window period of 23 years. In this method, for the prediction of rainfall for a reference year, data of 23 years just prior to the reference year was first used for PC analysis of the predictor data set. The first few PCs that explain 80%

Table 2 Details of the 8 predictors used for the new ensemble forecast system

No.	Predictor	Used for forecasts in	Correlation coefficient (1981–2010)	Significant level	Predictor association with Monsoon
1.	Europe land surface air temperature anomaly (January)	April	0.39	At and above 5% level	Associated with the strength and phase of the North Atlantic Oscillation (NAO) as well as with the anomalous winter snow cover (Rajeevan et al. 1998)
2.	Equatorial pacific warm water volume anomaly (February + March)	April	−0.36	At and above 5% level	Used as precursor for El Nino/La Nina development (Rajeevan and McPhaden 2004)
3.	SST gradient between Northwest Pacific and Northwest Atlantic (December + January)	April and June	0.43	At and above 1% level	Related to ENSO and NAO
4.	Equatorial SE India ocean SST (FEB)	April and June	0.55	At and above 1% level	The interannual variation in the SST anomalies over the Southeast Indian region can be due to the changes in the solar surface heating caused by the changes in the cloud cover or resulted from the changes in the low-level dynamical flow pattern

(continued)

of the total variability of the predictors set during these 23years (training period) were then related against the predictor series for the same period using the multiple linear regression method. Scores of the selected PCs for the reference year were then calculated using the PC loading matrix and predictor values for the reference year. These score values along with the coefficients of the trained regression equation were used for calculating the predictor value for the reference year.

Table 2 (continued)

No.	Predictor	Used for forecasts in	Correlation coefficient (1981–2010)	Significant level	Predictor association with Monsoon
5.	East Asia MSLP (FEB + MAR)	April and June	0.55	At and above 1% level	Shifting of the climatological low-pressure area over the north Pacific and indication of changes in circulation pattern and the typhoon tracks over the North Pacific
6.	NINO 3.4 SST Anom. Tendency MAM(O)-DJF (−1)	June	−0.41	At and above 5% level	Provide indication of El Nino/La Nina development
7.	North Atlantic MSLP (MAY)	June	−0.44	At and above 1% level	Influence of North Atlantic oscillation and Quasi-Biennial oscillation
8.	North Central Pacific zonal wind gradient 850 hPa (MAY)	June	−0.59	At and above 1% level	Activity of the pre-Meiyu front are good precursors for ISMR

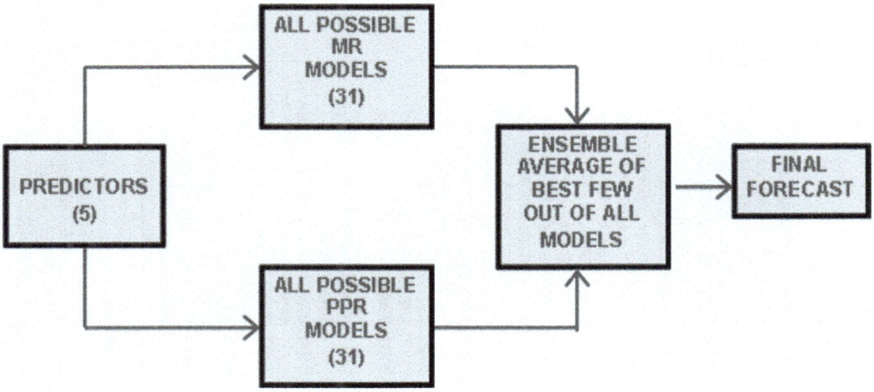

Fig. 2 The performace of the model used for first satage LRF issued in April

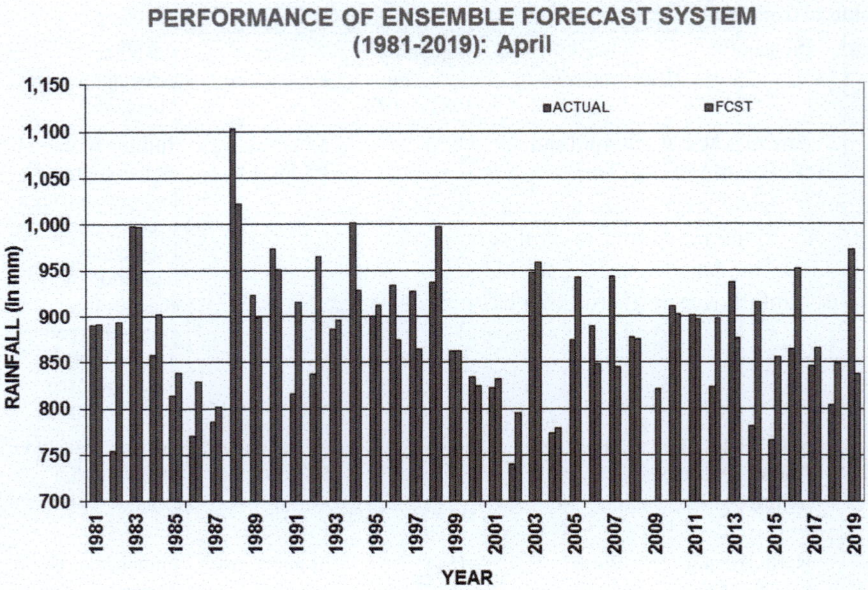

Fig. 3 The performace of the model usef for second stage forecast (issued in June)

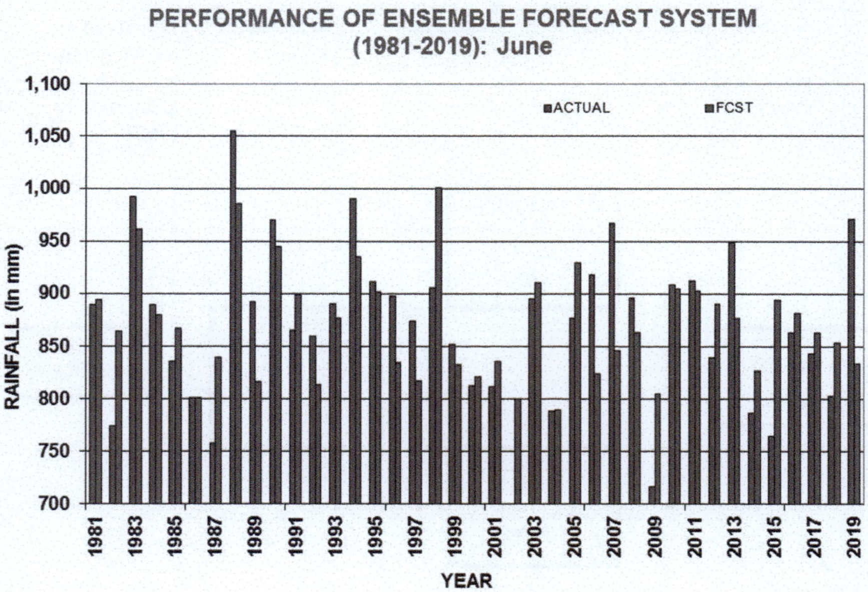

Fig. 4 The performance of operational forecast issued by IMD

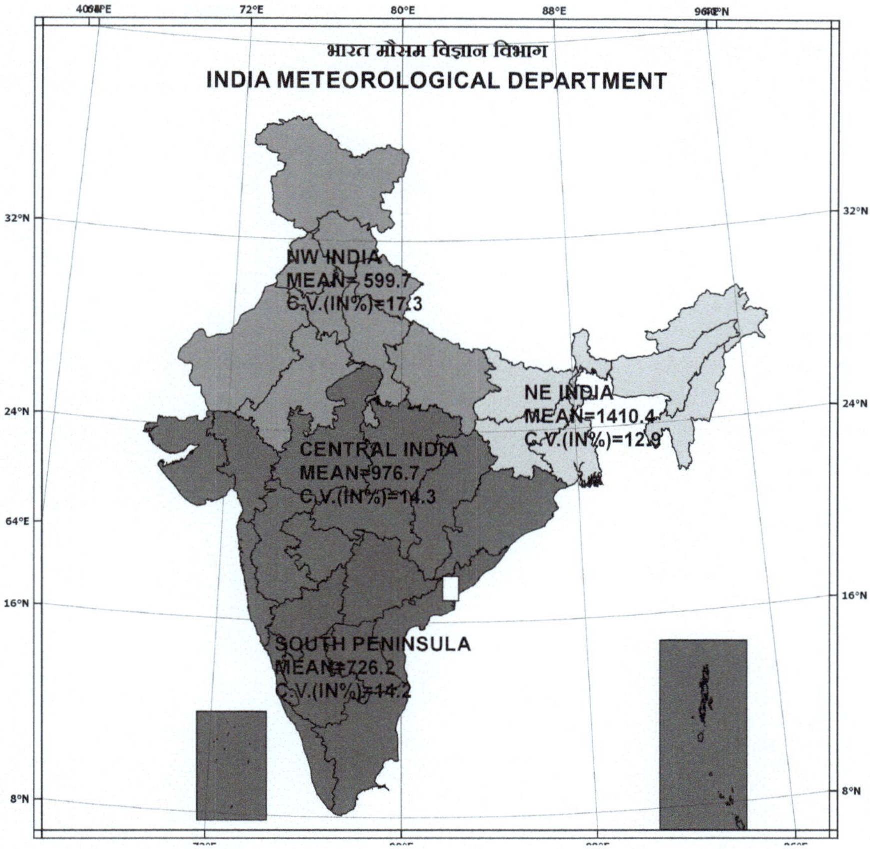

Fig. 5 The four homogeneous regions of India

Table 3 shows the performance of the four models for the last 15 years (2005–2019) along with RMSE during the same period.

4 Improvement in the IMD's Operational Seasonal Forecasting System for the All Indian Summer Monsoon Season Rainfall

The all India season rainfall forecast for the country as a whole was reintroduced in 1988 using 16-parameter power regression and parametric models after it was discontinued in 1923. IMD implemented new state-of-the-art LRF models in 2003–2007 following the review of an old forecasting system in 2002. The performance of the operational long-range forecast for the season rainfall over the country as a

Table 3 The performance of the four homogeneous regions for the last 15 years (2005–2019) along with RMSE during the same period

Year	NW India		Central India		NE India		South Peninsular India	
	Actual	Forecast	Actual	Forecast	Actual	Forecast	Actual	Forecast
2005	533	623	1092	1056	1197	1375	789	757
2006	606	534	1177	962	1247	1437	698	653
2007	501	590	1088	963	1691	1563	903	794
2008	648	545	975	866	1454	1393	695	679
2009	404	436	805	881	1117	1444	683	660
2010	658	603	1052	1045	1197	1449	821	743
2011	644	591	1092	904	1302	1409	710	647
2012	553	579	945	967	1418	1465	635	711
2013	677	505	1177	969	1088	1367	870	815
2014	510	570	885	1017	1246	1330	677	683
2015	509	544	827	943	1333	1362	598	744
2016	584	631	1035	1070	1291	1455	659	833
2017	551	602	920	935	1405	1374	715	651
2018	602	615	914	981	1080	1332	705	643
2019	592	538	1263	1012	1245	1262	842	677
RMSE (% of LPA) (2005–2019)	12.3		13.5		12.3		12.4	

whole for the period 1988–2019 is shown in Fig. 6. During the period, the absolute error was ≥10% of LPA in 8 years with the highest in 1994 (21%) followed by 2002 (20%).

The average absolute error (difference between forecast and actual rainfall) during the last 13 years (2007–2019) during which forecast was prepared using the new Statistical Ensemble Forecasting system (SEFS) was 6.57% of LPA compared to the average absolute error of 8.91% of LPA during the 13 years (1994–2006) just prior to that period. During 1994–2006, the forecast was within the ±8% of actual values for 8 years. Within these 8 years, forecast was within ±4% for 3 years. On the other hand, during 2007–2019, the forecast was within ±8% of actual values for 8 years with a forecast within ±4% during 5 years. Similarly, the CC between the actual and forecast rainfall for the periods (1994–2006) and (2007–2019) are −0.54 and 0.38, respectively. This clearly indicates improvement made in the operational forecast system in the recent 13 years' period compared to the earlier 13 years' period. It is not possible to have 100% success for forecasts based on statistical models. The problems with statistical models are inherent in this approach and are being faced by forecasters worldwide.

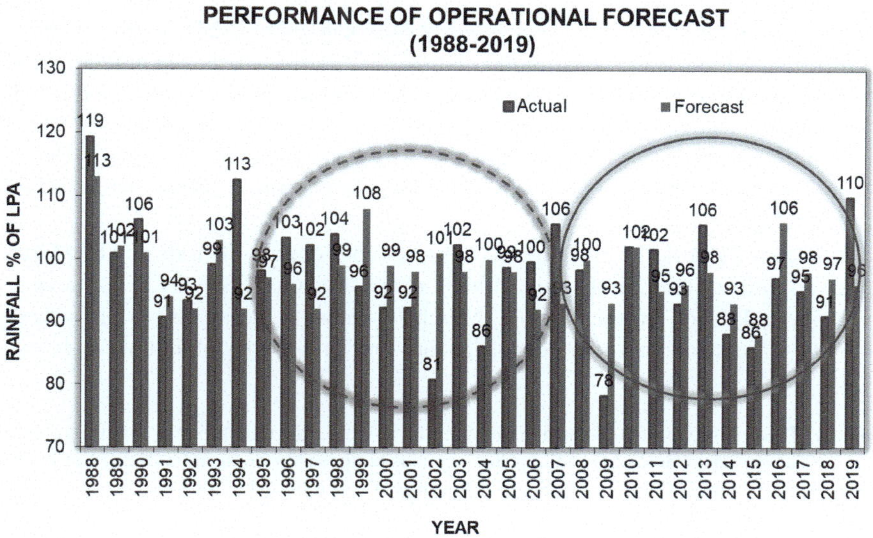

Fig. 6 Performance of IMD's Operational forecast for the season rainfall over the country as a whole for the period 1988–2019

5 Monsoon Mission Experimental Coupled Dynamical Model Forecasting System

The National Monsoon Mission (NMM) launched by the Ministry of Earth Sciences (MoES) in 2012 is a multi-institutional and inter-agency research program with the ultimate aim to develop a state-of-the-art dynamical prediction system for monsoon rainfall on different time scales (Rao et al. 2019). Indian Institute of Tropical Meteorology (IITM), Pune, coordinated and worked with different climate research centers from India and abroad on the development of a coupled model for forecasting the Indian summer monsoon rainfall (ISMR). Climate Forecast System version 2 (CFSv2) of the National Centers for Environmental Prediction (NCEP), USA, with a horizontal resolution of T126 (~100 km) was identified as the basic modeling framework for this purpose. IITM made improvement over the original version (Ramu et al. 2016). IMD, in 2017, adopted the latest high-resolution research version of the coupled model with a horizontal resolution of approximately 38km (T382).

The model climatology was prepared using retrospective forecasts generated for 28 years (1981–2008). The retrospective forecasts were prepared based on the average of 10 ensemble members with different initial conditions (ICs). The model skills (correlation coefficient (CC). and root mean square error (RMSE)) for the monthly and seasonal rainfall forecast over the country as a whole computed based on ICs pertaining to different months (January–May) is shown in the Table 4. CC and RMSE were computed between the observed all India rainfall and that derived from model hindcast for the period 1982–2008. Same skills are computed for the rainfall of four broad homogeneous regions of India (Northeast India (NEI), Northwest India (NWI), Central India (CI), and South Peninsula (SP)) also. The model hindcasts and forecasts were bias corrected using the z-score transformation (correction for both mean and variance) method (Hawkins et al. 2013). The model hindcasts performance of all India rainfall for the period 1982–2008 based on various ICs is shown in Fig. 7. Model forecasts of all India rainfall performance for the period 2009–2019 based on various ICs are shown in Fig. 8.

As seen in Table 4, the hindcast CC is highest for February IC followed by March IC for all India rainfall. In the case of homogenous regions, February IC is showing the highest skill for Northeast India, March IC showing the highest skill for both Northwest India and Southern Peninsula; and, for central India, all initial conditions

Table 4 The skill scores of the Monsoon Mission CFS model for the forecasting of seasonal rainfall over the country as a whole (AI) and four homogenous regions (Northeast India (NEI), Northwest India (NWI), Central India (CI), and South Peninsula (SP)) at different initial conditions

	JJAS									
	Model skill (CC)					RMSE (% of LPMA) (1981–2008)				
Initial conditions (IC)	AI	NEI	NWI	CI	SP	AI	NEI	NWI	CI	SP
January	0.30	0.23	0.21	0.30	0.37	9.9	11.7	17.4	14.4	14.5
February	0.48	0.31	0.41	0.09	0.39	9.7	11.4	16.6	18.4	14.4
March	0.46	0.29	0.51	0.15	0.61	10.6	11.6	15.3	19.3	12.3
April	0.37	0.23	0.34	0.06	0.40	9.9	11.6	17.3	17.0	14.3
May	0.23	0.16	0.15	0.06	0.25	10.1	11.7	19.2	15.6	15.3

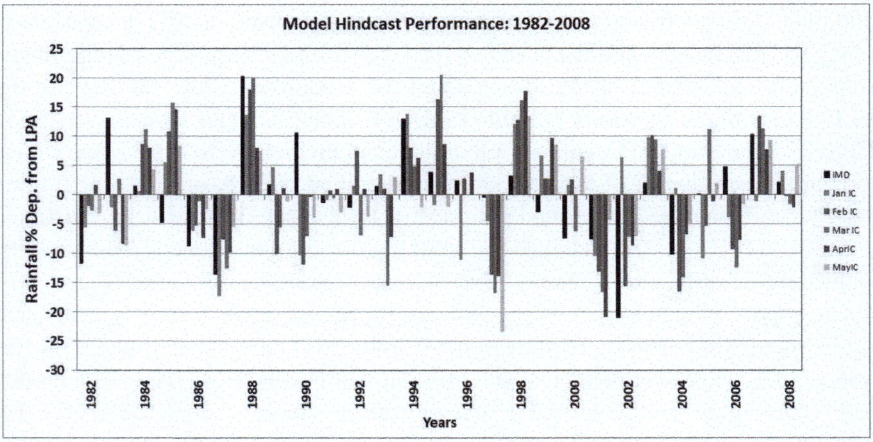

Fig. 7 Performance of the model hindcasts for the southwest monsoon season (June–September) rainfall over the country as a whole based on various initial conditions. The model hindcasts were bias corrected using the z-score transformation (correction for both mean and variance) method

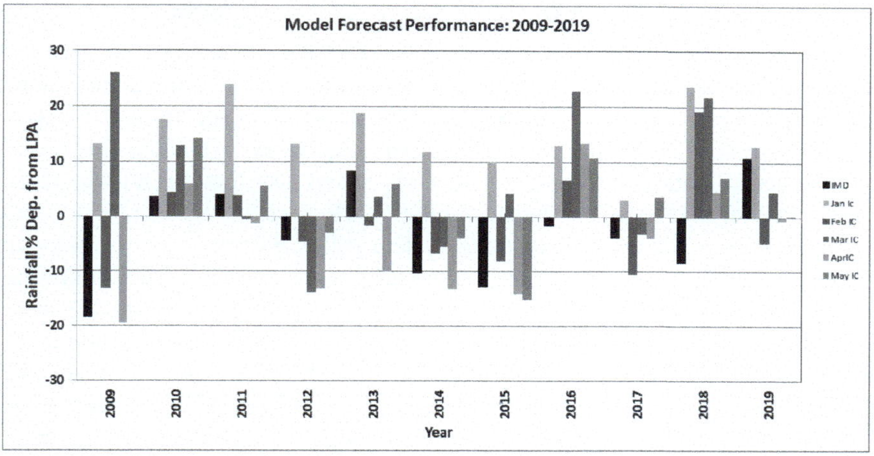

Fig. 8 Performance of the model forecasts for the southwest monsoon season (June–September) rainfall over the country as a whole based on various initial conditions. The model forecasts were bias corrected using the z-score transformation (correction for both mean and variance) method

are showing relatively poor skill except for January IC. The hindcast RMSE is the lowest for February IC followed by January IC and April IC for All India rainfall. In the case of homogeneous regions, hindcast RMSE are showing the lowest values for Northwest India and Southern Peninsula for March IC. The lowest RMSE value for Northeast India is from February IC and for Central India is from January IC. In Fig. 7, it is seen during 7(6) of 11 years 2009–2019, the February (March) IC-based forecast was able to indicate a correct sign of the observed rainfall anomaly. It was also observed that during the period 2007–2019, the absolute error of the all India model averaged rainfall based on February IC was 7.8% of LPA compared to 6.57% of the June SEFS system. Similarly, during the same period, the CC between the observed and forecasted all India rainfall based on MMCFS February IC (June SEFS) was 0.38 (0.4) indicating that in recent years SEFS performs better than the MMCFS.

Forecasts based on January's initial conditions of many years (2009, 2014, and 2015) were not able to predict the negative departures, whereas the years with positive departures (2010, 2011, 2013, and 2019) were predicted well. The model was able to predict the below-normal rainfall for the years 2009, 2014, and 2015 based on all initial conditions except for January.

6 Conclusions

IMD uses two approaches for seasonal forecasting of the ISMR, namely empirical statistical method and dynamical method. In the statistical method, the historical relationship between the ISMR and predictors is derived from the global atmosphere–ocean parameters (mainly derived from slowly varying boundary forcing), whereas, in the dynamical method, coupled General Circulation Models (GCM) of the atmosphere and oceans are used to simulate the summer monsoon circulation and associated rainfall. IMD does continuous in-house research and development in the statistical as well as dynamical models to achieve maximum accuracy in the operational forecasts provided for the country from time to time. Presently, operational forecasts for rainfall over the country as a whole and over the four homogeneous regions are prepared based on Statistical Ensemble Forecasting System (SEFS) and Monsoon Mission Coupled Forecasting System (MMCFS). As a part of in-house research, IMD implemented new state of art LRF models in 2003–2007 following the review of an old forecasting system in 2002. It is seen that after the implementation of the SEFS in 2007, there has been significant improvement in the accuracy of the seasonal forecasting of the all India southwest monsoon rainfall. The average absolute error in the all India forecast based on June SEFS (during the last 13 years (2007–2019) was 6.57% of LPA compared to the average absolute error of 8.91% of LPA during the 13 years (1994–2006) just prior to that period. The absolute error in the MMCFS simulations during the period 2007–2019 was 7.8% indicating that during the recent years, the SEFS performed better than MMCFS for all India rainfall seasonal forecasts. However, the advantage of the MMCFS has been that it can provide a forecast of spatial distribution of the rainfall and temperature for any period like month, bi-month, 3-month, or 4-month seasons, etc., with long lead time. Therefore, at present, IMD uses both the SEFS and MMCFS to generate operational seasonal forecasts for the country. Under monsoon mission project-II, various MoES institutions including IITM and IMD have been working for further improvement of the MMCFS. Seasonal forecasts of all India rainfall for the southwest monsoon season based on the MMCFS model have also shown comparatively good skill in predicting below-normal rainfall during El Nino events (e.g., 2014 and 2015). However, more systematic research is needed to further improve both statistical and dynamical approaches for more accurate seasonal rainfall forecasts.

References

Blanford HF (1884) On the connection of the Himalayan snow with dry winds and seasons of droughts in India. Proc Roy Soc Lond 37:3–22

Chan JCL, Shi J (1999) Prediction of the summer monsoon rainfall over South China. Int J Climatol 19:1255–1265

Charney JG, Shukla J (1981) Predictability of monsoons. In: Lighthill J, Pearce RP (eds) Monsoon dynamics. Cambridge University Press, pp 99–109

Friedman JH, Stuetzle W (1981) Projection pursuit regression. J Am Statist Assoc 76:817–823

Gadgil S, Rajeevan M, Nanjundiah R (2005) Monsoon prediction—why yet another failure? Curr Sci 88:1389–1400

Gadgil S (1996) Climate change and agriculture—an Indian perspective. In: Abrol YR, Gadgil S, Pant GB (eds) Climate variability and agriculture. Narosa, New Delhi, India, pp 1–18

Gowarikar V, Thapliyal V, Kulshrestha SM, Mandal GS, Sen Roy N, Sikka DR (1991) A power regression model for long range forecast of southwest monsoon rainfall over India. Mausam 42:125–130

Gowarikar V, Thapliyal V, Sarker RP, Mandal GS, Sikka DR (1989) Parametric and power regression models: New approach to long range forecasting of monsoon rainfall in India. Mausam 40:115–122

Hastenrath S (1995) Recent advances in tropical climate prediction. J Clim 8:1519–1532

Hawkins E, Osborne TM, Ho CK, Challinor AJ (2013) Calibration and bias correction of climate projections for crop modelling: an idealised case study over Europe. Agricult Forest Meteorol 170:19–31

Jagannathan P (1960) Seasonal forecasting in India: a review, FMU:1-80. India Meteorological Department, Pune, India

Kanamitsu M, Kumar A, Hann-Ming HJ, Jae-kyung S, Wanqui W, Fanglin Y, Song You H, Peitao P, Wilber C, Shrinivas M, Ming J (2002) NCEP dynamical seasonal forecast system 2000. Bull Amer Meteorol Soc 83:1019–1037

Krishna Kumar K, Soman MK, Rupa Kumar K (1995) Seasonal forecasting of Indian summer monsoon rainfall. Weather 50:449–467

Livezey RE, Masutani M, Ji M (1996) SST-forced seasonal simulation and prediction skill for versions of the NCEP/MRF model. Bull Amer Meteorol Soc 77:507–517

Normand C (1953) Monsoon seasonal forecasting. Q J R Meteorol Soc 79:463–473

Pai DS, Rao Suryachandra, Senroy Soma, Pradhan Maheswar, Pillai Prasanth A, Rajeevan M (2017) Performance of the operational and experimental long-range forecasts for the 2015 southwest monsoon rainfall. Curr Sci 112:68–75

Pai DS, Sreejith OP, Nargund SG, Musale M, Tyagi A (2011) Present operational long range forecasting system for southwest monsoon rainfall over India and its performance during 2010. Mausam 62, V2:179–196

Parthasarathy B, Diaz HF, Eischeid JK (1988) Prediction of all-India summer monsoon rainfall with regional and large-scale parameters. J Geophys Res 93(D5):5341–5350

Rajeevan M, Pai DS, Thapliyal V (1998) Spatial and temporal relationships between global land surface air temperature anomalies and Indian summer monsoon rainfall. Meteorol Atmos Phys 66(3):157–171

Rajeevan M (2001) Prediction of Indian summer monsoon: status, problems and prospects. Curr Sci 11:1451–1457

Rajeevan M, McPhaden MJ (2004) Tropical Pacific upper ocean heat content variations and Indian summer monsoon rainfall. Geophys Res Lett. https://doi.org/10.1029/2994GL020631

Rajeevan M, Pai DS, Dikshit SK, Kelkar RR (2004) IMD's new operational models for long range forecast of south-west monsoon rainfall over India and their verification for 2003. Curr Sci 86:422–431

Rajeevan M, Pai DS, Anil Kumar R, Lal B (2007) New Statistical models for long-range forecasting of southwest monsoon rainfall over India. Clim Dyn 28:813–828

Rajeevan M, Unnikrishnan CK, Preethi B (2012) Evaluation of ENSEMBLES multimodel seasonal forecasts of Indian Summer Monsoon variability. Clim Dyn 38:2257–2274

Ramu DA, Sabeerali C, Chattopadhyay R, Rao DN, George G, Dhakate A, Salunke K, Srivastava A, Rao SA (2016) Indian summer monsoon rainfall simulation and prediction skill in the CFSv2 coupled model: Impact of atmospheric horizontal resolution. J Geophy Res Atmos 1752–1775.:871. https://doi.org/10.1002/2015JD023538

Rao SA et al (2019) Monsoon mission: a targeted activity to improve monsoon prediction across scales. Bull Amer Meteorol Soc 100:2509–2532. https://doi.org/10.1175/BAMS-D-17-0330.1

Roads JO (2000) The Second International RSM workshop: meeting summary. Bull Amer Meteorol Soc 81(12):2979–2980

Shukla J, Wallace M (1983) Numerical simulation of the atmospheric response to equatorial Pacific sea surface temperature anomalies. J Atmos Sci 40:1613–1630

Thapliyal V, Kulshreshtha S (1992) Recent models for long-range forecasting of southwest monsoon rainfall over India. J Arid Environ 43:239–248

Walker GT (1923) Correlation in seasonal variations of weather. VIII. A preliminary study of world-weather. Mem Indian Meteorol Depart 24 (Part 4):75–131

Walker GT (1924) Correlation in seasonal variations of weather, IX. A Further Study of World Weather. Mem Indian Meteorol Depart 24(9):275–333

Webster PJ, Magana VO, Palmer TN, Shukla J, Thomas RA, Yanai M, Yasunari T (1998) Monsoons: processes, predictability and the prospects of prediction. J Geophys Res 103:14451–14510

Severe Weather Events Over the Indian Region: Insights from Ensemble Prediction System

Parthasarathi Mukhopadhyay, Tanmoy Goswami, Snehlata Tirkey, Radhika D. Kanase, R. Phani Murali Krishna, and Medha Deshpande

Abstract On the backdrop of increasing severe weather events (e.g. Lightning and thunderstorm, tropical cyclone, extreme rainfall, etc.) over India, a high-resolution (12.5 km) deterministic (GFS) and a 21-member ensemble-based (GEFS) global atmospheric general circulation models have been developed. It is found that, using the GEFS-based forecast, it is possible to predict occurrences of thunderstorms in 24–48-h lead time. The model is also able to provide skilfull forecast of tropical cyclones with 5 days of lead time. It is also observed that, using GEFS, we can predict the extreme rainfall probabilities in different locations over India. To avoid models' inability to simulate very high observed rain, a percentile-based extreme rainfall forecast methodology has been developed. This methodology provides the location of very heavy (>90th percentile) and extreme (>95th percentile rainfall) rainfall. These forecast products provide disaster managers a very crucial information to be able to take informed decisions on the event of an imminent disaster.

1 Introduction

Many previous studies have reported that there is an increase in the extreme rainfall events across various parts of the globe (Gordon et al. 1992; Hennessy et al. 1997; Allen and Ingram 2002; Trenberth et al. 2003). Goswami et al. (2006) and showed that there is an increase in rainfall of >100 and 150 mm day^{-1} categories over the central Indian region. A Study by Francis and Gadgil (2006) suggested that a fraction of extreme rain comes from the monsoon depressions and a large fraction of extreme rain comes from severe thunderstorm-like systems. Rajeevan et al. (2008) also reported that there is an increase in extreme rainfall over the central Indian region which is linked to rising sea surface temperature over the Indian Ocean. It was also reported that the increasing trend of extreme rain is associated with the increasing

P. Mukhopadhyay (✉) · T. Goswami · S. Tirkey · R. D. Kanase · R. Phani Murali Krishna · M. Deshpande
Ministry of Earth Sciences, Indian Institute of Tropical Meteorology, Dr. Homi Bhabha Road, Pune 411008, India
e-mail: mpartha@tropmet.res.in; parthasarathi64@gmail.com

V. K. Gahalaut and M. Rajeevan (eds.), *Social and Economic Impact of Earth Sciences*,
https://doi.org/10.1007/978-981-19-6929-4_3

trend of synoptic activity (Ajayamohan et al. 2010) over the Indian region. Singh et al. (2014) showed that there is a statistically significant increasing trend of wet spells over Indian landmass. A recent study by Roxy et al. (2017) reported that there is a threefold increase in extreme rain events over the central Indian region from 1950 to 2015. With the rising trend of extreme rainfall, Webster et al. (2005) reported that there is an increasing trend of cyclonic storms over the north Indian Ocean during the past three decades. In a recent study, Ray et al. (2021) showed the significant rise of extreme weather events (EWEs) and its impact based on 50 years of IMD data. Out of all EWEs, heavy rain and floods are reported to have the highest impact in terms of loss of lives (Ray et al. 2021) during the period 1970–2019.

Previous studies also suggest that extreme rainfall events are generally underestimated by current weather prediction models (GCMs) (Durman et al. 2001; Li et al. 2011). There are also many studies which have highlighted the use of different modelling frameworks to improve the prediction of extreme rainfall (DeMott et al. 2007; Li et al. 2012; Goswami et al. 2020).

Due to the increase in the frequency of severe weather events, it has become very important to develop and implement a modelling framework to predict severe weather events accurately with a reasonable lead time in order to avoid socioeconomic losses over the Indian region. The Ministry of Earth Sciences (MoES) under the Monsoon Mission has invested in the development of such state-of-the-art prediction framework in the recent years. These models as developed by scientists at the Indian Institute of Tropical Meteorology (IITM), Pune, are now made operational by the India Meteorological Department. Ensemble forecasting is a method used in numerical weather prediction. Instead of making a single forecast of the most likely weather, a set (or ensemble) of forecasts is produced. This set of forecasts aims to give an indication of the range of possible future state of the atmosphere.

In this paper, we will discuss the forecast skill of the Global Forecasting System (GFS) and the Global Ensemble Forecasting System (GEFS) models both of which are at a very high horizontal resolution (12.5 km). These models originally adopted from National Centers for Environmental Prediction, USA, were customized and modified at IITM, Pune, to implement a medium-range prediction system for 10 days of forecast. We have also discussed various forecast products developed from the output of these two models in the subsequent sections.

2 Model and Data

A very high-resolution deterministic Global Forecast System (GFS) with a spectral resolution of T1534 ($\sim$12.5 km) with 64 hybrid vertical levels (top layer around 0.27 hPa) has been implemented for daily operational forecast since June 2016. The global atmospheric model in GFS is a global spectral model (GSM) version 13.0.2, adopted from NCEP (http://www.emc.ncep.noaa.gov/GFS/doc.php). The dynamical core of GFS model is based on a two-time-level semi-implicit Semi-Lagrangian (SL) discretization approach (Sela 2010), while the physics is done in the linear, reduced

Gaussian grid in the horizontal space. It is for the first time that the SL dynamical core (previously Eulerian EL) is implemented in the GFS T1534 for operational forecast over India, equivalent to other global operational centres, namely ARPEGE (Meteo France), GEM (Environment Canada), GFS (NCEP, USA), GSM (Japan Meteorological Agency (JMA)), IFS (ECMWF), Met UM (United Kingdom Met. Office), etc. The major advantage of the SL framework over the EL approach is that it is an unconditionally stable scheme which accurately captures waves with high phase speed. It also saves a lot of computational time as compared to the EL framework due to longer time steps. A detailed description of the benefits of the SL approach is described in detail in the study by Staniforth and Côté (1991). The initial conditions for the forecasts are generated by the NCMRWF through the global data assimilation system (GDAS) cycle which has more Indian data in it. More details about the NCMRWF data assimilation system are documented by Prasad et al. (2016). The GFS T1534 model is being run daily for 10 days and the output is stored every 3-h interval. The model is being operationalized since June 2016 by the India Meteorological Department. Details of model physics are mentioned in Table 1.

To provide the probabilistic forecast at block level, Global Ensemble Forecast System (GEFS) has been developed at a resolution of 12.5 km with 21 ensemble members. This ensemble prediction system is being used to generate 10 days of probabilistic forecasts. Currently, this is the highest resolution operational global ensemble prediction system being used in India.

There are a variety of applications initiated based on the GFS/GEFS forecast at 12.5 km over the Indian region which are shown in Fig. 1.

3 Results and Discussions

Among severe weather systems that affect the lives and properties of people, prominent systems are thunderstorms/lightning during the pre-monsoon season, tropical cyclones during pre- and post-monsoon seasons and heavy precipitation events during the monsoon season. In the backdrop of climate change, numerous studies establish an increasing trend of extreme precipitation and severe weather events over the Indian region. In 2018, there was an unprecedented thunderstorm and severe weather activities during pre-monsoon months (March–April–May) causing widespread loss of public life and properties. This prompted the Ministry of Earth Sciences to initiate a time-bound targeted activity to build a thunderstorm prediction system for the country. As a part of this initiative, using the Global Ensemble Forecast System (GEFS), along with regional models, several indices have been developed to predict thunderstorms, gusty wind, heavy rain and lightning-related products such as Lightning Potential Index (LPI) from high-resolution mesoscale model. The global model provides an outlook of 2 days with a probability of thunderstorm and regional models provide 3 hourly outlooks of lightning probability over the Indian region. Figure 2a shows the probability of thunderstorm based on an index 'Super Cell Parameter' (SCP), indicating a high probability (>75%) of thunderstorm over Uttar Pradesh,

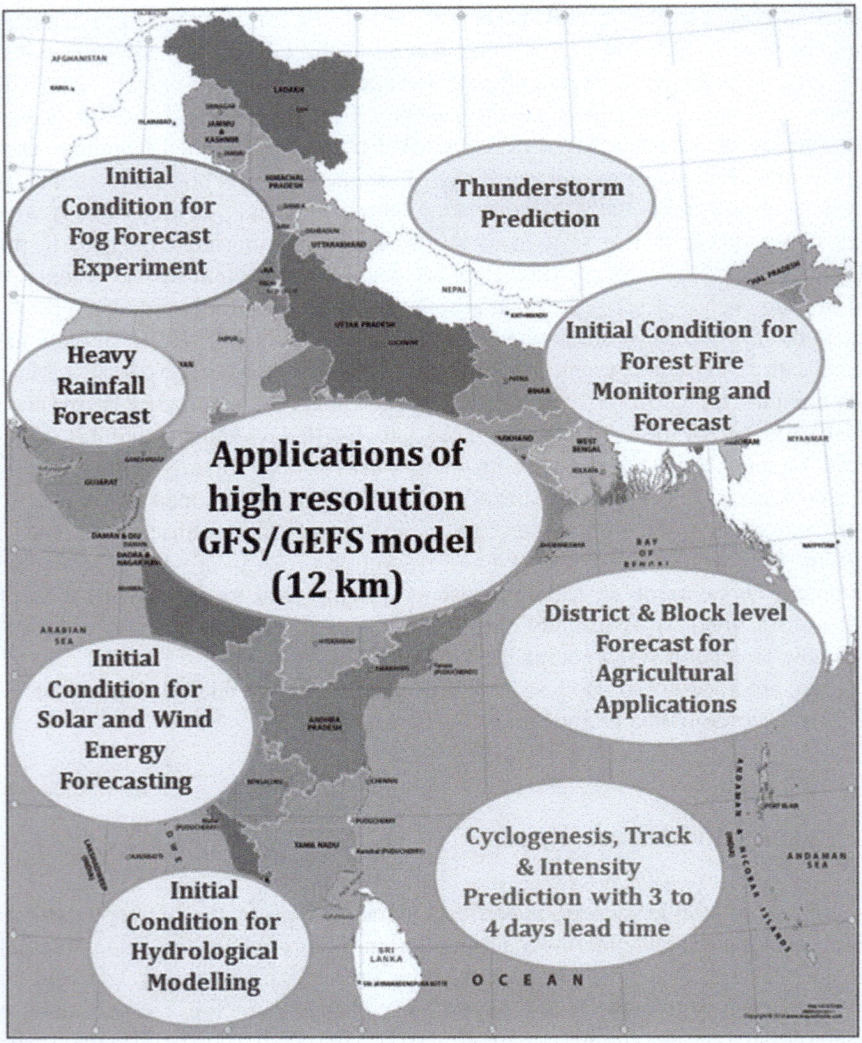

Fig. 1 High resolution GFS/GEFS model forecast for variety of societal applications including severe weather phenomenon

Bihar and the northern part of West Bengal around 15UTC (0830IST) with model being initialized at 0000 UTC of 25 June 2020. Figure 2b shows the probability of hail derived from GEFS forecast at 12.5 km indicating a higher probability of hail over some parts of eastern UP and Bihar states and many parts of northern districts of West Bengal. Figure 2b shows a higher probability of hail on 25 June with the model being initialized on 23 June 0000UTC with a lead time of 48 h. Such probabilistic severe weather forecasts have been implemented at IMD with the development of globally highest resolution short-range ensemble forecast system.

Table 1 Model physics in GFS T1534

Physics	Description
Convection	Scale-aware Revised Simplified Arakawa Schubert (RSAS) and mass flux-based SAS shallow convection scheme
Microphysics	Zhao–Carr–Moorthi microphysics formulation for grid-scale condensation and precipitation
Gravity wave drag	Orographic gravity wave drag, mountain drag and stationary convective gravity wave drag
Planetary boundary layer	Hybrid Eddy Diffusion Mass flux turbulence/vertical diffusion scheme
Radiation	Solar radiation and IR based on RRTM (originally from AER modified at EMC with Monte Carlo independent Column Approximation (McICA). Cloud fraction for radiation computed diagnostically from prognostic cloud condensate

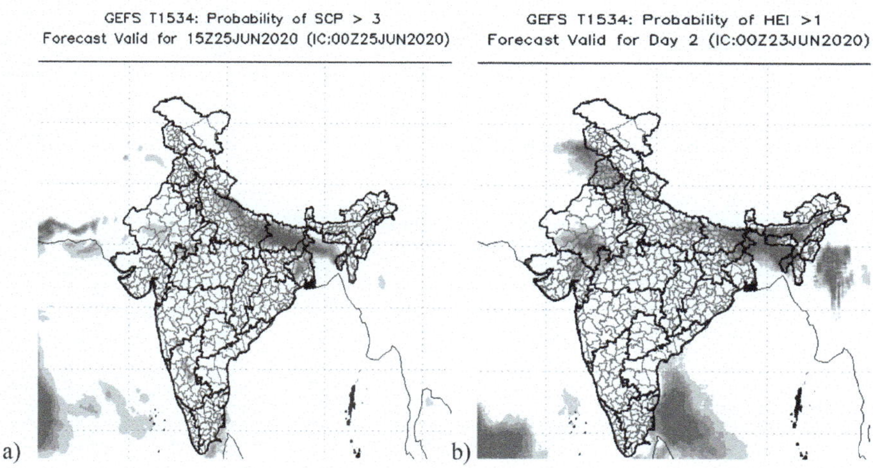

Fig. 2 **a** Forecast probability of thunderstorm based on Super Cell Parameter (SCP) index, **b** 48 h forecast probability of hail based on hail index derived from GEFS

Another severe weather system that affects the Indian coastal population is tropical cyclone. Though forecast of tropical cyclones has attained reasonable accuracy during the recent years, the gap was to build an indigenous system of ensemble-based cyclone track and landfall system. The implementation of GEFS has made it possible to address the uncertainty of track and landfall arising due to a single deterministic member of the model. The ensemble approach also provides the strike probability and tracks based on 21 members of the ensemble forecast with probability at a longer lead time. The probabilistic forecasts of tropical cyclones also help to distinguish between a region with lesser cyclone strike probability and against those with higher probability and this in turn helps the disaster managers for better preparedness and proper evacuation.

As mentioned in the introduction, episodes of intense rainfall are on the rise with the backdrop of climate change and prediction of these events is becoming increasingly challenging. However, the highest resolution GEFS ensemble prediction system-based probabilistic forecast is found to be useful. Figure 3 shows the prediction of a heavy rainfall episode over Mumbai on 10 June 2018. The predictions by 21 members are shown and the probability computed based on the ensemble prediction is also shown in Fig. 3 along with observation. The results suggest a realistic estimate of probabilities of extreme rainfall over Mumbai on 10 June.

To provide further insight into the ensemble prediction of heavy rainfall episodes, yet another example is shown below. The rainfall forecast averaged over the central Indian region (18°–25°N,73°–85°E) valid for 1 June–18 June 2020, is shown in Fig. 4 with 1–6-day lead time. It is interesting to note that the rainfall peak on 3 June and around 14 June 2020 have been reasonably predicted by all the members with a lead time longer than 5 days. This indicates that the GEFS probabilistic prediction system has a better skill of predicting heavy rain episodes which is the need of the hour. As majority of the members are able to capture the heavy rain episode, the prediction of heavy rain will have higher probability.

To objectively quantify the efficacy of a probabilistic forecast of rainfall, various skill scores such as Brier Score (BS), Reliability Diagram and Relative Operating Characteristic (ROC) are used. The Brier Score is an error score which is calculated as

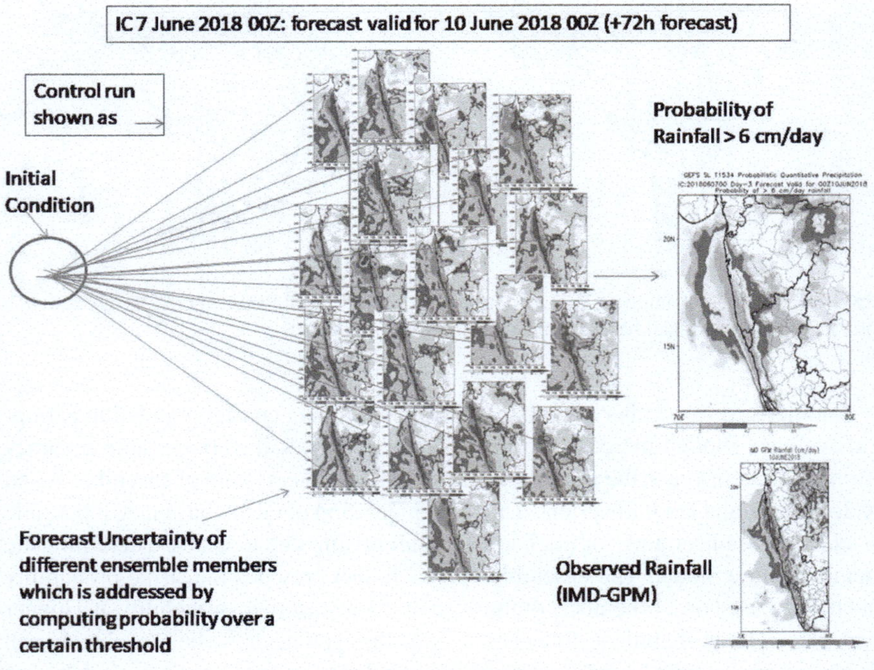

Fig. 3 GEFS forecast based on initial condition of 7 June 2018 0000UTC valid for 10 June 2018 00UTC (+72 h forecast)

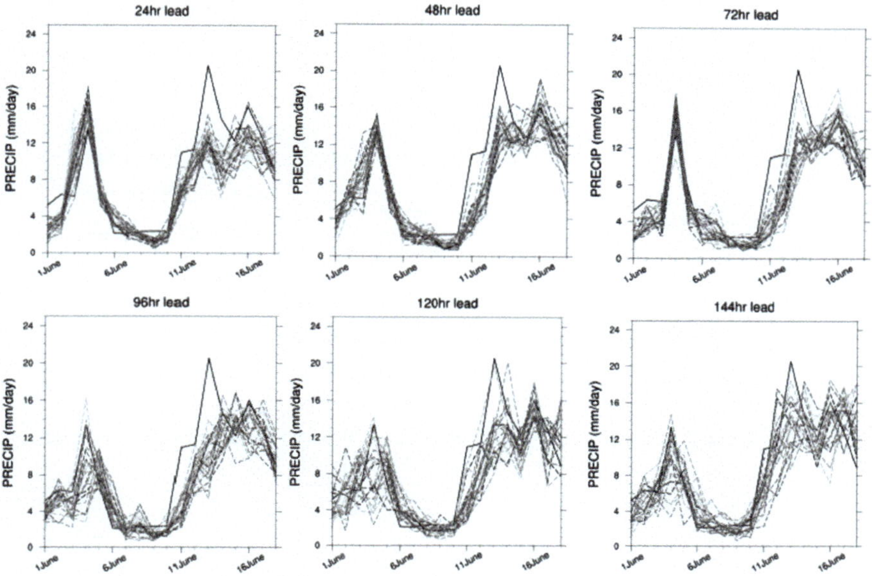

Fig. 4 Rainfall forecast valid for 1–18 June 2020 over the Central India (18°–25°N,73°–85°E) for Day 1–6. Black line is the observation (IMD-GPM merged rainfall). Coloured lines are ensemble members from GEFS T1534

the mean squared difference between the forecast probability and binary observation. Hence, a lower score indicates a better forecast. In Fig. 5, the BS is calculated for rainfall of JJAS 2020 over all India land points. A score of less than 0.1 is obtained for all thresholds with increasing lead time indicating skilful forecasts. Though a near-perfect score obtained for higher thresholds is due to a lesser frequency of such events.

The Reliability Diagram brings out how well the forecast probabilities agree with the observed frequency. Thus, a perfect forecast is indicated by a curve along the diagonal. Here, in Fig. 6, the Reliability Diagrams for Day 1 rainfall forecast with different thresholds are shown. It is evident that the model is able to generate skilful forecast though shows some over-confidence for higher forecast probability categories.

Another metric is the ROC which plots the hit rate against the false alarm rate with increasing probability thresholds. A perfect ROC curve extends from the bottom left to the top left and then to the top right thus higher the area under the curve more skilful is the forecast. Figure 7 shows the ROC for Day 1 rainfall forecast for JJAS 2020. The forecast shows good skill but the skill decreases for higher threshold.

The model has also shown very good skill in the track and intensity forecast of tropical cyclones which has been extensively assessed by Deshpande et al. (2021). Thus, such probabilistic forecast of the Indian Summer Monsoon (ISM) in general and extreme events like heavy rain and tropical cyclones in particular are essential

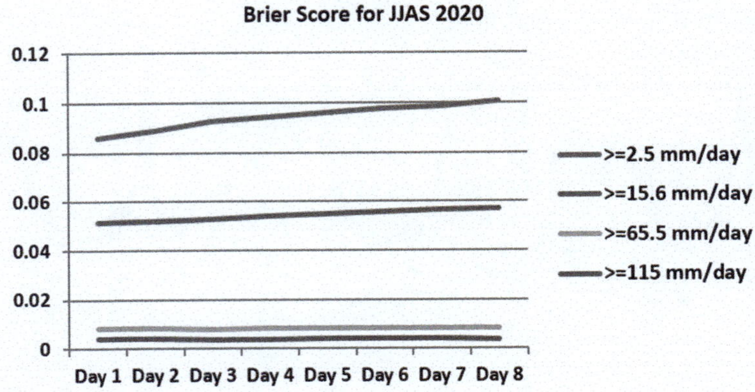

Fig. 5 Brier Score (BS) for rainfall forecast of JJAS 2020 from GEFS T1534 for all India land points

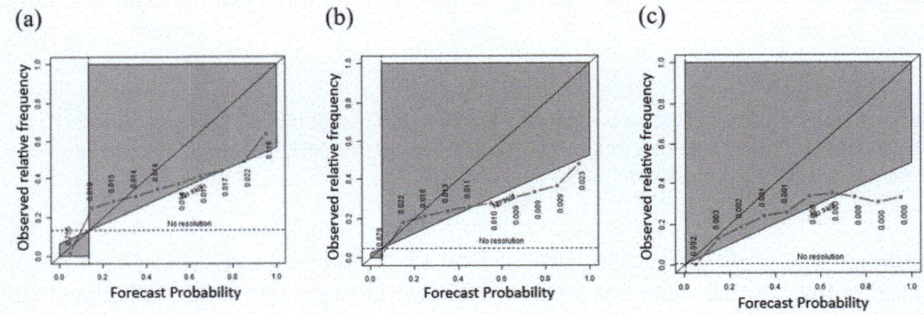

Fig. 6 Reliability diagram for Day 1 rainfall forecast of JJAS 2020 from GEFS T1534 for **a** 2.5 mm/day and **b** 15.6 mm/day and **c** 65.5 mm/day thresholds

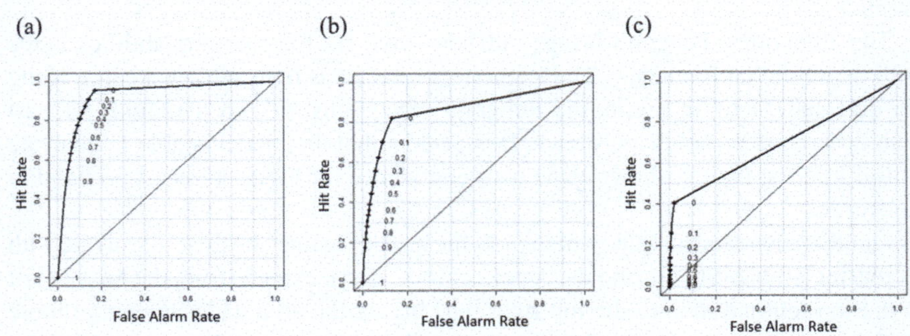

Fig. 7 ROC for Day 1 rainfall forecast of JJAS 2020 from GEFS T1534 for **a** 2.5 mm/day and **b** 15.6 mm/day and **c** 65.5 mm/day thresholds

for disaster managers and operational meteorologists for issuing reliable forecast to the common people.

To address GFS's inability to simulate very high observed rainfall, a percentile-based forecast for extreme rain and wind has been prepared. The 90th and 95th percentile values of rain and wind for each model grid and for each lead time of forecast till 5-day (120 h) forecast have been computed from the GFS model forecast climatology based on hindcasts. Based on the model hindcast, the extremes for these weather parameters are predicted. This has an added advantage as it depicts the region where the extreme events with respect to rain and wind are likely based on model climate. Figure 8 shows the percentile forecast valid for 14 June 2020 with 9 June initial conditions. It is evident that heavy rain over Maharashtra associated with strong monsoonal winds is well predicted with 5 days of lead time.

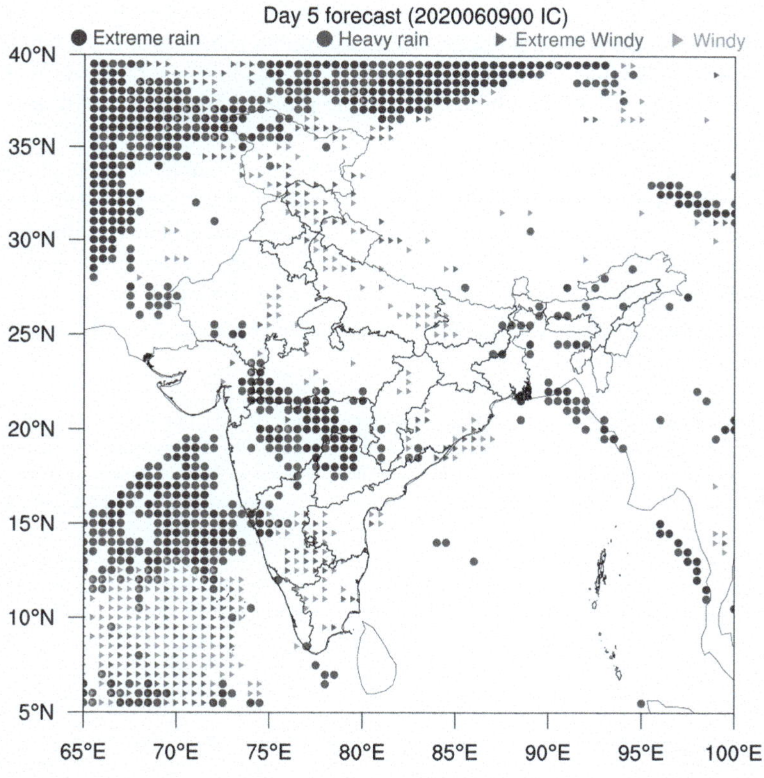

Fig. 8 Percentile based day 5 forecast of Extreme rain, heavy rain, extreme windy and windy conditions based by GFS valid for 14 June 2020 with initial condition of 9 June 2020 0000UTC

4 Conclusions

This paper discusses the current capability of India in providing skilful forecasts of severe weather systems, namely thunderstorms, tropical cyclones and heavy precipitation events using the globally highest resolution (12.5 km) ensemble prediction system based on the Global Forecast System (GFS). This ensemble forecast system is found capable of predicting probability of severe thunderstorms with good skill in 24–48-h lead time. The ensemble forecast has added much-needed value in improving tropical cyclone track and strike probability with longer lead times (~5–6 days) providing disaster managers and operational meteorologists sufficient time of issuing warnings and other evacuation-related actions. This as such is one of the reasons behind the reduced loss of public life and properties during the recent years.

In the backdrop of climate change, heavy precipitation events pose a threat to the community due to sudden flooding and devastation. The ensemble prediction system based on GEFS has shown high skill in forecasting the ISM rainfall as a whole and also in capturing the heavy rainfall events at a longer lead time with good reliability. Even the deterministic forecast has been made to provide enhanced skill through the use of percentile-based forecast.

Though the present prediction system provides reasonable skill in predicting severe weather systems with a lead time of 5–6 days, there is a need to enhance the lead time for even better management of disastrous events. The resolution of the GFS model needs to be increased further particularly to resolve the orographic effect of the Western Ghats and other mountainous regions. Research has already been initiated to improve the model resolution and take the existing ensemble prediction system of India to the next level.

Acknowledgements Indian Institute of Tropical Meteorology (IITM), Pune, is fully funded by the Ministry of Earth Sciences (MoES), Government of India, New Delhi. The model forecast presented here is being run in the 'PRATYUSH' high-power computation facility of MoES at IITM, Pune. The authors are grateful to the Director, IITM, Pune, for the encouragement for carrying out the studies.

References

Ajayamohan RS, Merry Beld WJ, Kharin VV (2010) Increasing trend of synoptic activity and its relationship with extreme rain events over central India. J Clim 23. https://doi.org/10.1175/2009jcli2918

Allen MR, Ingram WJ (2002) Constraints on future changesin climate and the hydrologic cycle. Nature 419:224

DeMott CA, Randall DA, Khairoutdinov M (2007) Convective precipitation variability as a tool for general circulation model analysis. J Clim 20:91–112. https://doi.org/10.1175/JCLI3991.1

Deshpande M, Kanase R, Krishna RP, Tirkey S, Mukhopadhyay P, Prasad VS, Johny CJ, Durai VR, Devi S, Mohapatra M (2021) Global Ensemble Forecast System (GEFS T1534) evaluation for tropical cyclone prediction over the North Indian Ocean. MAUSAM 72(1):119–128. https://doi.org/10.54302/mausam.v72i1.123

Durman CF, Gregory JM, Hassell DC, Jones RG, Murphy JM (2001) Comparison of extreme European daily precipitation simulated by global and a regional climate model for present and future climates. Quar J Royal Meteorol Soc 127:1005–1015. https://doi.org/10.1002/qj.49712757316

Francis PA and Gadgil S (2006) Intense rainfall events over the west coast of India. Meteorol Atmos Physcs 94:27–42. https://doi.org/10.1007/s00703-005-0167-2

Gordon HB, Whetton PH, Pittock AB, Fowler AM, Haylock MR (1992) Simulated changes in daily rainfall intensity due to the enhanced greenhouse eAect: implications for extreme rainfall events. Clim Dyn. https://doi.org/10.1007/BF00209165

Goswami BN, Venugopal V, Sengupta D, Madhusoodanan M Sand Prince K X (2006) Increasing trend of extreme rainevents over Indian in a warming environment. Science 314:1442–1445

Goswami T, Goswami BB, Phani RPM, Mukhopadhyay P (2020) Evaluation of SP-CAM and SP-CCSM in capturing the extremes of summer monsoon rainfall over Indian region. J Earth Syst Sci. https://doi.org/10.1007/s12040-020-1381-5

Hennessy KJ, Gregory JM, Mitchell JFB (1997) Changes in daily precipitation under enhanced greenhouse conditions. Clim Dyn 3:667–680

Kamaljit R, Giri RK, Ray SS, Dimri AP, Rajeevan M (2021) An assessment of long-term changes in mortalities due to extreme weather events in India: a study of 50 years' data, 1970–2019. Weather Clim Extr 32(2021):100315. https://doi.org/10.1016/j.wace.2021.100315

Li F, Collins WD, Wehner MF, Williamson DL, Olson JG (2011) Response to precipitation extreme to global warming in an aqua planet climate model: towards robust projection from regional to global scale. Tellus Ser A63(5):876–883. https://doi.org/10.1111/j.1600-0870.2011.00543.x

Li F, Rosa D, Collins WD, Wehner MF (2012) Superparameterization: a better way to simulate regional extreme precipitation? J Adv Model Earth Syst 4M04002. https://doi.org/10.1029/2011MS000106

Prasad VS, Jand JC, Sodhi JS (2016) Impact of 3D Var GSI-ENKF hybrid data assimilation system. J Earth Syst Sci 125(8):1509–1521

Rajeevan M, Bhate J, Jaswal AK (2008) Analysis ofvariability and trends of extreme rainfall events over India using 104 years of gridded daily rainfall data. Geophys Res Lett 35:L18707. https://doi.org/10.1029/2008GL035143

Roxy MK, Ghosh S, Pathak A, Athulya R, Mujumdar M, Murtugudde R, Terray P, Rajeevan M (2017) A threefold rise in widespread extreme rain events over central India. Nat Commun. https://doi.org/10.1038/s41467-017-00744-9

Sela JG (2010) The derivation of the sigma pressure hybrid coordinate Semi-Lagrangian model equations for the GFSNCEP Office Note 462, 31 pp. http://www.lib.ncep.noaa.gov/ncepofficenotes/files/on462.pdf

Singh D, Tsiang M, Rajaratnam B, Diffenbaugh NS (2014) Observed changes in extreme wet and dry spells during thesouth Asian summer monsoon season. Nat Clim Change. https://doi.org/10.1038/NCLIMATE2208

Staniforth A, Côté J (1991) Semi-Lagrangian integration schemes for atmospheric models—a review. Mon Weather Rev 119(9):2206–2223

Trenberth KE, Dai A, Rasmussen RM, Parsons DB (2003) The changing character of precipitation. Bull Amer Meteorol Soc 84:1205

Webster PJ, Holland GJ, Curry JA, Chang HR (2005) Changes in tropical cyclone number, duration, and intensity in a warming environment. Science 309:1844–1846

Monsoon Variability and Change

Ashwini Kulkarni and K. Koteswara Rao

Abstract India receives 75% of the annual rainfall during the summer monsoon season June through September. The summer monsoon (south-west monsoon) rainfall exhibits variations on all time scales, diurnal to multi-decadal. The year-to-year variability is characterized by excess and deficit monsoons which have a large impact on livelihood of the Indian population, agriculture, health, power sector, irrigation, etc. Hence, to study the variability of the Indian monsoon is of paramount importance. In this chapter, we discuss intra-seasonal and interannual variability of the Indian monsoon using century-long period data. The observed changes and trends in seasonal rainfall have also been examined. In the backdrop of unequivocal warming of the earth's surface, it is important to study the impact of warming on the amount and pattern of seasonal rainfall. We present spatial distribution of projected changes in seasonal rainfall using IPCC-Coupled Model Inter-comparison Project 5 (CMIP5) model simulations for two Representative Concentration Pathways (RCP), RCP4.5 and RCP8.5, in three time epochs; near term, middle of the century and long term.

1 Introduction

South Asian summer is dominated by the Indian monsoon, which arrives over the Indian landmass during the 4 months from June to September and contributes 80% of the annual rainfall. It supports the livelihood of about 17% of the world's population. India being an agrarian country, the food security of billions of people living in India depends on the rainfall received in these 4 months. A slight deviation in rainfall could have a major impact on water supply, food, energy and GDP. The variability of ISMR at different time scales, from intra-seasonal to multi-decadal, impacts the life of more than one billion people.

A. Kulkarni (✉)
Indian Institute of Tropical Meteorology (IITM-MoES), Pune, India
e-mail: ashwini@tropmet.res.in

K. Koteswara Rao
Azim Premji University, Bengaluru, India

V. K. Gahalaut and M. Rajeevan (eds.), *Social and Economic Impact of Earth Sciences*,
https://doi.org/10.1007/978-981-19-6929-4_4

Indian landmass also receives part of its annual rainfall in the post-monsoon season (October–December), known as the northeast monsoon. The north-east monsoon is an important source of water for south Peninsular India, covering about 22% area of the country. India gets about 10–20% of its annual rainfall during this monsoon, while the south Peninsula gets 30–60% of its annual rainfall. The southern Peninsular India with the adjoining coast on the Bay of Bengal receives about 50% of its annual rainfall during this season (Kulkarni et al. 2020). The north-east monsoon plays a vital role in the agricultural activities over seven meteorological subdivisions in India comprising coastal Andhra Pradesh, Rayalseema, Coastal Karnataka, North Interior Karnataka, South Interior Karnataka, Kerala and Tamilnadu.

It has been observed that the warming of the climate system is unequivocal since the 1950s (IPCC AR5). The combined land and ocean temperature has increased at an average rate of 0.07 °C per decade since 1880; however, the average rate of increase since 1981 (0.18 °C) is more than twice as great. The ten warmest years on record have all occurred since 1998, and nine of the ten have occurred since 2005. Changes in hydrological cycle and precipitation patterns are the most important outcome of unprecedented warming. The frequency, intensity, amount and type of precipitation have shown major changes over various regions of the globe. The land temperatures over India have also shown unequivocal warming. The annual average temperatures over the Indian landmass have shown an increasing trend of about 0.64° C (per 100 yrs) during the period 1901–2016 (Ross et al. 2018). The warming trend over India is more accelerated in post-1975 period which may alter the amount and pattern of seasonal rainfall.

The Indian summer monsoon rainfall series over the past century has remained stable in spite of the steady rise in global mean temperatures. Over the core monsoon zone, the contribution from increasing heavy rain events is offset by decreasing moderate events and, hence, in the long term, the series is trendless (Goswami et al. 2006). There is a climatic shift around the mid-1970s; hence, it is important to study the changes in monsoon variability with respect to the climate shift.

In this chapter, we discuss the observed changes in Indian monsoon variability on all time scales in nonglobal (1901–1975) and global (1976–2019) warming periods. Also, the projected future changes in monsoon rainfall are discussed using a suite of climate models. Section 2 gives data and methodology. The observed changes in the annual cycle and the subseasonal and interannual variability are discussed in Sect. 3. Section 4 describes the projected changes in mean monsoon using a suite of CMIP5 models. The socio-economic impacts of changing climate are given in Sect. 5.

2 Data and Methodology

(i) The high-resolution (0.25° × 0.25°) daily gridded rainfall data prepared by the India Meteorological Department (Pai et al. 2014) and updated thereafter for the period 1901–2019 have been used in the analysis. This data is the most up-to-date data. The tools such as Mann–Kendall trend, composite analysis, etc., have been applied.

(ii) A suite of 25 models from CMIP5 archive has been used. The simulations have been validated for the period 1986–2005 and the projections have been examined in three time slices 2016–2035, 2046–2065 and 2080–2099 for two scenarios RCP4.5 and RCP8.5. The models have different resolutions. For comparison, the models have been converted to uniform grid size 0.25 × 0.25 long/lat.

3 The Inter-Annual and Intra-Seasonal Variability

Monsoons are an integral part of the Earth's global climate system which is vital for the livelihood of two-thirds of the world's population. Monsoon systems are characterized by pronounced seasonal reversals of surface winds and precipitation which arise due to the annual cycle of solar heating and accompanying large-scale changes in the land–sea thermal contrasts (Ramage 1971; Webster et al. 1998). Monsoon is characterized by the reversal of winds, the change in rainfall amount with seasons is much more important for the people of the region, due to the dependence of the agricultural output, water resources, and economy on the amount and behavior of the rainfall during the season. Therefore, for the people of the region, monsoon means rainfall.

3.1 The Annual Cycle

The monsoon arrives over India every year but no two monsoons are same. Generally, monsoon starts around 1 June over the coast of Kerala and gradually travels north, north-westwards covering the entire country by 15 July. The onset of monsoon is quite abrupt; however, it withdraws gradually from north-west India to south-east. The annual cycle of precipitation over India is dominated by the seasonal reversal of winds associated with that of a larger scale climate system that covers India and the tropical Indian and Pacific Oceans (e.g. Ramage 1971; Webster et al. 1998; Goswami et al. 2006). The latitudinal variation of monthly mean rainfall averaged over Indian longitudes (75–95° E) is given in Fig. 1. July and August happen to be the rainiest months.

The mean pattern shows that, in Jan–Feb–Mar, there is a rainfall of the order of 2–4 mm/day over the northern parts of India. July–August are the main months of rainy season with maximum rainfall occurring over 18–24° N. As compared to 1901–1975, the maximum rainfall zone shrank in 1976–2019. Also, it is observed that during the later period the rainfall over southern parts has been reduced, while winter rainfall over northern India has increased.

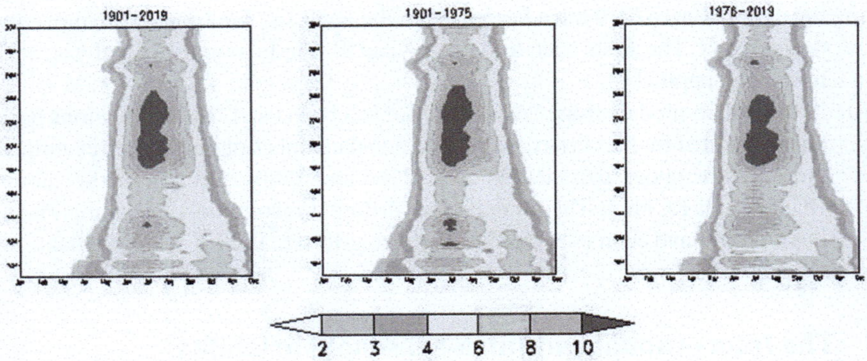

Fig. 1 Latitudinal variation of monthly mean rainfall (mm/day)over India for three periods 1901–2019 (Left), 1901–1975 (middle) and 1976–2019 (right)

3.2 Interannual Variability

The year-to-year variability of Indian summer monsoon rainfall, also known as interannual variability, is partly due to slowly varying surface boundary forcings and partly due to internal dynamics. The year-to-year variability of Indian monsoon rainfall is governed by the slowly varying surface features. El Nino conditions in the Pacific have a significant role in modulating the interannual variability of ISM rainfall (Sikka 1980; Pant and Parthasarathy 1981; Rasmusson and Carpenter 1983; Webster et al. 1998). Though not every ENSO leads to deficit monsoon over India almost 50% of the droughts are associated with ENSO; however, in the last few decades, the ENSO–Monsoon relationship has been weakened (e.g. Kripalani and Kulkarni 1997; Krishna Kumar et al. 1999), frequency and intensity of droughts have been increased and some of them are not associated with ENSO.

The coupled mode in the Indian Ocean (Indian Ocean Dipole; IOD; Saji et al. 1999) is also known to modulate interannual variability of ISM rainfall. Anomalously, high sea surface temperatures (SSTs) in the Western Equatorial Indian Ocean and lower SSTs in the Eastern Equatorial Indian Ocean are called positive IOD. It is observed that positive IOD leads to good monsoon in India (Ashok et al. 2004; Saji et al. 1999; Saji and Yamagata 2003). In addition to IOD and ENSO, there is a strong link between ISM rainfall and the Equatorial Indian Ocean oscillation (EQUINOO; Gadgil et al. 2004). In general, the positive phase of the EQUINOO is favourable for a good monsoon. The co-occurrence of EQUINOO and ENSO also determines the variations in ISM rainfall on the interannual time scale.

Eurasian snow cover also plays a major role in the year-to-year variability of ISM rainfall (Blanford 1884). Generally, positive Eurasian snow cover anomalies during winter and spring tend to be followed by an anomalous deficit rainfall over the Indian subcontinent in the subsequent summer monsoon season, while negative snow cover anomalies tend to be followed by abundant rainfall (Bhanu Kumar 1987; Bamzai and Kinter 1997). It has been observed that all non-ENSO related droughts over

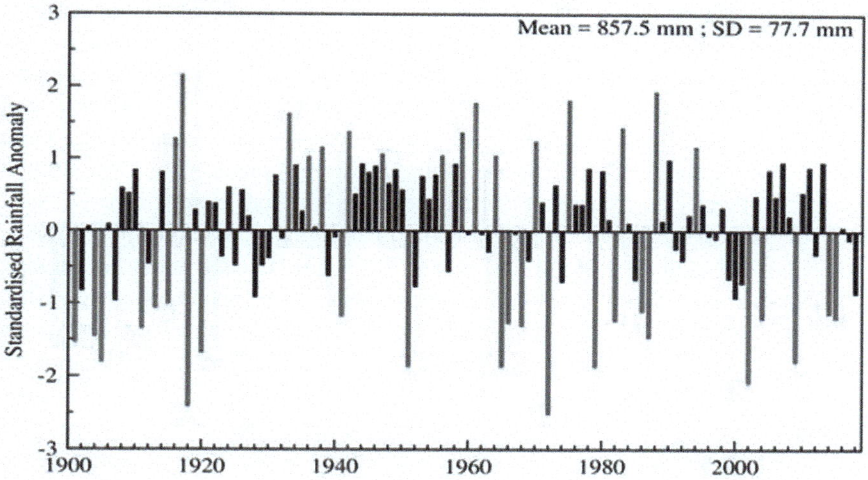

Fig. 2 Interannual variability of Indian summer monsoon rainfall based on 1901–2019

India have been associated with excessive snow depth over Eurasia (Kripalani and Kulkarni 1999).

Other than ENSO, IOD and Eurasian snow, the Atlantic, western North Pacific circulation changes (e.g. Chowdary et al. 2019; Srinivas et al. 2018) also play a role in monsoon interannual and decadal variability (Sankar et al. 2016; Yadav 2017).

The interannual variability of all India summer monsoon rainfall is characterized by excess (green) and deficit monsoons (red) as shown in Fig. 2. The all India monsoon rainfall time series has mean = 7.06 mm/day and SD = 0.64 mm/day based on 1901–2019. Figure 2 depicts the standardized rainfall anomalies for this time period. The rainfall is deficit (excess) if standardized rainfall anomaly is less (greater) than −1 (+1). The bottom panel shows decadal variability of ISMR.

During 119 years period (1901–2019), there are 23 deficit monsoons and 16 excess monsoons. During the period 1930–1960, when ISMR was in above normal epoch of its natural variability, there are only two deficit monsoons. However, 1901–1930 and 1961–1990 experienced quite frequent deficit monsoons when ISMR was in below normal phase. Also, the ENSO conditions have more impact on ISMR during below normal phase (Kripalani and Kulkarni 1997).

3.3 Regional and Seasonal Changes in Summer Monsoon and Trends

Indian summer monsoon rainfall series averaged over India landmass does not show any long-term trend on a century scale because the contribution from increasing heavy events is offset by decreasing moderate rain events (Goswami et al. 2006); however, there is a significant decreasing trend after the 1950s. This drying trend is evident in multiple datasets (Annamalai et al. 2013). The decreasing trend may be a result of multi-decadal epochal variability (Guhathakurta et al. 2015); due to the east-west shift in monsoon rainfall or due to anomalous warming of the Indo-Pacific warm pool (Annamalai et al. 2013) or weakening of land-ocean temperature gradient (Kulkarni 2012). Some studies attribute it to increasing anthropogenic aerosol concentration over the Northern Hemisphere which may cool the Northern Hemisphere and slow down the tropical meridional overturning circulation (e.g. Ramanathan et al. 2005; Chung and Ramanathan 2006; Bollasina et al. 2011, 2014); to significant weakening of monsoon low-level south-westerly winds, the upper tropospheric tropical easterlies from the outflow aloft, the large-scale monsoon meridional over-turning circulations (Rao et al. 2010; Joseph and Simon 2005; Sathiyamoorthy 2005; Fan et al. 2010; Krishnan et al. 2013) to significant increase in the duration and frequency of 'monsoon-breaks' (dry spells) over India since the 1970s (e.g. Ramesh Kumar et al. 2009; Turner and Hannachi 2010); to warming of the western Indian ocean and weakening of SST gradient (Roxy et al. 2015). Some studies state that changes in land use land cover may affect the evapotranspiration which in turn affects the monsoon (Niyogi et al. 2010; Pathak et al. 2014; Paul et al. 2016; Krishnan et al. 2016) (Fig. 3).

There is considerable spatial variability in precipitation changes. As compared to the period 1901–1975, the rainfall has been reduced by 1–5 mm/day over central parts of India which is the core monsoon zone, Kerala and extreme north-eastern parts; however, it is observed to increase in recent period over Jammu and Kashmir region as well as some parts of western India (Kulkarni et al. 2017). There is a statistically significant decreasing trend after the 1950s with a reduction of 10–20% rainfall over parts of northern and central India, and the Indo-Gangetic Planes which are mainly rainfed regions (Krishnan et al. 2013; Roxy et al. 2015).

Trends in Indian rainfall records have been extensively studied, but the subject remains complicated by the high spatiotemporal variability of rainfall arising from

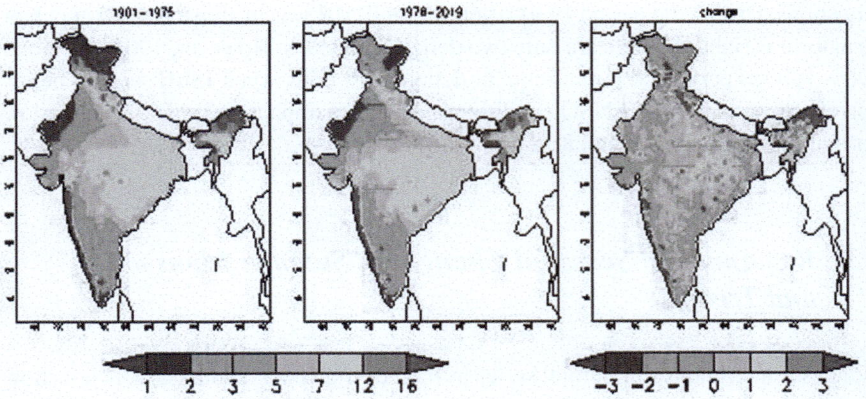

Fig. 3 Change in mean summer monsoon rainfall (mm/day) from 1901–1975 to 1976–2019

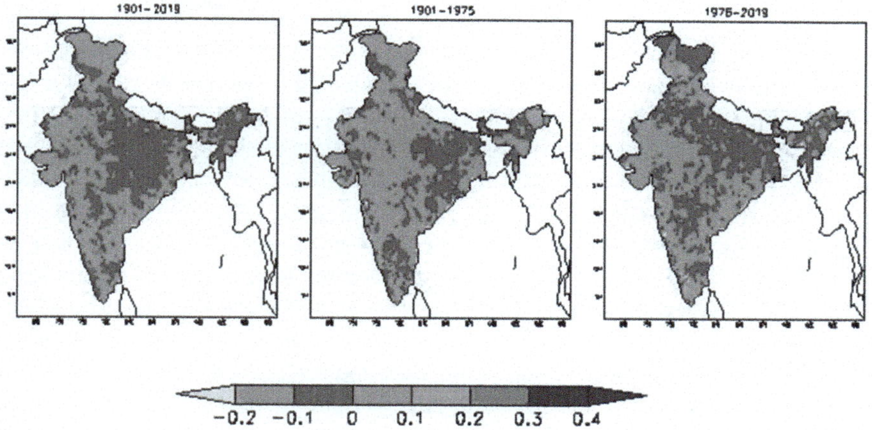

Fig. 4 Linear trend (mm/year) in Indian summer monsoon rainfall

complex atmospheric dynamics. The linear trend in annual, as well as seasonal rainfall, depict statistically significant decreasing trend over Jharkhand, Chattisgarh, Kerala and eight subdivisions, viz., Gangetic WB, West UP, Jammu and Kashmir, Konkan and Goa, Madhya Maharashtra subdivision, Rayalseema, Coastal AP and North Interior Karnataka showed significant increasing trends (Guhathakurta and Rajeevan 2008). Based on the high-resolution gridded data for 1901–2018, the linear trend in annual, as well as seasonal rainfall, depicts statistically significant decreasing trend over Kerala, the Western Ghats and some parts of central India in Uttar Pradesh, Madhya Pradesh and Chattisgarh as well as some parts of north-eastern states, whereas rainfall over Gujarat, Konkan–Goa, Jammu–Kashmir and east coast shows a significant increasing trend (Fig. 4). In the recent period, the entire northern parts of India depict a decreasing trend.

3.4 Intra-Seasonal Variability

The part of Indian summer monsoon rainfall variability contributed by its internal dynamics is more difficult to predict. Several recent modeling studies show that a significant fraction of the interannual variability of the seasonal mean Indian summer monsoon is governed by internal chaotic dynamics (e.g. Goswami 1998). This intermittent behaviour of rainfall is associated with a hierarchy of quasi-periods, namely 3–7 days (Bhalme and Parasnis 1975), 10–20 days (Krishnamurti and Bhalme 1976; Krishnamurti and Ardunay 1980) and the 30–60 days (Madden and Julian 1971, 1994; Sikka and Gadgil 1980; Yasunari 1980; Hartmann and Michelson 1989). The two periods the 10–20 days and the 30–60 days have been related with the active (wet spells) and break (dry spells) cycles of the monsoon rainfall over the Indian sub-continent, also known as monsoon intra-seasonal oscillations (MISO). MISO originate as an ocean atmosphere-coupled event in the tropical Indian Ocean, propagating northwards and, depending on the phase of the oscillation, results in the wet and dry spells of the monsoon. The 30–60 days of oscillations are also characterized by the northward movement of weather anomalies including rainfall and outgoing long-wave radiation over the Indian longitudes (Sikka and Gadgil 1980; Singh and Kripalani 1985, 1986, 1990; Kripalani et al. 1991, 1999). These fluctuations can cause extreme wet and dry regional conditions that adversely impact agricultural yields, water resources, infrastructure and human systems (Gadgil and Kumar 2006; Turner and Annamalai 2012). We applied Butterworth's band pass filter for 30–60 and 10–20 days bands to all India daily rainfall anomalies of the years 1901–2019. Figure 5 shows the daily rainfall anomalies (bars), the filtered anomalies for 30–60 days (red curve) and 10–20 days band (blue curve) for two representative monsoons; 1961 (excess), 1972 (deficit) and 1976 (normal). It is seen that the fast moving 10–20 days mode is enhanced in 1961 and 1976, while 30–60 days mode is enhanced in 1972. This is true for more than 90% of the years (Kulkarni et al. 2006). The active monsoon spells are strengthened when the positive phases of both the oscillations occur simultaneously, i.e. when they are in phase, while monsoon activity is suppressed when they are out of phase. In more than 70% of the flood years, the variance retained in 10–20 days mode is significantly more than that retained in 30–60 days mode (Kripalani et al. 2004). While in 40% of the droughts, the 30–60 days mode is dominant. Also, it is observed that, in drought years, 30–60 days oscillations are more organized.

It is observed that, during the period 1951–2011, the peak season precipitation has decreased over core monsoon region with increase in daily precipitation variability (Singh et al. 2014). Also, there is significant increase (decrease) in frequency (intensity) of dry spells. These changes in characteristics of extreme wet and dry spells are supported by increase in convective available potential energy and low-level moisture convergence along with changes in large scale circulation in upper level. The seasonal strength of Indian summer monsoon rainfall may depend on the frequency and duration of spells of break and active periods associated with the fluctuations of the ISOs. Thus, the predictability of the seasonal (June–September) mean Indian monsoon depends on the extent to which the ISOs could be predicted.

4 Projected Changes in Mean Monsoon

Global coupled models are the best tool to examine the future changes in monsoons. CMIP3 and CMIP5 models consistently project increases in the mean and variability of South Asian monsoon precipitation, despite weakening circulation at the end of the twenty-first century relative to the present (e.g. Kripalani et al. 2007; Kitoh et al. 2013). The CMIP5 models improve upon the simulation of global monsoon intensity and precipitation climatology compared to CMIP3 models; however, common model biases and large disagreement among the model projections persist. The rainfall

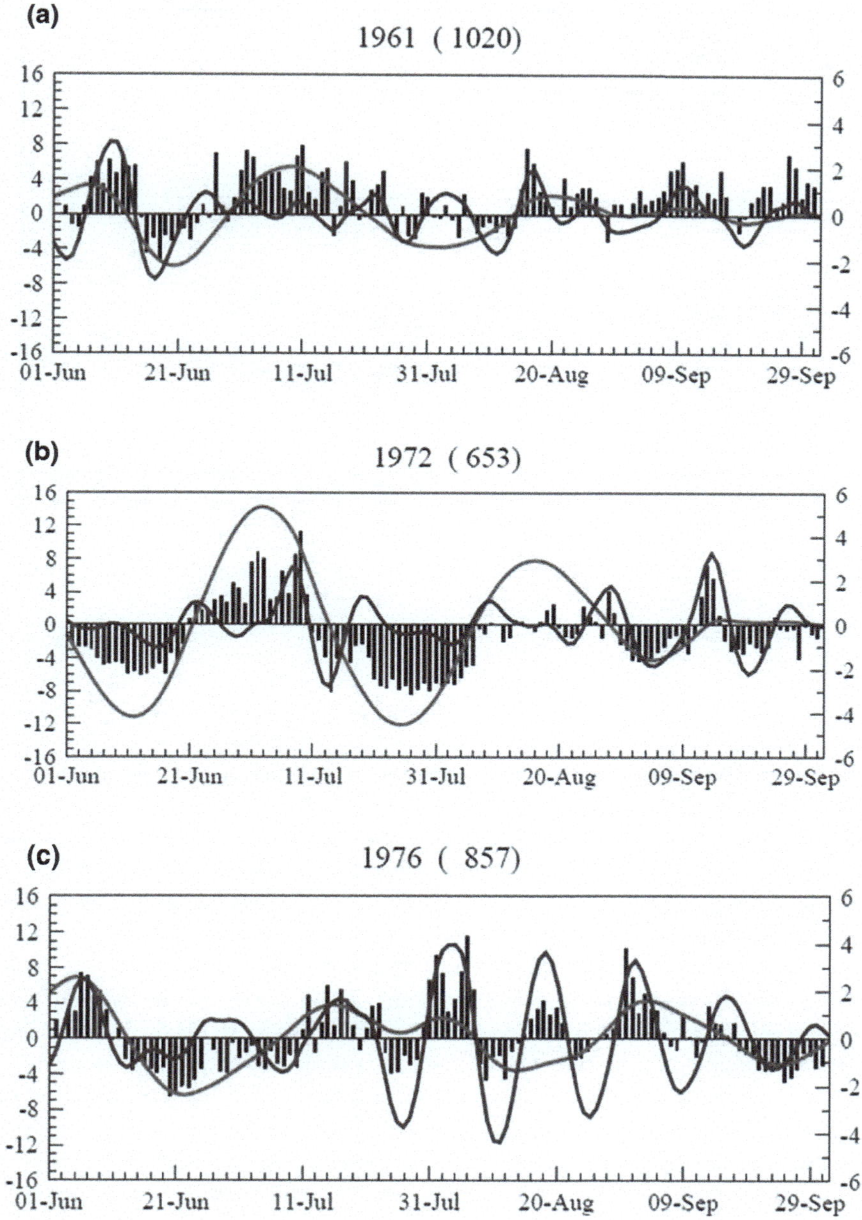

Fig. 5 All India daily rainfall anomalies (bars), 30–60 days (red curve) and 10–20 days (blue curve) filtered rainfall anomalies for 1961 (excess monsoon, top), 1972 (deficit monsoon, middle) and 1976 (normal monsoon, bottom). The numbers in bracket shows all India rainfall for that year

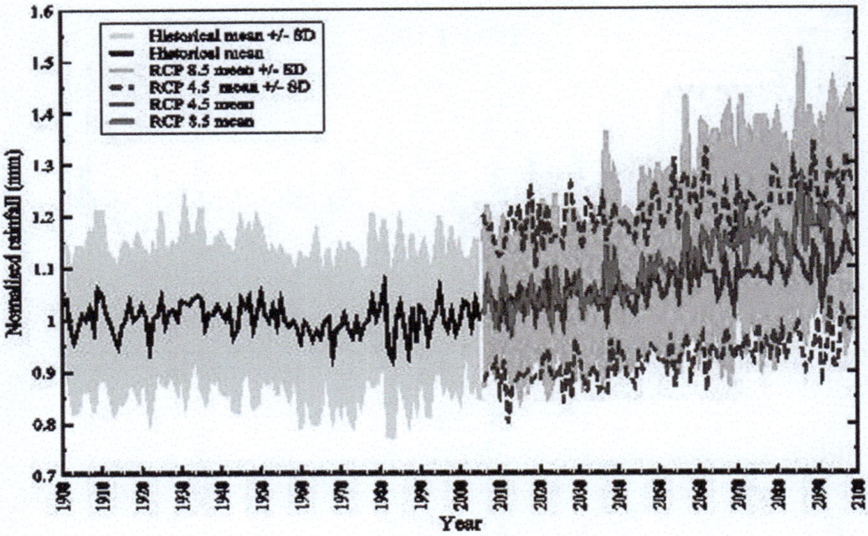

Fig. 6 Multi-model mean simulation and projection of year-to-year variability of Indian summer monsoon rainfall for 1901–2100. Y-axis shows normalized rainfall

variability over South Asian domain projects increase on all time scales, daily to decadal. Selected CMIP5 models project more severe floods and droughts in the future climate over South Asia (eg., Sharmila et al. 2015; Singh et al. 2019).

Figure 6 shows the interannual variability of simulations and projections of normalized rainfall anomaly by multi model ensemble of 25 CMIP5 models. The historical simulation 1901–2005 (black line) does not show any trend; however, the projections under RCP4.5 (blue) and RCP8.5 (red) both show significant increasing trend in summer monsoon rainfall over India. The background shaded show that there is a large uncertainty in model simulations as well as projections.

Figure 7 shows projected percent change in summer monsoon rainfall over India in three time slices (2016–2035, first column), (2046–2065, middle column) and (2080–2099, last column) with respect to 1986–2005) under RCP4.5 (first row) and RCP8.5 (second row). In general, the rainfall is projected to increase in all time epochs under both scenarios. The maximum increase is over the desert region of Rajasthan and the minimum increase is over east coast and north-eastern part. The increase is more towards the end of the century and also under RCP8.5.

5 Socio-Economic Impacts of Changing Climate on India

The observations show that the all India mean rainfall has decreased in global warming period (post-1975) and the reduction is mainly over Central India. The excess and deficit monsoons have become more severe.The daily distribution has

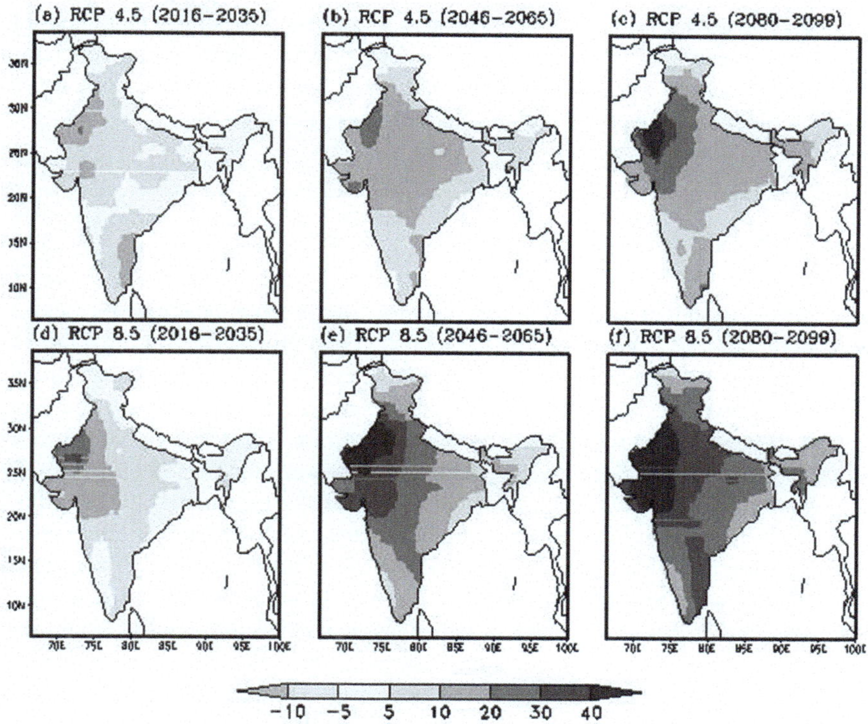

Fig. 7 Projected change(%) in mean summer monsoon rainfall during three periods with respect to 1976–2005 under RCP4.5 and RCP8.5 scenarios using multi-model mean of 25 CMIP5 models

become more erratic. There are few number of rainy days with heavy rainfall. This can have adverse impact on agriculture. The recent period is characterized by the long dry spell after the onset of monsoon which may cause drying of the seeds and reduction in net crop production. Since India has agrarian economy, the variation in monsoon rainfall has direct impact on Indian economy.

The rapid changes in Indian climate as given by climate models such as unequivocal consistent warming, increase in rainfall, increasing frequency and intensity of floods and droughts, increase in extreme weather events leading to heat/cold waves and flash floods, etc., may place serious stress on country's food security, water resources, health and ecosystem. Models suggest the increment in mean rainfall and its variability over India which may have adverse impacts on agricultural food production. The increasing frequency and severity of droughts and floods due to changing rainfall patterns may be detrimental to surface and groundwater recharge affecting water security adversely.

Higher temperatures and higher moisture may be conducive to vector-borne diseases such as malaria and dengue. Also, higher temperatures, higher climate variability and extreme weather events may be related with a risk of heat strokes, cardiovascular and neurological diseases and stress-related disorders. All these factors may affect the economically weaker section more severely. To combat these possible impacts of changing climate, we should develop proper mitigation and adaptation strategies. The climate models have large inter-model differences; hence, one should use the projections of weather parameters with utmost caution.

6 Summary

Since the entire Indian economy, water resources, agriculture, health and livelihood of the population in India depend on the monsoon rainfall received during 4 months from June to September, the study of observed and projected changes and trends in seasonal monsoon rainfall is of vital importance. In this chapter, we study the changes in mean monthly and seasonal monsoon rainfall, in the global warming period (1976–2019) with respect to the non-global warming period (1901–1975). It is seen that though July–August remain the months of maximum rainfall in both the periods, the region of maximum rainfall shrank in the global warming period. Also, it is observed that during the later period the rainfall over southern parts has been reduced, while winter rainfall over northern India has increased. Indian summer monsoon rainfall series averaged over Indian landmass does not show any long-term trend on a century scale but depicts multi-decadal variability. As compared to the period 1901–1975, the rainfall has been reduced by 1–5 mm/day over central parts of India, Kerala and extreme north-eastern parts; however, it is observed to increase in recent period over Jammu and Kashmir region as well as some parts of western India. There is a statistically significant decreasing trend after the 1950s with a reduction of 10–20% rainfall over parts of northern and central India, and the Indo-Gangetic Planes which are mainly rainfed regions. The CMIP5 model projections show a general increase in summer monsoon rainfall over India in all the three time slices (2016–2035), (2046–2065) and (2080–2099) with respect to (1986–2005) under both the scenarios RCP4.5 and RCP8.5.

References

Annamalai H, Hafner J, Sooraj KP, Piilai P (2013) Global warming shifts the monsoon circulation. Drying South Asia 26:2701–2718

Ashok K, Guan Z, Saji NH, Yamagata T (2004) Individual and Combined Influences of ENSO and the Indian Ocean Dipole on the Indian Summer Monsoon, 17:3141–3155

Bamzai AS, JL Kinter III (1997) Climatology and interannual variability of Northern Hemisphere snow cover and depth based on satellite observations. Center for Ocean Land Atmosphere Studies Report 52, COLA, 4041 Powder Mill Road, Suite 302, Calverton, MD 20705, USA, 48 pp

Bhalme HN, Parasnis SS (1975) 5–6 days oscillation in the pressure gradients over India during monsoon. Indian J Met Geophys 26:77–78

Bhanu Kumar OSRU (1987) Seasonal variation of Eurasian snow cover and its impact on the Indian summer monsoon. Largescale efects of seasonal snow cover. In: Proceedings of Vancouver Symposium, Publication 166, International Association of Hydrological Sciences, pp 51–60

Blanford HF (1884) On the connection of Himalayan snowfall and seasons of drought in India. Proc R Soc London 37(232–234):3–22. https://doi.org/10.1098/rspl.1884.0003
Bollasina M, Ming Y, Ramaswamy V (2011) Anthropogenic aerosols and the weakening of the South Asian summer monsoon. Science 334:502–504
Bollasina M, Ming Y, Ramaswamy V, Schwarzkopf MD, Naik V (2014) Contribution of local and remote anthropogenic aerosols to the 20th century weakening of the South Asian Monsoon. Geophys Res Lett 41(2):680–687. https://doi.org/10.1002/2013GL058183
Chowdary JS, Patekar D, Srinivas G, Gnanaseelan C, Parekh A (2019) Impact of the Indo-Western Pacific Ocean capacitor mode on South Asian summer monsoon rainfall. Clim Dyn 53:2327
Chung CE, Ramanathan V (2006) Weakening of North Indian SST gradients and the monsoon rainfall in India and the Sahel. J Clim 19:2036–2045
Fan F, Mann ME, Lee S, Evans JL (2010) Observed and modeled changes in the South Asian monsoon over the historical period. J Clim 23:5193–5205
Gadgil S, Vinayachandran PN, Francis PA (2004) Extremes of the Indian summer monsoon rainfall, ENSO and equatorial Indian ocean oscillation. Geophys Res Lett 31:L12213
Gadgil S, Rupa Kumar K (2006) The Asian monsoon — agriculture and economy. In: The Asian Monsoon. Springer Praxis Books. Springer, Berlin, Heidelberg. https://doi.org/10.1007/3-540-37722-0_18
Goswami B (1998) Interannual variations of Indian summer monsoon in a GCM: external conditions versus internal feedbacks. J Clim 11:501–522
Goswami BN, Venugopal V, Sengupta D, Mdhusoodanan MS, Xavier PK (2006) Increasing trend of extreme rain events over India in a warming environment. Science 314:1442–1445
Guhathakurta P, Rajeevan M (2008) Trends in rainfall patterns over India. Int J Climatol 28:1453–1469
Guhathakurta P, Rajeevan M, Sikka DR, Tyagi A (2015) Observed changes in southwest monsoon rainfall over India during 1901–2011. Int J Climatol 35:1881–1898
Hartmann DL, Michelson ML (1989) Intra-seasonal periodicities in Indian rainfall. J Atmos Sci 46:2838–2862
IPCC, 2013: Climate Change (2013) The physical science basis. Contribution of Working Group I to the Fifth Assessment Report of the Intergovernmental Panel on Climate Change [Stocker TF, Qin D, Plattner G-K, Tignor M, Allen SK, Boschung J, Nauels A, Xia Y, Bex V, Midgley PM (eds)] Cambridge University Press, Cambridge, United Kingdom and New York, NY, USA, 1535 pp
Joseph PV, Simon A (2005) Weakening trend of the southwest monsoon current through peninsular India from 1950 to the present. Curr Sci 89:687–694
Kitoh A, Endo H, Krishna Kumar K, Cavalcanti IFA, Goswami P, Zhou T (2013) Monsoons in a changing world: a regional perspective in a global context. JGR: Atmos 118:3053–3065
Kripalani RH, Kulkarni A (1997) Claimatic impact of El Nino/La Nina on the Indian monsoon: a new perspective. Weather 52:39–46
Kripalani RH, Kulkarni A (1999) Climatology and variability of historical Soviet snow depth data: some new perspectives in snow—Indian monsoon teleconnections. Clim Dyn 15:475–489
Kripalani RH, Singh SV, Arkin P (1991) Large-scale features of rainfall and outgoing long-wave radiation over Indian and adjoining regions. Contri Atmos Phys 64:159–168
Kripalani RH, Kulkarni A, Sabade SS (1999) Characteristic features of intra-seasonal Madden Julian Oscillations over the Indo-Pacific region as revealed by NCEP/NCAR Reanalysis data. In: Proceedings of the second international conference on reanalysis (Reading, UK, 23–27 August 1999) WCRP-109, WMO/TD-No. 985, pp 294–297
Kripalani RH, Kulkarni A, Sabade SS, Ravadekar JV, Patwardhan SK, Kulkarni JR (2004) Intraseasonal oscillations during monsoon 2002 and 2003. Current Sci 87(3):325–331
Kripalani R, Oh JH, Kulkarni A, Sabade SS, Chaudhari HS (2007) South Asian summer monsoon precipitation variability: coupled climate model simulations and projections under IPCC AR4. Theor Appl Climatol 90(3–4):133–159

Krishna Kumar K, Rajagopalan B, Cane MA (1999) On the weakening relationship between the Indian monsoon and ENSO. Science 284:2156–2159

Krishnamurti TN, Ardunay P (1980) The 10–20 day westward propagating mode and the breaks in the monsoon. Tellus 32:15–26

Krishnamurti TN, Bhalme HN (1976) Oscillation of a monsoon system. Part I: observational aspects. J Atmos Sci 33:1937–1954

Krishnan R, Sabin TP, Ayantika DC, Kitoh A, Sugi M, Murakami H, Turner AG, Slingo JM, Rajendran K (2013) Will the South Asian monsoon overturning circulation stabilize any further? Clim Dyn 40:187–211

Krishnan R, Sabin TP, Vellore R, Mujumdar M, Sanjay J, Goswami BN, Hourdin F, Dufresne J-L, Terray P (2016) Deciphering the desiccation trend of the South Asian monsoon hydroclimate in a warming world. Clim Dyn 47:1007–1027

Kulkarni A (2012) Weakening of Indian summer monsoon rainfall in warming environment. Theor Appl Climatol 109:447–459

Kulkarni A, Sabade SS, Kripalani RH (2006) Intraseasonal vagaries of the Indian summer monsoon rainfall. IITM Researh Report No RR114, 51 pp

Kulkarni A, Deshpande N, Kothawale DR, Sabade SS, RamaRao MVS, Sabin TP, Patwardhan S, Mujumdar M, Krishnan R (2017) Observed climate variability and change over India. In: Krishnan R, Sanjay J (eds) Climate change over India—an interium report. Center for Climate Change Research, IITM, Pune

Kulkarni A, Sabin TP, Chowdary JS, Rao KK, Priya P, Gandhi N, Bhaskar P, Buri VK, Sabade SS, Pai DS, Ashok K (2020) Precipitation changes in India. Assessment of climate change over the Indian region. Springer, Singapore, pp 47–72. https://doi.org/10.1007/978-981-15-4327-2_3

Madden RA, Julian PR (1971) Detection of a 40–50 day oscillation in the zonal wind in the tropical Pacific. J Atmos Sci 28:702–708

Madden RA, Julian PR (1994) Observations of the 40–50 day tropical oscillation—a review. Mon Wea Rev 122:814–837

Niyogi D, Kishtawal C, Tripathi S, Govindraju RS (2010) Observational evidence that agricultural intensification and land use change may be reducing the Indian summer monsoon rainfall. Water Res Res 46:W03533

Pai DS, Shridhar L, Rajeevan M, Sreejith OP, Satbhai NS, Mukhopadhyay B (2014) Development of a new high spatial resolution (0.25° × 0.25°) long period (1901–2010) daily gridded rainfall data set over India and its comparison with existing data sets over the region. Mausam 65:1–18

Pathak A, Ghosh S, Kumar P (2014) Precipitation recycling in the indian subcontinent during summer monsoon. J Hydrometeor 15:2050–2066

Pant GB, Parthasarathy B (1981) Some aspects of an association between the southern oscillation and Indian summer monsoon. Arch Meteorol Geophys Bioclim B29:245–252

Paul S, Ghosh S, Oglesby R, Pathak A, Chadrasekharan A, Ramasankaran R (2016) Weakening of Indian summer monsoon rainfall due to changes in land use land cover. Sci Rep 6:10

Ramage CS (1971) Monsoon meteorology. International geophysics series, vol 15. Academic Press, San Diego, California, pp 296

Ramanathan VC, Chung DK, Bettge T, Buja L, Kiehl JT, Washington WM, Fu Q, Sikka DR, Wild M (2005) Atmospheric brown clouds: impact on South Asian climate and hydrologic cycle. Proc Natl Acad Sci USA 102:5326–5333

Ramesh Kumar MR, Krishnan R, Syam S, Unnikrishnan AS, Pai DS (2009) Increasing trend of "break-monsoon" conditions over India-role of ocean-atmosphere processes in the Indian Ocean. IEEE Geosci Rem Sens Lett 6:332–336

Rao SA, Choudhari H, Pokhrel S, Goswami BN (2010) Unusual Central Indian drought of summer monsoon 2008: role of southern tropical Indian ocean warming. J Clim 23:5164–5173

Rasmusson EM, Carpenter TH (1983) The relationship between eastern equatorial Pacific SSTs and rainfall over India and Sri Lanka. Mon Wea Rev 111:517–528

Ross SR, Krishnamurti TN, Pattnaik S, Pai DS (2018) Decadal surface temperature trends in India based on a new high resolution data set. Sci Rep 8(7452):1–10

Sikka DR (1980) Some aspects of the large scale fluctuations of summer monsoon rainfall over India in relation to fluctuations in the planetary and regional scale circulation parameters. Proc Ind Acad Sci (Earth Planetary Sci) 89:179–195

Roxy MK, Kapoor R, Terray P, Murtugudde R, Ashok K, Goswami BN (2015) Drying of Indian subcontinent by rapid Indian Ocean warming and a weakening land-sea thermal gradient. Nat Commun 6(7423)

Saji H, Yamagata T (2003) Possible impacts of Indian ocean dipole mode events on global climate. Clim Res 25(2):151–169

Saji NH, Goswami BN, Vinayachandran PN, Yamagata T (1999) A dipole mode in the tropical Indian Ocean. Nature 401:360–363

Sankar S, Svendsen L, Gokulapalan B, Joseph PV, Johannessen OM (2016) The relationship between Indian summer monsoon rainfall and Atlantic multidecadal variability over the last 500 years. Tellus A 68:1–14

Sathiyamoorthy V (2005) Large scale reduction in the size of the tropical easterly jet. Geophys Res Lett 32:L14802

Sharmila S, Joseph S, Sahai AK, Abhilash S, Chattopadhyay R (2015) Future projection of Indian summer monsoon variability under climate change scenario: an assessment from CMIP5 climate models. Glob Planet Chang 124:62–78

Sikka DR, Gadgil S (1980) On the maximum cloud zone and the ITCZ over India longitude during the southwest monsoon. Mon Wea Rev 108:1840–1853

Singh SV, Kripalani RH (1985) The south to north progression of rainfall anomalies across during the summer monsoon season. Pure Appl Geophys 123:634–637

Singh SV, Kripalani RH (1986) Application of the extended empirical orthogonal function analysis to inter-relationships and sequential evolution of monsoon fields. Mon Wea Rev 114:1603–1610

Singh SV, Kripalani RH (1990) Low frequency intra-seasonal oscillation in Indian rainfall and out-going long-wave radiation. Mausam 41:217–222

Singh D, Tsiang M, Rajaratnam B, Diffenbaugh NS (2014) Observed changes in extreme wet and dry spells during the South Asian summer monsoon season. Nat Clim Chang, pp 1–6

Singh D, Ghosh S, Roxy MK, McDermid S (2019) Indian summer monsoon: extreme events, historical changes, and role of anthropogenic forcings. Wires Clim Change 2019(e571):1–35

Srinivas G, Chowdary JS, Kosaka Y, Gnanaseelan C, Parekh A, Prasad K (2018) Influence of the Pacific-Japan pattern on Indian summer monsoon rainfall. J Clim 31(10):3943–3958

Turner AG, Hannachi A (2010) Is there regime behavior in monsoon convection in the late 20th century? Geophys Res Lett 37:L16706

Turner A, Annamalai H (2012) Climate change and the South Asian summer monsoon. Nat Clim Chang 2:587–595

Webster PJ, Magana VO, Palmer TN, Shukla J, Tomas RA, Yanai M, Yasunari T (1998) Monsoons: processes, predictability, and the prospects for prediction. J Geophys Res 103(14):451–510

Yadav RK (2017) On the relationship between east equatorial Atlantic SST and ISM through Eurasian wave. Clim Dyn 48:281–295

Yasunari T (1980) A quasi-stationary appearance of the 30–40 day period in the cloudiness fluctuations during the summer monsoon over India. J Meteor Soc Japan 58:225–229

Use of Remote Sensing in Weather and Climate Forecasts

Ashim Kumar Mitra

Abstract Satellites basically measure the radiance coming from the earth's surface and cloud tops. By making such measurements at appropriate wavelengths and applying physical and statistical techniques, it is possible to compute a wide range of products for weather monitoring and forecasting. Further, the satellite meteorological data on a global scale are vital inputs in Numerical Weather Prediction (NWP) models as initial conditions. For a tropical country like India where high-impact convective events are very common, it is necessary to have good quality high density observations both in spatial and temporal scale to ingest into an assimilation cycle of NWP models. In view of the importance of satellite data, accurate estimation of satellite observations could be the only viable alternative solution for data sparse regions. Availability of multi-spectral imager channels in INSAT3D/3DR satellites with the staggering mode of temporal frequency at every 15 min provides us a new way to look at the weather events over the Indian region. The most effective ways to utilize these products are by combining the channels using red–green–blue (RGB) composites of INSAT3D/3DR satellite with its microphysical properties and RAPID SCAN mode. The RAPID scan mode has a higher temporal resolution, i.e., approximately 4 min unlike 30-min of conventional mode, for analyzing weather phenomena particularly intensification, propagation, and decay of various types of weather systems including nowcasting applications. Further, INSAT3D/3DR sounder provides continuous upper level temperature and moisture profiles with a spatial resolution of 10 × 10 km with temporal resolution of half an hour. With its sounder payload, the amount of the water vapor present in the atmospheric column in the form of total precipitable water (TPW) is being retrieved operationally and can be considered as a precursor for mesoscale activities.

A. K. Mitra (✉)
Satellite Meteorology Division, India Meteorological Department, New Delhi 110003, India
e-mail: ak.mitra@imd.gov.in

V. K. Gahalaut and M. Rajeevan (eds.), *Social and Economic Impact of Earth Sciences*,
https://doi.org/10.1007/978-981-19-6929-4_5

1 Introduction

Meteorological services are essential for all countries to safeguard people and to support sectors like agriculture, water resources, and power. These services include weather forecasts, decision support system, warnings, and providing information for the purposes of protection and safety. Weather forecasting and severe weather warnings are essential to the success of the public and private sectors, including business and commerce, agriculture, forestry, marine and fisheries, the airline industry, military applications, and urban infrastructure management. Timely collection of data related to current weather is essential for generating weather forecasts. Indeed, a meteorologist's forecast will only be as good as the data and knowledge that is available.

Satellite measurements of meteorological parameters have better advantages over in situ observations. INSAT, the Indian National Satellite System is a series of multipurpose geostationary satellites launched by ISRO (Indian Space Research Organization) to meet the requirement for telecommunication, broadcasting, meteorology, and search and rescue operations. INSAT is the largest domestic communication system in the Asia Pacific Region and was commissioned in 1983. The Indian National Satellite (INSAT) system was commissioned with the launch of INSAT-1B in August. INSAT system ushered in a revolution in India's television and radio broadcasting, telecommunications, and meteorological sectors. Some INSATs also carry instruments for meteorological observations and data relays for providing meteorological services. KALPANA-1 was an exclusive meteorological satellite. The history of INSAT satellites since 1982 for meteorological use is shown in Table 1. The successful launch of indigenous geostationary satellite INSAT3D on July 26, 2013 and INSAT3DR on September 8, 2016 with advanced meteorological payloads onboard has provided a new opportunity to Indian meteorologists. Currently, these two satellites are operational at India Meteorological Department. Both of them carry four payloads: Imager (Six Channels), Sounder (Nineteen Channels), Data Relay Transponder (DRT) & Satellite Aided Search and Rescue (SAS & R). The Channel specifications of INSAT-3D/3DR IMAGER are given in Table 2.

The first satellite completely dedicated to satellite meteorology was launched on April 1, 1960. TIROS 1 (Television and Infrared Observational Satellite), the twenty-second successful launched satellite, was harbox shaped, about 57 cm in height and 107 cm in diameter. Nine additional satellites were launched in the TIROS series; the last, TIROS 10, was launched on July 2, 1965. The first 5 years of satellite meteorology are also documented by Hubert and Lehr (1967), to which the readers can get all the details related to references.

The vertical structure of temperature and water vapor plays an important role in the meteorological processes of the atmosphere. For years the radiosonde network has been the primary observing system for monitoring tropospheric temperature and water vapor. Routine observations are very difficult over the oceanic regions due to logistics problems and high cost factors. The radiosonde networks are also limited over land regions. Therefore, satellites-based observation for obtaining these

Table 1 INSAT satellites since 1982

Satellite	Period	Payload	Channels	Spatial resolution	Temporal resolution
INSAT-1A	(1982–1982)	VHRR	Visible (0.55–0.75 μm)	2.75 km	3 hourly
INSAT-1B	(1983–1993)		Infrared (10.5–12.5 μm)	11 km	
INSAT-1C	(1988–1989)				
INSAT-1D	(1990–2002)				
INSAT-2A	(1992–2002)	VHRR	Visible (0.55–0.75 μm)	2 km	3 hourly
INSAT-2B	(1993–2004)		Infrared (10.5–12.5 μm)	8 km	
INSAT-2C	(1995–2002)				
INSAT-2D	(1997–1997)				
INSAT-2E	(1999–2012)	VHRR DRT	Visible (0.55–0.75 μm)	2 km	3 hourly
			Infrared (10.5–12.5 μm)	8 km	
			Water vapor (5.7–7.1 μm)	8 km	
KALPNA-1	(2002–2017)	VHRR DRT	Visible (0.55–0.75 μm)	2 km	3 hourly with 3 triplet (00,06,12 UTC) during 2002–2005 Hourly (00,06,12 UTC) during 2005–2008 Half Hourly 2008–17
			Infrared (10.5–12.5 μm)	8 km	
			Water vapor (5.7–7.1 μm)	8 km	
INSAT-3A	(2003–2016)	VHRR	Visible (0.55–0.75 μm)	2 km	3 Hourly (2003–2008) Hourly (2008–16)
			Infrared (10.5–12.5 μm)	8 km	
			Water vapor (5.7–7.1 μm)	8 km	
		CCD	Visible (0.62–0.68 μm)	1 km	Six times during day (03,05,06,07,09 and11 UTC)
			NIR (0.77–0.86 μm)	1 km	
			SWIR (1.55–1.69 μm)	1 km	
		DRT	Transponder (24 channels)		

Table 2 INSAT-3D and INSAT-3DR imager specification

Channel no.	Channel ID	Channel name	Spectral range (μm)	Ground resolution (Km)	Purpose
1	VIS	Visible	0.55–0.75	1	Clouds, surface features
2	SWIR	Short wave infrared	1.55–1.70	1	Snow, Ice, and water phase in clouds
3	MIR	Medium wave infrared	3.7–3.9	4	Clouds, fog, fire
4	WV	Water vapor	6.5–7.1	8	Upper-troposphere moisture
5	TIR1	Long wave infrared	10.3–11.3	4	Cloud top and surface temperature
6	TIR2	Split	11.5–12.5	4	Lower-troposphere moisture

parameters are the best source of information because they provide observations on a global scale with improved temporal resolution. With the launch of first Advanced Microwave Sounding Unit (AMSU) sounder, onboard the National Oceanic and Atmospheric Administration's (NOAA) polar-orbiting satellites, observations at middle and upper levels have been possible from the surface to an altitude of about 45 km in most situations. With the advancement of new technology, several other satellites, Atmospheric Infrared Sounder (AIRS), Microwave Limb Sounder (MLS) and GPS Radio Occultation provide profiles of temperature and water vapor with reasonable accuracies. At this juncture, India's advanced weather satellite, INSAT-3D, the first geostationary sounder system over the Indian Ocean, was launched (located at 83°E) on July 26, 2013, for improved understanding of mesoscale systems. It was the second of its kind after the Geostationary Operational Environmental Satellites (GOES).

Currently, the sounding system of INSAT3D/3DR satellite provides vertical profiles of temperature (40 levels from surface to ~70 km), humidity (21 levels from surface to ~15 km) and integrated ozone from surface to top of the atmosphere. The sounder retrieval scheme at India Meteorological Department (IMD), New Delhi, is adapted from the operational High resolution Infrared Radiation Sounder (HIRS) processing scheme and Geostationary Operational Environmental Satellites (GOES) algorithms developed by Cooperative Institute for Meteorological Satellite Studies (CIMSS), University of Wisconsin, USA (Ma et al. 1999; Li et al. 2008), in which physical and regression-based retrievals are employed. Spectral bands in and around the CO_2 and H_2O absorbing bands are designed to yield information about the vertical structure of atmospheric temperature and moisture and are important in weather monitoring and forecasting in the study of mesoscale weather phenomena.

2 INSAT3DR RAPID SCAN Mode

To improve forecasting accuracy for the detection of severe weather and deep convection phenomena, a more flexible and powerful observing strategy is needed. International satellite operators and researchers typically define rapid scanning as taking images at intervals shorter than the routine scan interval of the same area considered as baseline for a given satellite. For example, the Meteosat Second Generation (MSG) operational Rapid Scanning Service (RSS) currently delivers 5-min scans over Europe at the same spatial resolution as the routine 15-min scans. The Geostationary Operational Environmental Satellite-14 (GOES-14) imager was operated by the National Oceanic and Atmospheric Administration (NOAA) in an experimental rapid scan 1-min mode during parts of the summers of 2012 and 2013. This scan mode, known as the super rapid scan operations for GOES-R (SRSOR), emulates high-temporal-resolution sampling of the mesoscale region scanning of the Advanced Baseline Imager (ABI) on the next-generation GOES-R series (Schmit et al. 2013).

INSAT-3D/3DR IMAGER payload can be operated in three modes, namely, full frame mode, programmed normal scan mode, and programmed sector scan (RAPID SCAN) mode. The details of these scan modes are provided in Table 3. The motivation and intent behind carrying out the Rapid Scan from INSAT-3DR IMAGER payload are to understand the rapidly evolving systems and their physical processes involved in the formation of various weather phenomena. More importantly, the temporal resolution achieved in rapid scan is quite high, i.e., approximately 4 min and 30 s. While achieving this high-temporal resolution, the spatial resolution is maintained, i.e., VIS & SWIR scanning are done at 1 km resolution and TIR1/TIR1/MIR at 4 km and WV at 8 km. Therefore, when INSAT-3DR is operated in rapid scan mode as shown in Fig. 1, satellite data are available to forecasters at an interval of every 4 min and 30 s in native resolution similar to what SEVIRI instrument onboard the MSG satellites and GOES ABI (Advanced Baseline Imager) scan of 5-min intervals. With this time range, it is possible to forecast small features such as individual storms with reasonable accuracy. A forecaster using the latest radar, satellite and observational data is able to make analysis of the small-scale features present in a small area such as a city and make an accurate forecast for the following few hours. It is, therefore, a powerful tool in warning the public of hazardous, high-impact weather including tropical cyclones, thunderstorms, and tornados which cause flash floods, lightning strikes, and destructive winds. During rapid scan operation, which is even better than that of Doppler weather radars which has a temporal resolution of 10 min. Unlike Doppler weather radars, satellite observations do not suffer from line of sight constrain which reduce their effective range to 250 kms. Hence, rapid scan satellite acquisition will be a huge boost for the nowcasting of severe weather events.

Table 3 INSAT3D and INSAT3DR scanning mode

Mode of operation	Time of coverage (min)	Coverage area
Full frame mode	26	18 × 18°
Programmed normal scan mode	23	14 × 18°
Programmed sector scan (RAPID SCAN) mode	6	4° in NS and 18° in EW

Fig. 1 Extent of coverage: 9 blocks (4.5° coverage in 351 lines), number of repetitions: 4, time required: 27 min (9 blocks with 4 repetitions)

3 INSAT 3D/3DR RGBs and Its Utilization

The Red–Green–Blue or RGB processing in the satellite data offers a unique solution from different signatures of spectral channels into meaningful inference for severe mesoscale phenomena than any individual IR and VIS grayscale images or traditional single-channel color enhancement techniques.

RGB with multi-spectral channels are false color images designed to enhance a specific feature such as fog, snow, dust, and convections. These images are created by combining band or band differences into each of the red, green, and blue components with a defined recipe. The advantage of RGB products is the ability to look at a single image to identify a feature by its color code scheme instead of analyzing multiple single channels. In the INSAT-3D RGB, the image manipulations refer to reflectance factor (RF) values for solar channels and equivalent brightness temperature (EBT) values for IR channels. The INSAT3D/3DR enhancement operations expand to the full range of display values (0–1023) and a limited range (MIN, MAX) of RF or EBT values, applying subsequently a gamma correction for possible non-linear expansion (Mitra et al. 2018).

$$INSAT = 3D(R, G, B) = 1023 * \left[\frac{(BT, \Delta BT, R, \Delta R)}{MAX - MIN}\right]^{\left(\frac{1}{\gamma\, R,G,B}\right)}$$

where BT is the brightness temperature, ΔBT is the brightness temperature difference, R is the reflectance, ΔR is the reflectance difference, and γ is the gamma enhancement.

In the early 2000s, European Meteorological Satellite (EUMETSAT) developed a set of RGB recipes or best practices following the launch of Meteosat Second Generation with the SEVIRI instrument aboard. The standards identification of features is expressed in terms of combinations of channels or channel differences, attribution to the RGB color planes and enhancement. Over the Indian region, the technique of Lensky and Rosenfeld (2008) (operational in EUMESAT) has been fine-tuned using INSAT-3D multi-spectral channel specifications and implemented in INSAT Meteorological Data Processing System (IMDPS) at IMD in 2015. In this study, here fine-tuning is basically the adjustment of EUMETSAT RGB recipes to retain legacy coloring due to differing spectral characteristics between SEVIRI and INSAT-3D channels. The reflectance of solar channel and BT of INSAT-3D thermal channels are key ingredients for the preparation of the daytime and nighttime microphysical products.

With the availability of the spectral channels in the day and nighttime, Daytime microphysics (DTMP) using shortwave thermal infrared (1.6 μm) (SWIR), visible (0.5 μm) and thermal IR channels (10.8 μm) and nighttime microphysics (NTMP) using the brightness temperature (BT) differences between 10.8, 12.0, and 3.9 μm is prepared as shown in Table 4 and Table 5, respectively.

Table 4 DTMP RGB composition color scheme

Beam	Channel	Range	Gamma
Red	VIS (0.55–0.75 um)	0…+100%	1.0
Green	SWIR (1.67 um)	0…+60%	1.0
Blue	IR (10.8 um)	+ 203…+323 °K	1.0

Table 5 NTMP RGB composition color scheme

Beam	Channel	Range	Gamma
Red	IR 12.0 um–IR 10.0 um (TIR2–TIR1)	−4…+2 K	1.0
Green	IR 10.8 um–IR 3.9 um (TIR1–MIR)	−4…+6 K	1.0
Blue	IR 10.8 um (TIR1)	+243…+293 K	1.0

4 Identification of Snow/Fog/Clouds and Dust

To identify snow/fog/cloud and dust regions, weather forecasters sometimes rely upon comparisons between visible, shortwave IR, and thermal infrared satellite imagery during daytime. When these phenomena appear in the scene at the same time, interpretation via single-channel imagery comparisons becomes extremely difficult. Here, DTMP RGB scheme intelligently results in a more definitive characterization of the cloud/snow distribution. In Fig. 2a, snow and ice clouds appear red because they strongly absorb in 1.6 μm. The small particles of ice cloud signature appear orange in (the western disturbance), whereas large particle ice cloud [Mature Cumulonimbus (CB) cell] appears with the greater red component. Snow grains are usually larger than cloud ice particles and over the ground it appears as full red color. Generally, ice particles that form by mixed phase process in a supercooled water cloud grow quickly to much larger sizes than crystals forming by vapor deposition in ice-only clouds. This helps to separate convective precipitating clouds from non-precipitating or layer ice clouds (Lensky and Rosenfeld 2008). Ocean appears dark blue because of high thermal emission. In contrast to daytime, the NTMP RGB scheme, Fig. 2b, designed to monitor the evolution of nighttime fog/low stratus and different types of clouds in the nighttime. Apart from this, other applications are detection of nighttime fog, detection of fires, low-level moisture boundaries, cloud classification and dust movement. In NTMP, the BT difference between the 12.0 and 10.8 μm channels (12.0−10.8) is a measure of the clouds' opaqueness (Inoue 1987) and it is displayed in red whereas the BT difference between 10.8 and 3.9 μm channels (10.8−3.9) modulates the green beam. The BT difference in these two bands distinguishes stratus clouds consisting of water droplets versus cirrus clouds as well as cloud-free regions. Lensky and Rosenfeld (2003a) showed how sensitive is the BT difference between 10.8 and 3.9 μm channels to particle size, and used this information to delineate precipitation (Lensky and Rosenfeld 2003b). Large difference between these two channels indicates clouds with small drops. Radiation difference between the cloud and fog is due to the spectral response differences which provide the basis of fog detection. Due to the limitation of a single IR (10.8 μm) channel during fog, poor discrimination of low clouds/fog/stratus (Eyre et al. 1984; Bader et al. 1995) is resulted. Therefore, to identify the fog extent (horizontal) during night, the differences in the radiative properties of fog and stratus in the 3.9 and 11 μ channels result in temperature contrasts will allow discriminating the fog. This technique is based on the principle that the emissivity of water cloud at 3.9 μ is less than at 10.7 μ (Ellrod 1995). In NTMP, fog with small drops or shallow clouds appears in white. These spectral signatures of the snow/fog and clouds can be well analyzed with RGBs and shown in Fig. 2c. On January 21, 2021, a western disturbance with induced cyclonic circulation was approaching toward western Himalayas and northern parts of the country. At 10:30 IST, DTMP RGB shows 3 different color schemes, i.e., clouds over northern Afghanistan, snow over Jammu and Kashmir and fog over Indo Genetic plains. Transect plot of these events can be seen in Fig. 2c. High cirrus cloud exhibits lower thermodynamic temperature than a snow-covered background. However, in the visible (0.5 μm) region, highly reflective snow cover contrasts fog and cloud features (VIS > SWIR), whereas the lower absorption (SWIR > VIS) properties of SWIR (1.6 μm), clearly distinguishing fog from cloud and snow. To identify the dust, because of the larger reflectance in the SWIR (1.6 μm) and visible (0.5 μm), dust appears bright cyan. Here, unlike snow and cloud, dust exhibits higher temperatures from the TIR band. A transect plot of dust on October 13, 2020 across northern parts of Arabian sea marked as A–D is shown in Fig. 2d. It can be seen that in all the cases (A–D), reflectance from SWIR (40–55%) is higher than the visible (15–18%) (SWIR > VIS) along with higher TIR (10.8 μm) values(>293 K). It is because dust is less absorptive at the 10.8 μm wavelength and largely influenced by the underlying surface emissions, therefore have a warmer brightness temperature.

5 Convective System Identification from RGBs

On March 6, 2020, the western disturbance (WD) was observed over north Pakistan and neighborhood at middle-tropospheric levels with associated trough at higher levels. The induced low pressure area (LPA) was observed over central parts of West Rajasthan and neighborhood. In Fig. 3a, DTMP RGB shows the well-depicted WD and LPA from 0715 to 1215 UTC. Due to synoptic weather systems, there was a strong convective activity over West Rajasthan and adjoining areas, which developed at 0715UTC. DTMP RGB imagery demonstrates significant convective clouds over the same area. The associated convective clouds in the DTMP images suggest the occurrence of a thunderstorm over Rajasthan, which subsequently moving eastwards. Afterward, as seen in the IMD Delhi radar, the squall line with reflectivity up to 50 dBz moving northeastwards across Delhi has started causing moderate rain and thunderstorm with the gusty wind over western parts of Delhi and is likely to cause moderate rain and TS with gusty winds over remaining parts of Delhi. From the actual IMD observations (from 0830 h IST of 06/03/2020 to 0830 h IST of 07/03/2020), thunderstorms were reported over many places over Uttar Pradesh, Haryana, Chandigarh and Delhi and Hailstorm reported over Nainital, Sonipat, Rohtak, Lucknow, Hardoi, Ajmer, Alwar, and Jaipur. Both INSAT-3D RGB and radar observations have been found in good agreement in terms of delineating areas of deep convection. This exercise suggests that thunderstorm events can be more effectively analyzed by the INSAT-3D RGB technique at every 15-min interval, with the maximum zone of reflectivity by the radar observations.

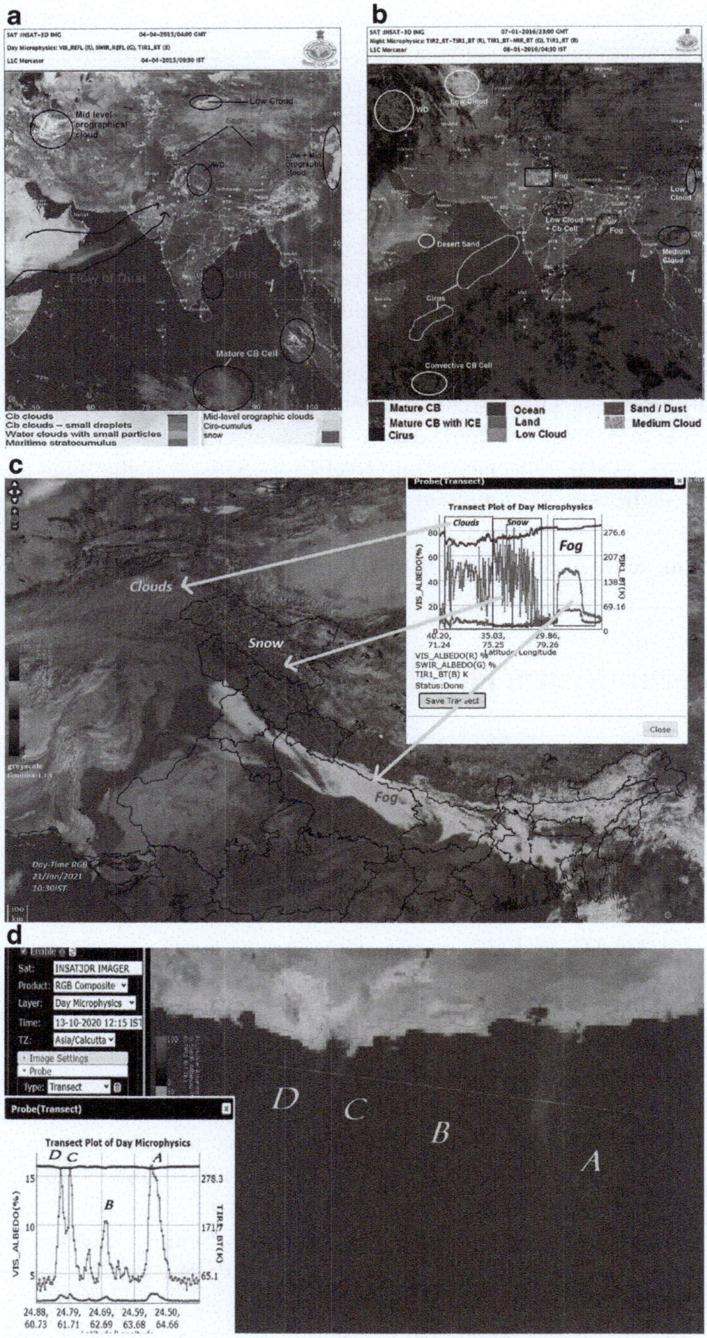

Fig. 2 INSAT-3D RGB composition of the **a** DTMP color scheme and **b** NTMP color scheme **c** Spectral signature of Snow/Fog and cloud and **d** Dust

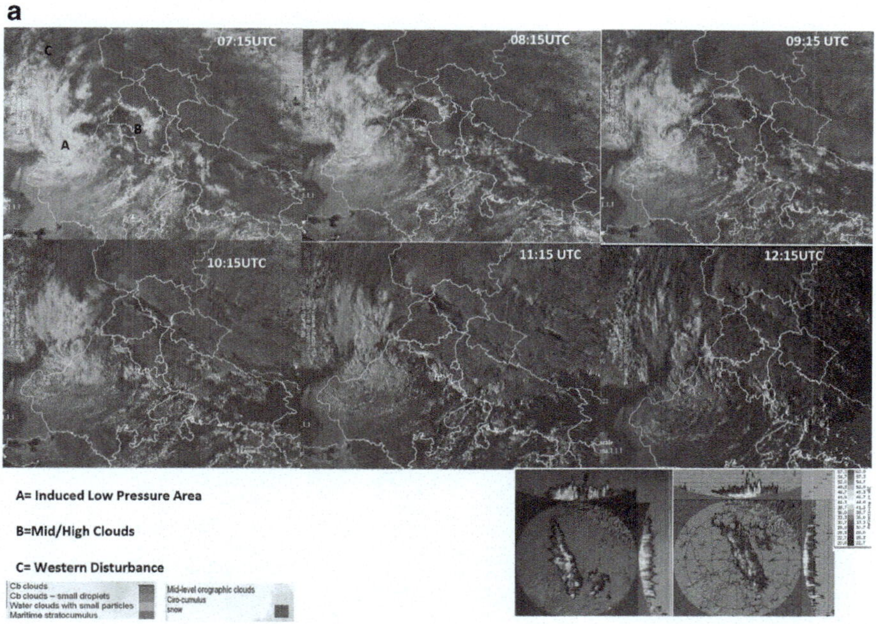

Fig. 3 **a** DTMP RGB and Radar reflectivity (dBZ) of March 06, 2020. **b** INSAT-3DR visible imageries during life cycle of AMPHAN. **c** INSAT-3DR Cloud Top Brightness Temperature (CTBT) imageries during life cycle of AMPHAN. **d** SCAT SAT imageries during life cycle of AMPHAN. **e** ASCAT imageries during life cycle of AMPHAN

INSAT-3D DTMP and NTMP schemes are available for the forecasters and user community on a real-time basis for the identification of weather events. The thresholds technique was also developed and verified separately (Mitra et al. 2018) for both the RGB products for the identification of fog, snow, and low clouds. It was found that the RGBs, i.e., DTMP and NTMP have a reasonable agreement with ground-based observations and MODIS satellite data. Mitra et al. (2018) found a very good probability of fog detection more than 94 and 85% with acceptable false alarm conditions <8 and 10% for DTMP and NTMP, respectively. This technique has significantly minimized misclassification between low clouds, snow, and fog. This method is found to be very useful for day-to-day weather forecasting and is being used by forecasters on operational purpose. For the severe weather phenomenon particularly thunderstorms, initial hours are critical to recognize the location and expanding the cautioning lead times. Traditionally, satellite instruments only observe the highest cloud tops, blocking the internal structure of the storm (Setvak et al. 2010). Considering the advantages of RGB products of the high temporal sampling of INSAT-3D/3DR satellite and its multi-spectral channels capability, this may be taken as a measure of convective activity.

b

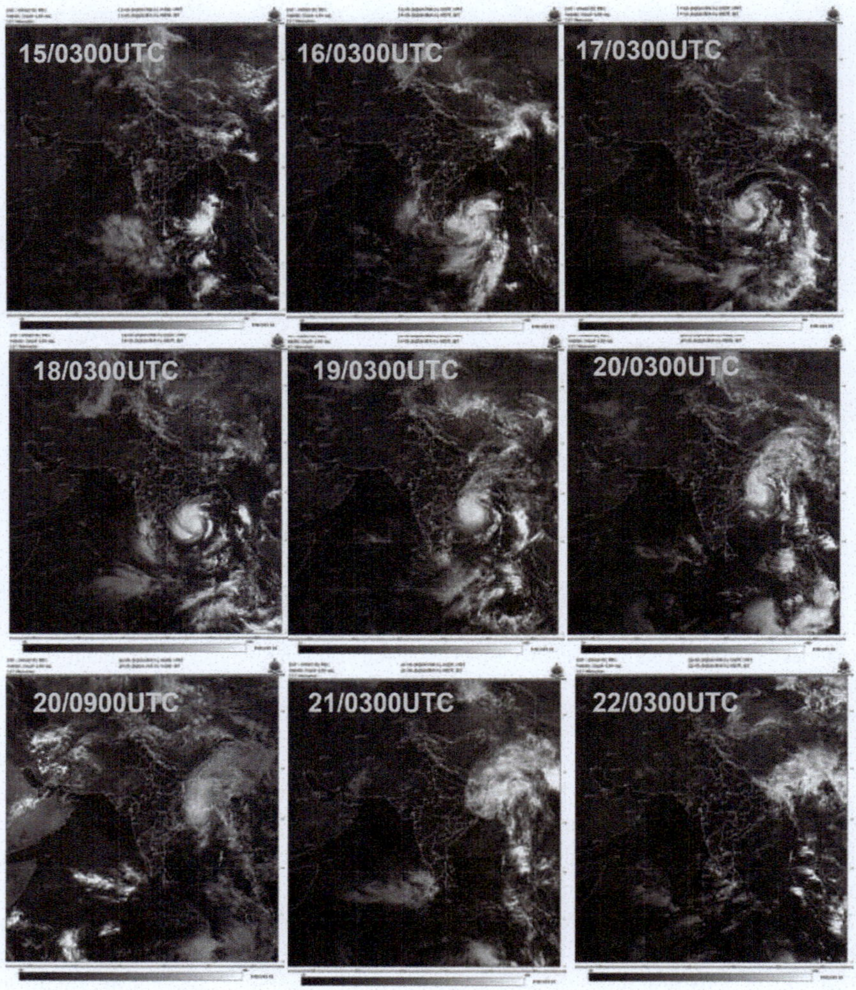

Fig. 3 (continued)

6 Super Cyclonic Storm "AMPHAN" Over Southeast Bay of Bengal (May 16, 2022–May 21, 2020)

The Bay of Bengal's fiercest storm this century, super-cyclone "Amphan," slammed into the coast of eastern India and Bangladesh on May 20, 2020, afternoon, bringing heavy gales and threat of deadly storm surges and flooding. Various numerical weather prediction models run by Ministry of Earth Sciences (MoES) institutions, and IMD's dynamical-statistical models developed in-house were utilized to predict

c

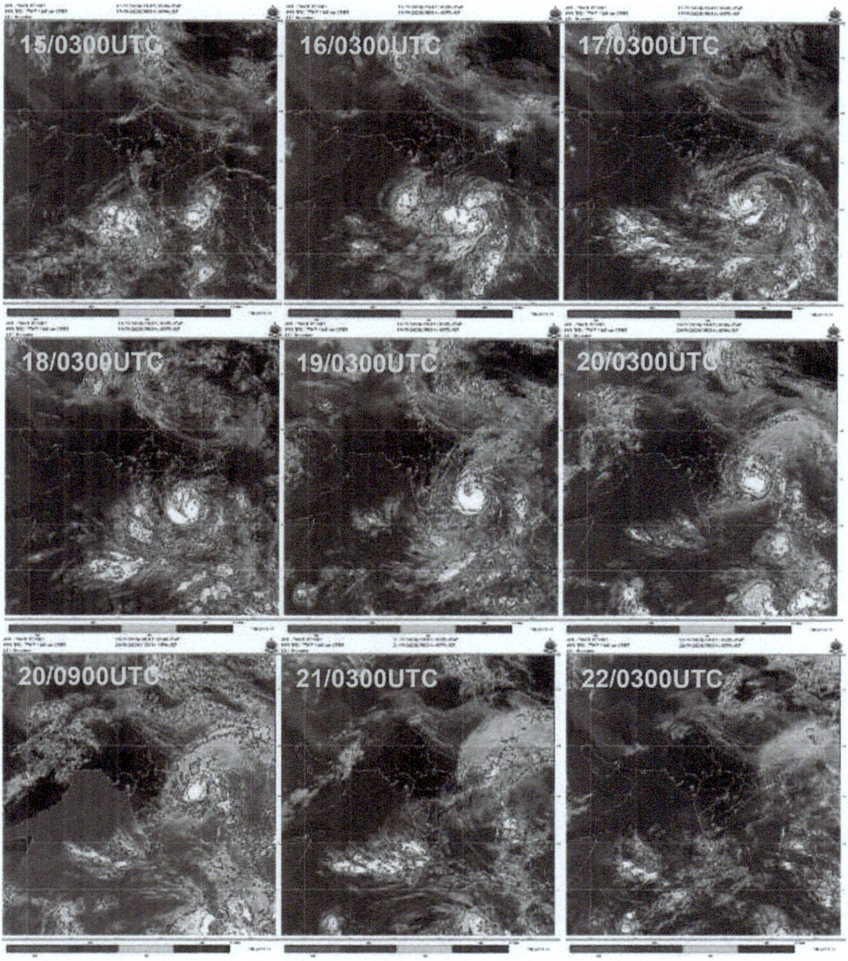

Fig. 3 (continued)

the genesis, track, landfall, and intensity of the cyclone. In addition to that, it was also extensively monitored with the help of available satellite observations from INSAT3D, INSAT3DR, and polar-orbiting satellites including SCATSAT, ASCAT. The available satellite observations during the life cycle of AMPHAN are shown in Fig. 3b–e.

At 0000 UTC of the 16th, the intensity of the system was T1.5. Associated broken low to medium clouds with embedded intense to very intense convection lay over the Bay of Bengal as seen from the figure. Minimum CTT was minus 93 °C. Convection

d

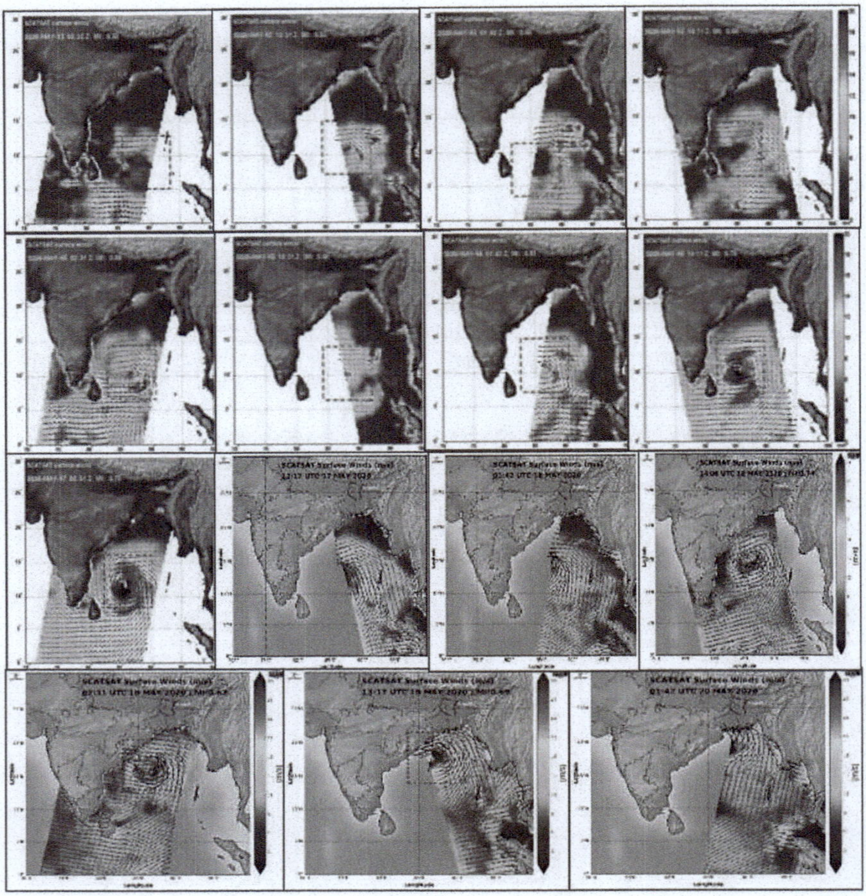

Fig. 3 (continued)

and organization had increased further depicted in visible imageries in Fig. 3b. At 0300 UTC of 17th, intensity of the system was T3.0 associated with curved band pattern. Minimum CTT was minus 93 °C with associated broken low to medium clouds with embedded intense to very intense convections. This can be seen from the cloud top brightness temperature imageries (CTBT) in Fig. 3c. In Fig. 3d, SCAT SAT imagery at 1411 UTC of 16th May was indicating winds around 40 knots to the south of system center, associated with a banding eye pattern in the visible imagery. At 0600 UTC of 18th, the system further intensified and the current intensity of the system was T6.5. Eye was clearly visible with circular pattern and it was continuing with a diameter of 15 km. Eye had become colder with temperature minus 21 °C. Wall cloud temperature was minus 93 °C. Minimum cloud top temperature was minus 93 °C with associated broken low to medium clouds with embedded intense

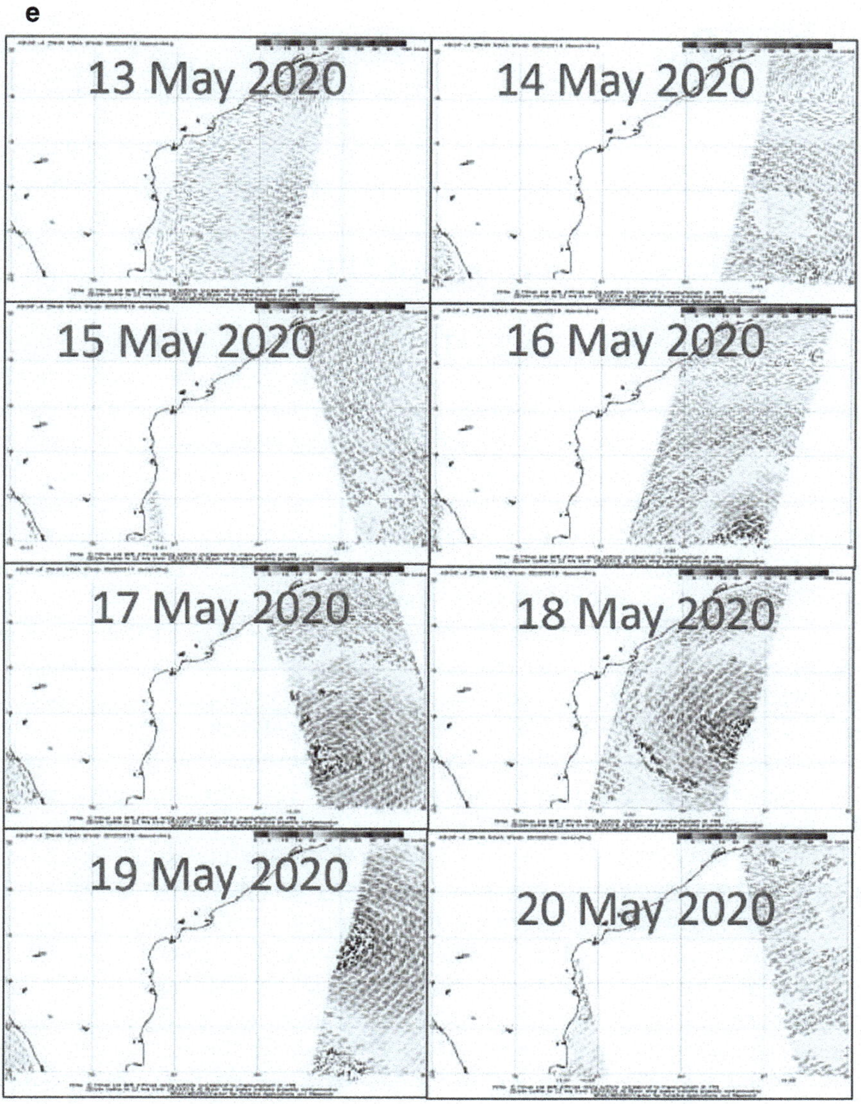

Fig. 3 (continued)

to very intense convection in the visible imagery and ASCAT winds in Fig. 3e. With accurate track and forecast of super cyclonic storm "AMPHAN" round the clock, IMD earned a lot of appreciations from various national and international agencies.

Table 6 Specification of INSAT sounder

Channels (spectral range microns)	Resolution
Visible (0.67)	10 km × 10 km
SWIR (3.67)	10 km × 10 km
MIR (6.38)	10 km × 10 km
LWIR (11.66)	10 km × 10 km

7 INSAT3D/3DR Sounder Derived TPW

First time in Indian satellite, i.e., INSAT3D and INSAT3DR, atmospheric sounder payload, has been introduced to observe vertical profiles of temperature and humidity in the atmosphere over the Indian sub-continent. With staggering mode, it is possible to obtain continuous upper level temperature and moisture profiles with a spatial resolution of 10 × 10 km and temporal resolution of half an hour. INSAT3D/3DR sounder has one visible spectral channel and 18 channels in shortwave infrared (SWIR), middle infrared (MIR), and longwave infrared (LIR) regions. Further details of the INSAT3D sounder can be found in Mitra et al. (2015). Table 6 shows the channel specifications of INSAT3D/3DR sounder.

One of the important weather monitoring capabilities of the INSAT-3D/3DR sounder is the estimation of water vapor in the atmosphere in the form of total precipitable water (TPW). TPW observations are essential for weather, climate modeling, and prediction. The TPW may be used for monitoring the mesoscale to synoptic-scale convective activity, monsoonal activities, and moisture gradients. Yuan et al. (1993) showed an increment of ~8 mm in the tropical TPW resulting from doubling of atmospheric CO_2. Water vapor varies in time and in space (both vertically and horizontally) and the gaps in the observations make its use impossible for climate and weather forecasting/nowcasting related studies (Trenberth and Olson 1988). This could be possible with higher temporal and spatial resolution of accurate temperature and moisture profiles, either from in situ observations or remotely sensed data. Parihar et al. (2018) and Mitra et al. (2015) showed the usefulness of INSAT3D sounder-derived TPW and described that the heavy and heavy-to-very heavy rainfall corresponds well with the higher INSAT3D TPW values. This indicates the reliability of using the TPW product to forecast monsoon precipitation over the Indian region.

8 A Case Study with Validation of TPW

Keeping in mind the potential impact of TPW, hourly INSAT-3D sounder-derived TPW, and GNSS TPW were analyzed for thunderstorm events that occurred in Shimla, Trivandrum, and Dibrugarh during June 2017 and shown in Fig. 4. The purpose of using the IMD GNSS instrument is that it has a maximum coverage of India and access to multiple satellites, redundancy, and availability at all times.

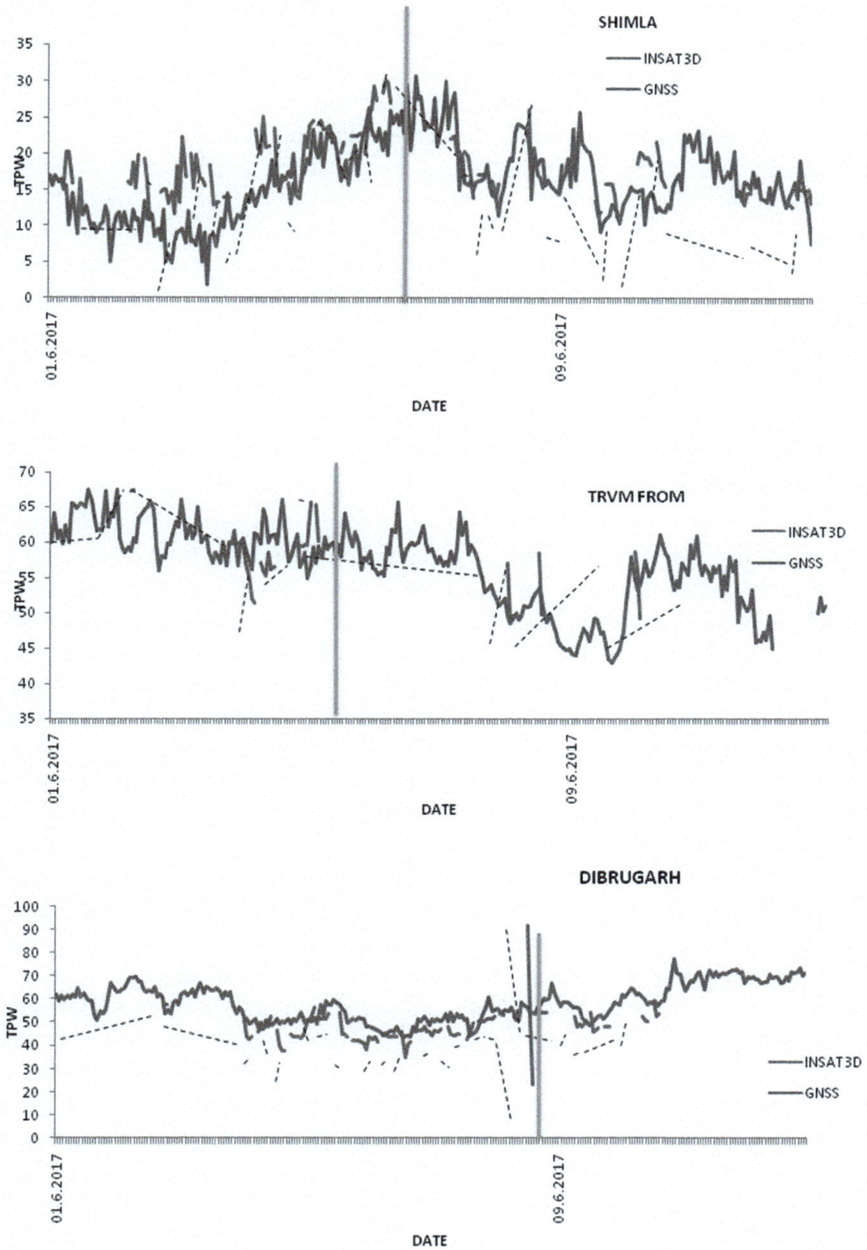

Fig. 4 INSAT-3D and GNSS TPW and hourly realized thunderstorms events at Shimla, Trivandrum and Dibrugarh in June 1–10, 2017). Green line is the time of occurrence of thunderstorms. Dotted lines are cloudy scene of INSAT3D

The case studies show that during the thunderstorm events, INSAT-3D-derived TPW compares reasonably well with GNSS TPW observations. It can be seen that prior to rainstorms, thunderstorms, and hailstorms there is a rapid increase in the value of TPW in nearly 1–3 h. Along with other meteorological parameters (e.g., CAPE; convective available potential energy), instability indices with INSAT-3D TPW and the mesoscale activities can be very easily detected and utilized for weather forecasts. However, the above case studies confirm the usefulness of INSAT-3D-derived TPW prior to the event and it can be considered one of the precursors for convective activities. Nowadays, TPW is also considered as one of the important input parameters in numerical weather prediction for nowcasting in any severe weather events.

9 Summary and Future Scope for Climate and Weather Prediction Using Remote Sensing

There is an increasing demand for higher spatio-temporal resolution datasets for regular day-to-day monitoring and prediction of atmospheric and oceanic states as well as to address several important science questions related to weather and climate. Since there is a limitation for ground-based observations, especially over the oceans and difficult terrains, there is a lot of scope for improving the current Earth Observations (EO) systems using Indian satellites. Along with the growing demand for monitoring different components of the earth systems, several technological advancements in payload instruments have taken place. To take full advantages from available improved observational capabilities an entirely new suit of payloads from geostationary (GEO) is recommended.

There is also a critical requirement to undertake the atmosphere and ocean process study that will lead toward the improvements in modeling of physical processes by considering water vapor, cloud, precipitation, aerosol, radiation, atmospheric circulation, and associated feedbacks. Thus, it is very much required to enhance our understanding of the physical processes that are still unresolved in the numerical weather prediction models either due to the complexity of these processes or due to their spatio-temporal scales. Intensive observational efforts will be required in near future to understand these processes that are critical to our ability to predict the weather at desired space time scales. It is also important to reduce the uncertainty in future climate projections by the climate models.

Ministry of Earth Sciences (MoES) under its long-term vision plan is gearing toward the development of an advanced weather prediction system to provide advisories at higher spatial scale, i.e., at administrative block level with improved skill. This will be used to develop advisories for sectors like agriculture, disaster management, water resources, power, tourism and pilgrimage, smart cities, renewable energy sector, and transport. This information will help toward the generation of clean energy (solar, wind, and tidal) and its transmission and management, depending upon users' requirements. Because it requires a proper assessment of potential energy generation sources and devising early contingency plans. Therefore, new advanced INSAT series data must be used in real time for critical life and property forecasting and warning applications primarily by IMD/MoES. The summary of the new payloads for future INSAT meteorological series of satellites in a single or multiple platforms based on their applications potential are may be recommended as follows:

- 16-Ch Advanced Imager (ABI, AHI, FCI) with high spatial (1 km) and temporal resolution (<15 min)
- Hyper-spectral infrared sounder (HES/GIFTS, IRS)
- Lightning Imager (GLM, LI)
- Radiation budget monitor
- Space weather monitoring Instruments (GOES-R Space Environment In-Situ Suite SEISS)
- Microwave Radiometer for all weather capability (2 or 3 over tropical region based on AM and PM)

These proposed additional payloads will provide us with valuable observations for a better understanding of various atmospheric and oceanic processes. Our capability for weather and climate prediction is also expected to improve as these observations are very useful for data assimilation so that initial conditions are improved and also for model forecast validation. There has been a paradigm shift in the role of IMD in the forecasting of weather and climate in the past 5 years over the Indian region with the help of INSAT3D and INSAT3DR, SCATSAT satellites with no synoptic-scale weather hazard go undetected and unpredicted. There has been significant improvement in forecast accuracy of various severe weather events including tropical cyclones, heavy rainfall, fog, heat wave, cold wave, thunderstorm, etc. The accuracy has increased by about 15–40% for different severe weather events forecast during the last 5 years. Further, upcoming GEO Imaging Satellite (GISAT) missions and Oceansat-III are also being planned. GISAT is an Indian geo-imaging satellite for providing images quickly during disasters with very high spectral resolutions.

With the establishment of Multi-mission Meteorological Data Receiving and Processing System (MMDRPS) at Satellite Division of IMD, New Delhi, which is an end-to-end near real-time data processing system to acquire and process all meteorological data transmitted by the Imager, Sounder, and DRT payloads of INSAT-3D, INSAT-3DR, and INSAT-3DS (yet to be launched) satellites, round the clock surveillance of weather systems including severe weather events around the Indian region will be achieved successfully and make India self-reliant for satellite data utilization in weather forecasting.

Acknowledgements I am grateful to Dr. M. Mohapatra, Director General of Meteorology IMD for his valuable suggestions and providing all facilities to complete the work. I am also very much grateful to SAC, Ahmadabad team for their technical, software expertise and implementation of "RAPID" tool at IMDPS, New Delhi. Finally, I am thankful to Dr. M. Rajeevan, Secretary, MoES for his great support and encouragement.

References

Bader MJ, Forbes GS, Grant JR, Lilly RBE, Waters J (1995) Images in weather forecasting. Cambridge University Press, 493 pp

Ellord GP (1995) Advances in the detection and analysis of fog at night using GOES multi spectral infrared imagery. Weather Forecast 10:606–619

Eyre JR, Brownscombe JL, Allam RJ (1984) Detection of fog at night using advanced very high resolution radiometer. Meteorol Mag 113:266–271

Hubert LF, Lehr PE (1967) Weather satellites. Published: Blaisdell Publication Company

Inoue T (1987) An instantaneous delineation of convective rainfall areas using split window data of NOAA-7 AVHRR. J Meteor Soc Japan 65:469–481

Lensky IM, Rosenfeld D (2003a) Satellite-based insights into precipitation formation processes in continental and maritime convective clouds at nighttime. J Appl Meteor 42:1227–1233

Lensky IM, Rosenfeld D (2003b) A night rain delineation algorithm for infrared satellite data based on microphysical considerations. J Appl Meteor 42:1218–1226

Lensky IM, Rosenfeld D (2008) Clouds-aerosols-precipitation satellite analysis tool (CAPSAT). Atmos Chem Phys 8:6739–6753

Li Z, Li J, Menzel WP, Timothy JS, James PN, Daniels J, Ackerman SA (2008) GOES sounding improvement and applications to severe storm now casting. Geophys Res Lett 35:3

Ma XL, Schmit TJ, Smith WL (1999) A nonlinear physical retrieval algorithm—its application to the GOES-8/9 sounder. J Appl Meteorol 38:501–513

Mitra A, Bhan S, Sharma A, Kaushik N, Parihar S, Mahandru R, Kundu PK (2015) INSAT-3D vertical profile retrievals at IMDPS, New Delhi. Mausam 66:687–694

Mitra AK, Shailesh P, Bhatla R, Ramesh KJ (2018) Identification of weather events from INSAT-3D RGB scheme using RAPID tool. Curr Sci 115(7), 10 October 2018

Parihar S, Mitra AK, Mohapatra M, Bhatla R (2018) Potential of INSAT-3D sounder-derived total precipitable water product for weather forecast. Atmos Measure Tech 11:6003–6012. https://doi.org/10.5194/amt-11-6003-2018

Setvak M, Lindsey DT, Novak P, Wang PK, Radová M, Kerkmann J, Grasso L, Su S-H, Rabin RM, Šť'ástka J, Charvát Z (2010) Satellite-observed cold-ring-shaped features atop deep convective clouds. Atmos Res 97:80–96. https://doi.org/10.1016/j.atmosres.2010.03.009

Schmit TJ, Goodman SJ, Lindsey DT. Rabin RM, Bedka KM, Gunshor MM, Cintineo JL, Velden CS, Scott Bachmeier A, Lindstrom SS, Schmidt CC (2013) Geostationary operational environmental satellite (GOES)-14 super rapid scan operations to prepare for GOES-R. J Appl Remote Sens 7:073462

Trenberth KE, Olson JG (1988) An evaluation and intercomparison of global analyses from the National Meteorological Center and the European centre for medium range weather forecasts. AMS 28:477–1520

Yuan L, Anthes R, Ware R, Rocken C, Bonner W, Bevis M, Businger S (1993) Sensing climate change using the global positioning systems. J Geophys Res 98:25–30

Forecasting of Severe Weather Events Over India

Raghavendra Ashrit, Anumeha Dube, Kuldeep Sharma, Harvir Singh, Sushant Kumar Aditi Singh, Saji Mohandas, and S. Karunasagar

Abstract Forecasting severe weather events is of immense value since it can help alleviate (if not mitigate) the disastrous impacts to a certain extent. While accurate and precise forecasting is a great challenge, the advances in recent decades in the NWP in India have shown improved accuracy and reliability in the forecasting capabilities, especially for severe weather events. Improved observations, monitoring, modelling and forecasting capabilities in recent years in India have helped in early detection and warning for severe weather events. This Chapter documents the recent developments in India, and NCMRWF in particular, in forecasting (i) Heavy Rainfall (ii) Tropical Cyclones (iii) Thunderstorms and Hailstorms (iv) Heat and Cold Waves (v) Fog and Visibility.

1 Introduction

Severe weather is an extreme meteorological event or phenomenon, which represents a real hazard (to human life and property). The definition of severe weather is most often impacted based, and usually defined by "local/regional" thresholds that are related to the inability of the "local/regional" populations to safely conduct their normal business, to the point of being life threatening. Naturally, the set of severe weather phenomena is different in different geophysical environments (different physical processes and antecedent conditions are in play). In general, severe weather includes events like (category 1) heavy rain, strong wind/ wind gusts, hail, lightning, tornadoes, flash floods and extreme temperature. Some of the more localized events are suggested as (category 2) snow storms, thunderstorms, dust/sand storms, sea swell/ tsunamis/storm surges and extended areas of fog for transport (aviation especially). In this chapter, an overview of some of the severe weather events in India is presented with a brief discussion on the prediction or early warning capability and know-how used in India.

R. Ashrit (✉) · A. Dube · K. Sharma · H. Singh · S. K. A. Singh · S. Mohandas · S. Karunasagar
National Centre for Medium Range Weather Forecasting, Ministry of Earth Sciences, 50, Institutional Area, Phase-II, Sector-62, Noida, Uttar Pradesh, India
e-mail: raghu@ncmrwf.gov.in

V. K. Gahalaut and M. Rajeevan (eds.), *Social and Economic Impact of Earth Sciences*,
https://doi.org/10.1007/978-981-19-6929-4_6

2 Heavy Rainfall Events

Heavy rains are the most common natural disasters which affect the human society, the environment and the economy. They are generally associated with deep convective systems and produce copious rainfall in a short period of time (a few hours). As per the IMD definitions (Table 1), heavy rainfall is reported at a place when the accumulated rainfall in a day exceeds 7 cm. If the rainfall in a day exceeds 12 cm, then it is reported as very heavy rainfall.

The Indian subcontinent is highly vulnerable to heavy rainfall events. Most of the heavy and extreme rainfall events occur during the southwest monsoon season (June to September, JJAS), the Western Ghats (WGs), North-Eastern (NE) states (Assam, Meghalaya, Mizoram, Arunachal Pradesh, Sikkim, Manipur and Tripura) of India and central India are the most prominent regions of heavy rainfall. Central India receives rainfall mainly due to the Low-Pressure Systems (LPS) and Monsoon Depressions (MD) that form over the Bay of Bengal (BoB) and move towards the west north-westward during JJAS (and only on very few occasions do these LPS and MDs move towards northward and produce a significant amount of rainfall over the NE states. The WGs and the NE states of India are the regions which are characterized by steep orography and therefore the heavy rainfall in these regions is often due to the forced ascent of air parcels over the mountains.

Recently, India has witnessed some unprecedented heavy rainfall events that caused flooding and landslides. The heavy rainfall in Mumbai (26 July 2005), Kedarnath (Uttarakhand) in June 2013, heavy rains and flooding in Chennai (2 Dec 2015), flooding in Madhya Pradesh during July–August 2016 and Mount Abu in Gujarat (July 2017) are some of the examples of severe weather/extreme rainfall events. The most recent unprecedented heavy rainfall occurred over Kerala during August 2018 (1–20 August) resulting in large flooding with the death of more than 400 people and affecting millions of people. According to the India Meteorological Department

Table 1 IMD classification of 24-h accumulated rainfall

	Terminology	Rainfall Range in (cm)	Percentile
1	Very light rainfall	Trace-0.25	
2	Light rainfall	Up to 1	Up to 65
3	Moderate rainfall	2–6	65.0–95.0
4	Heavy rainfall	7–11	95.0–99.0
5	Very heavy rainfall	12–20	99.0–99.9
6	Extremely heavy rainfall	≤21	>99.9
7	Exceptionally heavy rainfall	When the amount is a value near about the highest recorded rainfall at or near the station for the month or season. However, this term will be used only when the actual rainfall amount exceeds 12 cm	

(IMD), rainfall over Kerala during August (1–19 August) was 164% above normal thus making it the fourth highest record August rainfall (+94%). Similarly in some of the neighbouring districts of Karnataka state, heavy rains caused widespread flooding and landslides. Figure 1a shows the spatial distribution of the rainfall along the west coast of India, accumulated during 7–17 August 2018 based on the satellite and gauge merged rainfall analysis. The IMD-NCMRWF satellite + gauge merged (NSGM) rainfall analysis shows the highest observed rainfall amounts exceeding 100 cm over several areas in Kerala and coastal Karnataka (Fig. 1a).

NCMRWF provides real-time forecast based on NWP models for up to 10 days. Figure 1b shows accumulated rainfall during 7–17 August 2018 as predicted by NCMRWF's NWP model. A comparison with observations (Fig. 1a) suggests that the model (Fig. 1b) successfully predicted the increased rainfall amount over Kerala and coastal Karnataka. A detailed analysis of the forecasting and validation for the Kerala event is presented by Ashrit et al. (2020).

Although the prediction of these heavy rainfall events and scientific understanding of the processes responsible for these events have been improved continuously, there are still many challenges associated with its prediction like where (location), when (time), and how much (intensity) precipitation will occur. The reported improvement is attributed to increased horizontal and vertical resolutions as well as improved physics parameterization schemes in NWP models. However, the major credit for the substantial improvements in weather forecasting goes to the sophisticated data assimilation (DA) systems which utilize satellite data.

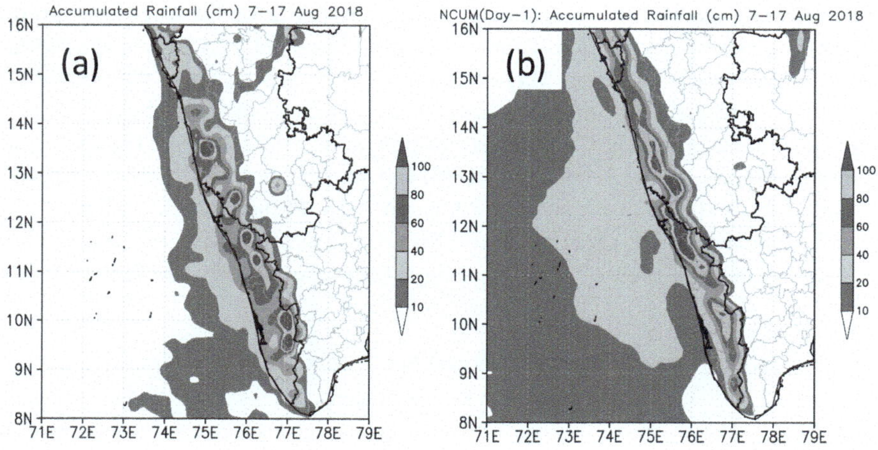

Fig. 1 Accumulated rainfall during 7–17 Aug 2018 over coastal Karnataka and Kerala. **a** Observed. **b** NCMRWF Model forecast

3 Tropical Cyclones in the North Indian Ocean

Tropical cyclones are intense circular storms that originate over warm tropical oceans and are characterized by low pressure at the centre of the storm, string winds, and heavy rainfall. These weather systems draw most of their energy from the sea surface and maintain their strength as long as it remains over warm waters (>26 °C). As soon as a cyclone makes landfall it tends to lose its strength and gradually dissipates. Cyclonic systems usually move from east to west due to the upper level steering flow. They are the major natural hazards that affect the coastal region of India. In view of the serious risk posed to life, property and infrastructure, the availability of reliable and timely information about the probability of formation of tropical cyclones, their location, movement, intensification, and landfall become very crucial for disaster management and mitigation.

Tropical cyclones over the North Indian Ocean are categorized as (Table 2):

In the coastal areas with the approach of a tropical cyclone, there are strong winds, torrential rains and storm surges which is an elevation of the sea surface that can reach up to 6 m (20 ft) above normal levels. Such a combination of high winds and water makes cyclones a serious hazard for coastal areas in tropical and subtropical areas of the world.

The Indian subcontinent, with a total coastline of about 7516 km and almost 40% of the population living within 100 km of this coastline, is exposed to nearly 10% of the world's tropical cyclones making it one of the worst cyclone-affected areas. On an average, the Bay of Bengal and Arabian sea are affected by about 5–6 tropical cyclones each year out of which 2–3 cyclones may reach the intensity of a severe cyclonic storm. There are two definite seasons of tropical cyclones in the North Indian Ocean (NIO), i.e. from May to June and from September to the middle of December. In recent years, there have been many instances of severe to very severe tropical cyclones affecting both the Bay of Bengal as well as the Arabian Sea, for example, 'Phailin' in 2013, 'Hudhud' in 2014, 'Chapala' in 2015, 'Vardah' in 2016, 'Mora' and 'Ockhi' in 2017, and 'Mekunu' in 2018. In the year 2018 alone there were 7 tropical cyclones observed in the NIO out of which 5 were in the severe to very severe category. Figure 2 shows the plot of tropical cyclones in the Indian Ocean

Table 2 Tropical cyclone categories in NIO based on maximum sustained windspeed

Category	Wind speed
Depressions	(31 km/h $\leq$ wind $\leq$ 50 km/h)
Deep depressions	(51 km/h $\leq$ wind $\leq$ 62 km/h)
Cyclonic storms	(63 km/h $\leq$ wind $\leq$ 88 km/h)
Severe cyclonic storms	(89 km/h $\leq$ wind $\leq$ 117 km/h)
Very severe cyclonic Storms	(118 km/h $\leq$ wind $\leq$ 165 km/h)
Extremely severe cyclonic storms	(166 km/h $\leq$ wind $\leq$ 220 km/h)
Super cyclones	(221 km/h $\leq$ wind)

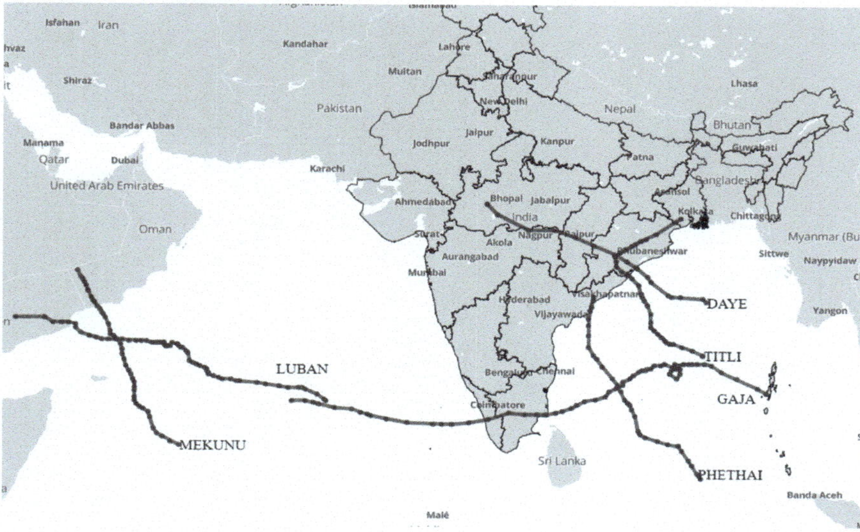

Fig. 2 Tropical Cyclones affecting the North Indian Ocean during 2018

during 2018. In India, IMD is the operational agency which gives all cyclone-related warnings and forecasts for all the countries adjoining the North Indian Ocean.

3.1 Forecasting Tropical Cyclones at NCMRWF

The loss of life and property caused due to tropical cyclones is particularly high in densely populated areas, usually when the forecasts are inaccurate or are not given well ahead of time. In the last decade or so there have been many advances in the field of numerical weather prediction (NWP) which have resulted in an improvement in the forecasts of tropical cyclone tracks. All over the world, there are several numerical weather prediction models that are used by different meteorological agencies in giving forecasts (usually up to 5 days) for cyclones.

At the National Centre for Medium Range Weather Forecasting (NCMRWF), the operational weather prediction model is adapted from the Unified Model (UK Met Office) and is known as NCUM, which has a horizontal resolution of 12 km. The output of this model comes in the form of winds, mean sea level pressure, temperature, rainfall, etc. These basic variables are then used to compute the current and future locations of a tropical cyclone as well as the various hazards associated with it like the increase in wind speed as well as heavy rainfall. Output from this model is also used to calculate the landfall time and location for a given cyclone, which is further used in issuing out the evacuation or preparedness warnings to the general public by the IMD. Due to the high horizontal resolution of the NCUM model, it is also very useful in predicting the strength/intensity of a cyclone. This intensity is usually

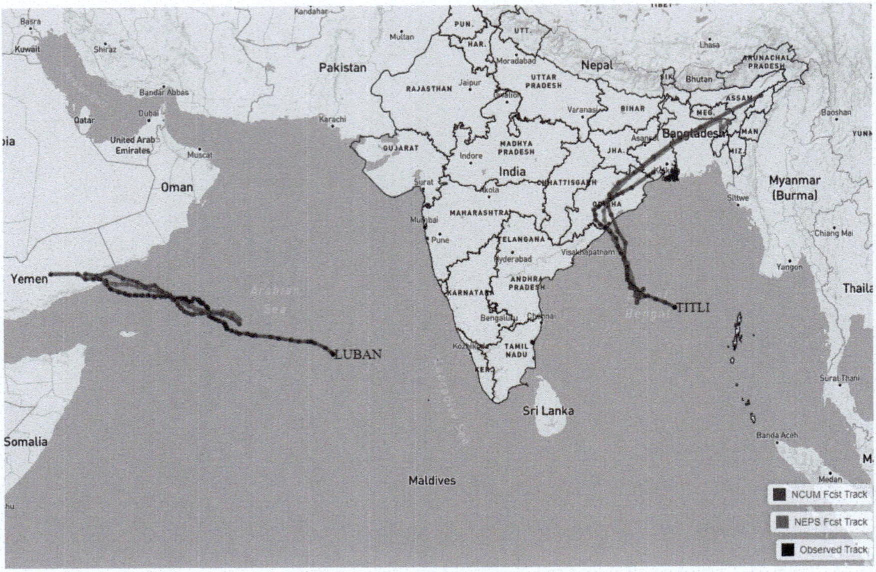

Fig. 3 Observed and forecast tracks for two very severe cyclones affecting the North Indian Ocean during 2018 (Titli and Luban)

measured in terms of the maximum winds that are seen within the cyclonic storm. Figure 3 shows the observed and forecast tracks for 2 very severe cyclonic storms Titli (8–12 October 2018) and Luban (6–15 October 2018) observed in 2018. The forecast tracks are from NCUM and the Ensemble prediction System, NEPS (mean track), two global models operational at NCMRWF.

3.2 Forecast Track Errors

For the two cyclonic systems, the NCUM and NEPS forecast track errors in Km are shown in Tables 3 and 4. The errors are computed against the IMDs' best track data. For the cases of VSCS Titli (Table 3), it is found that the lowest initial position error is 54 km in NEPS mean and slightly higher value of 58 km in NCUM. In NCUM all 00 and 12 UTC analyses show initial position errors lower than 70 km except for 12 UTC analyses on 8th Oct where the initial position errors is 189 km. The NCUM and NEPS mean track errors are comparable up to 48 h. For the Arabian Sea VSCS Luban, the NCUM, and NEPS (ensemble mean) forecast track errors suggest that the mean initial position error in NCUM is 43 km and in NEPS it is 67 km. The lowest initial position error is 11 km in NCUM and 12 km in NEPS mean (00UTC 8th Oct 2018). NEPS mean shows a very large initial position error (>350 km) on 6th Oct 12 UTC. Similarly, NCUM also shows a large error of 12 UTC on 8th Oct (109 km)

Table 3 Forecast track errors (in Km) for VSCS Luban from 3 to 13 Oct 2018. NCUM and NEPS track errors are based on 00UTC and 12UTC (numbers in the brackets indicate the number of cases)

	0	24	48	72	96	120
NCUM	43(12)	60(14)	125(17)	148(17)	186(17)	285(14)
NEPS	67(13)	56(14)	97(14)	138(13)	171(10)	245(9)

Table 4 Forecast track Errors (in Km) for VSCS Titli from 6 to 12 Oct 2018. NCUM and NEPS track errors are based on 00UTC and 12UTC (numbers in the brackets indicate the number of cases)

	0	24	48	72	96	120
NCUM	58(8)	99(8)	153(9)	142(8)	148(6)	183(3)
NEPS	54(7)	99(9)	135(9)	172(8)	165(6)	237(4)

and 00 UTC on 9 Oct 2018 (90 km). The NEPS mean track errors are lower at all lead times after 48 h.

3.3 Forecasting Landfall

The 850 hPa winds at the time of landfall of VSCS Titli over Odisha are shown in Fig. 4 for analysis and Day-2 forecast (top panels). Intense winds (>40kts) near the coast are well predicted in the Day-2 forecast. Similarly, the rainfall forecast (bottom) indicates the accurate location of peak rainfall over the coast. The regions affected by heavy rains >8 and 16 cm/day over Odisha, parts of AP and West Bengal are all accurately represented in the Day-2 forecast. Model tends to overestimate the rainfall amount; however, the location is accurately predicted. A similar analysis for the case of VSCS Luban is presented in Fig. 5.

The VSCS Luban made landfall at around 5:30–6:00 UTC on 14 Oct 2018). The forecasts based on 11th Oct predicted landfall 6 h early while the forecast based on 12th and 13th Oct had almost no landfall time error. For VSCS Titli, which made landfall at around 23UTC of 10th-00 UTC of 11 Oct 2018, the landfall time errors in the forecast based on 8th, 9th and 10th Oct is about 12 h, + 6 h and 0 h, respectively.

3.4 Four-Stage Warning Strategy of India Meteorological Department

Cyclone Warning Division (CWD) of IMD, New Delhi is the nodal agency in India for issuing the cyclone-related alerts and warnings. Additionally, the Area Cyclone Warning Centres (ACWC) and Cyclone Warning Centres (CWCs) play a crucial role

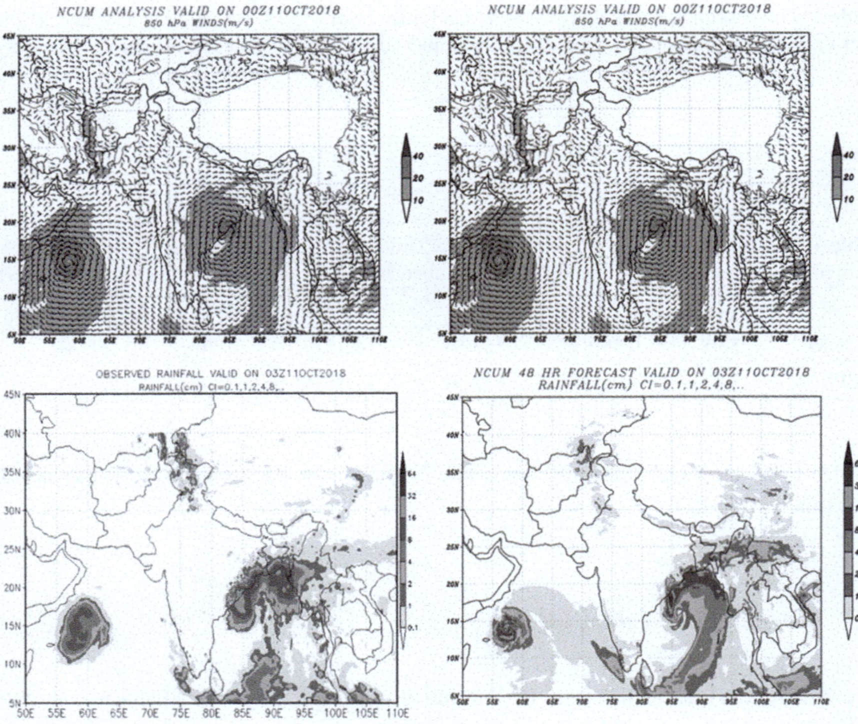

Fig. 4 Observed (left) and Day-2 forecast (right) 850 hPa winds (top) and rainfall (bottom) valid at the time of the landfall over Odisha for VSCS Titli on 11th Oct 2018

in the dissemination of regional and local warnings. The cyclone warnings are issued to state government officials in four stages.

Different colour codes are being used since post-monsoon season of 2006 to indicate the different stages of the cyclone warning in the bulletins as desired by the NDMA. Cyclone Alert (Yellow), Cyclone Warning (Orange) and Post-Landfall Outlook (Red). During disturbed weather over the Bay of Bengal and the Arabian Sea, the ports likely to be affected are warned by concerned ACWCs/CWCs by advising the port authorities through port warnings to hoist appropriate Storm Warning Signals. IMD also issues "**Fleet Forecast**" for the Indian Navy, Coastal Bulletins for Indian coastal areas covering up to 75 km from the coastline and sea area bulletins for the sea areas beyond 75 km. The special warnings are issued for fishermen four times a day in normal weather and every three hours in accordance with the four-stage warning in case of disturbed weather.

For the general public, the coastal residents and fishermen are warned by State Government officials and broadcast warnings through All India Radio and National Television (Doordarshan) telecast programmes in national and regional hook-up (Table 5).

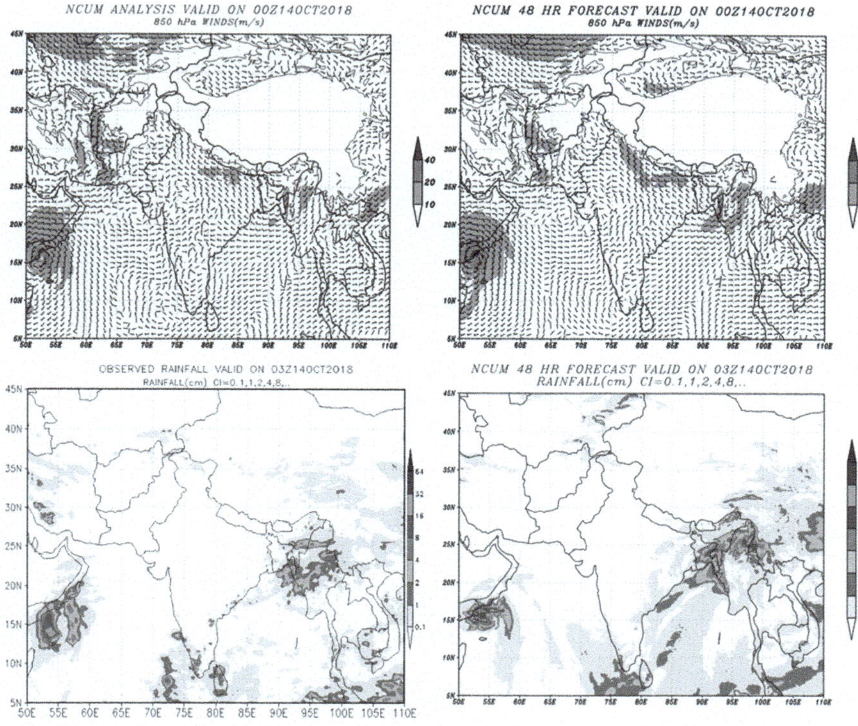

Fig. 5 Observed (left) and Day-2 forecast (right) 850 hPa winds (top) and rainfall (bottom) valid at the time of the landfall over Yamen for VSCS Luban on 14th Oct 2018

3.5 *Tropical Cyclone Monitoring/Warning Using Satellites*

Satellite information is very important for tracking and determining the intensity trends of tropical cyclones. When a system is well offshore and out of effective radar range, forecasters use satellite imagery to continuously track the storm's movement and development. Figure 6 shows the Meteosat-7 visible imagery of VSCS 'Phailin' on 12 Oct 2013 06UTC (left) On the right Meteosat-7 IR image with 12.5 km ASCAT winds overlay for Cyclone Hudhud on 10 October 03:00 UTC is shown. The satellite imagery gives information about the top of the storm and also about the wind speeds over the ocean surface. These are crucial for monitoring the formation and development of storms and are used to pinpoint the storm centre. Tropical Cyclone forecasters use both visible and infrared satellite imagery to track the motion and cloud patterns of cyclones and infrared to monitor cloud-top temperatures. The forecasters also pay attention to the sea surface temperature. Cyclones passing over warmer (*cooler*) water pick up energy and intensify (*weaken*). For a cyclone to form the sea surface temperature needs to be at least 26 °C.

As per the IMDs Standard Operating Procedure (SOP; IMD 2013), 1-hourly Satellite Bulletins (based on satellite cloud imagery data and derived products) are made at

Table 5 Four stages of cyclone warning strategy operational in IMD

Stages	Warning	Information	Issued by	Target agency
First stage 72 h in advance	Pre-cyclone Watch	Early warning about the development of a cyclonic disturbance in the NIO and its likely intensification into a tropical cyclone and the coastal belt likely to experience adverse weather	Director General of Meteorology	Cabinet secretary and other senior officers of the Government of India including the Chief Secretaries of concerned maritime states
Second stage 48 h in advance	Cyclone alert	information on the location and intensity of the storm likely direction of its movement, intensification, coastal districts likely to experience adverse weather and advice to fishermen, general public, media and disaster managers	ACWCs/CWCs and CWD at HQ	
Third stage 24 h in advance	Cyclone warning	Landfall point is forecast at 3 hourly interval giving the latest position of cyclone and its intensity, likely point and time of landfall, associated heavy rainfall, strong wind and storm surge along with their impact	ACWCs/CWCs/and CWD at HQ	Advice to the general public, media, fishermen and disaster managers
Fourth stage 12 h in advance	Post-landfall outlook	Likely direction of movement of the cyclone after its landfall and adverse weather likely to be experienced in the interior areas at least 12 h in advance of the expected time of landfall	ACWCs/CWCs/and CWD at HQ	

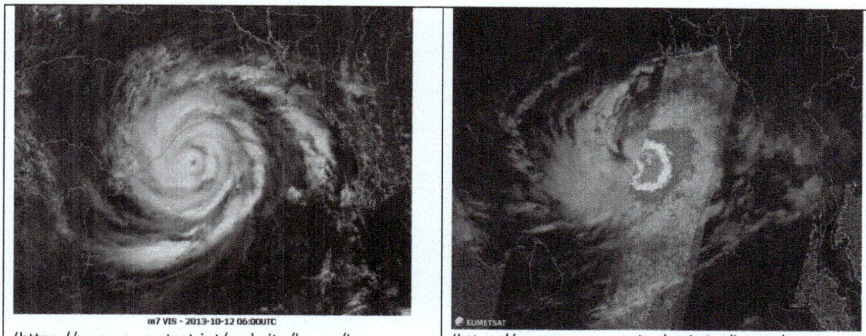

(https://www.eumetsat.int/website/home/Images/ImageLibrary/DAT_2081889.html)

(https://www.eumetsat.int/website/home/Images/ImageLibrary/DAT_2354888.html)

Fig. 6 On left Bay of Bengal VSCS 'Phailin' visible imagery from METEOSAT-7 on 12th Oct 2013 06UTC. On the right Meteosat-7 IR image with 12.5 km ASCAT winds overlay for Cyclone Hudhud on 10 October 03:00 UTC. Imagery shows the area with hurricane force winds in the centre along with tight circulation centre

the Satellite Meteorology Division of IMD, New Delhi. In these bulletins, the centre and intensity of a cyclone as estimated by "Dvorak Technique", its past motion and amount of convection associated with tropical cyclones and other characteristic features are described.

Ultimately, the assimilation of data from Indian satellites (e.g. INSAT-3D, Megha Tropiques, SARAL) in the numerical weather prediction (NWP) models carried out for improved representation of tropical cyclone environment in the initial conditions which yield improved prediction of cyclone track, intensity, rainfall in the NWP models.

3.6 Use of RADARS for Monitoring and Warning Tropical Cyclones

IMD has a network of over 27 radars spread all over the country as in Fig. 7. They are categorized mainly as:

- S and C Band Radars for cyclone detection are installed along the coast. Eleven numbers of S-Band high power radars are located along the east and west coasts of India and are used primarily for the detection of cyclones approaching the Indian Coasts. Some of these radars are also used for the detection of storms and other severe weather phenomena for use in local forecasting. The effective range of these radars is 400 km.
- X Band radars for storm detection are installed at airports for the detection of localized phenomena such as thunderstorms, dust storms and squalls. The effective range is 150 km.

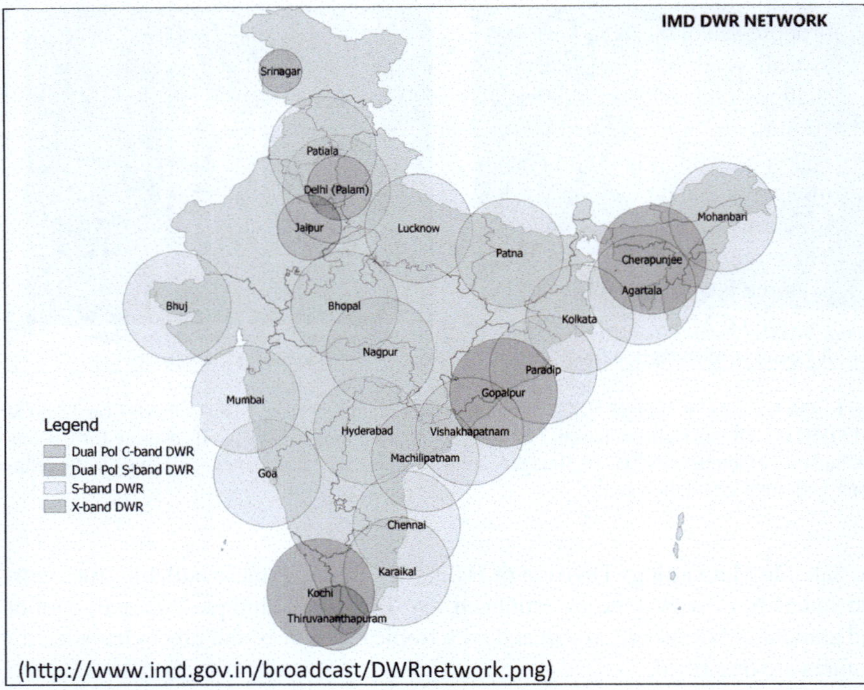

Fig. 7 IMD's Doppler Weather Radar network over India

- IMD also has a network of multipurpose radars known as MULTIMET Radars operating in the X Band. These were dual-purpose radars used for tracking the radiosonde balloon as well as detecting storms.

The radar network of IMD is now being upgraded with doppler weather radar technology replacing conventional radars. IMD is also installing C band polarimetric doppler weather radars and the first two are installed in DELHI and JAIPUR. Eventually, the network in India will comprise 55 doppler weather radars covering entire India.

For the case of VSCS Hudhud, the DWR Reflectivity is shown in Fig. 8 as the system is crossing the coast on 12 Oct 2014 over the city. The eye of the cyclone is clearly seen in the DWR imagery. This image from Doppler Radar of Vizag (taken 02:20) shows Vizag at the (+) mark. The image shows the eyewall and eye (the place where the strongest winds and heaviest rainfall take place). So, the eyewall is about to hit Vizag and the coastal areas around it. The image shows the VSCS Hudhud crossing the coast. This will be the period when the rains will be stopping, almost (the image below shows a blue-shaded region which is the eye where rains vanish), and winds will also decrease until the eye passes. However, just after this eye passes, the weather once again turns severe with a complete reversal of wind.

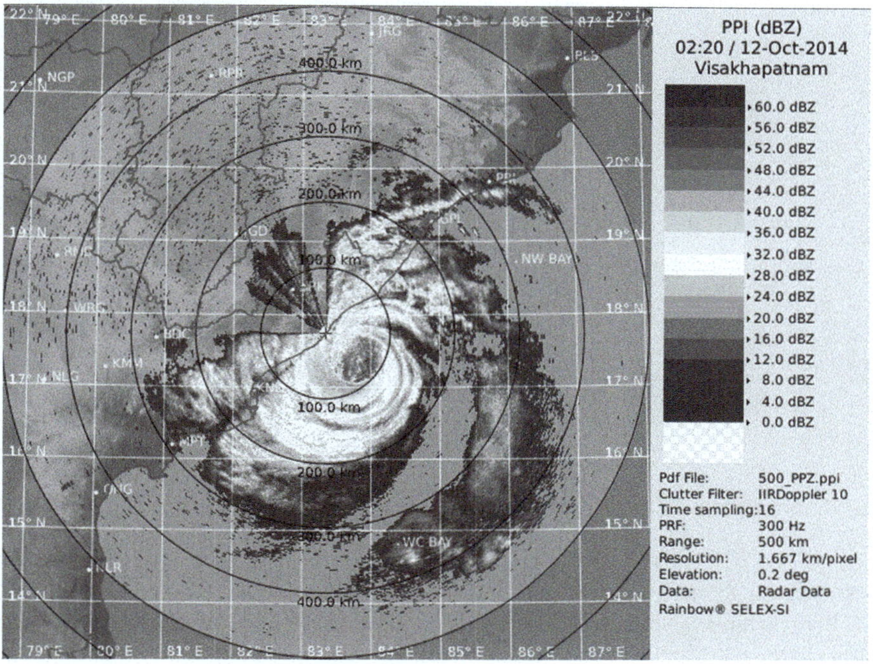

Fig. 8 Visakhapatnam DWR reflectivity during Cyclone Hudhud while crossing the coast over city on 12th Oct 2014

4 Thunderstorms and Hailstorms

A thunderstorm is a violent short-lived weather disturbance. It is associated with high-impact weather phenomena like lightning, thunder, dense clouds, heavy rain or hail, and strong gusty winds. These cause loss of life and damage to crops and property. Recently, India has witnessed many thunderstorm events a few to be named are, a severe thunderstorm over north India on 2 May 2018 in which more than 130 people died and on 21 April 2015 over Bihar which took the lives of 65 people. National Crime Records Bureau (NCRB) indicates that 9.86% of the total number of deaths in India attributed to nature (317,259) is due to lightning alone. Similar significant fraction of loss of life and damage to crops and the aviation sector are associated with other phenomena associated with thunderstorms, such as downbursts, high winds and hailstorms. During pre-monsoon season (March–May) Gangetic West Bengal and surrounding areas get severe thunderstorms called Nor'westers, which are locally called as 'Kal-Baisaki'. Northwest India gets convective dust storms called locally as 'Andhi'. These dust storms are generally associated with strong dry winds and reduced visibility (almost nil). When downpours associated with thunderstorms are very heavy and of very short duration it is called 'cloud burst'. It is defined as rainfall of more than 10 cm in 1 h.

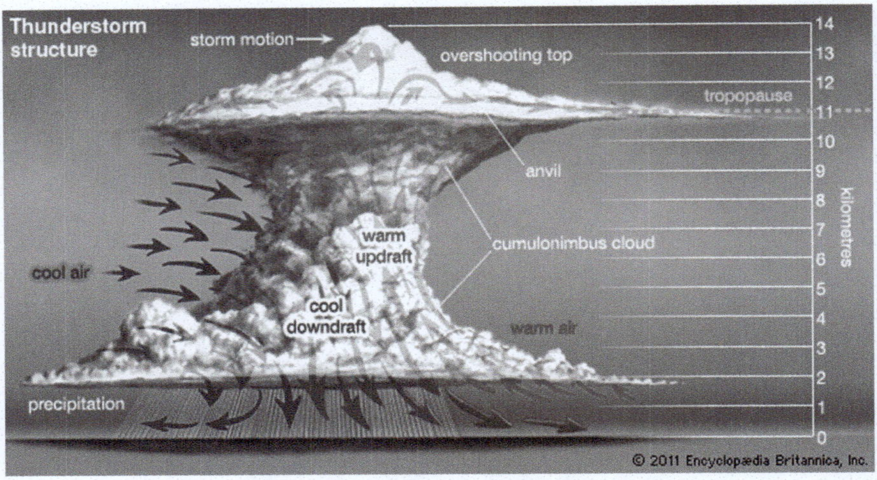

Fig. 9 Structure of a fully mature thunderstorm

In general, thunderstorms have a spatial extent of a few kilometres and a life span of less than an hour. Thunderstorms originate mainly during the summer season, warm days of the season, and warm hours of the day. Warm, moist and rising unstable air is the most important factor in the development of thunderstorms. Besides the convective mechanism, warm and moist winds also rise and become unstable due to orographic obstacles (Fig. 9).

4.1 Forecasting Thunderstorm

A natural calamity is beyond human control but numerical weather prediction (NWP) models can be used to provide forecasting guidance a few days in advance to minimize the losses incurred due to it. NCMRWF being a numerical weather forecasting centre runs Unified Model up to 10 days in advance to provide guidance to weather forecasters to advice the general public. Depending on the weather phenomenon and its severity, suitable weather parameters are selected for this purpose. For forecasting thunderstorms atmospheric instability parameters like Total Totals Index (TTI > 45), Showalter Index (SWI < 0), KI (>35), lifted index (LI < −1), convective available potential energy (CAPE high values) and convective inhibition (CIN low values) are useful. These parameters are calculated from NWP model running at NCMRWF. On 7 February 2019, Delhi, NCR and adjoining areas witnessed a severe hailstorm (Fig. 10a). In Fig. 10b IMD's weather warning issued for different types of severe weather over north India has been highlighted. NCMRWF model predicted values of TTI, KI, CAPE and CIN as 57, 36, 1469 and 15, respectively, (Fig. 10c and d) indicate severe thunderstorm activity over a large part of the north and northwest India.

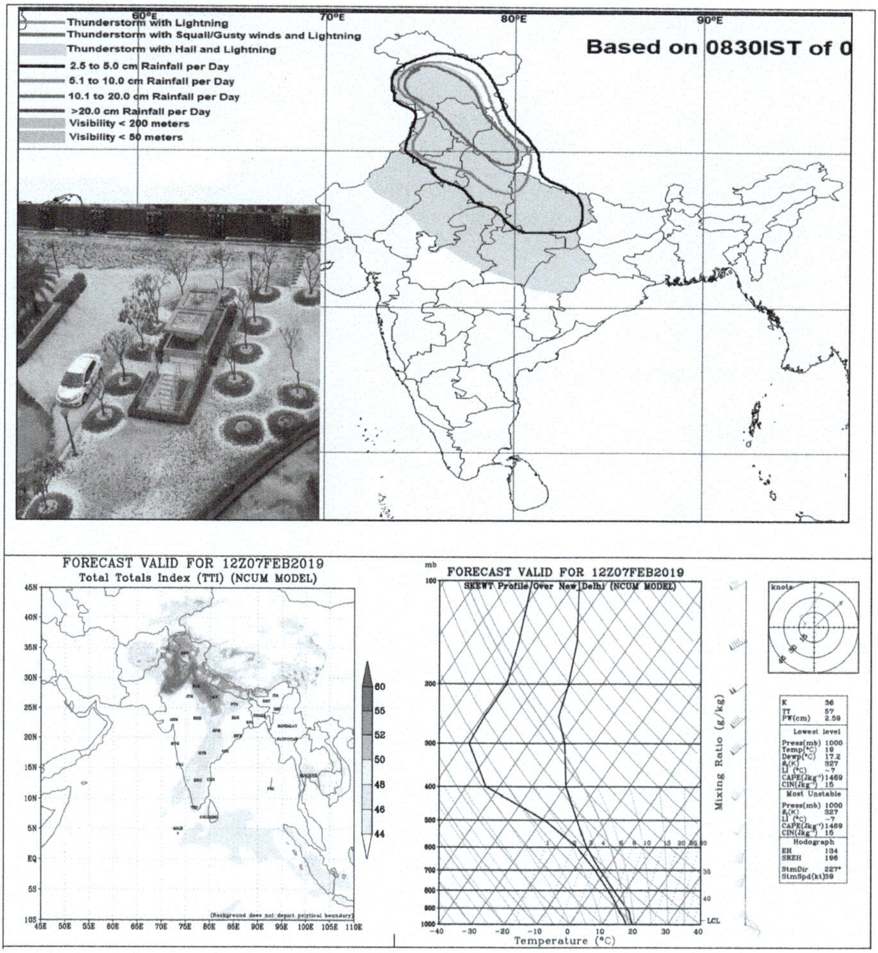

Fig. 10 **a** A glimpse of hailstorm event on February 7, 2019 (photo credit: Manisha Pandey/India Today). **b** IMD weather warning. **c** and **d** NCMRWF model predicted TTI and Skew-T chart

4.2 *Forecasting Lightning*

Lightning strikes are the biggest natural killers in India, causing more than 2,000 deaths each year, according to the top experts from the IMD and the NDMA. India has witnessed an increasing death toll and damages due to lightning bolts over the past few years. Despite improved understanding, monitoring and prediction capabilities brought about by scientific and technological progress, lightning and thunderstorms still cause widespread loss of life and property every year mainly due to a communication gap which does not result in last mile connectivity. With the joint initiative of scientists from the IMD, IITM and NCMRWF, a lightning warning system was

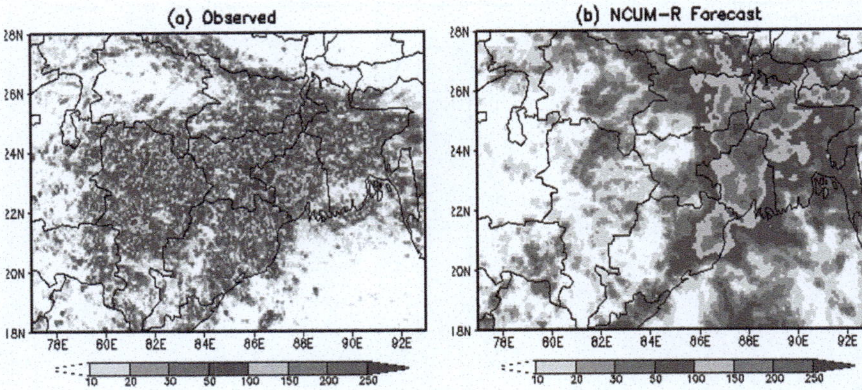

Fig. 11 Observed (left) and NCUM-R forecast (right) 24 h total lightening flash counts during May 2021

developed in 2018, having a location-specific forecast of up to 48 h about the occurrence of thunderstorms, lightning strikes, squalls, gusty winds or hailstorms. The country is now capable of having 'real-time' information about lightning updated every 5 min to alert the people about potential threats. The models developed by the Ministry of Earth Sciences are very definite and are utilized by forecasters around the country to provide detailed information for each district in the country every three hours. Figure 11 shows the comparison of observed and NCMRWF model (NCUM-R) predicted 24 h lightning count over eastern India covering Bihar, Jharkhand, Chhattisgarh, Odisha and West Bengal. The enhanced lightning activity over the West Bengal region is accurately predicted in the model. Lightning activity is underestimated in Chhattisgarh. The forecasters use these products for early warning of lightning.

5 Heat/Cold Waves

Temperature conditions, which vary from extremely hot to chilly weather, differ from place to place. In fact, people living in a place for a long get acclimatized to the normal weather/climate of that place/region. For instance, in the cold climate region (extratropical regions or high-level stations), a day temperature of more than 30 °C may be a hot day for the local residents whereas this temperature in the warm climate region may present the feeling of comfortable weather. In the equatorial and tropical regions, locals are accustomed to warm/hot weather. Comfortable temperature range for inland and coastal stations of the same climate regimes may also differ substantially due to varying humidity levels and wind flow.

Similarly, people in cold climate conditions are acclimatized to rather low temperatures and threshold values for cold wave conditions in those regions are much lower than for people living in warm climates. Thus, health-related problems due to heat/cold waves in the extra-tropical countries are different, compared to those in the tropics. India Meteorological Department (IMD) is mandated to provide heat/cold wave advisories to help decision-making by the various sectors to minimize these risks and potential damages.

5.1 Heat Waves

Heat wave is a pervasive meteorological phenomenon resulting in the mortality of people, agricultural losses, and increases in many risks such as health-related risks, wildfires, power and water shortage and accidents, amongst others. As per the WMO guidelines (WMO 2018).

A period of marked unusual hot weather (maximum, minimum and daily average temperature) over a region persisting at least three consecutive days during the warm period of the year based on local (station-based) climatological conditions, with thermal conditions recorded above given thresholds.*

As per the IMD terminology, heat wave is considered if the maximum temperature of a station reaches at least 40 °C or more for plains and at least 30 °C or more for Hilly regions. When actual TMAX remains 45 °C or more irrespective of normal TMAX, heat waves occur (Nair et al., this book).

As per WMO (2018), a heat wave differs from warm spells. Similar to heat waves, warm spells are defined as a persistent period of abnormally warm weather. A warm spell can similarly be defined in terms of the 90th or 95th percentile of daily maximum temperature (T_{max}). A warm spell occurs at any time of the year, whereas heat waves can only occur in the warm season. Figure 12 shows the spatial distributions of IMD's observed and NCUM Model forecast T_{max} valid on 28 May 2018. This figure is an example of the prevailing heat wave conditions over parts of Rajasthan, Madhya Pradesh, Uttar Pradesh, Delhi, Haryana, Punjab and some parts of Maharashtra. The observed maxima of T_{max} (>42 °C) are seen over a wide region over central and western parts of India. NWP Models do feature the capability to forecast severe heat wave conditions. They can be effective tools of guidance for forecasters.

NCMRWF Model forecast of maximum temperatures for all cases of T_{max} >40 C is evaluated for an entire MAM season during 2017 (Fig. 13). The forecast verification is presented in the form of a spatial map of the Probability of Detection (POD) and Equitable Threat Score (ETS) in Fig. 13. The appendix provides the detailed formula for the verification metrics. POD gives the fraction of the observed events that were correctly forecasted. It varies from 0 to 1 with 1 representing perfect forecast. On the other hand, ETS measures the fraction of observed events that were correctly predicted, adjusted for hits associated with random chance (for example, it is easier to correctly forecast rain occurrence in a wet climate than in a dry climate). The POD values excess of 0.6 can be considered skilful over large parts of India. The ETS values also indicate reasonable skill in predicting high temperatures during the MAM season.

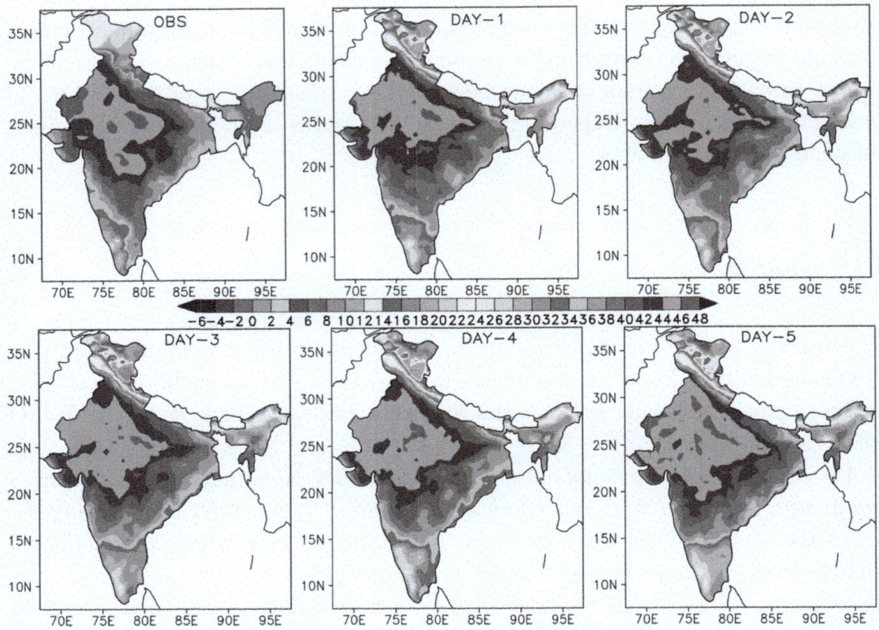

Fig. 12 Observed and model forecast T_{max} on 28th May 2018

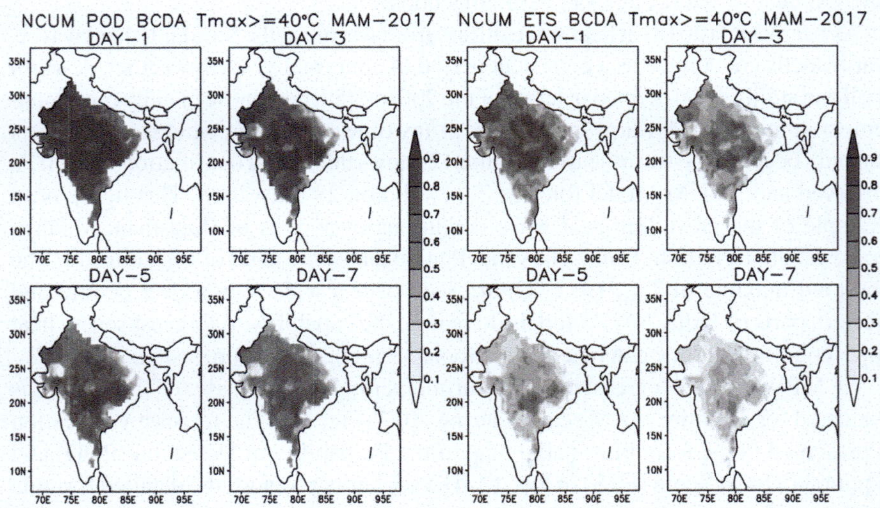

Fig. 13 Probability of Detection (POD; left) and Equitable threat Score (ETS; right) for forecast T_{max} > 40C during MAM-2017

Higher daily peak temperatures and longer, more intense heat waves are becomingly increasingly frequent globally due to climate change. India too is feeling the impact of climate change in terms of increased instances of heat waves which are more intense in nature with each passing year, and have a devastating impact on human health thereby increasing the number of heat wave casualties. As per NDMA (2016), there were over 1000 reported heat wave-related deaths in India every year from 2010 to 2015 and over 22,000 deaths during 1992–2015.

5.2 *Heat Action Plan (HAP)*

Many states are affected during the Heat wave season, such as the state of Andhra Pradesh, Telangana, Odisha, Gujarat, Rajasthan, Madhya Pradesh, Uttar Pradesh, Vidarbha region of Maharashtra, Bihar, Jharkhand and Delhi. NDMA's Heat Action Plan (HAP) provides a framework for Indian cities to protect their citizens from extreme heat. Ahmedabad was among the first city to prepare a heat wave action plan in 2015. The heat wave action plan aims to provide a framework for implementation, coordination and evaluation of extreme heat response activities in cities/towns in India that reduces the negative impact of extreme heat. The Plan"s primary objective is to alert those populations at risk of heat-related illness in places where extreme heat conditions either exist or are imminent and to take appropriate precautions, which are at high risk. Preventive heat management and administrative action need to be taken by city administration authorities.

The heat wave action plan is intended to mobilize individuals and communities to help protect their neighbours, friends, relatives, and themselves against avoidable health problems during spells of very hot weather. Broadcast media and alerting agencies may also find this plan useful. Severe and extended heat waves can also disrupt general, social and economic services. For this reason, Government agencies will have a critical role to play in preparing and responding to heat waves at a local level, working closely with health and other related departments on long-term strategic plans. Details of strategies can be found in NDMA (2016). Some of the key components include:

- Establish Early Warning System and Inter-Agency Coordination to alert residents on predicted high and extreme temperatures.
- Capacity building/training programme for health care professionals at the local level to recognize and respond to heat-related illnesses, particularly during extreme heat events.
- Public Awareness and community outreach
- Collaboration with non-government and civil society organizations.

5.3 Criteria for Cold Wave

A cold wave is a meteorological event generally characterized by a sharp drop of air temperature near the surface, leading to extremely low values, steep rise of pressure and strengthening of wind speed, or associated with hazardous weather like frost and icing. It often causes severe impacts on human health, agriculture and high heating costs, and can even result in mortality for human beings and livestock.

As per the IMD terminology, a cold wave is considered when the minimum temperature is 10 °C or less for plains and 0 °C or less for Hilly regions (Smitha A Nair et al., this book).

In January 2017 entire north India was devastated by a cold wave. This occurrence had a severe effect on several North Indian states, including Himachal Pradesh, Jammu and Kashmir, Punjab, Harayana, Rajasthan, and Uttar Pradesh. The lowest temperature in Gulmarg due to the cold wave was recorded at − 12.4 C. The banks of Dal Lake in Srinagar froze. Keylong of Himachal Pradesh and Kargil of Jammu and Kashmir witnessed low temperatures of − 13.9 °C and − 15.6 °C, respectively. At least 40 people died as a result of the January 2017 cold wave.

During 10–15 Jan 2017, the entire northern part of India was under the grip of a severe cold wave spell. Figure 14 shows the observed and model predicted Tmin valid on 13 Jan 2017. As in the observations, model forecasts show a large part of northern and central India featuring the Tmin values lower than about 5 °C. The models could predict the cold waves 3–4 days in advance. Such model-based products are crucial for the forecasters to monitor and forecast severe cold wave events.

The NCMRWF Model forecast of maximum temperatures for all cases of Tmin < 10 °C is evaluated for an entire DJF season during 2016–17 (Fig. 15). The forecast verification is presented in the form of a spatial map of the Probability of Detection (POD) and Equitable Threat Score (ETS) in Fig. 15. The POD values higher than 0.6 clearly indicate high POD over large parts of central and northern India. The ETS values are relatively lower indicating relatively lower skill in predicting the lower Tmin values during winter.

6 Fog/Visibility

The impact of fog on the daily lives of human beings has significantly increased during recent decades due to increasing air, marine and road traffic. The financial and human losses related to fog and low visibility become comparable to the losses from other weather events such as cyclones. Fog and associated low visibility conditions over any region cause significant impacts on the human activities and adversely affect the economy (Gultepe et al. 2007). Fog events lead to deaths, damage to property and delays worldwide. The number of deaths in the USA in fog-related traffic accidents in the period 1982–1991 was more than twice that due to the combination of flash floods, hurricanes, lightning and tornadoes (Rosenfled 1996). The British Airports Authority

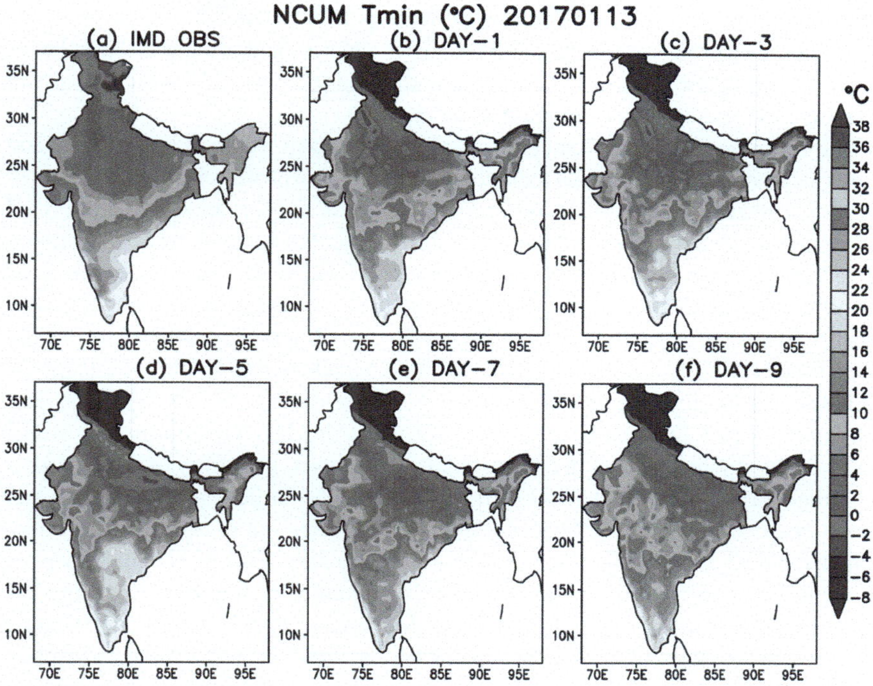

Fig. 14 Observed and model forecast T_{min} on 13th Jan 2017

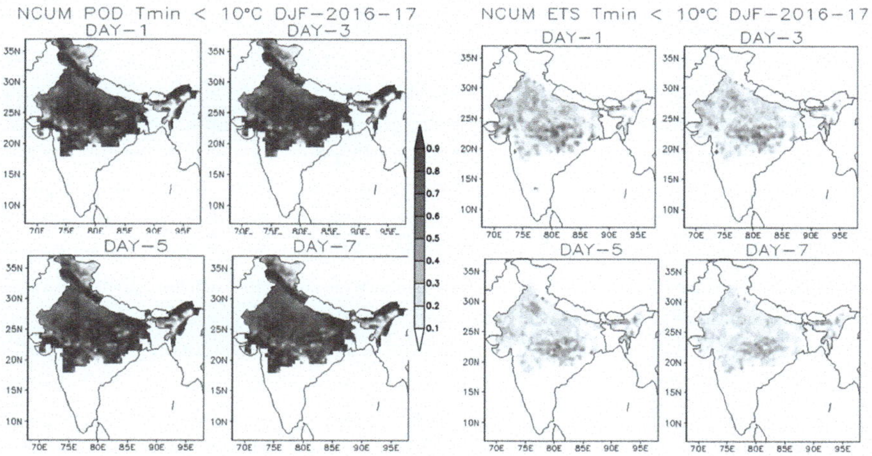

Fig. 15 Probability of Detection (POD;) and Equitable threat Score (ETS; right) for forecast T_{min} < 10C during DJF-2016–17

cancelled nearly half the flights at Heathrow Airport in December 2006. The Indira Gandhi International Airport (IGIA) in New Delhi, experienced the highest number of fog hours during the months of December and January 2008–2009, which disrupted aviation services severely (Jenamani and Tyagi 2011).

The techniques for fog predictions can be categorized as manual, statistical and numerical modelling. In manual techniques, atmospheric profiles from radiosonde or numerical weather prediction (NWP) model and surface observations are interpreted at forecasting location to estimate fog point. Statistical techniques rely on a training period of data which connects forecast profiles, local observations and observed visibility. The prediction of fog with NWP models has long been a challenge and success is also very limited mainly due to the complex processes involved in fog formation. However, continuous efforts have been made in improving models for the accurate prediction of fog and visibility.

A number of 1-D models such as COBEL and Air Mass Transformation Model have achieved some success in forecasting fog. However, the utility of these models is limited in practice as they do not incorporate the large-scale meteorological conditions that affect the genesis, evolution and dissipation of fog. Thus, a three-dimensional (3-D) model which incorporates processes like horizontal pressure gradient, advection and diffusion is required for predicting fog. Mesoscale models such as Fifth-Generation NCAR/Penn State Mesoscale Model (MM5) and Weather Research Forecasting (WRF) were developed and used to forecast fog, especially the low visibility in fog by different researchers. These models use parameterization schemes to forecast visibility as they do not have detailed microphysical processes. The liquid water content (LWC)/ice water content (IWC), temperature or relative humidity is often used to parameterize visibility in these models.

There are operational models globally which predict visibility. The HIRLAM model at Danish Meteorological Institute (DMI) (Peterson and Neilson 2000) is one such model which predicts visibility at 2 m height using the variables such as relative humidity, temperature and cloud cover. The Rapid Update Cycle (RUC) model of the USA, calculates visibility from prognostic relative humidity, cloud and hydrometeor fields (Smironova et al. 2000). However, the visibility of any region also depends on the aerosol content of that region. In HIRLAM-DMI constant aerosol concentration is assumed whereas the RUC model neglects aerosol. The operational global model of NCMRWF known as NCUM also predicts visibility.

Visibility forecasts from NCUM are shared with India Meteorological Department (IMD) every year under the Forecast Demonstration Project (FDP) during the winter season (December–February). The spatial extent of fog over any area is identified using the satellite observations. Ground-based and conventional observational networks use horizontal visibility (Vis) to identify fog. If observed visibility at a particular location is $\leq$ 1 km, then it is considered as foggy (WMO manual 2003). Figure 16 shows the comparison of the visibility forecast from NCUM with INSAT-3D fog image and METAR visibility observations for 28 December 2014. Different types of fog are identified based on the visibility values. Visibility forecasts at Amritsar, Delhi, Lucknow and Varanasi are verified with METAR observations in Fig. 17.

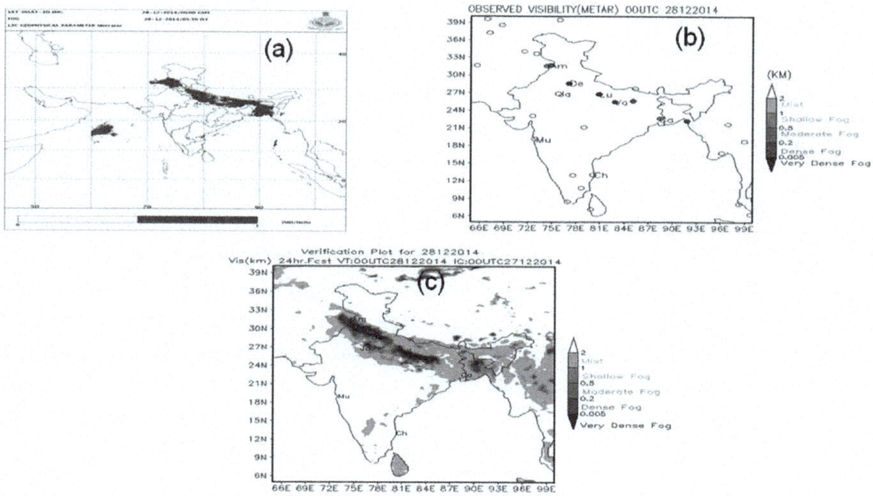

Fig. 16 Obsrved fog from a INSAT-3D, b Visibility from METARS and c Day-1 forecast of visibility from NCUM on 28th Dec 2014

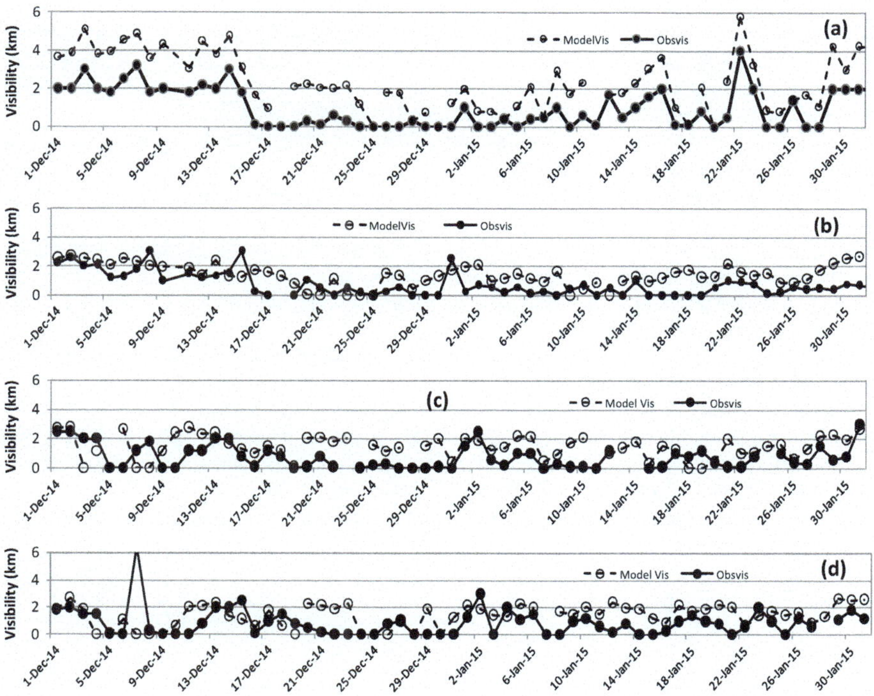

Fig. 17 Comparison of Day-1 forecasts of Visibility from NCUM with Observations at a Amritsar, b Delhi, c Lucknow and d Varanasi during Dec 2014–Jan 2015

7 Conclusions

Severe weather events often have devastating impacts on "local/regional" populations and their normal businesses. Understandably they are different in different geophysical environments Generally severe weather includes events like heavy rain, strong wind/wind gusts, hail, lightning, tornadoes, flash floods and extreme temperature. Some of the more localized events are snow storms, thunderstorms, dust/sand storms, sea swell/tsunamis/storm surge and extended area of fog for transport (aviation especially). Forecasting for severe weather events is of immense value since it can help alleviate (if not mitigate) the disastrous impacts to a certain extent. While accurate and precise forecasting is a great challenge, the advances in the recent decade in the NWP in India have shown improved accuracy and reliability in the forecasting capabilities, especially for severe weather events. This chapter has documented the glimpses of successful forecasting of severe weather events over India.

References

Ashrit R, Sharma K, Kumar S, Dube A, Karunasagar S, Arulalan T, Mamgain A, Chakraborty P, Kumar S, Lodh A, Dutta D, Momin I, Bushair MT, Prakash BJ, Jayakumar A, Rajagopal EN (2020) Prediction of the August 2018 heavy rainfall events over Kerala with high resolution NWP models. Meteorol Appl 27(2):1–14. https://doi.org/10.1002/met.1906

Gultepe I, Pagowski M, Reid J (2007) Using surface data to validate a satellite based fog detection scheme. Weather Forecast 22:444–456

India Meteorological Department 2013: Cyclone Warning in India, Standard Operating Procedure, July 2013

Jenamani RK, Tyagi A (2011) Monitoring fog at IGI airport and analysis of its runway wise spatio-temporal variations using Meso-RVR network. Curr Sci 100

National Disaster Management Authority 2016: Guidelines for Preparation of Action Plan—Prevention and Management of Heat-Wave

Petersen C, Nielsen NW (2000) Diagnosis of visibility in DMIHIRLAM. Scientific Report 00-11. DMI, Copenhagen, Denmark

Rosenfeld J (1996) Cars vs. the weather. A century of progress. Weatherwise 49:14–23

Smirnova TG, Benjamin SG, Brown JM (2000) Case study verification of RUC/MAPS fog and visibility forecasts. In: Preprints for 9th conference on aviation, range, and aerospace meteorology, Orlando, FL. American Meteorological Society, Boston, pp 31–36

World Meteorological Organization, 2016: Guide to Climatological Practices, 2nd edn. WMO-182

Weather and Climate Modeling

Saji Mohandas

Abstract The current manuscript documents a summary of the evolution of Numerical Weather Prediction (NWP) as an inevitable component of modern meteorology and its growth and development toward future climate and earth system models. NWP is essentially an initial-boundary value problem and the basic assumption of atmospheric continuum led to the development of subgrid scale physical parameterization schemes. The errors are inherent to the initial conditions and science formulations in the deterministic models and the natural chaotic behavior of the atmosphere brought out the importance of the ensemble approach to assess the uncertainty in forecasts and a probability-based narrative. The growth of recent high-resolution global models necessitates resolving more and more intricate processes, which were so far relevant only in climate models, and signaling toward a merger with seamless weather and climate prediction system across all scales and earth system models. Society at large needs to adapt and respond to the climate change impacts and to the natural transformation processes in the earth's climate system. High-resolution global ensemble-based data assimilation-forecast systems can better resolve the scale interactions and improve the skill of the NWP models in simulating the under-resolved components of the coupled atmosphere-earth-ocean-cryosphere systems. There is a significant uncertainty in the forcings and the actual internal variability of the climate system, posing the problem of a large spread in the climate projections of global warming by the models from different leading centers. The treatment of cloud processes, scale-aware parameterizations, atmosphere-land-vegetation coupling, and aerosol processes are the immediate major challenges in weather, climate, and earth system modeling. Carbon and methane cycles are important for the realistic estimation of greenhouse gas forcing, compounded by the dynamic vegetation, ocean biogeochemistry, hydrology and water cycle, ecosystem, land–ocean-ice coupling, and the representation of the intrinsic modes of intra-seasonal to decadal variability that play crucial roles in the projection of future climate.

S. Mohandas (✉)
National Centre for Medium Range Weather Forecasting (NCMRWF) (Govt. of India), A-50, Sector-62, Noida 201309, UP, India
e-mail: saji@ncmrwf.gov.in

V. K. Gahalaut and M. Rajeevan (eds.), *Social and Economic Impact of Earth Sciences*,
https://doi.org/10.1007/978-981-19-6929-4_7

1 Introduction to Numerical Weather Prediction

The evolution of meteorology as an interdisciplinary science was remarkable after the first numerical prediction model in history was described by Charney et al. (1950). Numerical Weather Prediction (NWP) is now the most vibrant branch of meteorology, which uses mathematical models for the prediction of future weather conditions based on the current weather conditions. Based on the laws of physics, fluid mechanics, and chemistry applied at the scale of planetary motions, a closed set of governing equations are formulated, which describes the various meteorologically significant weather processes like general circulation, heat and energy transfer, insolation and radiative processes, moisture processes and transport as well as surface hydrological processes. These equations are basically the conservation of energy, mass, and momentum. The two most important approximations are hydrostatic and incompressible fluid assumptions. Thus we have momentum equations, hydrostatic equations, thermodynamic energy equations, continuity equations, equation of state, and moisture conservation equations (see Holton 1992; Gill 1982). The prognostic variables are wind components, surface pressure, temperature, and tracers while density is mostly a diagnostic parameter derived from the continuity equation. These sets of partial differential equations are converted into difference equations for numerical treatment (Refer Holton 1992; Haltiner and Williams 1980). Thus the assumption of the continuum is compromised as the fluid is not sampled with a grid at the scale of the continuum. The state of the atmosphere is thus represented as averaged in space and time. The model dynamics basically deal with the part of the equations containing averaged parameters. The subgrid-scale effects are parameterized by the model physical parameterization schemes.

Generally, the lower part of the atmosphere like the troposphere and part of the stratosphere is included in the general circulation models, which is meteorologically vital. The model atmosphere is segregated into different layers and each layer is treated separately as a two-dimensional surface. All the meteorological variables are represented in each layer using some form of horizontal discretization scheme. The continuous problem is replaced by a discrete problem, whose solutions are the best possible approximations of the continuous problem. Examples of horizontal discretization methods are finite difference, Galerkin techniques like finite element and spherical harmonics, and adaptive grids. The grid structure can be uniform latitude–longitude or staggered grid (Arakawa A, B, C, D, and E), Icasohedral, skipped, variable resolution, adaptive grid, Gaussian or double-Fourier, depending on the scheme and the nature of the domain of the models. See Fig. 1 for various types of grids and staggering. Vertical discretization is generally a finite difference with non-uniform resolution. The most preferred vertical coordinates are Sigma (terrain following), Eta (step-coordinates), and hybrid (terrain following in the lower levels and transforming to height or pressure coordinates at the upper levels) (Fig. 2). Time integration schemes are normally explicit, semi-implicit, or semi-Lagrangian or a combination of these.

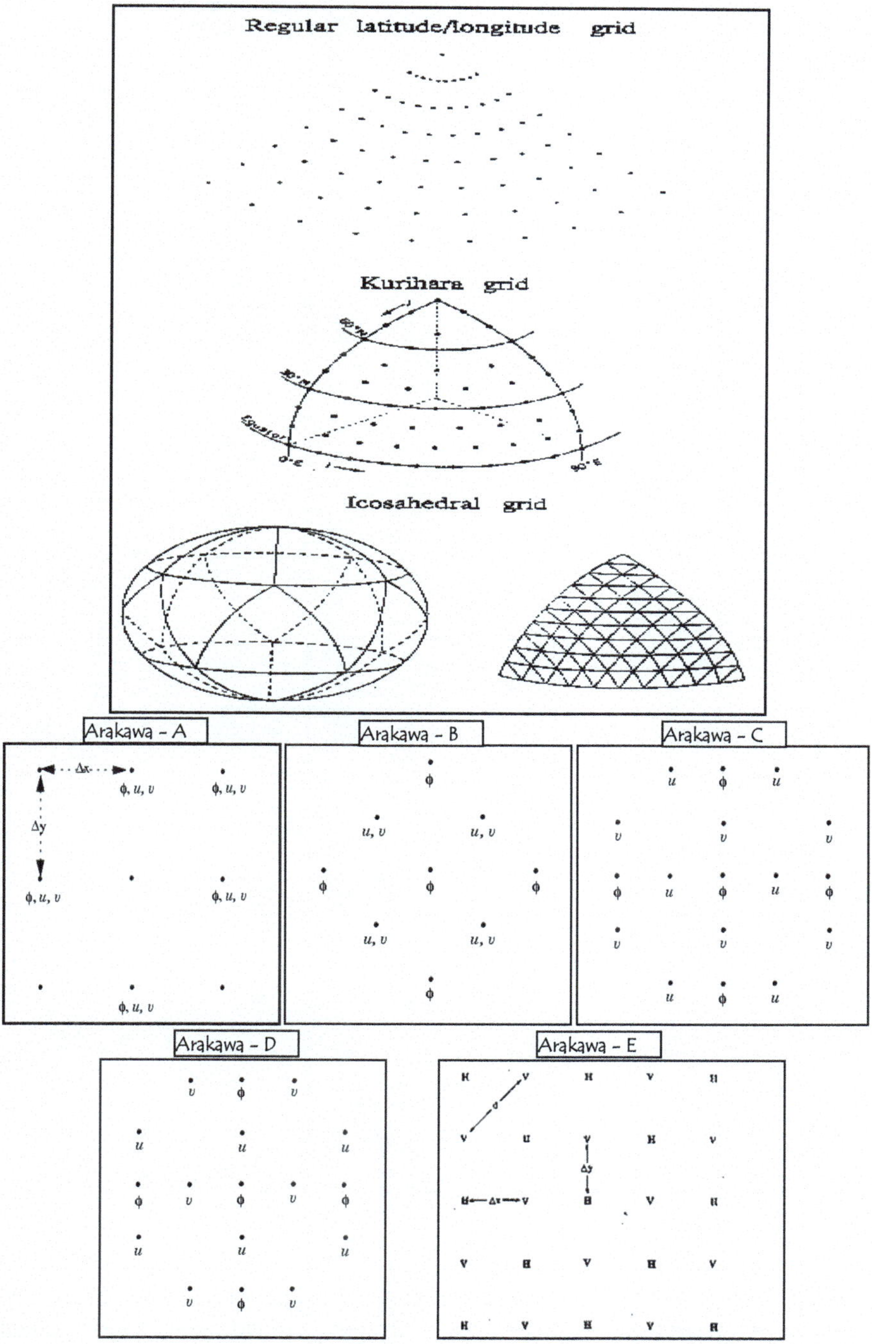

Fig. 1 Different types of grids and different types of staggering

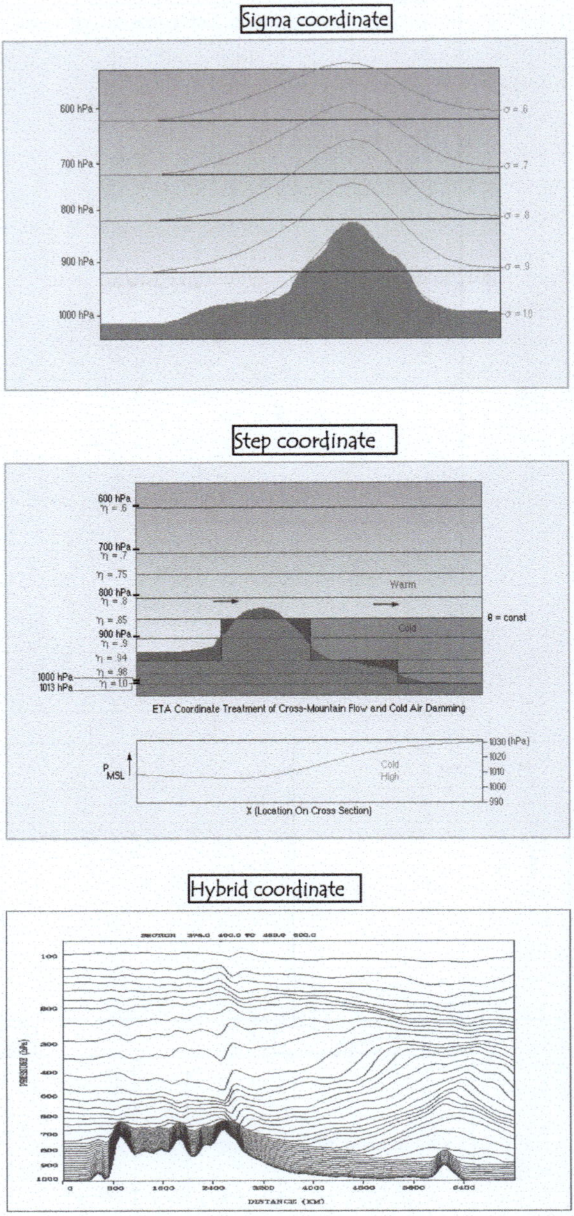

Fig. 2 Examples of vertical coordinate systems

The components of NWP models are the governing equations, physical processes, numerical procedures and approximations, and the initial conditions. From the initial period of the equivalent barotropic model and filtered equations, the governing equations have become primitive equations of momentum, continuity, hydrostatic, perfect gas law, thermodynamic energy, and mass continuity equations. The right-hand side of the prognostic equations contains the terms of pressure gradient, gravity, friction, adiabatic processes, and other subgrid-scale parameterizations. There are hydrostatic or non-hydrostatic assumptions and Eulerian or Lagrangian frame of references. The conservation principles to be followed are that of mass, momentum, thermodynamic energy, water in various forms like vapor, cloud water, rain, snow, ice, graupel, etc., and conservation of other gaseous and aerosol materials. The meteorological variables on real earth are to be cast on a plane surface using some type of map projections. The map projections normally used are Lambert-conformal, Polar Stereographic, Mercator, and Gaussian and the equations are to be accounted for these projections sometimes with extra terms in it.

NWP is an initial-boundary value problem (Bjerknes 1904; Pielke et al. 1999; Goosse 2015). An initial value problem is a differential equation together with a value at a given point of time called initial conditions. A boundary value problem is a differential equation together with additional restraints called boundary conditions, which also should be satisfied for any solution of the differential equations. When the domain is global, we don't have to worry about the lateral boundary conditions. However, the surface and upper boundary conditions have to be specified. The surface conditions are easier to specify as there is a material surface and the surface properties can be specified over the entire domain. The top boundary has no material surface and if specified will create an additional equation to be solved. Specifying pressure at the top as zero makes it vertically unbound and is undesirable for many applications. In practice, it can be assumed to be with zero vertical velocity and free slip boundary conditions for other variables. But this will cause wave reflection at the top. The radiation boundary condition is based on the diagnostic relationship between pressure and vertical velocity at the top. But these are formulated in terms of vertical wave numbers and frequencies and are hence difficult to implement. Yet another approach that can prevent spurious vertical reflection from the top of the atmosphere is the absorbing layer approach, which can be explicit or implicit. The explicit absorbing layer is to have increasing horizontal diffusion when approaching the top. The implicit absorbing layer is coarsening of the vertical resolution while approaching the top and is simple to implement.

The set of governing equations is partial differential equations (PDEs) which need to be converted into difference equations for a numerical solution. Classical PDEs are Laplace, Poisson, advection equation, heat, and wave equation. The expressions for classical PDEs are given below.

$$\frac{\partial^2\varphi}{\partial x^2} + \frac{\partial^2\varphi}{\partial y^2} = 0\text{----Laplace;} \quad \frac{\partial\varphi}{\partial t} = \alpha\frac{\partial^2\varphi}{\partial x^2}\text{----Heat equation}$$

$$\frac{\partial^2\varphi}{\partial x^2} + \frac{\partial^2\varphi}{\partial y^2} = f(x)\text{----Poisson;} \quad \frac{\partial^2\varphi}{\partial t^2} = C^2\frac{\partial^2\varphi}{\partial x^2}\text{----Wave equation}$$

$$\frac{\partial\varphi}{\partial x} + U\frac{\partial\varphi}{\partial y} = 0\text{----Advection;}$$

Nonlinear equations are linearized in the vicinity of a known solution and are converted into some form of classic PDEs to apply a standard solution. A well-posed problem is one for which the solution exists and is unique. Another assumption is made that the equations describe well enough the physical problem in our hand to assure that the problem is well posed. There are a number of important considerations to take into account when a numerical method is chosen.

- Consistency: The discrete operator converges toward the continuous operator of the PDE for finer and finer resolutions. At a given time the truncation error may vanish but the discretization error will anyway grow with time.
- Accuracy: Order decides how well the method approximates the solution.
- Stability: Errors are damped with time.
- Convergence: The numerical solution converges toward the continuous solution when the resolutions are finer. Consistency and stability ensure the convergence.
- Efficiency: The number of operations per timestep.
- Monotonicity and conservation.
- Controlling the computational mode.
- Pole treatment.
- Aliasing and nonlinear instability: Arises from the advection term.

The finite difference methods use the truncated Taylor series expansion for space and time derivatives. It can be two-level, three-level, multi-stage, higher-order, time-splitting, semi-implicit, and semi-Langrangian. For a one-dimensional case, let the grid points be represented as

$$x_j = (j-1)\Delta x, \quad j = 1, 2, \ldots N+1$$

By Taylor series expansion we have

$$\varphi_{j-1} = \varphi(x_j - \Delta x) = \varphi_j - \varphi_j'\Delta x + \varphi_j''\left(\frac{\Delta x^2}{2!}\right) - \varphi_{j+\theta_2}'''\left(\frac{\Delta x^3}{3!}\right)$$

$$\varphi_{j+1} = \varphi(x_j + \Delta x) = \varphi_j + \varphi_j'\Delta x + \varphi_j''\left(\frac{\Delta x^2}{2!}\right) + \varphi_{j+\theta_1}'''\left(\frac{\Delta x^3}{3!}\right)$$

Forward and backward difference formulations are

$$\varphi_j' = \frac{\varphi_{j+1} - \varphi_j}{\Delta x} + E; \quad E = -\varphi_j''\left(\frac{\Delta x}{2!}\right) - \varphi_{j+\theta_1}'''\left(\frac{\Delta x^2}{3!}\right)$$

$$\varphi_j' = \frac{\varphi_j - \varphi_{j-1}}{\Delta x} + E; \quad E = \varphi_j''\left(\frac{\Delta x}{2!}\right) - \varphi_{j+\theta_2}'''\left(\frac{\Delta x^2}{3!}\right)$$

Centered difference is

$$\varphi_j' = \frac{\varphi_{j+1} - \varphi_{j-1}}{2\Delta x} + E; \quad E = \left(\frac{\Delta x^2}{3!2}\right)\left(\varphi_{j+\theta_1}''' + \varphi_{j+\theta_2}'''\right)$$

Dispersion and round off error by the centered scheme are given by

$$\varphi_j'' = \frac{\varphi_{j+1} - 2\varphi_j + \varphi_{j+1}}{\Delta x^2} + O(\Delta x^2)$$

For finite difference in time, similarly we can define the grid points as a function of time,

$$t_n = n\Delta t_{n=0,1,\ldots}$$

Forward in time is given by

$$\left(\frac{\partial \varphi}{\partial t}\right)_j^n \to \frac{\varphi_j^{n+1} - \varphi_j^n}{\Delta t} + O(\Delta t)$$

Also centered in time (second order) is

$$\left(\frac{\partial \varphi}{\partial t}\right)_j^n \to 2\frac{\varphi_j^{n+1} - \varphi_j^{n-1}}{2\Delta t} + O(\Delta t^2)$$

Linear advection equation can be written as first order in time and space as

$$\frac{\partial \varphi}{\partial t} + u_0\frac{\partial \varphi}{\partial x} = O \to \frac{\varphi_j^{n+1} - \varphi_j^n}{\Delta t} + u_0\left(\frac{\varphi_j^n - \varphi_{j-1}^n}{\Delta x}\right) = 0;\ (u_0 > 0)$$

Explicit solution will have only one time level in the RHS,

$$\varphi_j^{n+1} = \varphi_j^n - \alpha\left(\varphi_j^n - \varphi_{j-1}^n\right);\ \alpha = \frac{u_0\Delta t}{\Delta x}$$

Implicit solution is a two-time level iteration approach,

$$\varphi_j^{n+1} + \frac{\alpha}{4}\left(\varphi_{j+1}^{n+1} - \varphi_{j-1}^{n+1}\right) = \varphi_j^n - \frac{\alpha}{4}(\varphi_{j+1}^n - \varphi_{j-1}^n)$$

Let us consider first order in time and second order in space (explicit approach),

$$\varphi_j^{n+1} = \varphi_j^n - \frac{\alpha}{2}(\varphi_{j+1}^n - \varphi_{j-1}^n)$$

This needs to satisfy the CFL criterion for numerical stability,

$$\propto = \frac{u_0\Delta t}{\Delta x} \le 1$$

Leapfrog is an example of second order in both time and space,

$$\varphi_j^{n+1} = \varphi_j^{n-1} - \alpha\left(\varphi_{j+1}^n - \varphi_{j-1}^n\right)$$

When applied wave solution,

$$\varphi_j^n = \varphi_0 \lambda_k^n \exp(ikj\Delta x)$$

The solution will have two modes; physical mode which is the real solution and the computational mode which is the noise due to the approximation.

$$\lambda = ip + \sqrt{1 - p^2} \xrightarrow[\Delta x \to 0;\, \Delta t \to 0]{} 1 \quad \text{physical mode}$$

$$\lambda = ip - \sqrt{1 - p^2} \xrightarrow[\Delta x \to 0;\, \Delta t \to 0]{} -1 \quad \text{computational mode}$$

Thus there will be phase errors and computational dispersion due to different wave numbers. Hence some filtering or damping of numerical solutions are to be worked out. Implicit schemes are stable, but more costly. Ideally, the numerical scheme should be damping the fast-moving sound waves but resolving the slow-moving meteorological waves like Rossby waves and gravity waves. Semi-implicit and split-explicit schemes are mostly popular in weather and climate models.

A separate issue arises for the advection equations which involves the nonlinear interaction of different wave numbers. The shortest wavelength which can be resolved is twice the grid distance (2*dx*). When there is a product term of two variables, some of the resulting structures are too small to be described by the grid resolution or truncation. Thus these structures are falsely interpreted as large-scale structures. This leads to a changing of energy over the range (2*dx*–4*dx*) and cascade of energy in this window (Fig. 3). This misrepresentation is called aliasing which occurs only during the nonlinear interaction and which leads to the fast growth of some wave numbers. The solutions for this problem involve filtering of unstable wavelengths, numerical smoothing or diffusion, usage of energy-conserving or built-in-diffusion numerical discretization schemes, and introduction of the Galerkin technique or semi-Lagrangian scheme of advection.

The current NWP models include all the state-of-the-art physical parameterizations, like radiation, clouds, convection, microphysics, gravity wave drag, surface drag and fluxes, land-surface processes, and vertical diffusion and boundary-layer

Fig. 3 Aliasing

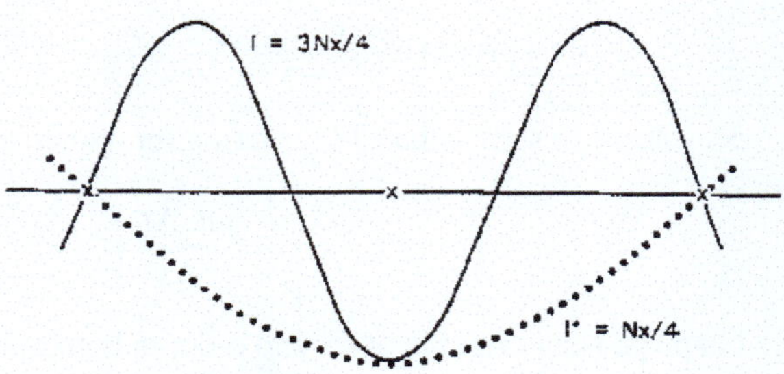

turbulent processes. Figure 4 shows the day-7 forecasts of wind and geopotential (at 850 hPa) and daily accumulated precipitations by NCMRWF (National Centre for Medium-Range Weather Forecasting) Unified Model (NCUM) compared against the corresponding analyses and day-1 rainfall forecasts valid for 15 August 2019 during the peak day of the extreme rainfall during Kerala flood event. It shows fairly good accuracy of forecasts in general circulation details while underpredicting the quantitative locations of rainfall peaks in Kerala. Whatever be the sophistication and development in NWP, the deterministic forecasts are prone to some errors, mainly because of two types of errors; (1) The error due to imperfect initial conditions and (2) The errors due to the imperfect model formulations. Hence to account for the chaotic behavior of the governing equations of the atmosphere, an ensemble of forecasts is made to assess the uncertainty involved in the forecasts and to indicate a range of future possibilities. It is a form of Monte Carlo analysis with a set of forecasts having slightly perturbed initial conditions or slightly differing initial time of integration. A good ensemble prediction system provides the probability of a particular event by running as many ensemble members in parallel with the control run and the number of ensembles can be a function of computational resources available and the ability of the model to provide enough range of possibilities.

Recently more and more earth system components are being coupled and bundled together with the atmospheric models as with the advances in science and resolution, the newer micro-level processes are becoming more and more important and are being better resolved. At higher resolutions, chemistry and various kinds of aerosols have a significant impact on all the weather parameters including clouds. Beyond the medium range, the slowly changing ocean surface characteristics and sea ice also become a more and more important component of the earth system in controlling the weather and climate, and the coupled models become essential tools for extended range forecasts. Earth system models (EMSs) include physical, chemical, and biological processes, unlike the global climate models (GCM), which just represented the physical atmospheric and oceanic processes only. At its core, ESMs have added representations of the global carbon cycle, dynamic vegetation, atmospheric chemistry, ocean bio-geo-chemistry, and even continental ice sheets. Instead of using predetermined inputs of atmospheric components such as aerosols and greenhouse gases, an ESM can simulate how these components change over time in response to anthropogenic activity and changing climate conditions. Therefore, a changing climate will affect the concentration of greenhouse gases and aerosols which will in turn feedback to the climate. Hence the first step toward achieving the goal of unified weather-climate modeling is the development of a strategy toward the seamless prediction system suited for all the space and time scales and modeling of more and more intrinsic processes related to both the regimes like the policy adopted by some global meteorological institutions.

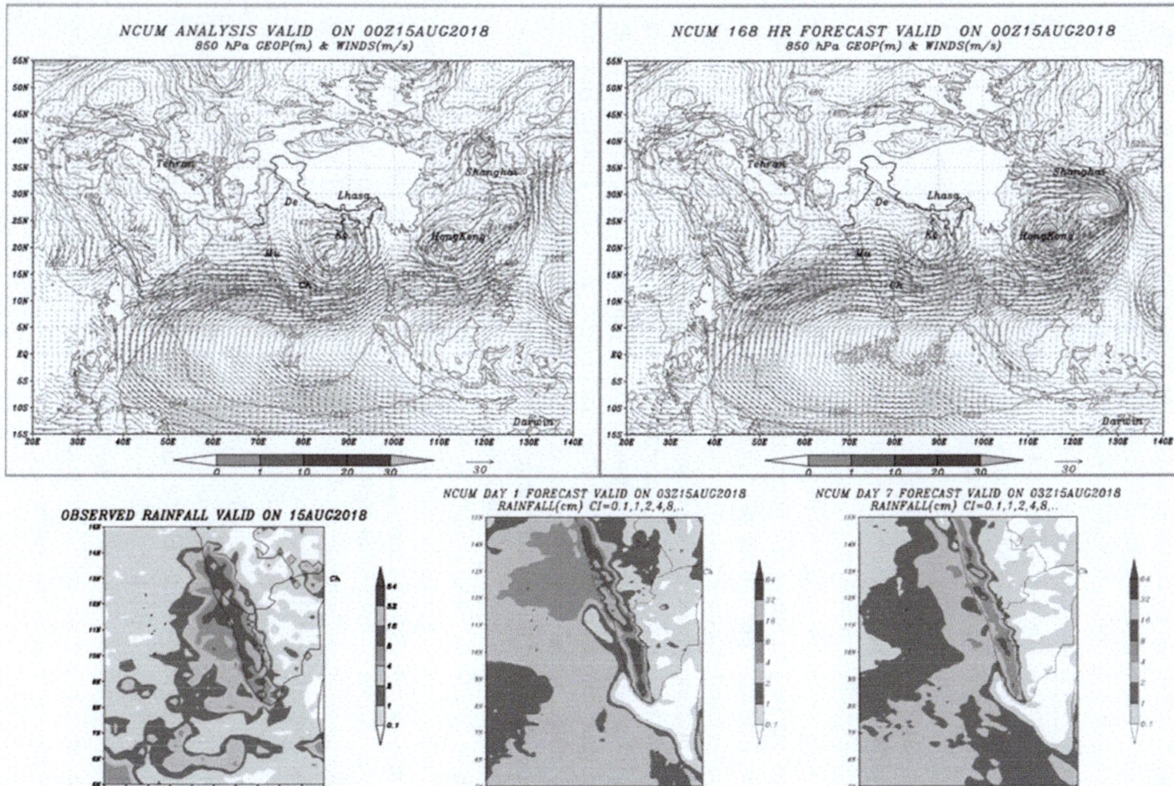

Fig. 4 Day-7 NCUM forecasts of wind (ms-1) and geopotential (m) (top right) valid on August 15, 2019 are compared against NCUM analysis (top left). The day-7 daily accumulated rainfall forecasts (cm per day) (bottom rightmost) are compared against the day-1 forecast (bottom middle) and IMERG rainfall analysis (bottom left)

2 Unified Modeling Strategy

The resolution of NWP models has been increasing with the advances in computational power so that currently the global models have become of the order of 10 km or less so that it will have to account for various new factors. The hydrostatic and shallow water assumptions adopted for the simplification of numerical equations become no more valid and the compressibility of the atmosphere becomes more critical. The Nonhydrostatic, fully-compressible, deep atmosphere, unapproximated primitive equations in the spherical coordinate system were used for very high-resolution operational global modeling systems for the first time by UK Met Office. Historically, the so-called hydrostatic models, which were used for decades for the weather and climate predictions basically consist of the following assumptions.

- Spherical approximation to spheroidal geometry.
- Hydrostatic approximation (Fully-compressible $dw/dt = 0$).
- Shallow atmosphere approximation ($r << a$; and some curvature terms are neglected).
- Traditional approximation (Coriolis effects or cos ϕ terms are neglected).

There are four combinations of approximations used in the hierarchical development of modern day's non-hydrostatic models (see Staniforth 2001).

1. Deep non-hydrostatic Euler equations (Davies et al. 2005)
2. Shallow Nonhydrostatic primitive equations (Tanguay et al. 1990)
3. Deep Quasihydrostatic equations (White and Bromley 1995)
4. Shallow hydrostatic primitive equations (Kasahara 1974).

The current state-of-the-art models which are not following the shallow hydrostatic primitive equations fall into the following two types.

(1) The hydrostatic primitive equations but with the vertical acceleration term, Dw/Dt, restored in the vertical momentum equation (see Tanguay et al. 1990), together with Laprise's (1992) terrain-following "hydrostatic pressure" as the vertical coordinate. The shallow-atmosphere equations may be obtained from the deep-atmosphere Euler ones by (a) setting the spherical polar coordinate r to the earth's mean radius a wherever it appears undifferentiated and (b) dropping all of the $2\Omega \cos\phi$ terms in the components of the momentum equation, as well as all metric terms not involving $\tan\phi$. The first step corresponds to the shallow-atmosphere approximation per se, whereas step b corresponds to the "traditional" approximation (e.g., Phillips 1973; White and Bromley 1995) and is necessary in order to obtain conservation principles for energy, axial angular momentum, and potential vorticity, analogous to those of the deep-atmosphere equations.

(2) The more complete deep-atmosphere Euler equations (Cullen et al. 1997). An example is a scheme adopted by the UK Met Office for the dynamical core of its Unified Model (UM) for a wide range of scales (Davies et al. 2005). UM is a grid-point model with non-hydrostatic, fully compressible, deep-atmosphere

formulation using a terrain-following height-based vertical coordinate. The grid-point formulation facilitates UM to be run as a global or regional configuration.

The deep-atmosphere equations in spherical coordinate systems can be written as

$$\frac{Du}{Dt} - \frac{uv\tan\phi}{r} + \frac{uw}{r} - 2\Omega v \sin\Phi + 2\Omega w \cos\phi + \frac{1}{\rho r \cos\phi}\frac{\partial p}{\partial \lambda} = F^u,$$

$$\frac{Dv}{Dt} + \frac{u^2\tan\phi}{r} + \frac{uw}{r} + 2\Omega u \sin\Phi + \frac{1}{\rho r}\frac{\partial p}{\partial \phi} = F^v,$$

$$\delta_v\frac{Dw}{Dt} - \frac{(u^2+v^2)}{r} - 2\Omega u \cos\Phi + g + \frac{1}{\rho}\frac{\partial p}{\partial r} = \delta_v F^w,$$

$$\frac{D\rho}{Dt} + \rho\left[\frac{1}{r\cos\phi}\frac{\partial u}{\partial \lambda} + \frac{1}{r\cos\phi}\frac{\partial}{\partial \phi}(v\cos\phi) + \frac{1}{r^2}\frac{\partial}{\partial r}(r^2 w)\right] = 0$$

$$\frac{D(c_p T)}{Dt} - \frac{1}{\rho}\frac{Dp}{Dt} = F^T,$$

$$p = \rho R T,$$

where (F^u, F^υ, F^w) and F^T are any parametrized source/sink terms, $g \equiv d\Phi/dr$ is the gravitational acceleration, and $\Phi = \Phi(r)$ is the geopotential. A "vertical acceleration" switch has been introduced in the UK Met Office Unified Model, where vertical acceleration is retained or dropped according to whether δ_V is, respectively, set to unity or zero. Also,

$$\frac{D}{Dt} = \frac{\partial}{\partial t} + \frac{u}{r\cos\phi}\frac{\partial}{\partial \lambda} + \frac{v}{r}\frac{\partial}{\partial \phi} + w\frac{\partial}{\partial r},$$

$$R = C_p - C_v$$

Currently, the difference between weather and climate models is narrowing down and efforts are underway to merge these two branches of development to form a unified system for all scales. By unifying the Global atmosphere and Climate versions, future enhancements will only be included only if accepted across all timescales. Scientific integrity is ensured due to verification against real-time observations, analyses, and climatic long-term datasets. The possibility also arises for a better understanding of the underlying processes due to the combined framework. Robust and scientifically sound solutions are the other advantages of the unified approach. Coupling with the other earth system components and data assimilation systems, across all timescales, this strategy will provide a trusted configuration and a seamless prediction system that can be more versatile in terms of code size and version control. NWP models are the backbone of the modern weather forecasting and are commonly used for many applications of extreme weather warnings like heavy rain and floods, tropical cyclones, active and break monsoons, gusty winds, dust/hail storms, heat/cold waves, pollution, and fog.

3 Introduction to Climate Modeling

Today climate information is useful to various sectors and planners. The users of climate information have different needs with different time horizons and with different tolerances for uncertainty. Society will need to adapt and respond to climatic impacts, such as sea-level rise, an ice-depleting Arctic, and large-scale ecosystem changes. Information about the future climate comes from computer models that simulate the physical, chemical, and biological processes in Earth's climate system. The climate models are all derived from the fundamental physical laws such as Newton's laws of motion and the chemistry and thermodynamics of gases, liquids, solids, and electromagnetic radiation. These are supplemented by empirical relationships determined from observations of complex processes such as turbulent mixing, ice crystal formation in clouds and waves in both air and water, biological processes, sea-ice growth, and glacier movement. The main components within a climate model are the atmosphere, land surface, ocean, and sea ice.

Predictability in weather forecasts is intrinsically limited by chaos, for periods beyond 1–2 weeks. The climate models also consist of their atmospheric part identical to a weather forecast model, but run for longer periods to simulate interactions between atmosphere, land, ocean, and cryosphere ever on time scales of months to millennia. Climate projections are not expected to match individually simulated weather systems, but only statistics of the simulated weather such as the mean and year-to-year range of annual rainfall. The history of improvements in the resolution and quality of climate models is synonymous with the history of enhancements in the computational power of the high-performance machines. With higher and higher resolutions, the future climate models will be able to shed more and more light on the regional impact of the climate change and the divide between the weather and climate model formulations will narrow down.

Unified weather-climate prediction models can be tested in a "weather forecast" mode, with initial conditions at a particular time of the data assimilation cycle. Weather forecast mode will evaluate rapidly evolving processes that are routinely observed. Such shorter simulations are used to test model performance over a range of grid resolutions that are mostly valid for prospective climate simulation capabilities. This approach is being successfully used by the UK Met Office, a leading international modeling center and involves substantial effort and infrastructure. The same approach is the basis of the NCUM model, for the advancement toward a seamless prediction system, with the use of the same dynamic framework and mostly the same science settings for all the scales from cloud-resolving to climate scale.

Recently NCUM adopted a strategy for a very high-resolution global deterministic system (12 km grid size) along with the same resolution of ensemble prediction system (referred to as NEPS) to predict the tropical weather systems in the medium range (currently with 10 days lead time). The hybrid four-dimensional variational assimilation system takes into account the high-resolution NEPS outputs and proved to be a game changer in significantly enhancing the accuracy of the medium-range forecasts to become comparable with the global leaders. Madden–Julian Oscillations (MJOs) being the primary mode of intra-seasonal oscillations over the tropics, the skill of the deterministic control and ensembles are being evaluated at the intra-seasonal scale for MJO predictions. Figure 5 shows a typical verification of MJO Index forecasts by NCUM control and ensemble systems, which shows reasonable skill. Apart from this, a coarser resolution coupled modeling system (CNCUM) is also being developed at NCMRWF for experimental monthly and extended range forecasts.

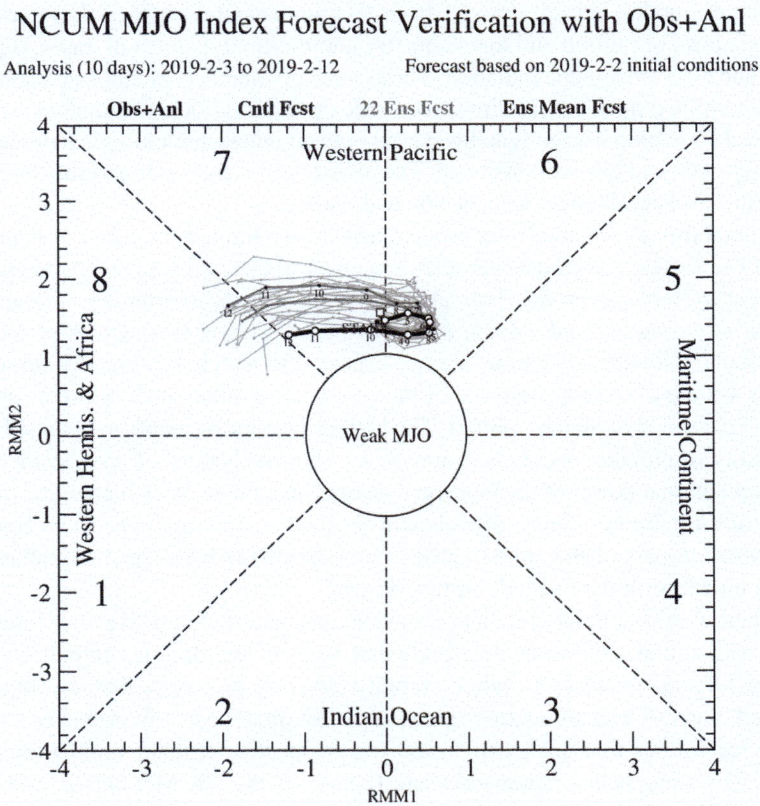

Fig. 5 Verification of MJO Index with observation + analysis (black), control forecast (red), and ensemble forecasts from 22 NCUM ensembles during February 2–12, 2019

Global warming to date due to greenhouse gas emissions has been partially compensated by an uncertain amount of cooling caused by human-induced enhancement of light scattering by aerosols and by their effect on clouds; this compensation has been estimated to be from 20 to 70% (IPCC 2007). Over the twenty-first century, global aerosol emissions are expected to not increase further, but greenhouse gas emissions are likely to accelerate for at least the next few decades, so this compensation will become less significant. There are wide variations among the climate models, with a range of approximately 30%, regarding the simulated global average warming.

A significant source of uncertainty in the future climate projections arises from the lack of knowledge or imperfect knowledge about specific quantities or the behavior of a system. Future greenhouse gas and aerosol emissions and natural processes such as volcanic eruptions and solar variability are used as inputs to the models—both pose significant uncertainty in the forcing on the climate system. Apart from forcings, there is uncertainty from the natural internal variability of the climate system. Another uncertainty arises from incomplete representations of known but complicated and small-scale processes (such as cumulus clouds) and of poorly understood processes (such as ice nucleation in clouds) and the uncertainty from "unknown unknowns". In short, the recently observed trend in global climate change far surpasses the projections by the climate community. However comprehensive and with finer grid size, the models become, there are significant spread in the projections across the leading models.

4 Climate Modeling Strategy

The evolution of climate modeling can be traced back from simple energy-balance models and single-column radiation-convection equilibrium models to current high-end atmosphere–ocean coupled general circulations models (Fig. 6). Addition of submodels of terrestrial vegetation, ecosystems, and biogeochemical cycles such as the carbon cycle, role of aerosols and atmospheric chemistry and inclusion of representations of processes related to the formation of the Antarctic ozone hole and of aerosol effects on climate (e.g., their interaction with clouds) qualifies this comprehensive system to be referred to as *Earth system models* (ESMs).

Uncoupled component models, often run at a higher resolution or with idealized configuration, allow a more controlled focus on individual processes such as clouds, vegetation feedbacks, or ocean mixing. Process models such as *single-column models* and *cloud-resolving models* were developed as a bridge between observations and parameterized representation of unresolved processes needed in GCMs (Arakawa and Wu 2013). *Superparameterized climate models* use a high-resolution cloud-resolving model on each grid column (Khairoutdinov et al. 2005; Grabowsky 2001; Khairoutdinov and Randall 2001) and are computationally demanding and are hence very limited in their use to research so far. *Nested regional atmospheric and oceanic models* forced at the lateral boundaries address the problems of locally high

Fig. 6 Illustration of increasing complexity and diversity of elements incorporated into models used in the Intergovernmental Panel on Climate Change (IPCC) process over the decades. Evolution of the resolution (left side) and physical complexity (right side) of climate models used to inform IPCC reports from the mid-1970s to 2007 (FAR—First Assessment Report, SAR—Second Assessment Report, TAR—Third Assessment Report, AR4—Fourth Assessment Report) (*Source* NRC 2012)

spatial resolutions, such as orographic snowfall, and runoff, or oceanic eddies and coastal upwelling. Regional atmospheric models (Giorgi and Mearns 1999) are used to further the understanding of regional climate processes, and to provide dynamical downscaling of GCM simulations at seasonal or inter-annual time scales. *Stretched* or *variable grids* are used to simulate the entire globe but telescope too much finer resolution within a region of interest.

Earth system models of intermediate complexity (EMICs) use highly simplified process representations but add slowly evolving components such as ice-sheet and dynamic vegetation models that are critical on centennial to millennial scales. Climate models have also started to include some representations of human systems, like air quality, water quality, irrigation, and water management in land-surface models. These so-called *integrated assessment models* coupled with integrated models of energy economics, energy systems, and land-ice with ESMs could be useful tools for exploring climate mitigation and adaptation. Advances in ESMs, refinement in model physics and resolutions, and improved coupling of global and regional models are the keys to address these research frontiers in climate modeling. Models should have a seasonal and decadal predictive skill, should be able to replicate historical trends and modes of variability (e.g., El Niño/Southern Oscillation [ENSO]; decadal-scale Atlantic and Pacific variability), and should be able to capture the processes and feedbacks involved in major paleoclimate events, such as the last glacial cycle and decadal-scale climate transitions that occurred during the glacial period.

Some of this vision may not be realistic because of intrinsic limits in predictability and practical limits to resolution, physical understanding, and observational constraints. Substantial improvements in model resolution are expected and important, but the challenges of simulating climate physics are not magically resolved as models go to high resolution and increased complexity. It takes time to add and properly validate new processes and components to a model. Extensive testing and sensitivity experiments are required, involving hierarchical regional climate models and global climate models with a variety of scale-sensitive parameterizations. However, it is difficult to foresee the advances in technology, observational capacity, and process understanding that will extend modeling capability in the coming decades. There needs to be a strategic research agenda for climate science, observations, and modeling as the climate modeling community keeps pace with the information needs of a changing climate system, while at the same time improving climate model capabilities and skills.

5 Challenges for Climate Models

Climate change is expected to affect society in many ways, including impacts on health, infrastructure, food and water security, ecological integrity, and geopolitical stability. Advances in climate modeling are required in a number of research fronts to improve the information that climate models can provide. A question on climate sensitivity and global warming deals with how much will the planet warm this century? If the climate models cannot capture the mean state and main features of the atmospheric and ocean circulation, they cannot provide meaningful insight regarding regional details. There are uncertainties in the magnitude and rate of expected global warming for a given radiative forcing across the climate models. This uncertainty in climate sensitivity is due to a range of internal feedbacks in different climate models, particularly with respect to the treatment of cloud processes, the carbon cycle, and aerosols within climate models.

Simulation of clouds and how they will respond to future greenhouse gas and aerosol changes is a central challenge in climate modeling. The complex interactions between the physics parameterizations and the low cloud feedbacks from marine boundary-layer clouds in the lowest 1–2 km of the atmosphere are the largest sources of the spread between the climate models (Soden and Held 2006; Soden and Vecchi 2011). Low clouds are particularly sensitive to human-induced aerosol increases, which change their typical droplet size and albedo. In particular, the boundary-layer cloud and cloud-aerosol uncertainties in climate models will not automatically go away in atmospheric models of the cloud-resolving resolution, although they may become easier to reduce. Although these are short-term processes, they have a potentially large spatial and cumulative effect on modeled tropical circulation; systematic biases can influence overall climate sensitivity in decadal to centennial predictions in climate models.

The cumulative extent of greenhouse gas emissions, primarily the amount of carbon dioxide (CO_2) and methane (CH_4) released into the atmosphere, is of first-order importance to future climate. About half the CO_2 from fossil fuel combustion remains in the atmosphere and is the principal forcing of climate change; the remainder is absorbed by the land and oceans. There are numerous feedbacks in the carbon cycle, however, both positive and negative, that influence the amounts of CO_2 and CH_4 that remain in the atmosphere versus those which are taken up in the ocean and the land surface. These carbon sinks need to be included in climate models to provide the best possible estimate of future greenhouse gas forcing in the atmosphere. The photosynthetic uptake in vegetation can increase in a high-CO_2 environment, providing negative feedback to CO_2 accumulation in the atmosphere. Imbalances between photosynthesis and decomposition processes can affect the net carbon storage on land.

Ecosystem models predict the distribution of natural land cover on the basis of local temperature, precipitation, and other factors. These ecosystem models are now being coupled with general circulation models (GCMs) to produce feedback on the loops involving the biogeochemical cycle of carbon. Earth system models for the next decade will include multiple processes that interact with carbon cycling including the major biogeochemical cycles providing nutrients important for life (e.g., nitrogen and phosphorus). The establishment and mortality within ecosystems will have to account for other factors like age, health, and disturbances beyond fires and land use, like pests, infestation, and other processes. A major advance in the next decade must be in the representation of carbon-climate feedback via subsurface processes, most importantly the soil water. Models of next-generation should include the dynamics of carbon-rich permafrost soils as well as functional classification of microbe communities and mechanistic representation of soil biogeochemistry.

CO_2 exchange between the oceans and the atmosphere is driven by the difference in CO_2 partial pressures in the surface waters and the atmosphere, with the oceanic value dependent on ocean circulation, marine biology, and carbonate chemistry. Ocean biogeochemistry is a central determinant of the uptake of CO_2 from the atmosphere and will change as the climate and ocean change. Ocean biogeochemistry models currently include climate-sensitive carbonate chemistry, a rudimentary representation of different classes of phytoplankton and zooplankton, and multiple nutrient cycles (nitrogen, phosphorus, silica, and iron). New modeling directions need to include the cascading impacts on the entire marine biota from ocean acidification and purposeful and inadvertent additions of macronutrients (e.g., from rivers) and micronutrients such as iron, and their impact on surface CO_2 concentrations. Better resolution of coastal circulation and biogeochemistry will be helpful, as well as improved coupling with continental hydrology models.

Most climate models now include an interactive simulation of aerosols to describe aerosol-climate interactions, but the underlying chemistry and microphysics are only crudely parameterized. This limitation introduces uncertainty in the model quantification of aerosol radiative forcing and its dependence on the hydrologic cycle, both through hygroscopic growth and precipitation scavenging. In addition, atmospheric oxidant and nitrogen chemistry are generally not described in climate models, and this limitation stymies a proper description of simple chemical feedbacks, such as methane-hydroxyl radical (OH) coupling, and more complicated feedbacks involving the effects of changing land cover on atmospheric composition.

Atmospheric aerosols are greatly sensitive to land cover and vegetation. Increased desertification associated with the drying of the subtropics could represent an important source of dust. Changes in ecosystem structure and function would affect the supply of organic aerosol produced by the oxidation of biogenic volatile organic compounds (VOCs). The resulting climate feedback loops are potentially important, and they could be either positive or negative depending on the poorly understood radiative properties of dust and the climate dependence of biogenic VOC emissions. The latter emissions depend in a complicated way on vegetation type, temperature, water availability, leaf phenology, and CO_2. Aerosol yields from biogenic VOCs may also depend on the preexisting supply of anthropogenic aerosols, further complicating the feedback loops.

Atmospheric chemistry plays a critical role in aerosol formation and contributes to other climate-chemistry feedbacks driven by changes in land cover. Deposition of reactive nitrogen (nitrate, ammonium) may significantly affect carbon uptake by ecosystems, and climate change in turn will affect the terrestrial emission and atmospheric chemistry of nitrogen oxides and ammonia. Understanding these effects requires coupling of the sophisticated, dynamic ecosystem and land-surface models. Aerosol chemistry, through direct and indirect effects on atmospheric absorption and scattering, is one of the greatest sources of intermodel climate variability. The advance of coupled land-surface, vegetation, boundary-layer, and aerosol chemistry models promises to be an exciting frontier that may transform aspects of climate modeling.

Climate change and its effects on the water cycle, water availability, and food security require model skill at regional scales. Accurate simulations of regional precipitation patterns and their trends are difficult and are closely related to the hydrological cycle, land-surface processes, ecosystems, ice-ocean interactions, and severe weather. Climate models consistently agree the mean precipitation will increase over the warmest parts of the tropics and poleward of 45° latitude, which may be a consequence of the increased water-holding capacity of a warmer atmosphere as well as increased rates of evaporation (IPCC 2007). Orography and frontal forcing are some of the precipitation mechanisms where increasing resolution is beneficial. Improved model fidelity at regional scales is essential to assessment of water resource and agricultural stress and to drought and flood hazards, which are also an elements of climate extremes. Another challenge for global and regional climate models is their representation of patterns or modes of variability, such as ENSO, Arctic Oscillation (AO), North Atlantic Oscillation (NAO), Southern Annual Mode, Northern Atlantic

Oscillations, and West Pacific decadal variability. Work is needed to better understand modes of decadal variability, the underlying ocean–atmosphere feedbacks, and their representations in models.

Projection—years in advance—of severe weather events as a function of the mean climate state are useful in assessments of climate hazards and climate change adaptation strategies (Klein-Tank et al. 2009). As these events are a function of the mean climate state, statistical probabilities for extreme weather events may be possible to project. Experience is growing in the application of advanced statistical methods to the assessment of climate hazards and climate change adaptation strategies (Klein-Tank et al. 2009). Some coupled models are now able to simulate inter-annual variations in the frequency and intensity of tropical cyclones. Drought persistence is another example, involving feedbacks between soil moisture, evapotranspiration, atmospheric and surface temperatures, dust aerosols, cloud condensation nuclei, and interactions between regional and synoptic circulation patterns (i.e., blocking). Simulation of these feedbacks requires multiscale modeling with an interactive and sophisticated treatment of land-surface and boundary-layer processes.

References

Arakawa A, Wu CM (2013) A unified representation of deep moist convection in numerical modeling of the atmosphere. Part I. J Atmos Sci. https://doi.org/10.1175/JAS-D-12-0330.1

Bjerknes V (1904) Das Problem der Wettervorhersage, betrachtet vom Standpunkte der Mechanik und der Physik. Meteor Z 21:1–7 (Weather forecasting as a problem in mechanics and physics. English translation by Mintz Y, Mimeographed, Los Angeles 1954)

Charney J, Fjørtoft R, von Neumann J (1950) Numerical integration of the Barotropic vorticity equation. Tellus 2(4):237

Cullen MJP, Davies T, Mawson MH, James JA, Coulter SC, Malcolm A (1997) An overview of numerical methods for the next generation UK NWP and climate model. In: Lin C, Laprise R, Ritchie H (eds) Numerical methods in atmosphere and ocean modelling, The Andre Robert memorial volume. Canadian Meteorological and Oceanographical Society, Ottawa, pp 425–444

Davies T, Cullen MJP, Malcolm AJ, Mawson MH, Staniforth A, White AA, Wood N (2005) A new dynamical core for the Met Office's global and regional modelling of the atmosphere. Q J R Meteorol Soc 131:1759–1782

Gill AE (1982) Atmosphere-ocean dynamics, vol 30. Academic Press

Giorgi F, Mearns LO (1999) Introduction to special section: regional climate modeling revisited. J Gerontol Ser A Biol Med Sci 104(D6):6335–6352

Goosse H (2015) Climate system dynamics and modelling. Cambridge University Press, Science, 358 pp

Grabowsky WW (2001) Coupling cloud processes with the large-scale dynamics using the cloud-resolving convection parameterization (CRCP). J Atmos Sci 58(9):978–997

Haltiner GJ, Williams RT (1980) Numerical prediction and dynamic meteorology. Wiley

Holton JR (1992) Dynamic meteorology, 3rd edn. Academic Press

Intergovernmental Panel on Climate Change (2007) Summary for policymakers. Contribution of working group II to the fourth assessment report of the intergovernmental panel on climate change. In: Parry ML, Canziani OF, Palutikof JP, van der Linden PJ, Hanson CE (eds) Climate change 2007: impacts, adaptation and vulnerability. Cambridge University Press, Cambridge

Kasahara A (1974) Various vertical coordinate systems used for numerical weather prediction. Mon Weather Rev 102:509–522

Khairoutdinov MF, Randall DA (2001) A cloud resolving model as a cloud parameterization in the NCAR community climate system model: preliminary results. Geophys Res Lett 28(18):3617–3620

Khairoutdinov M, Randall DA, DeMott C (2005) Simulation of the atmospheric general circulation using a cloud resolving model as a super-parameterization of physical processes. J Atmos Sci 62:2136–2154

Klein-Tank AMG, Zwiers FW, Zhang X (2009) Guidelines on analysis of extremes in a changing climate in support of informed decisions for adaptation. Climate data and monitoring WCDMP-No. 72. World Meteorological Organization, Geneva

Laprise R (1992) The Euler equations of motion with hydrostatic pressure as an independent variable. Mon Weather Rev 120:197–207

National Research Council (2012) A national strategy for advancing climate modeling. The National Academies Press, Washington, DC. https://doi.org/10.17226/13430

Phillips NA (1973) Principles of large-scale numerical weather prediction. In: Morel P (ed) Dynamic meteorology. Reidel, Dordrecht, pp 1–96

Pielke RA, Liston GE, Eastman JE, Lu LX (1999) Seasonal weather prediction as an initial value problem. J Geophys Res 104:19463–19479

Soden BJ, Held IM (2006) An assessment of climate feedbacks in coupled ocean-atmosphere models. J Clim 19:3354–3360

Soden BJ, Vecchi GA (2011) The vertical distribution of cloud feedback in coupled ocean-atmosphere models. Geophys Res Lett 38

Staniforth A (2001) Developing efficient unified nonhydrostatic models. In: Spekat A (ed) Proceedings of the symposium on the 50th anniversary of NWP, Potsdam, Germany, 9–10 March 2000. Deutsche MeteorologischeGesellschafte.V., Berlin, Germany, pp 185–200

Tanguay M, Robert A, Laprise R (1990) Semi-implicit semi-Lagrangian fully compressible regional forecast model. Mon Weather Rev 118:1970–1980

White AA, Bromley RA (1995) Dynamically consistent, quasi-hydrostatic equations for global models with a complete representation of the Coriolis force. Q J R Meteorol Soc 121:399–418

Operational Extended Range Forecast of Weather and Climate over India and the Applications

D. R. Pattanaik, Rajib Chattopadhyay, and A. K. Sahai

Abstract Monsoon rainfall over India during June to September shows intra-seasonal variability with spells of above-normal rainfall (active) and subdued rainfall (break) cycles. India Meteorological Department (IMD) with the support from other sister's organizations of the Ministry of Earth Sciences (MoES) has implemented the operational extended range forecast (ERF) system based on the Climate Forecast System (CFS) version 2 coupled model for extended range forecasts valid up to 3–4 weeks. The operational ERF of rainfall during southwest and northeast monsoon season, maximum and minimum temperature, cyclogenesis potential, etc. has demonstrated useful skill. The ERF for the southwest monsoon season clearly captured the intra-seasonal variability of monsoon including delay/early onset of monsoon, active/break spells of monsoon, and also withdrawal of monsoon at least 2 weeks in advance. The forecast at met-subdivision level is being used for application in Agriculture. Similarly, the cyclogenesis potential in case of Super Cyclone AMPHAN also captured very well 2 weeks in advance, which has the potential application in the disaster management sector. In addition to the regular ERF products for application in agriculture, additional products are being prepared, viz., (i) active phase of monsoon rainfall having potential of heavy rainfall events; (ii) land-surface hydrology products like soil moisture and runoff; (iii) heat wave/cold wave; (iv) transmission window products for vector-borne diseases, etc. for applications in disaster management, hydrology, energy, and health sectors, respectively.

Keywords Coupled model · Extended range forecast · Indian monsoon · CFSv2 · Sectoral applications

D. R. Pattanaik (✉)
India Meteorological Department, New Delhi, India
e-mail: dr.pattanaik@imd.gov.in; drpattanaik@gmail.com

R. Chattopadhyay · A. K. Sahai
Indian Institute of Tropical Meteorology, Pune, India

V. K. Gahalaut and M. Rajeevan (eds.), *Social and Economic Impact of Earth Sciences*,
https://doi.org/10.1007/978-981-19-6929-4_8

1 Introduction

Indian summer monsoon exhibits strong sub-seasonal variability. The spectral signature of sub-seasonal variability is clear from Fig. 1, which is power spectral plot of rainfall over the Indian region. It is clear that the sub-seasonal low-frequency variations of monsoon rainfall range from 2 to 90 days. The spectral signatures are classified as synoptic-scale variability, super synoptic-scale variability and monsoon intra-seasonal variability (MISO). The spectral signatures are identified physically as different types of rain-bearing systems such as low pressure, depressions, and northward propagating rain-bearing systems as MISOs. These systems and their propagations bring convectively active cloud clusters over the Indian region and cause rainfall over different regions of India. Further, the vast Indian land mass is surrounded by the Indian Ocean to the south and the Himalaya to the north. During the summer monsoon season from June to September the land is heated over the vast Asian (covering India) landmass giving rise to low-pressure belt and the cool southern hemisphere is associated with high-pressure region leading to north–south pressure gradient with wind blowing from southern hemisphere and becoming southwesterly after crossing equator (Ramage 1971; Rao 1976). The reversal of wind happened in the winter monsoon with northeasterly wind mainly over India associated with high pressure in the northern hemisphere over Siberian region and low pressure over southern hemisphere. Associated with this reversal of wind the seasons over India are classified with peak monsoon rainfall occurring during June to September and decreased during other seasons. Like the temporal variability, there also exists large spatial variability with more rainfall over the western coast and northeastern states and less rainfall over northwest India and rain shadow regions of Tamil Nadu during JJAS (Rao 1976). Thus, the rainfall distribution is not homogeneous and varies widely from month to month and season to season over India.

During the summer monsoon season, the west coast of peninsular India gets heavy rainfall which has elevated orography (the Western Ghats) parallel to the coast (Rao 1976; Gadgil 2000; Sikka 2006; Fransis and Gadgil 2006). Similarly, the northeastern regions and foothills of Himalayas get plenty of orographic rain. A large part of the southern peninsula belongs to rain shadow area with several districts in this region getting low as well as erratic rainfall during the season. Several analyses have shown that sub-seasonal variability of monsoon has two preferred location on a broader spatial scale, a strong continental convergence zone associated with convection over the land region (between 10° and 25° N) and the other over the eastern equatorial Indian Ocean. The intra-seasonal variability can be defined as the see-saw pattern of the two convergence systems oscillating out of phase with one another (Yasunari 1979; Sikka and Gadgil 1980; Gadgil 2000; Krishnamurti and Subrahmanyam 1982). The oscillation is accompanied by a northward phase propagation of rainfall and other circulation feature anomalies. Hence, the MISO is associated with an explicit northward propagation of positive or negative precipitation (or convection) anomalies (Sikka and Gadgil 1980). Such oscillations bring a sequence of active monsoon and break monsoon situations, which are spells of dry and wet conditions, that often last for 1–2 week or more. Sub-seasonal variability of

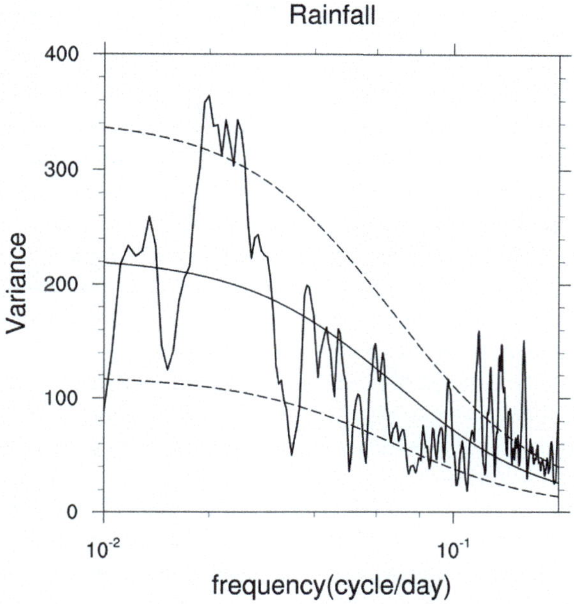

Fig. 1 Power spectral plot of rainfall over Indian region showing the synoptic and low-frequency components of monsoon sub-seasonal variability

monsoon rainfall has dominant variance associated with 30–60-day periodicity, and has a common mode of variability with the seasonal mean, which is hypothesized to be strengthening (weakening) the seasonal mean in its active (break) phases and the large-scale structure of active/break phases, 30–60-day mode and seasonal mean are often similar (Goswami 2012). The intra-seasonal variability can modulate synoptic activity and cause spatial and temporal clustering of lows and depressions (Goswami et al. 2003; Rajeevan et al. 2010) which can impact the regional rainfall patterns.

Forecasting of intra-seasonal oscillations and the synoptic variability is a great challenge and it is an integral part of India Meteorological Department's operational forecasting strategy. The forecast of intra-seasonal oscillations provide forewarning and outlook in different time scales and hence it is important for several stakeholder applications. IMD has adopted a multi-scale seamless forecast strategy in which operational forecasts in the extended (i.e., 2–3 weeks) range are part of the MoES/IMD's "get set go" strategy to improve the operational hydro-meteorological service, as several stakeholder from different sectors, e.g., agriculture, hydrology, energy, and health use the service and is in great demand. Extended range rainfall forecast can provide valuable forewarning of extreme events as well as decision support outlook in the sub-monthly scale. The field of application of extended range forecast is vast and it can help several planning from the country level to state level to block-level scale. Advance planning of several activities, e.g., crop sowing, application of pesticide, planning of irrigation (Chattopadhyay et al. 2018), dam water management (Sahai et al. 2019; Pattanaik and Das 2015), health emergency warning associated with the heatwave and cold waves (Mandal et al. 2019), vector-borne disease (Sahai et al. 2020), and several other applications can be based on extended range outlooks.

2 Sub-seasonal Variability and Its Predictability

Most challenging task of weather forecasting is its prediction in the extended range time scale, which is neither a complete initial value problem nor a complete boundary value problem (Pattanaik et al. 2019). The sub-seasonal variability of monsoon rainfall over India consists of active and break phases of monsoon along with its transition from active to break and vice versa. The time series of daily observed and normal rainfall during three recent monsoons over India (2017, 2018, and 2019) as shown in Fig. 2a–c clearly show active, break, and transition from active-break phase and vice versa, although the phase and timing of the active and break phases are different in different years.

The concept of predictability as well as real-time prediction of sub-seasonal variability is comparatively new compared to short range and seasonal forecasting. The monsoon sub-seasonal variability can be potentially decomposed into few quasi-periodic modes. It is known that the evolution of these modes is associated with distinct dynamical features, multi-component teleconnections in the local and global scale impacting the spatiotemporal distribution of rainfall and other fields. The low-frequency mode evolves on a seasonal mean in a way that it provides a low-frequency systematic phase evolution background and hence it can have better predictability in the sense that the patterns are more or less repeated within a season with some degree of homogeneity. Of course, the repetition of patterns does not guarantee an infinite predictability like a sinusoidal situations, but it can help in improving the forecast quality in the sub-seasonal scale beyond the weekly (or chaos dominated) scale.

This hypothesis has been stated in several versions and a work by (van Den Dool and Saha 1990) can be used as a background in developing a statistical and dynamical forecast system for the monsoon intra-seasonal oscillation. Several recent studies have shown that the prediction skill essentially depends on model fidelity, initial conditions, and boundary conditions. With the improvement in modeling with improved dynamics and physics in last two to three decades, the prediction skill of the low-frequency mode is higher and hence extended range forecast in the 2–3-week scale was seen to be useful in predicting the active-break spells and their transitions (Chattopadhyay et al. 2019; Pattanaik et al. 2019), which is an integral part of monsoon intra-seasonal oscillations. It is to be noted that, however, the predictability of monsoon active and break periods differs (Sahai et al. 2017) though both statistical and dynamical models confirm that the operational predictability could be extended to 2–3 weeks (Sahai et al. 2019).

3 The Current Strategy for Dynamical Extended Range Prediction

The high-resolution models adopted for the operational use by IMD have shown improved skill of monsoon forecasting in short-to-medium range time scale (Sharma

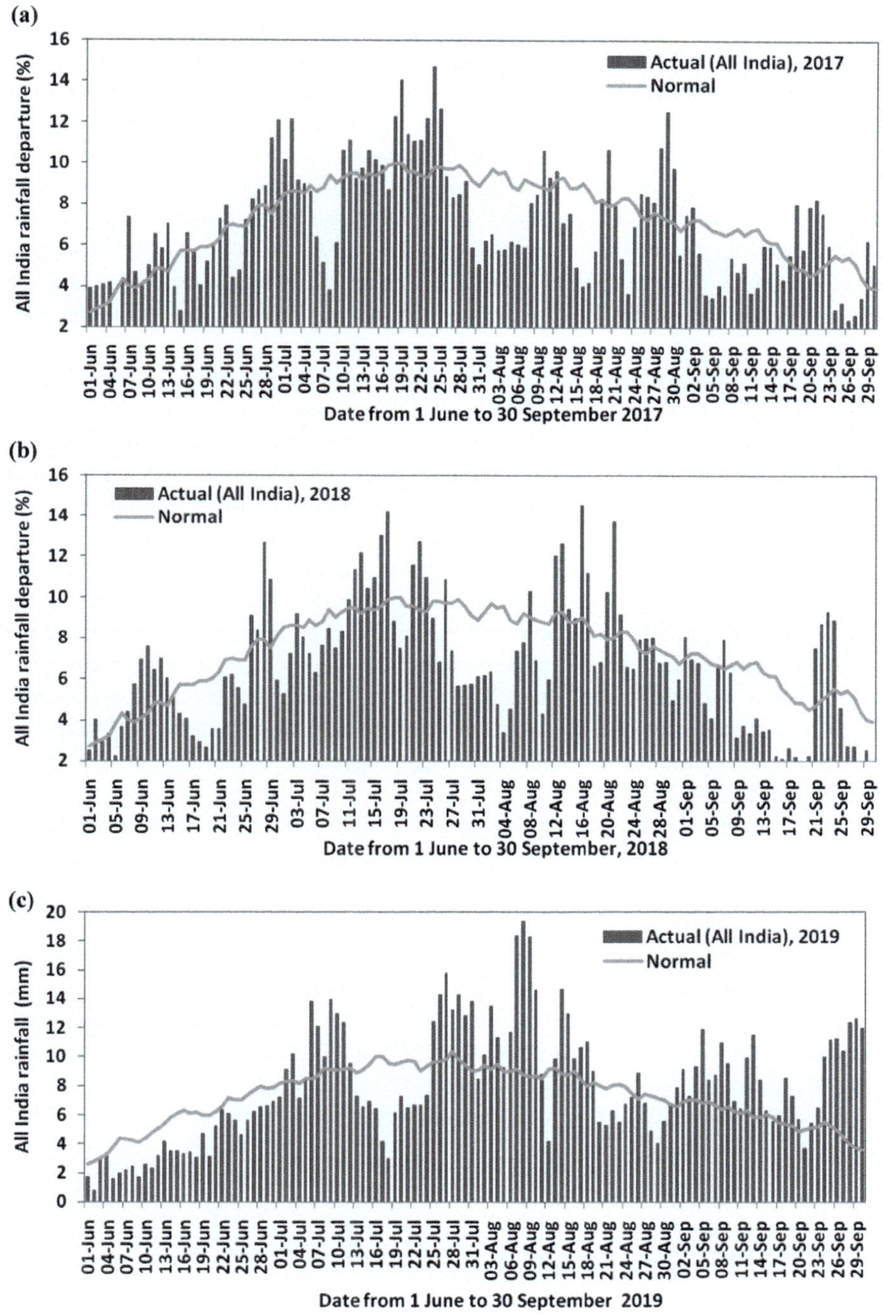

Fig. 2 Daily actual and normal rainfall time series during monsoon seasons of 2017, 2018, and 2019

et al. 2019; Chakraborty et al. 2020, 2021). The current extended range forecasting technique exploits the low-frequency predictability hypothesis by considering the time scales involved in slowly varying low-frequency oscillation forecasts. It is shown by several studies that the predictability on this time scale is controlled both by initial conditions and boundary forcing (Krishnamurti et al. 1992; Fu et al. 2008). Predictability (hence error growth) is limited by the uncertainty in the initial condition and uncertainty in the dynamical forecast mainly arises from the uncertainty in the initial conditions, boundary forcing as well as from the incomplete representation of the model physics. Such problems have led to the acceptance of an ensemble forecast approach which is usually used to sample these errors (Palmer 1993; Harrison et al. 1999).

India Meteorological Department (IMD) issues a real-time weekly averaged operational forecast of the monsoon intra-seasonal oscillations and active/break spells of monsoon up to 4 weeks lead time using an indigenously developed ensemble-prediction-based approach (Through collaborative efforts with IITM, Pune and NCMRWF, Noida) using a version of the dynamical model from National Centers for Environmental Prediction (NCEP) Climate Forecast System version 2 (CFSv2). The ensemble is created based on a perturbation scheme, which has been introduced indigenously to develop reliable initial conditions in a coupled model framework. At IITM, as part of National Monsoon Mission (NMM), the initial testing of the in-house developed ensemble prediction system is carried out using the CFS in different versions of the model (Pattnaik et al. 2013). There are several approaches to generate ensembles of different initial conditions, the operational version at IMD uses an approach which is similar to the "complex-and-same-model environment group" as classified in Buizza et al. (2008).

Other than initial conditions, the key question on how to tackle model uncertainty due to imperfect model dynamics and physics has been addressed by introducing four slightly different versions of same modeling framework with similar dynamics and physics. A suit of slightly different versions of four models is used and the ensemble forecasts are generated using the perturbed initial conditions. These four models are based on two resolutions of CFSv2 coupled model (Saha et al. 2014) and atmospheric component of the coupled model, i.e., GFS runs at T126 and T382 resolutions. Each sub-model is run with 4 initial conditions. These 16 member initialized runs are currently generated once in a week (every Wednesday and forecast is given for 4 weeks). A schematic of the model run and its probable applications are described in Fig. 3.

4 Skill of Extended Range Forecast (ERF)

The skill of the extended range forecast is evaluated based on 11 years of hindcast data for different homogeneous regions of India. The same skill is shown in Fig. 4. As it can be seen that the Multi-model Ensemble (MME)-based ERF skill is better than the individual component models, a multi-model and multi-ensemble strategy

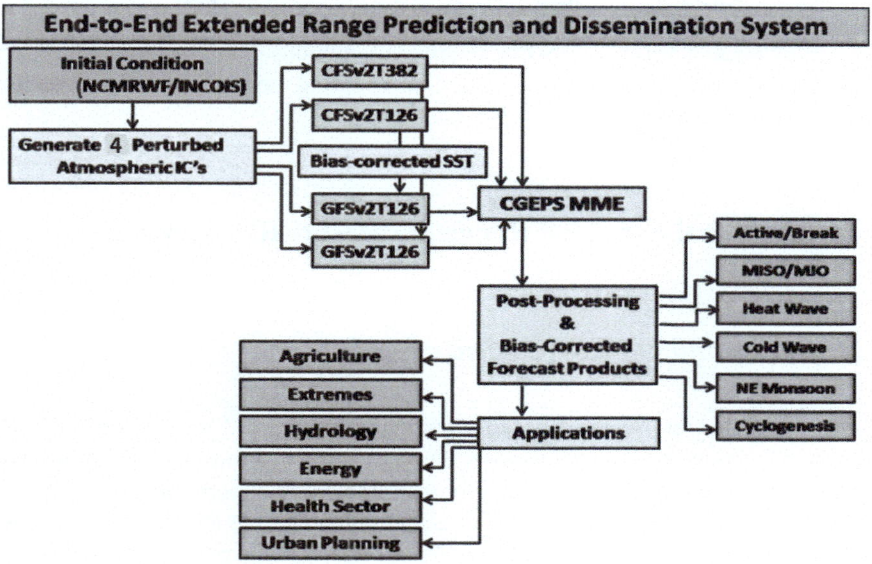

Fig. 3 Extended range forecast system and its applications

is recommended for the extended range outlook. The skill over monsoon zone and central India is more useful for operational applications as compared to Northeast India where the skill is modest.

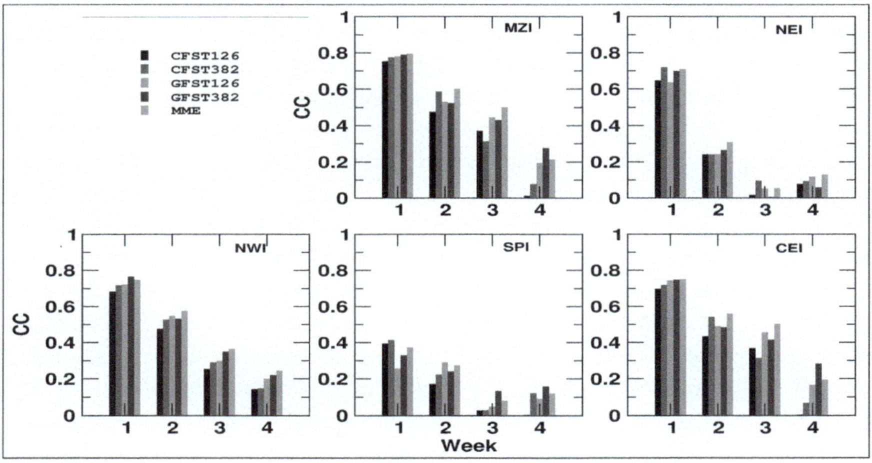

Fig. 4 Weekly lead-time averaged skill of extended range forecast for different homogeneous regions of India (Monsoon zone of India (MZI), Northeast India (NEI), Northwest India (NWI); South Peninsula India (SPI), and Central India (CEI)). The skill of multi-model ensemble (MME) and individual model ensemble mean is shown separately. The figure is adapted from Sahai et al. (2019)

The above discussion suggests that there is potential for application of extended range forecast in several sectors that require sub-monthly scale forewarning. Over the Indian region, the extended range prediction during the monsoon season would potentially improve the agriculture yield outlook.

5 Operational Extended Range Forecast (ERF) System of IMD

5.1 Background of ERF System of IMD

Considering the time scale of weather prediction (up to 3–4 weeks) the ERF of weather using numerical model requires the role of ocean–atmosphere coupling and thereby a coupled model is best suited for the same. For the forecasting of monsoon on this time scale, the accuracy of ERF using numerical models depends on how correctly it simulates the statistics (amplitude, phase propagation, and frequency spectra) of the Intra-seasonal Oscillation (ISO). The operational ERF of IMD started very recently and a modest beginning occurred in IMD following the 2 drought years 2002 and 2004 toward the beginning of twenty-first century, which was associated with one or more long dry spells (Kalsi et al. 2004; Pattanaik and Rajeevan 2007). These 2 drought years in quick succession associated with long dry spells of monsoon were instrumental for IMD to make a modest beginning about the prediction of active-break cycle of monsoon in the real time. However, considering the importance and demand of ERF, IMD started to generate operationally the forecast products of precipitation and temperature based on available model products from different centers in India and abroad from 2008 onward. Initially, some empirical models were used and subsequently the available dynamical model outputs were also used for the same. The two groups of products have been documented in the review articles by Pattanaik (2015) and Pattanaik et al. (2019). The two groups of products as highlighted in the review articles are empirical models and dynamical models. After the initial few years during 2008 and 2009, the improved dynamical models like the ECMWF monthly forecast systems, NCEP CFSv2 climate model, and the JMA Ensemble Prediction System (EPS) are used for the preparation of real-time MME-based ERF during the period from 2010 to 2016. The intra-seasonal monsoon forecast during the drought year of 2009 and other recent years showed useful skill in MME (Tyagi and Pattanaik 2010; Pattanaik et al. 2013a; Pattanaik 2014). The MMEERF products are also prepared on smaller spatial domains (4 homogeneous regions of India and 36 met-subdivisions) during the southwest monsoon season from June to September for the preparation of Agro-met advisory (Pattanaik 2014).

5.2 Current Operational ERF System of IMD Since 2017

At present, the ERF system at IMD is running operationally once a week on every Wednesday and the forecast is generated for 4 weeks starting from subsequent Friday to Thursday and so on. The current operational ERF modeling system is a suite of models at different resolutions based on the CFSv2 coupled model adopted from NCEP as shown in Fig. 3. The hindcasts and forecasts are generated with the atmospheric initial conditions of Global Data Assimilation System (GDAS) run at NCMRWF and oceanic initial conditions of Global Ocean Data Assimilation System (GODAS) run at INCOIS. The above four suite of models are run operationally for 32 days based on every Wednesday initial condition with four ensemble members (1 control and 3 perturbed) each for CFSv2T382, CFSv2T126, GFSbcT382, and GFSbcT126. This is based on the Ensemble Prediction System (EPS) of IITM developed by Abhilash et al. (2014, 2015). For 2018 operational forecast, the hindcast run is performed for 15 years (2003–2017). Similarly, the year 2018 is added into the hindcast run for the forecast of 2019. A verification of intra-seasonal forecasts for 2017 and 2018 monsoon has been discussed in a recent paper by Pattanaik et al. (2019), which witnessed many dry spells of monsoon rainfall as shown in Fig. 2a–b.

6 Some Typical Cases of ERF of Monsoon 2019

2019 monsoon is an excess monsoon year with seasonal departure of monsoon during June to September which is found to be +10%. The main features of the ISO during 2019 monsoon as shown in Fig. 2c are characterized by delayed onset over southern tip of India and subdued monsoon rainfall during June, very active monsoon during last week of July to first 3 weeks of August, and also very active September.

During 2019, the monsoon hits Kerala coast 7 days later than usual on June 8. A day later, a low-pressure system developed over the Arabian Sea, which intensified into the severe cyclonic storm "Vayu" and gradually weakened into a well-marked low-pressure area over northeast Arabian Sea and adjoining Saurashtra and Kutch and crossed Gujarat coast (western coastal state of India) as a weak system on 18 June 2019. The cyclonic system pulled in the moisture-rich westerly winds, halting the advancement of the monsoon. In order to see the ERF of onset of monsoon over Kerala, the extended range forecast of weekly mean wind and rainfall anomaly for 4 weeks based on the initial conditions of 08 and 15 May 2019 are shown in Fig. 5a–d. As seen from Fig. 5a, b, the mean wind forecast for 4 weeks based on IC of 8 May, the 4 weeks forecast of wind, and rainfall anomaly also indicated anti-cyclonic flow over the Arabian Sea and adjoining western coast of India throughout 4 weeks ending on 06 June 2019 associated with negative rainfall anomaly over southern parts of India including Kerala coast. This clearly indicates a likely delay of onset of monsoon over Kerala and it is unlikely to occur till 06 June 2019. Based on IC of 15 May (Fig. 5c, d) the forecast wind indicated a slight change pattern in week 4 valid for

07–13 June with dissipation of anticyclone over the Arabian Sea and strengthening of southwesterly wind over the eastern Arabian Sea and adjoining west coast of India (Fig. 5c). The corresponding forecast rainfall anomaly for 4 weeks based on IC of 15 May indicated positive anomaly over southern parts of Arabian Sea only during week 4 forecast valid for the period from 07 to 13 June 2019 (Fig. 5d). Thus, the circulation patterns indicated favorable patterns for monsoon onset only during the period 07–14 June, thereby a clear indication of delayed onset was indicated in the extended range forecast. Thus, the delayed onset was well predicted in the real-time ERF.

The delayed onset of monsoon, coupled with the developing cyclone over the Arabian Sea, has resulted in a sluggish progress of monsoon till almost end of June, which resulted into an all India rainfall departure of −13% in the month of June (IMD 2020). As shown in Fig. 2c, the daily rainfall averaged over India indicates below-normal rainfall over the country till the end of June. In order to see the performance of ERF forecast for this weak phase of monsoon, the 3-week forecast rainfall anomaly along with the observed rainfall anomaly for the target weeks of 07–13 June, 14–20 June, and 21–27 June 2019 are shown in Fig. 6a–c with 3-week lead-time forecasts. The observed gridded rainfall used for the verification purpose is obtained from the merged rainfall analysis obtained from stations observations over land region and satellite estimates over the ocean region (Mitra et al. 2009). As shown in Fig. 6, the weak phase of monsoon rainfall as seen in the observed rainfall during 07–27 June 2019 is well captured in the extended range forecast with 3-week lead time based on ICs of 05 June, 12 June, and 19 June, respectively.

Monsoon during the second half of August was almost with normal monsoon rainfall, followed by very active September. As indicated earlier, the September month witnessed unprecedented rainfall with a departure of about +52%, which has broken the record of 102 years. The active phase of September can be seen both in the daily rainfall plot shown in Fig. 2c. The active September in 2019 over different parts of India is associated with increase in convective activity over the Arabian Sea. It is to be mentioned here that during the season there were many synoptic-scale systems formed over the Arabian Sea and contributed to the rainfall over different parts of India like a monsoon depression over east-central and adjoining northeast Arabian Sea during third week of September, Very Severe Cyclonic Storm "HIKAA" over the Arabian Sea during 22–25 September 2019, etc. (IMD 2020). These systems in the Arabian Sea have contributed to the above normal over different regions of the country. The active week during 20–26 September 2019 is also well captured in week 1 and week 2 forecast, whereas week 3 forecast is slightly underestimated (Fig. 7a). The active week during 27 September–03 October (Fig. 7b) is also very well captured in the extended range forecast with lead time of 3 weeks as seen in observed and forecast anomaly. This led to the delay in the commencement of withdrawal of monsoon from northwest India. The withdrawal of monsoon began only on 09 October in 2019 (IMD 2020).

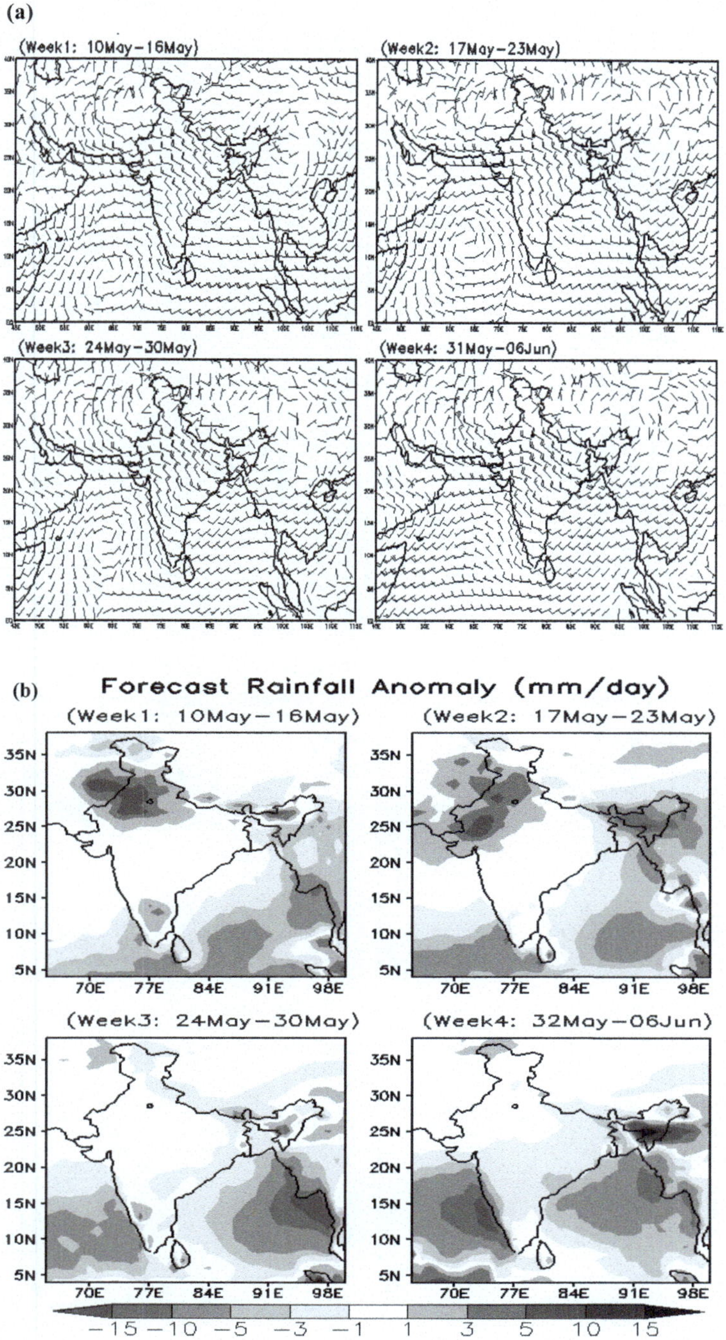

Fig. 5 **a** and **b** Four-week extended range forecasts of weekly 850 hPa mean wind and rainfall anomaly, respectively, based on the initial condition of 08 May 2019. **c** to **d** Same as (**a**) to (**b**) but with ICs of 15 May

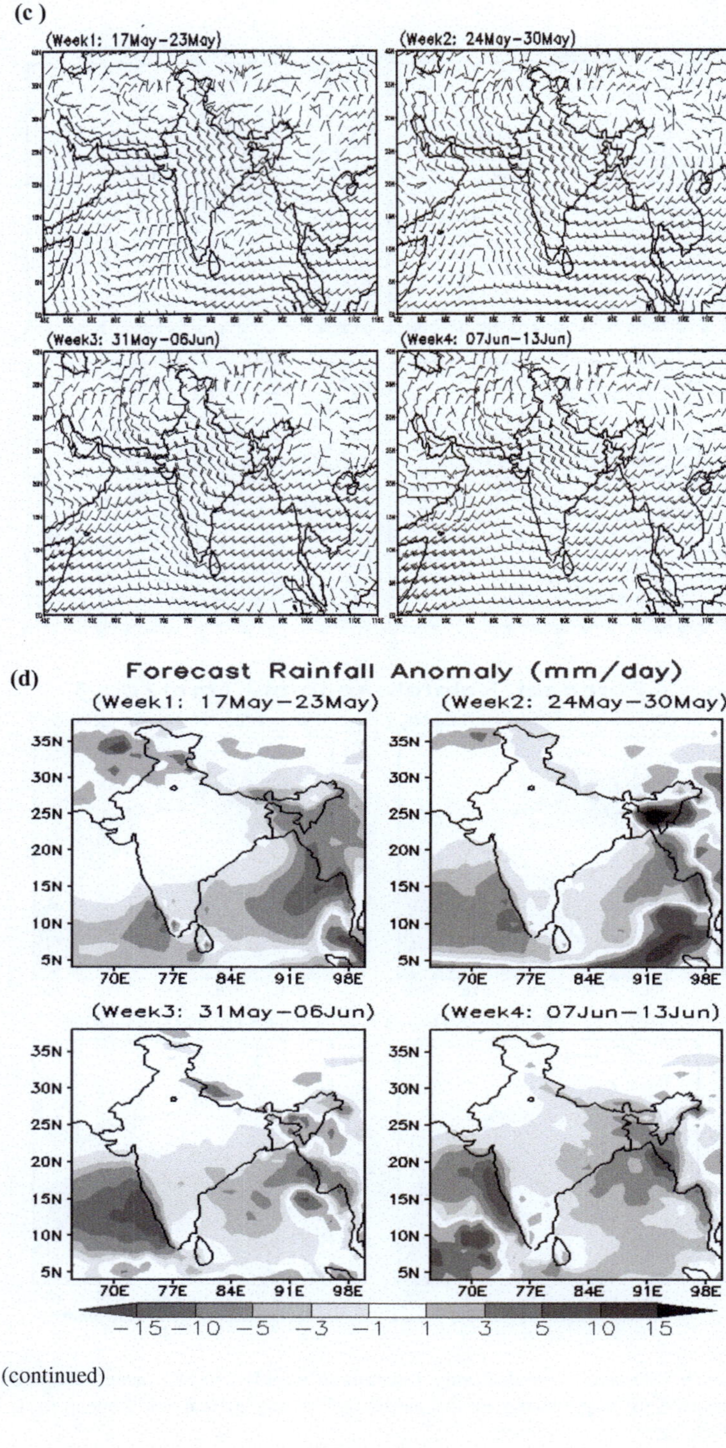

Fig. 5 (continued)

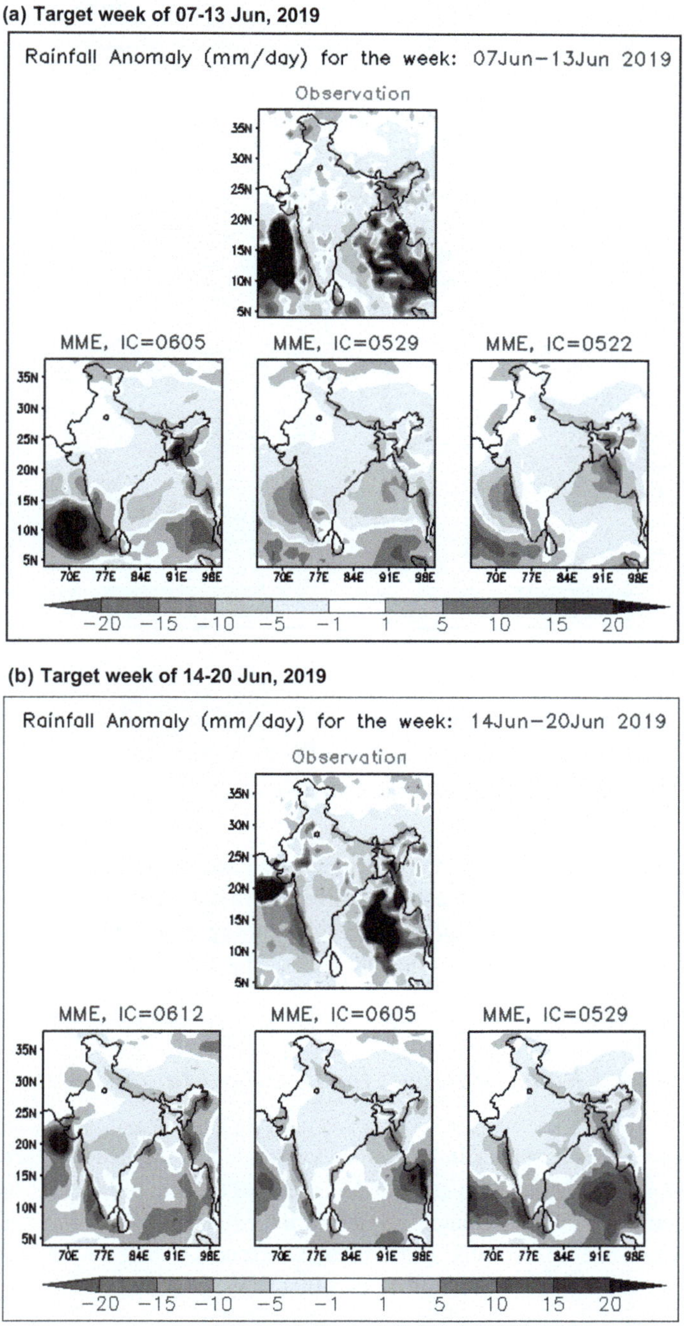

Fig. 6 **a** Observed weekly rainfall anomaly for the period 07–13 June 2019 and 3-week ERF rainfall anomaly for the same target week based on ICs of 05 June, 29 May, and 22 May. **b** to **c** Same as "(**a**)" but for the target weeks of 14–20 and 21–27 June 2019, respectively

(c) Target week of 21-27 Jun, 2019

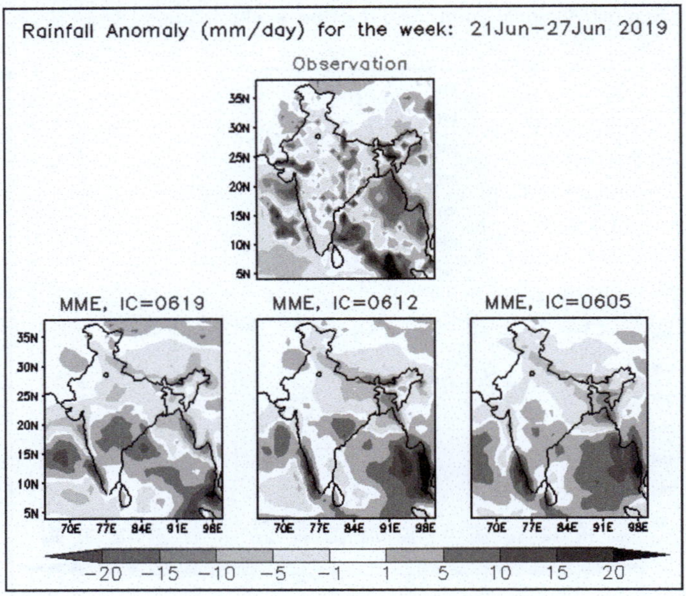

Fig. 6 (continued)

7 Application of ERF for Agro-advisory Services and Hence Economy of the Country

As mentioned earlier, the extremes of temperature and rainfall have broad and far-reaching impacts such as significant loss of life, health issues, and increased economic costs in transportation, agricultural production, energy, and infrastructure. Hence, a timely and accurate forecast can alleviate the adverse effects to a great extent.

7.1 Extended Range Forecast for Disaster Management

The extensive coastal belt of India is very vulnerable to the deadly storms known as cyclonic storms (Mohapatra et al. 2013). The forecasting of the genesis of Tropical Cyclone (TCs) and associated rainfall in the extended range time scale (about 2 weeks in advance) is also very useful in many respects. Using the latest generation global numerical weather prediction models there have been some efforts to study the predictability of TCs of the North Indian Ocean. Pattanaik et al. (2013b) and Pattanaik and Mohapatra (2014) in their recent studies have demonstrated that the prediction of genesis of TCs in extended range time scale using present generation coupled models has got useful skill over the north Indian Ocean. The current ERF system is also being

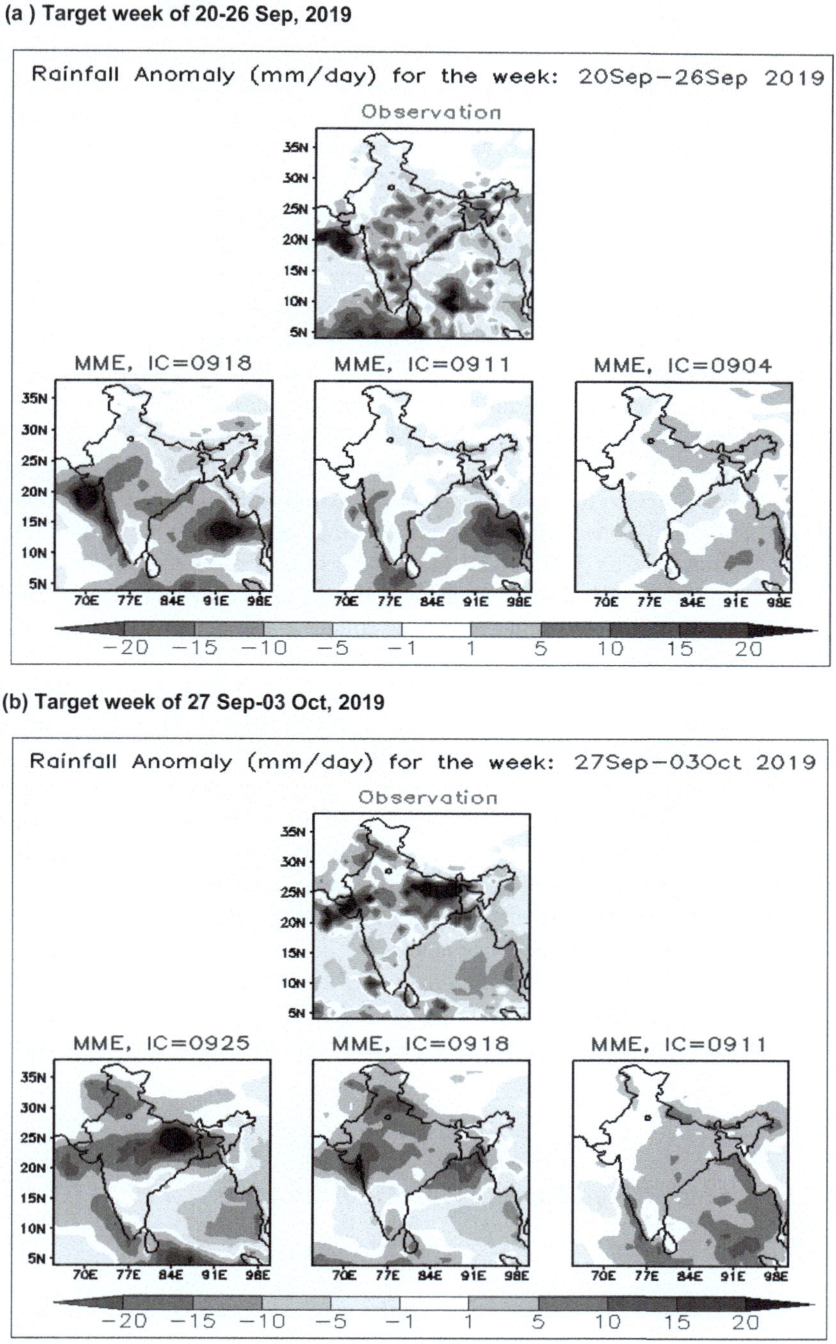

Fig. 7 **a** Observed weekly rainfall anomaly for the period 20–26 September 2019 and 3-week ERF rainfall anomaly for the same target week. **b** Same as "(**a**)" but for the target week of 27 September–03 October 2019

used for cyclogenesis potential probability operationally Over the Bay of Bengal the Super Cyclonic Storm (SuCS) "AMPHAN" initially formed as a low-pressure area over southeast Bay of Bengal and adjoining south Andaman Sea, in the morning of 13 May 2020. It gradually intensified into a depression, cyclonic storm, and finally into a Super Cyclone around noon of 18 May 2020, which can be seen from the observed track and satellite image (Fig. 8a, b). The SuCS "Amphan" over northwest Bay of Bengal moved north-northeast ward and crossed West Bengal-Bangladesh coasts as a Very Severe Cyclonic Storm in the afternoon of 20 May 2020. IMD predicted accurately the landfall point and time, track and intensity as well as associated adverse weather like wind, rainfall, and storm surge due to SuCS "Amphan". The operational ERF system of IMD was used to provide the cyclogenesis potential probability. The 850 hPa extended range forecast wind anomalies clearly indicated the presence of anomalous cyclonic circulation indicating the formation of a system over the Bay of Bengal during the week from 15 to 21 May 2020 valid for week 2 and week 1 forecasts with initial conditions of 6 (top panel in Fig. 9) and 13 May (Bottom panel in Fig. 9) 2020, respectively. The cyclogenesis potential probability is also calculated based on the ERF by considering all 16 ensemble members separately. The GPP is calculated by considering a set of forecast variables like low-level vorticity, vertical wind shear, mid-level humidity, etc. similar to that discussed by Kotal et al. (2009). However, in the present case, the GPP is modified as discussed in Saranya et al. (2019). The real-time GPP based on ERF of 06 May and 13 May initial conditions is shown in Fig. 10a, b, respectively. It indicated the genesis of the system in "Week 1" and "Week 2" forecast and also its re-curvature northeastward. Thus, the extended range outlook had predicted the genesis of the system and also indicated that the system would intensify further and move initially north-northwestward and recurve north-northeastward thereafter toward north Bay of Bengal without crossing eastern coastal states of India. Considering the huge damage potential of a cyclone this early information about the genesis of the system can save huge loss of life and property by mitigating the disaster through early warning and by taking other appropriate measures by the disaster managers to reduce the gravity of damage (Ganesh et al. 2019). Similarly, the early warning on impending heavy rainfall events given by the ERP system can also improve (Pattanaik et al. 2015) the preparedness of disaster management authority so that the loss to life and property can be averted to a great extent.

7.2 *Extended Range Forecast for Application in Health Sector*

Heat wave and cold wave forecast based on the real-time ERF has been very useful in minimizing the loss of life and property (Pattanaik et al. 2016, 2019). In May 2015, India was also severely impacted by a severe heat wave with casualties of more than 2,400 people in many meteorological subdivisions over central, eastern

(a)

(b)

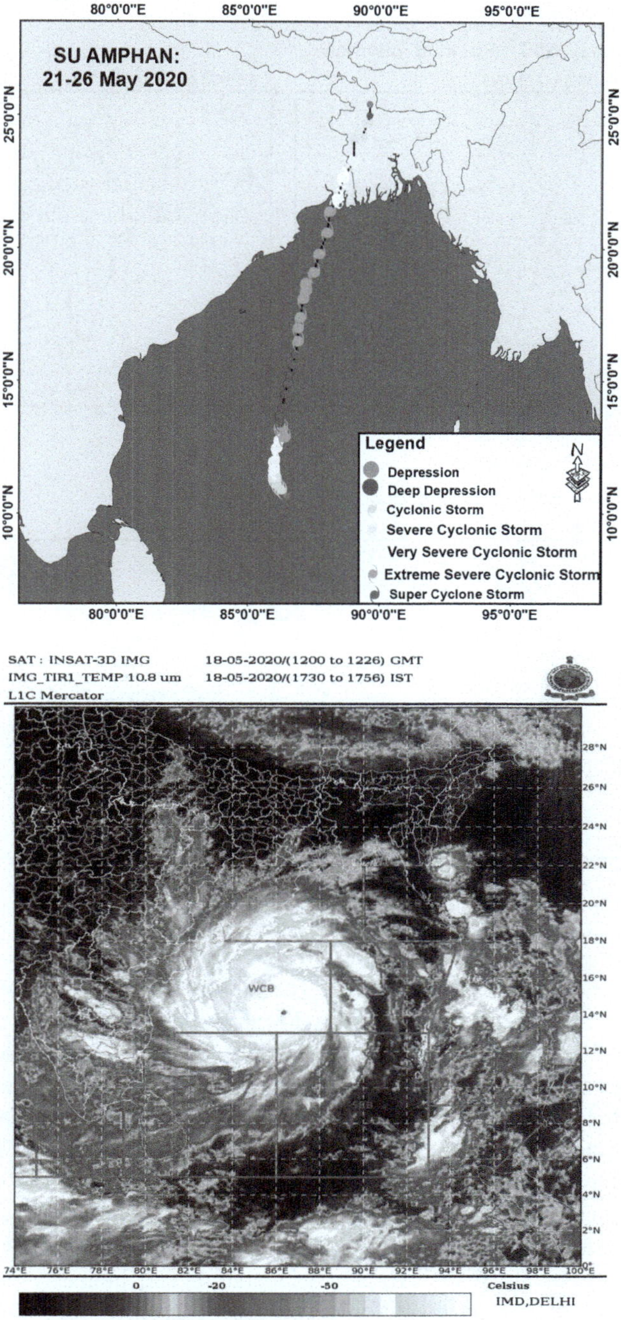

Fig. 8 **a** Operational observed track of SuCS "Amphan" during 16–21 May 2020. **b** INSAT 3D satellite image of SuCS "Amphan" during 1230 h IST of 18 May 2020 during super cyclone stage

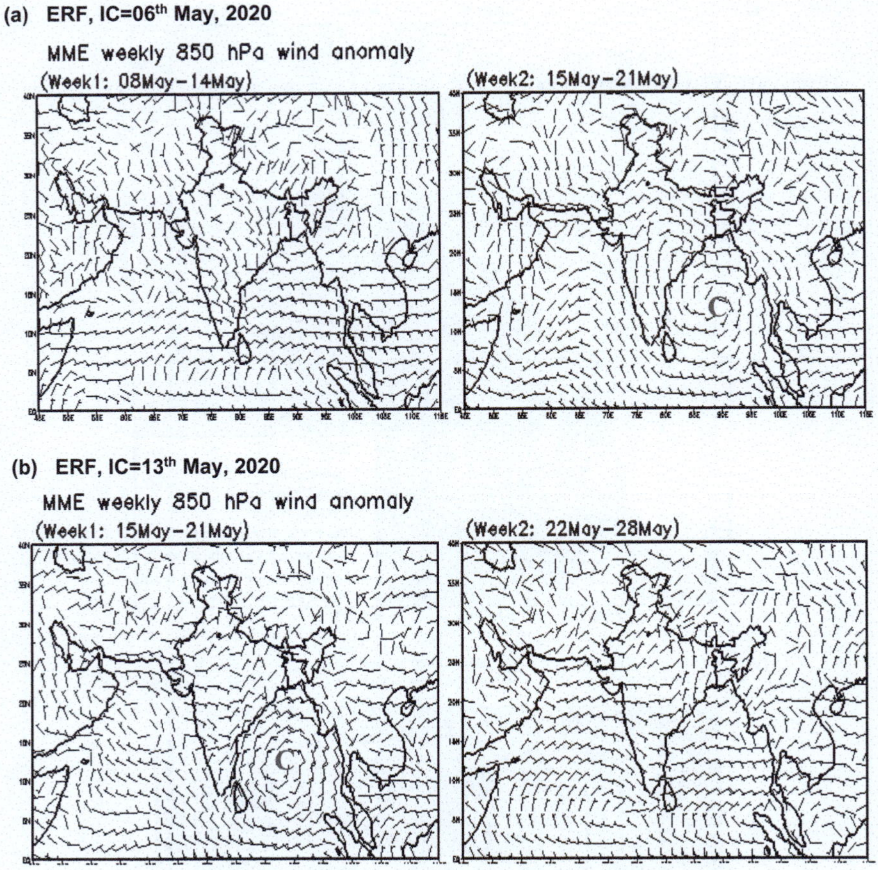

Fig. 9 **a**, **b** The 850 hPa forecast wind anomaly for 2 weeks (IC of 6 and 13 May)

coastal regions, north and northwestern parts of the country (Pattanaik et al. 2016). In 2016, the number of casualties came down drastically from 1111 deaths in 2016 (reduced 46%) as compared to deaths that were recorded in 2015. Heat waves also caused death of wildlife, birds, poultry in states, and most of the zoos in India (source: NDMA). However, after the heat action plans implemented over many states of the country, the number of deaths related to heat waves is decreased substantially in India from 2016 onward. Similarly, the ERF of cold wave in winter is also very skillful for various applications including that in the health sector (Pattanaik et al. 2019).

Further, it has been found from many recent studies that certain favorable weather conditions associated with favorable temperature and humidity can have some impacts on human health conditions. Thus, in addition to heat wave, cold wave, and high heat index (combined effect of high temperature and high humidity) as discussed in Pattanaik et al. (2013c), certain combination of minimum temperature, maximum temperature, and humidity is also more suitable for the transmission

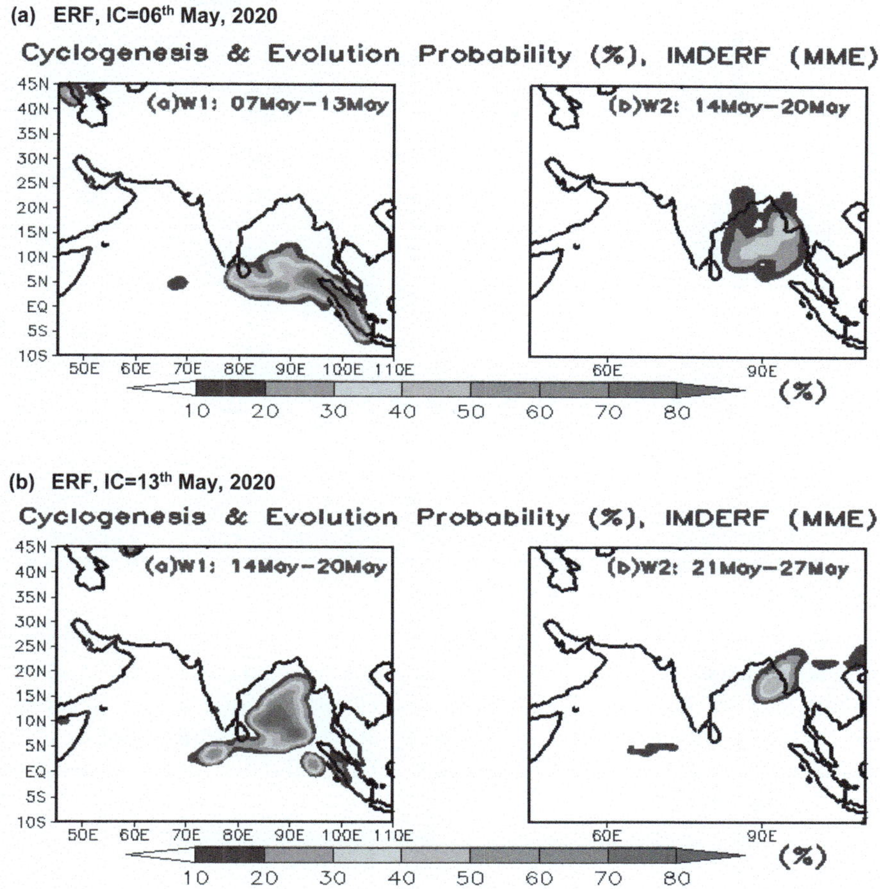

Fig. 10 **a**, **b** ERF of cyclogenesis potential probability based on 06 (top) and 13 May (bottom), respectively, valid for 2 weeks

of vector-borne diseases like Malaria, Dengue, etc. Pattanaik and Mukhopadhyay (2012) have discussed the vulnerability analysis of two transmission window (TW1 and TW2) products for vector-borne diseases over India during different seasons based on climatological data of meteorological variables. The two windows are defined as ***TW1; Tmax*** $\le$ ***35; Tmin*** $\ge$ ***20; RH*** $\ge$ ***55%*** and ***TW2; 25*** $\le$ ***T*** $\le$ ***30 °C; 60*** $\le$ ***RH*** $\le$ ***80%***. Recently, IMD has started preparing the climate information products for the health sector experimentally based on ERF. The probabilistic outlook for prevalence of transmission window for vector-borne diseases for Malaria with threshold maximum temperature (Tmax) between 33 and 39 °C and threshold minimum temperature (Tmin) between 16 and 19 °C is prepared and the real-time experimental ERF valid for week 2 forecast (02–08 October 2020) based on IC of 23 September 2020 is shown in Fig. 11a, b. As seen from Fig. 9, the probability of exceeding 75% for the transmission window for Malaria is indicated over major

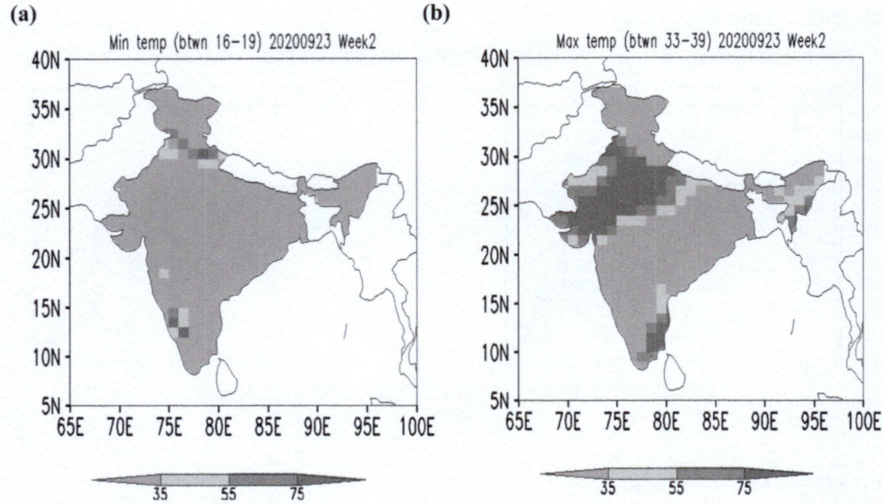

Fig. 11 **a** Probability of minimum temperature (Tmin) to be between 16 and 19 °C and **b** probability of maximum temperature (Tmax) to be between 33 and 39 °C during week 2 forecast valid for the period based on IC of 30 September 2020

districts of Punjab, Haryana, Rajasthan, Gujarat, some districts of Uttar Pradesh, Madhya Pradesh, Karnataka, Tamil Nadu and few districts of Manipur. Recently, an early health warning system is developed by the nonlinear clustering of weather parameters from the ERP outputs (Sahai et al. 2020), and initial results suggest that the system has promising skill.

7.3 *Extended Range Forecast for Hydrological Application*

In hydrology, ERP of the rainfall in a river catchment area and the duration and intensity of hot/dry spells would be helpful for dam managers to decide on the amount of water to be held in or released from the dams to evade the risks of floods or water shortage (Pattanaik and Das 2015; Sahai et al. 2017). The hydrologic prediction based on the ERP data and the variable infiltration capacity (VIC) model can provide a basis for predicting both meteorological and hydrological anomalies and the information can be provided to farmers and water managers. The forecast of root-zone soil moisture along with precipitation and temperature anomalies can be used for irrigation planning. IMD in collaboration with IIT Gandhinagar is preparing the soil moisture and runoff forecasts based on the ERF by using the VIC model in near-real time to estimate the agricultural and hydrological droughts severity and areal extent (Shah et al. 2017; Mishra et al. 2018; Pattanaik et al. 2019). The anticipated weekly soil moisture change for the 2 weeks during 04–10 September 2020 and 11–17 September 2020 with respect to observed week from 28 August to 03 September 2020 is shown in Fig. 12.

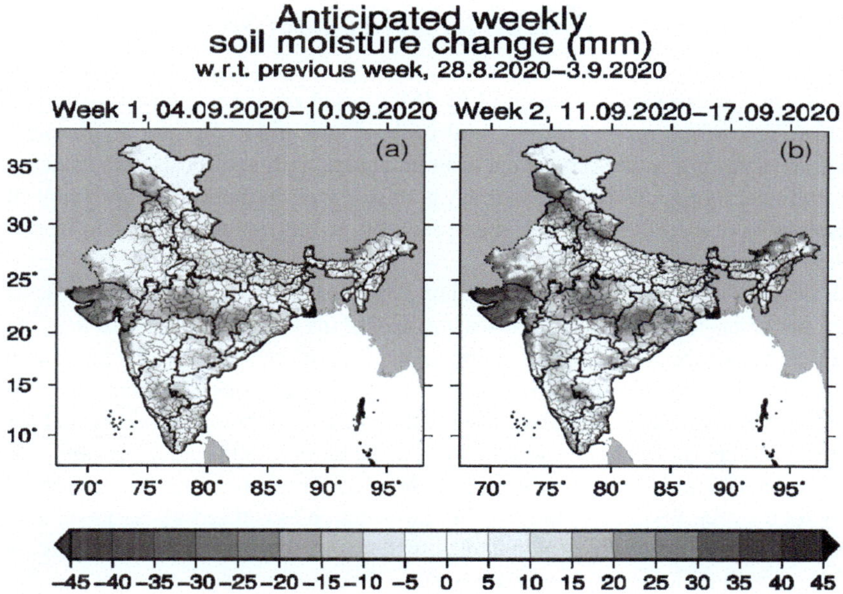

Fig. 12 Anticipated weekly soil moisture change (mm) for two forecasted weeks **a** 04–10 September 2020 and **b** 11–17 September 2020 with respect to observed week 28 August–03 September 2020

Further, an effort has been made to compute various drought indices (SPI, SPEI, and PDSI) using ERP forecasts (Shrivastava et al. 2018) and it was found that the probabilistic forecasts and the drought indices from the prediction system are quite useful to identify droughts over central India 20 days in advance.

7.4 Extended Range Forecast for Energy Sector

The energy sector significantly depends on the weather and climate, which impacts demand and supply (Dubus et al. 2018). If it is too hot or cold outside, the power consumption increases and the energy providers must ensure the availability of this extra power at the right time. Therefore, ERF of extreme temperatures can contribute to the preparedness in the energy sector. An outlook on impending heavy rainfall events, storms, etc. also helps them to manage the resources wisely. The initiatives undertaken under the monsoon mission for wind and solar energy will be further enhanced under MoES programs and the improved forecast would be provided to the stakeholders for strengthening the wind and solar energy sectors.

7.5 *Extended Range Forecast for Agricultural Sector*

The agro-met advisories are prepared by IMD based on the weather forecast on a real-time basis and proved to be useful in the decision-making process in every farm operation, rain water harvesting, crop planning, control of pests and diseases, irrigation scheduling, etc. (Chattopadhyay et al. 2018). A number of studies showed the usefulness of medium range weather forecast in Indian agriculture (Rathore et al. 2009; Chattopadhyay et al. 2016).

As shown clearly in a paper published by Chattopadhyay et al. (2018) in case of intermittent flash flood during Kharif season of 2015 over Assam, different situations (flooding period) were created depending on land situation, soil types, and nearness of the crop field to the source of the flood. As a real-time response, by using medium range weather forecast, different agro-met advisories were prepared and issued for different flooding situations and disseminated to the farmers of the agro-climatic zone in this region and those farmers who used these advisories were immensely benefited by saving the crop loss. Thus, the weather information (weather forecast) available at that time was very much useful in managing the situation arising due to the excess rainfall through real-time agro-met advisories. Though there has been considerable skill in medium range weather forecast which has already been demonstrated, there was mixed feeling in the skill of sub-seasonal-to-seasonal forecast in the country. It was noticed that various climate research centers in India and abroad using statistical and dynamical models could not predict the extent of deficiency of 2009 seasonal monsoon rainfall during June to September. However, a proper real-time monitoring of intra-seasonal fluctuation of monsoon rainfall during 2009 monsoon season by IMD was quite useful in assessing the extent and gravity of drought situation of the country (Tyagi and Pattanaik 2010). Similarly, the dry spells during the recent monsoon seasons of 2017 and 2018 are also very well captured in the real time (Pattanaik et al. 2020), which have tremendous economic value for the country like India. A recent study by Robertson et al. (2019) demonstrated the 2-week extended range forecast issued in real time during June–September 2018 monsoon period for the four districts of the state of Bihar. Successful monsoon onset and break phase forecasts in 2018 over Bihar are related to episodes of the Madden-Julian Oscillation, the model of which is shown to capture quite well at 1–2-week lead. A comprehensive third-party assessment of socio-economic benefits of Agro-met Services was carried out by reputed National Council of Applied Economic Research [Sharma A. (NCAER 2015)], Delhi and in their report the Council pointed out that the farming community of the country is using Agro-Meteorological Advisory Service products for critical actions during their farm operations, viz.,

(i) Management of sowing in case of delayed onset of rains.
(ii) Shifting to short-duration crop varieties in case of a long-term delay in rainfall.
(iii) Deferring of spraying of pesticides for disease control on forecast of occurrence of rainfall in near future.
(iv) Managing (curtailing) artificial irrigation in case of heavy rainfall forecast.

The study suggests that the Agro-met Advisory Project of IMD has the potential of generating net economic benefit up to Rs. 3.3 lakh crores on the four principal crops alone when Agro-Meteorological advisory Service is fully utilized by 95.4 million agriculture-dependent households. But, there was a need to extend the services to planner's communities especially for dry land agriculture planners by advising well in advance (at least 15 days) so as to plan agricultural activities, prepare for contingency. To mitigate this requirement, the Agricultural Meteorology Division, IMD, based on real-time extended range forecast started issuing Agro-met Advisory Service bulletin every week. For the agro-advisory purpose, the ERF is converted into met-sub-divisional category forecasts (normal, above normal, and below normal) and the advisory is issued for 2 weeks. It is also proposed to issue the advisories at district level from existing meteorological subdivision level for 2 weeks. This can be seen from Fig. 13a, b for week 1 forecast based on IC of 25 September and valid for the period from 25 September to 01 October 2020 indicating the withdrawal of southwest monsoon from northwest India associated with below-normal rainfall. It is not only the agricultural sector, which can be benefited financially with the accurate guidance to farmers, but also many more sectors like the energy (power) sector that can be benefited with inputs from ERF in managing the generation and distribution of power as per the increase and decrease of the demand.

8 Summary

IMD has recently implemented a Climate Forecast System (Version 2)-based coupled modeling system for operational extended range forecast (forecast up to 3–4 weeks). Dynamical model-based extended range forecast is the best tool suited to provide the scientific outlook beyond the weather time scale. The extended range forecast in the 2–3-week time scale based on the CFSv2 has improved our understanding of the monsoon system in general. Specifically, this technique has optimized the skill of the monsoon prediction system by improving the monitoring and prediction of tropical intra-seasonal oscillations. The dynamical models reflect our knowledge of the physical system and therefore their improvement (by improving various parameterizations) indicates an overall improvement in our understanding of the various physical processes. The skillful ERF can provide services to many important sectors like Agriculture, Water, Health, Energy, and Disaster Management.

The MME-based ERF provides useful guidance up to 2–3 weeks about the onset, withdrawal, active-to-break, and break-to-active transitions of monsoon during the southwest monsoon season, cyclogenesis potential probability during the cyclone season from October to December over the north Indian Ocean, heat wave in summer and cold wave during winter season. Similarly, IMD has also started preparing the climate information products for the health sector experimentally based on extended range forecasts.

With regard to the disaster management the longer lead-time forecast of the genesis of a tropical cyclone with huge damage potential can save loss of life and property

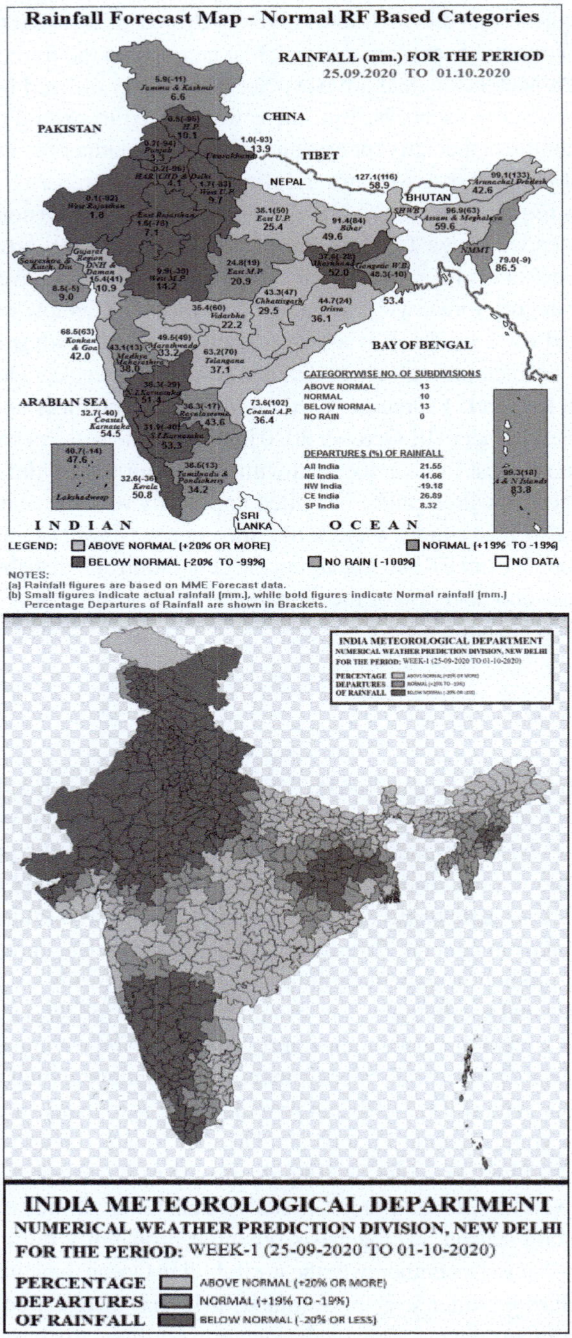

Fig. 13 **a** Met-subdivision-wise week 1 forecast based on IC of 25 September and valid for the period from 25 September–01 October 2020 indicating the withdrawal of southwest monsoon. **b** Same as “(**a**)” but for district-wise forecasts

by properly mitigating the disaster. The ERF of the genesis and evolution of cyclonic systems would serve as guidance to the disaster management authority to reduce the gravity of damage. Similarly, the early warning on impending heavy rainfall events can also improve the preparedness of disaster management authority so that the loss to life and property can be averted to a great extent.

For application in the agricultural sector, the meteorological sub-division-wise forecasts for 36 meteorological subdivisions for 2 weeks are being used for preparing agro-met advisory to farmers. It is suggested that the Agro-met Advisory Services of IMD have the potential of generating huge economic benefit when the Agro-Meteorological Advisory Service is fully utilized to all agriculture-dependent households. It is not only the agricultural sector, which can be benefited financially with the accurate guidance to farmers, but the energy sector will also be benefited with inputs from extended range forecast in managing the generation and distribution of power as per the demand.

Acknowledgements The authors are very much thankful to Dr. M. Rajeevan, Secretary, MoES and Dr. M. Mohapatra, Director General, IMD for their encouragement and for providing all support to work together with other institutes. Thanks are also due to CRS Pune for making available the gridded data. The real-time extended range forecast capability of IMD is being strengthened through the collaborative efforts of the MoES institutions, viz., the IITM, NCMRWF, and INCOIS. The CFSv2/GFSbc customized at IITM was implemented in IMD for the operational run.

References

Abhilash S, Sahai AK, Pattnaik S, Goswami BN, Kumar A (2014) Extended range prediction of active-break spells of Indian summer monsoon rainfall using an ensemble prediction system in NCEP climate forecast system. Int J Climatol 34:98–113. https://doi.org/10.1002/joc.3668

Abhilash S, Sahai AK, Borah N, Joseph S, Chattopadhyay R, Sharmila S, Rajeevan M, Mapes BE, Kumar A (2015) Improved spread-error relationship and probabilistic prediction from the CFS-based grand ensemble prediction system. J Appl Meteor Climatol 54:1569–1578

Buizza R, Leutbecher M, Isaksen L (2008) Potential use of an ensemble of analyses in the ECMWF ensemble prediction system. Q J R Meteorol Soc 134:2051–2066. https://doi.org/10.1002/qj.346

Chakraborty P, Sarkar A, Kumar S, George JP, Rajagopal EN, Bhatla R (2020) Assessment of NCMRWF global ensemble system with differing ensemble populations for tropical cyclones prediction. Atmos Res 244(1–20):105077. https://doi.org/10.1016/j.atmosres.2020.105077

Chakraborty P, Sarkar A, Bhatla R, Singh RS (2021) Assessing the skill of NCMRWF global ensemble prediction system in predicting Indian summer monsoon during 2018. Atmos Res 248(1–22):105255. https://doi.org/10.1016/j.atmosres.2020.105255

Chattopadhyay N, Bhowmik, Roy SK, Singh KK, Ghosh K, Malathi K (2016) Verification of district level weather forecast. Mausam 67(4):829–840

Chattopadhyay N, Rao KV, Sahai AK et al (2018) Usability of extended range and seasonal weather forecast in Indian agriculture. Mausam 69:29–44

Chattopadhyay R, Joseph S, Abhilash S et al (2019) Understanding the intraseasonal variability over Indian region and development of an operational extended range prediction system. Mausam 70:31–56

Dubus L, Muralidharan S, Troccoli A (2018) What does the energy industry require from meteorology? In: Troccoli A (ed) Weather & climate services for the energy industry. Palgrave Macmillan, Cham. https://doi.org/10.1007/978-3-319-68418-5_4

Francis PA, Gadgil S (2006) Intense rainfall events over the west coast of India. Meteorol Atmos Phys 94: 27–42

Fu X, Yang B, Bao Q, Wang B (2008) Sea surface temperature feedback extends the predictability of tropical intraseasonal oscillation. Mon Weather Rev 136:577–597. https://doi.org/10.1175/2007MWR2172.1

Gadgil S (2000) Monsoon-ocean coupling. Curr Sci 78:309–332

Ganesh SS, Abhilash S, Sahai AK, Joseph S, Chattopadhyay R, Mandal R, Dey A, Krishna RPM (2019) Genesis and track prediction of pre-monsoon cyclonic storms over North Indian Ocean in a multi-model ensemble framework. Nat Hazards 95:823–843. https://doi.org/10.1007/s11069-018-3522-6

Goswami BN (2012) South Asian monsoon. Intraseasonal variability in the atmosphere-ocean climate system. Springer, Berlin, Heidelberg, pp 21–71

Goswami BN, Ajayamohan RS, Xavier PK, Sengupta D (2003) Clustering of synoptic activity by Indian summer monsoon intraseasonal oscillations. Geophys Res Lett 30:1431. https://doi.org/10.1029/2002GL016734

Harrison MSJ, Palmer TN, Richardson DS, Buizza R (1999) Analysis and model dependencies in medium-range ensembles: two transplant case-studies. Q J R Meteorol Soc 125:2487–2515. https://doi.org/10.1002/qj.49712555908

India Meteorological Department (2020) Southwest monsoon 2019: a report. In: Sreejith OP, Pai DS, Devi Sunitha (eds) IMD Met Monograph: ESSO/IMD/Synoptic Met/02(2019)/24, pp 1–243

Kalsi SR, Hatwar HR, Jayanthi N, Subramanian SK, Shyamala B, Rajeevan M, Jenamani RK (2004) Various aspects of unusual behaviour of monsoon 2002. IMD Met. Monograph, Synoptic Meteorology No. 2/2004, National Climate Centre, India Meteorology Department, Lodi Road, New Delhi, India

Kotal SD, Kundu PK, Roy Bhowmik SK (2009) Analysis of cyclogenesis parameter for developing and non-developing low-pressure systems over the Indian Sea. Nat Hazards 50:389–402

Krishnamurti TN, Subrahmanyam D (1982) The 30–50 day mode at 850 mb during MONEX. J Atmos Sci 39:2088–2095

Krishnamurti TN, Subramaniam M, Daughenbaugh G et al (1992) One-month forecasts of wet and dry spells of the monsoon. Mon Weather Rev 120:1191–1223. https://doi.org/10.1175/1520-0493(1992)120%3c1191:OMFOWA%3e2.0.CO;2

Mandal R, Joseph S, Sahai AK et al (2019) Real time extended range prediction of heat waves over India. Sci Rep 9:9008. https://doi.org/10.1038/s41598-019-45430-6

Mishra V, Shah R, Azhar S, Shah H, Modi P, Kumar R (2018) Reconstruction of droughts in India using multiple land-surface models (1951–2015). Hydrol Earth Syst Sci 22:2269–2284. https://doi.org/10.5194/hess-22-2269-2018

Mitra AK, Bohra AK, Rajeevan MN, Krishnamurti TN (2009) Daily Indian precipitation analysis formed from a merge of rain–gauge data with the TRMM TMPA satellite-derived rainfall estimates. J Meteor Soc Jpn Ser II 87A:265–279

Mohapatra M, Nayak DP, Sharma RP, Bandyopadhyay BK (2013) Evaluation of official tropical cyclone track forecast over North Indian Ocean issued by India Meteorological Department. J Earth System Sci 122(3):589–601

Palmer TN (1993) Extended-range atmospheric prediction and the Lorenz model. Bull Am Meteor Soc 74:49–66. https://doi.org/10.1175/1520-0477(1993)

Pattanaik DR (2014) Meteorological subdivisional-level extended range forecast over India during southwest monsoon 2012. Meteorol Atmos Phys. https://doi.org/10.1007/s00703-014-0308-6

Pattanaik DR (2015) Operational extended range forecast activity in IMD and its application in different sectors. Vayu Mandal 40:44–70

Pattanaik DR, Das AK (2015) Prospect of application of extended range forecast in water resource management: a case study over the Mahanadi River basin. Nat Hazards 77:575–595

Pattanaik DR, Mohapatra M (2014) Multi-model ensemble based extended range forecast of tropical cyclogenesis over the north Indian Ocean. In: Mohanty UC, Mohapatra M, Singh OP, Bandopadhyay BK, Rathore LS (eds) Monitoring and prediction of tropical cyclones in the Indian Ocean and climate change, pp 203–216. https://doi.org/10.1007/978-94-007-7720-0_17

Pattanaik DR, Mukhopadhyay B (2012) Indian climatology in the context of human health. In: Dogra N, Srivastava S (eds) Climate change and disease dynamics in India. TERI Press, TERI, New Delhi, pp 71–110. ISBN: 9788179934128

Pattanaik DR, Rajeevan M (2007) North-west pacific tropical cyclone activity and July rainfall over India. Meteorol Atmos Phys 95:63–72

Pattanaik DR, Rathore LS, Kumar A (2013a) Observed and forecasted intraseasonal activity of southwest monsoon rainfall over India during 2010, 2011 and 2012. Pure Appl Geophys. https://doi.org/10.1007/s00024-013-0670-1

Pattanaik DR, Mohapatra M, Mukhopadhyay B, Tyagi A (2013b) A preliminary study about the prospects of extended range forecast of tropical cyclogenesis over the North Indian Ocean during 2010 post-monsoon season. Mausam 64:171–188

Pattanaik DR, Kumar G, Bhatla R, Soni VK, Bist S, Bhan SC, Mukhopadhyay B (2013c) Heat index outlooks over the Indian region using general circulation model outputs. IMS Bulletin (Vayu Mandal) 39:12–24

Pattnaik S, Abhilash S, De S et al (2013d) Influence of convective parameterization on the systematic errors of climate forecast system (CFS) model over the Indian monsoon region from an extended range forecast perspective. Clim Dyn 41:341–365. https://doi.org/10.1007/s00382-013-1662-7

Pattanaik DR, Pai DS, Mulhopadhyay B (2015) Rapid northward progress of monsoon over India and associated heavy rainfall over Uttarakhand: a diagnostic study and real time extended range forecast. Mausam 66(3):551–568

Pattanaik DR, Mohapatra M, Srivastava AK, Kumar A (2016) Heat wave over India during summer 2015: an assessment of real time extended range forecast. Meteorol Atmos Phys. https://doi.org/10.1007/s00703-016-0469-6

Pattanaik DR, Sahai AK, Mandal R et al (2019) Evolution of operational extended range forecast system of IMD: prospects of its applications in different sectors. Mausam 70:233–644

Pattanaik DR, Sahai AK, Phani Muralikrishna R, Mandal R, Dey A (2020) Active-break transitions of monsoons over India as predicted by coupled model ensembles. Pure Appl Geo Phys 177:4391–4422

Rajeevan M, Gadgil S, Bhate J (2010) Active and break spells of the Indian summer monsoon. J Earth Sys Sci 119:229–247

Ramage CS (1971) Monsoon meteorology. Academic Press, New York, p 296

Rao YP (1976) Southwest monsoon. Met Monograph, Synoptic Meteorology. India Meteorological Department, Delhi, pp 107–185

Rathore LS, Bhowmik, Roy SK, Chattopadhyay N (2009) Integrated agromet advisory services in India. In: Challenges and opportunities in agrometeorology. Springer Publication, New York, pp 195–205

Robertson AW, Acharya N, Lisa Goddard DR, Pattanaik AK, Sahai KK, Singh KG, Agarwal A, Buizer JL (2019) Subseasonal forecasts of the 2018 Indian summer monsoon over Bihar. J Geophys Res (Atmosphere) 124:13861–13875

Saha S, Moorthi S, Wu X, Wang J, Nadiga S, Tripp P, Pan H-L, Behringer D, Hou Y-T, Chuang H-y, Iredell M, Ek M, Meng J, Yang R, van den Dool H, Zhang Q, Wang W, Chen M (2014) The NCEP climate forecast system version 2. J Clim 27:2185–2208. https://doi.org/10.1175/JCLI-D-12-00823.1

Sahai AK, Sharmila S, Chattopadhyay R et al (2017) Potential predictability of wet/dry spells transitions during extreme monsoon years: optimism for dynamical extended range prediction. Nat Hazards 88:853–865. https://doi.org/10.1007/s11069-017-2895-2

Sahai AK, Chattopadhyay R, Joseph S et al (2019) Chapter 20—Seamless prediction of monsoon onset and active/break phases. In: Robertson AW, Vitart F (eds) Sub-seasonal to seasonal prediction. Elsevier, pp 421–438

Sahai AK, Mandal R, Joseph S, Saha S, Awate P, Dutta S, Dey A, Chattopadhyay R, Krishna RPM, Pattanaik DR, Deshpande S (2020) Development of a probabilistic early health warning system based on meteorological parameters. Sci Rep. https://doi.org/10.1038/s41598-020-71668-6

Saranya Ganesh S, Sahai AK, Abhilash S, Joseph S, Dey A, Mandal R, Chattopadhyay R, Phani R (2019) A new approach to improve the track prediction of tropical cyclones over North Indian Ocean, vol 2018, pp 7781–7789

Shah R, Sahai AK, Mishra V (2017) Short to sub-seasonal hydrological forecast to manage water and agricultural resources in India. Hydrol Earth Syst Sci 21:707–720

Sharma A (NCAER) (2015) National Council of Applied Economic Research, Report on "Economic benefits of dynamic weather and ocean information and advisory services in India and cost prising of customized products and services of ESSO-NCMRWF & ESSO-INCOIS"

Sharma K, Ashrit R, Ebert E, Ashis Mitra R, Bhatla GI, Rajagopal EN (2019) Assessment of met office unified model (UM) quantitative precipitation forecasts during the Indian summer monsoon: contiguous rain area (CRA) approach. J Earth Syst Sci 128(1):1–17. https://doi.org/10.1007/s12040-018-1023-3

Shrivastava S, Kar SC, Sahai AK, Sahai, Sharma AR (2018) Identification of drought occurrences using ensemble predictions up to 20-days in advance. Water Resour Manag. https://doi.org/10.1007/s11269-018-1921-9

Sikka DR (2006) A study on the monsoon low pressure systems over the Indian region and their relationship with drought and excess monsoon seasonal rainfall. COLA technical report 217. Centre for Ocean-Land Atmospheric Studies, Claverton, MD

Sikka DR, Gadgil S (1980) On the maximum cloud zone and the ITCZ over Indian longitudes during the southwest monsoon. Mon Weather Rev 108:1840–1853

Tyagi A, Pattanaik DR (2010) Real time monitoring and forecasting of intra-seasonal monsoon rainfall activity over India during 2009. IMD Met. Monograph, Synoptic Met. No. 10/2010, pp 1–45

van Den Dool HM, Saha S (1990) Frequency dependence in forecast skill. Mon Weather Rev 118:128–137

Yasunari T (1979) Cloudiness fluctuations associated with the northern hemisphere summer monsoon. J Meteor Soc Jpn 57:227–242

Frequency and Magnitude of Heat and Cold Waves over India

Smitha Nair, Mahima, Monika Sanghwahia, and D. S. Pai

Abstract Heat waves (HWs) and cold waves (CWs) are well-known forms of extreme weather events. The observed frequency, magnitude, and duration of heat waves (HWs) and cold waves (CWs) over the Indian main land have been studied. For this daily maximum temperature of 103 stations during the hot weather (AMJ (April–June)) season for the period 1961–2020 and daily minimum temperature data of 86 stations during the cold weather (DJF (December of previous year to current year February) season for the period 1971–2020 were used. The trends in the seasonal frequency of these extreme temperature events as well as their association with the ENSO events have also been examined. During the AMJ season, HWs are generally experienced over the north, northwest, central, east India, and north-east Peninsula [together called core HW zone (CHZ)] with the highest frequency during May. Noticeable increase (decrease) in the frequency and spatial coverage of HW/SHW days compared to their climatological values were observed during the El-Nino (La-Nina) years. There are significant increasing trends in the HW/SHW days at most of the stations from CHZ. The total number of HW/SHW days over CHZ showed noticeable increase during the recent decades 1991–2000, 2001–2010, and 2011–2020 compared to the previous three decades. During DJF, the CWs are generally experienced in the core CW zone (CCZ) that is nearly the same as CHZ with the highest frequency during January. Noticeable decrease (increase) in the frequency and spatial coverage of CW/SCW days were observed during the El-Nino (La-Nina) years compared to their climatological values. The total number of CW/SCW days over CCZ decreased during the recent decades, namely, 1991–2000, 2001–2010, and 2011–2020 as compared to the previous two decades. However, the areas of CW days of ≥8 days on an average increased over north along the plains of Himalayas and central India in the latest decade (2011–2020) as compared to the previous decade (2001–2010). This study coincided with the fact that the latest three decades were the warmest decades for the country as well as for the globe. Associated with intense and persistent extreme temperature events, large human mortality was

S. Nair (✉) · Mahima · M. Sanghwahia · D. S. Pai
India Meteorological Department, Pune, India
e-mail: smitha.nair@imd.gov.in

V. K. Gahalaut and M. Rajeevan (eds.), *Social and Economic Impact of Earth Sciences*,
https://doi.org/10.1007/978-981-19-6929-4_9

reported during some years of the study period. India Meteorological Department has recently established a seamless forecasting strategy from short-range forecasts to seasonal forecasts for HWs and CWs. With this new development, IMD is able to forewarn the agencies and people and save thousands of lives.

Keywords Heat waves · Cold waves · Trends · Climatology · El Nino/La Nina

1 Introduction

The earth is warming due to natural and anthropogenic factors which in turn has impacted Earth's climate in many ways. As per the IPCC (2019) special report on "Climate Change and Land", the average temperature over land for the period 2006–2015 was 1.53 °C higher than for the period 1850–1900. Since the pre-industrial period, the land surface air temperature has increased nearly twice as much as the global average temperatures. Rupa Kumar et al. (1994) have pointed out that the mean temperature trends over India were similar to the global and hemispheric trends. Kothawale and Roop Kumar (2005) concluded that the all-India mean annual temperature for the period 1971–2003 showed significant warming trend of 0.22 °C/10 yr. The 2020 annual mean land surface air temperature for the country was +0.29 °C above the 1981–2010 period average, thus making the year 2020 as the eighth warmest year on record since 1901 (IMD, Annual Climate Statement 2020). The data of recent 50 years (1971–2020) have shown that the increase in minimum temperatures is slightly more than that in the maximum temperature which is in tune with the earlier studies (Attri and Tyagi 2010; Rathore et al. 2013). The changes in mean values of temperatures cause increase in extreme temperatures events (Trenberth et al. 2007). These abnormal temperature events have high impacts on various sectors like health, agriculture, and economy (Meehl and Tebaldi 2004; Coumou and Rahmstorf 2012). On the health front, extreme heat can lead to dehydration, heat cramps, and heat strokes while extreme cold temperatures can lead to hypothermia, incoherence, and frostbite.

India experience HWs during March to July and CWs during November to March (Bedekar et al. 1974) with peaks during the middle 3 months; hot weather season of April to June (AMJ) and cold weather season of December of previous year to current year February (DJF), respectively. Due to significant impact of HWs and CWs on the human health and observed changes in their frequency, intensity, and persistency over various parts of the globe, there have been many studies on these extreme temperature events over India and their impact on human mortality (Raghavan 1966, 1967; Natarajan 1964; Bedekar et al. 1974; Subbaramayya and Surya Rao 1976; Chaudhury et al. 2000; De 2001; De et al. 2005; Pai et al. 2004, 2013; Rohini et al. 2016; Ratnam et al. 2016a). However, each of these studies used different threshold temperature values and data sets to describe the HWs and CWs.

Here, we describe various characteristics such as frequency, persistency, and spatial coverage of HWs and CWs in the country and the associated monthly and decadal variations. In addition, trends in the frequencies of these extreme temperature events and changes associated with the two ENSO phases (El Nino and La Nina) have also been highlighted.

2 Definition of HW and CW

There is no universal definition for HW (CW). However, an interval of hotter (colder) than normal weather conditions by a certain threshold value is generally used to describe HW (CW). The criteria (IMD 2002) used by the India Meteorological Department (IMD) for defining the HW/CW and Severe HW (SHW)/Severe CW(SCW) conditions over India since 2002 are given in Table 1. It is seen that the HW(CW) conditions signify increase (decrease) in the daily maximum (minimum) temperature at a station by a certain threshold compared to a climatological (normal) value. The climatological values were computed for the base period of 1971–2000. HW (CW) of relatively higher intensity is classified as severe HW (severe CW) or SHW (SCW).

Table 1 Criteria used in this study for defining heat and cold waves

Criteria for declaring heat wave based on maximum temperature (Tmax)
Heat wave over a station is declared only when the actual Tmax of the station is ≥40 °C for plains and ≥30 °C for hilly regions. However, when the Tmax is ≥40 °C for coastal stations and ≥45 °C for other stations, conditions are declared as heat wave
The following criteria are used for defining severity of the heat wave: When normal Tmax ≤ 40 °C and i. if (actual Tmax – normal Tmax) is 5–6 °C: heat wave ii. if (actual Tmax – normal Tmax) is ≥7 °C: severe heat wave When normal Tmax ≥ 40 °C and i. if (actual Tmax – normal Tmax) is 4–5 °C: heat wave ii. if (actual Tmax – normal Tmax) is ≥6 °C: severe heat wave
Criteria for declaring cold wave based on minimum temperature (Tmin)
When the Tmin is ≤10 °C for coastal stations, conditions are declared as cold wave The following criteria are used for defining severity of the cold wave over all the stations: When normal Tmin ≥ 10 °C and i. if (actual Tmin – normal Tmin) is –5 to –6 °C: cold wave ii. if (actual Tmin – normal Tmin) is ≤–7 °C: severe cold wave When normal Tmin ≤ 10 °C and i. if (actual Tmin – normal Tmin) is –4 to –5 °C: cold wave ii. if (actual Tmin – normal Tmin) is ≤–6 °C: severe cold wave

3 Data and Methodology

The daily maximum and minimum temperature data from the archives of the Climate Services Division of IMD, Pune have been used to derive the HW (CW)/SHW (SCW) information. Observations from 103 meteorological stations over the period 1961–2020 was used to derive the HW/SHW information and 95 out of the 103 stations had observations for at least 90% of the total days. Similarly, observations from 86 stations for the period 1971–2020 were used to compute the CW/SCW information. Out of these 86 stations, 80 had minimum temperature observations available for at least 90% of the total days. It has been found that approximately 99% of temperature values fall within the range of means ± three Standard Deviation (SD). Any outliers (i.e., values beyond three SD from the mean) or abnormally large difference in the maximum/minimum temperatures between two consecutive days in the data set were scrutinized by taking current weather into consideration and were accordingly assimilated into the data. The daily normal of maximum and minimum temperatures for the stations under study was calculated. Data from the hot (AMJ)/cold (DJF) seasons between 1971 and 2000 was used to perform this calculation. The anomalies in conjunction with the criteria in Table 1 were used to identify the days satisfying HW/CW conditions for each station. The data from the entire period of study, viz., 1961–2020 for AMJ and 1971–2020 for DJF have been used to prepare the climatological maps. To examine the link between ENSO and spatial distribution of the HWs/CWs over India, composite spatial maps of mean number of HW/CW (including SHW/SCW) days during the hot/cold weather season were prepared for El-Nino and La-Nina years.

For defining El-Nino/La-Nina events, criteria of Climate Prediction Centre (CPC), USA was used which is based on the Oceanic Nino Index (ONI). ONI is computed as the ERSST.v5 Sea surface temperature (SST) anomalies averaged over the Nino 3.4 region (5° N–5° S, 120° W–170° W). Means of ONI computed over running 3-month periods were used with a threshold of ±0.5 °C for defining El-Nino/La-Nina events. The events are declared when the threshold is met for a minimum of five consecutive overlapping in 3-month periods. In this study, for HW (CW) case, a reference year was classified as an El-Nino/La-Nina year, when any of the three 3-month seasons of March–May, April–June, and May–July (November of previous year to January of reference year, December of previous year to February of reference year, and January to March of reference year) is part of the five consecutive overlapping seasons. As per this definition, for HWs during the period 1961–2020, 17 years (1963, 1965, 1966, 1969, 1972, 1982, 1983, 1987, 1991, 1992, 1997, 1998, 2002, 2010, 2015, 2016, and 2019) satisfied the El-Nino criteria and 13 years (1964, 1971, 1973, 1974, 1975, 1976, 1985, 1988, 1989, 1999, 2000, 2008, and 2011) satisfied the La-Nina criteria. Similarly, for the CWs case, during 1971–2020, 18 years (1973, 1977, 1978, 1980, 1983, 1987, 1988, 1992, 1995, 1998, 2003, 2005, 2007, 2010, 2015, 2016, 2019, and 2020) satisfied El-Nino criteria and 19 years (1971, 1972, 1974, 1975, 1976, 1984, 1985, 1989, 1996, 1999, 2000, 2001, 2006, 2008, 2009, 2011, 2012,

2017, and 2018) satisfied the La-Nina criteria. Differences in the spatial distribution of HWs/CWs associated with the El-Nino and La-Nina events were examined by preparing composite spatial maps of frequency of HWs/CWs for the respective El-Nino/La-Nina years. The nonparametric Mann–Kendall test (Gilbert 1987) was employed to test the monotonic increasing or decreasing trends in the frequency and spatial coverage of the HWs/CWs. The Mann–Kendall test assumes the trends to be monotonic and thus no seasonal or other cycle is present in the data.

4 Heat Wave (HW)/Severe Heat Wave SHW Days

4.1 Climatology

Frequency of the HW/SHW Days

The season-wise spatial variation of mean number (frequency) of HW and SHW days experienced over India expressed as days per season during the hot weather season (April–June) for the period 1961–2020 is shown in Fig. 1a and b, respectively. It is seen in Fig. 1a that most areas of the country except northeast India and some parts of south Peninsula have experienced on an average $\geq$2 HW days per season. The Core HW Zone (CHZ) covers the states of Punjab, Himachal Pradesh, Uttarakhand, Delhi, Haryana, Rajasthan, Uttar Pradesh, Gujarat, Madhya Pradesh, Chhattisgarh, Bihar, Jharkhand, West Bengal, Orissa and Telangana and Marathawada, Madhya Maharashtra, and coastal Andhra Pradesh. SHWs of 1–3 days are mainly experienced over isolated areas of north and eastern parts of the CHZ (Fig. 1b).

Figure 2a–c shows monthly distribution of average HW days experienced during the period 1961–2020 for the 3 months of April–June, respectively. Among the 3 months of the season, average number of HW days experienced was relatively more during the month of May. The monthly distribution of average SHW days showed that about 1–2 days are experienced during May and June. The decade-wise spatial distribution of mean HW (Fig. 3a–f) and SHW days for the AMJ season shows that there was an increase in the average number of HW and SHW days over the CHZ during the recent three decades as compared to the previous three decades.

The observed spatial distribution of average HW/SHW days can be associated to annual path of the sun and difference in the distribution of moisture over the country due to the presence of seas on both sides of the Indian Peninsula and arrival and advance of monsoon over the country from June onward. During May and June, the Sun is located over central and north India and these regions also experience clear skies and dry weather. In April, thermal convective activity starts over the southern parts of the Peninsula and northeast India due to incursion of the moisture from the neighboring seas which reduces heating over these regions. By June, with the arrival of monsoon over the Peninsula, the moisture level increases further. As a result, the HW/SHW events are not frequently experienced over Peninsula except over northern parts of the east coast where intense prolonged HW/SHW events occur occasionally

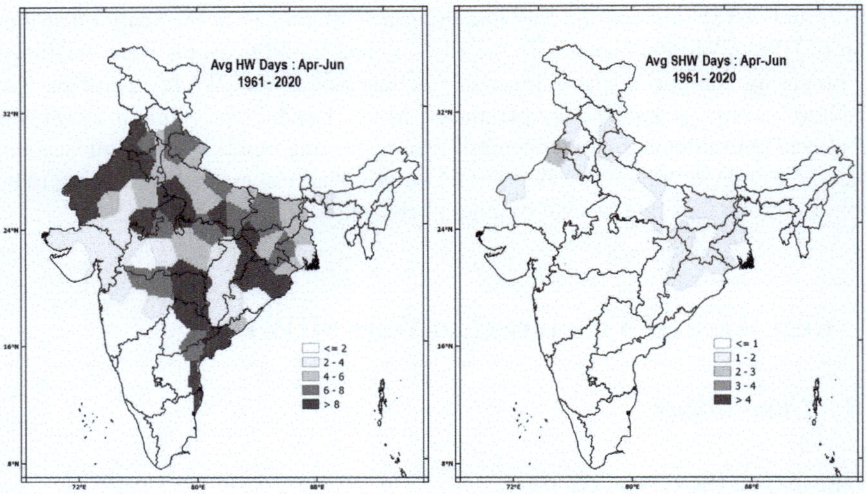

Fig. 1 **a** Average number of HW days over India during the hot weather season (AMJ) (1961–2020) **b** Same as (**a**) but for SHW days

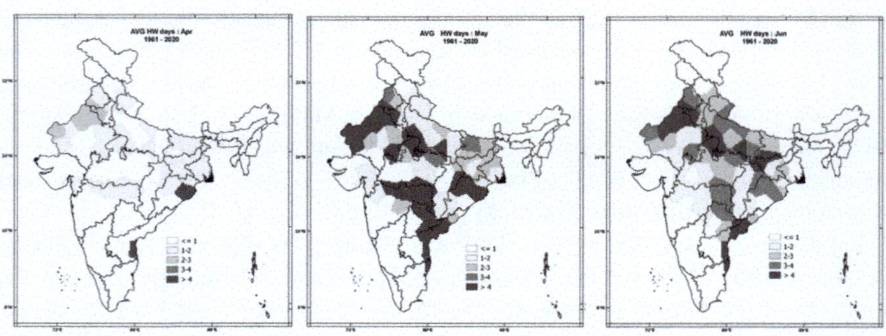

Fig. 2 Monthly climatology maps of average number of HW days for the 3 months **a** April, **b** May, and **c** June over India for the period 1961–2020

due to synoptic situations that oppose the sea breeze from the Bay of Bengal. In years (like 1998, 2002, 2012) when monsoon advance over north India was delayed, prolonged HW conditions were reported over many areas of north and northwest India.

Persistency of HW/SHW Days

The duration and intensity of HW events determine the amount of thermal stress on the human body. It is observed that most of the stations experienced HW spells of only 1–2 days frequently. This is expected as the HW/SHW conditions are very high-frequency temperature extreme events. However, a few spells of HW lasted for very long period, i.e., more than 10–15 days continuous and few spells of SHW lasted

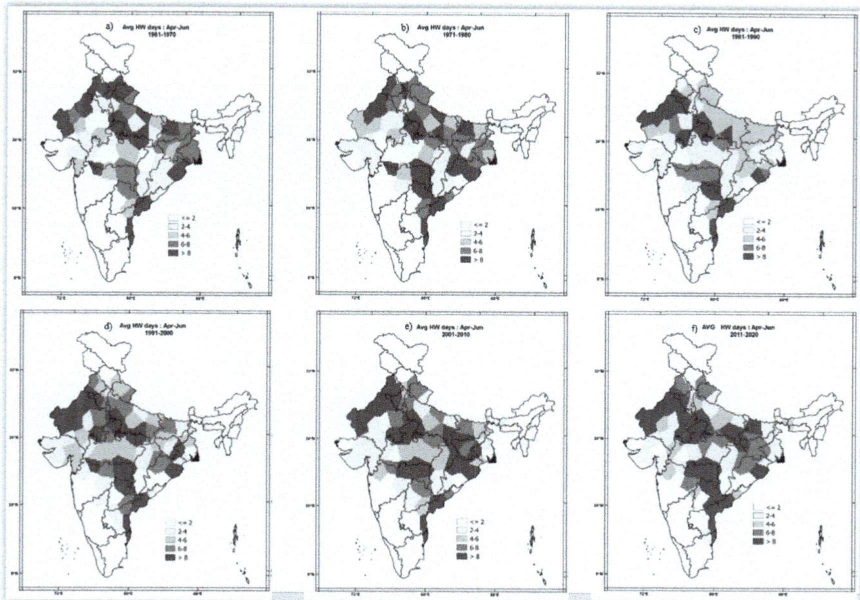

Fig. 3 Mean number of HW days over India during the hot weather season (AMJ) for the six decades, viz., **a** 1961–70, **b** 1971–80, **c** 1981–90, **d** 1991–2000, **e** 2001–2010, **f** 2011–2020

for more than 5 days continuous over some stations. There has been an increase in the persistency of HW conditions during the recent decades.

4.2 HW Days and ENSO

The composite spatial maps of mean HW days over India for the El-Nino and La-Nina cases are given in Fig. 4a, b. The composite map for El Nino shows higher frequency of HW days as compared to the climatology map (Fig. 1a) with large areas of CHZ experiencing ≥8 HW days. However, La-Nina case shows that most of the areas have experienced significantly less number of HW days than the climatology (Fig. 1b). Thus, areas of 8 HW days or more on an average were maximum during El-Nino case and minimum during the La-Nina case. The mean SHW days over India for the El-Nino and La-Nina cases (figures not given) also showed the same result. This is in tune with the finding of Revadekar et al. (2009) that temperatures over the Indian region are higher during El-Nino years than the La-Nina years.

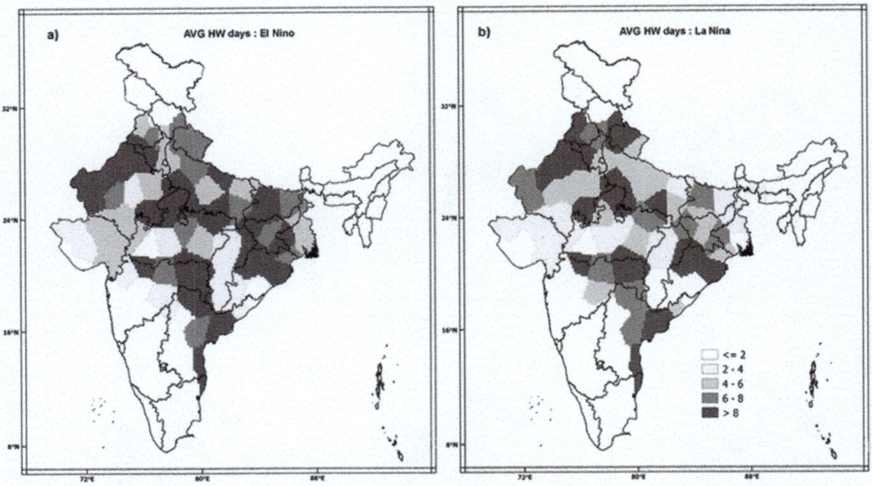

Fig. 4 Composite of average number of HW days during the AMJ season for **a** 17 El-Nino years, **b** 13 La-Nina years

4.3 *Trends in the Station-Wise Seasonal HW Days*

In Fig. 5, red rising arrows (blue falling arrows) are used to represent stations with increasing (decreasing) trend in the HW days with filled arrows showing trends significant at 95% confidence level. About 19 stations from north, northwest, east coast, and central India and 4 stations from peninsular India showed significant increasing trends. These are also the stations that experience highest HWs during the season (Fig. 1a). However, two stations from the east coast (Kolkata and Gopalpur) and two stations from North India (Ambala, Barelley, Bhariach, and Gorakpur) showed significant decreasing trends. In case of SHW (figure not shown), significant increasing trends were observed at five stations from northwest India (Amritsar, Hissar, Ganganagar, Phalodi, and Jaipur). However, no stations showed significant decreasing trend.

5 Cold Wave (CW)/Severe Cold Wave SCW Days

5.1 *Climatology*

Frequency of the CW/SCW Days

Figure 6a shows the spatial variation of seasonal climatology of CW days experienced over the country expressed as average CW days per season. Figure 6b shows the same but for SCWs. The Core CW Zone (CCZ) is almost exactly the same as the CHZ with

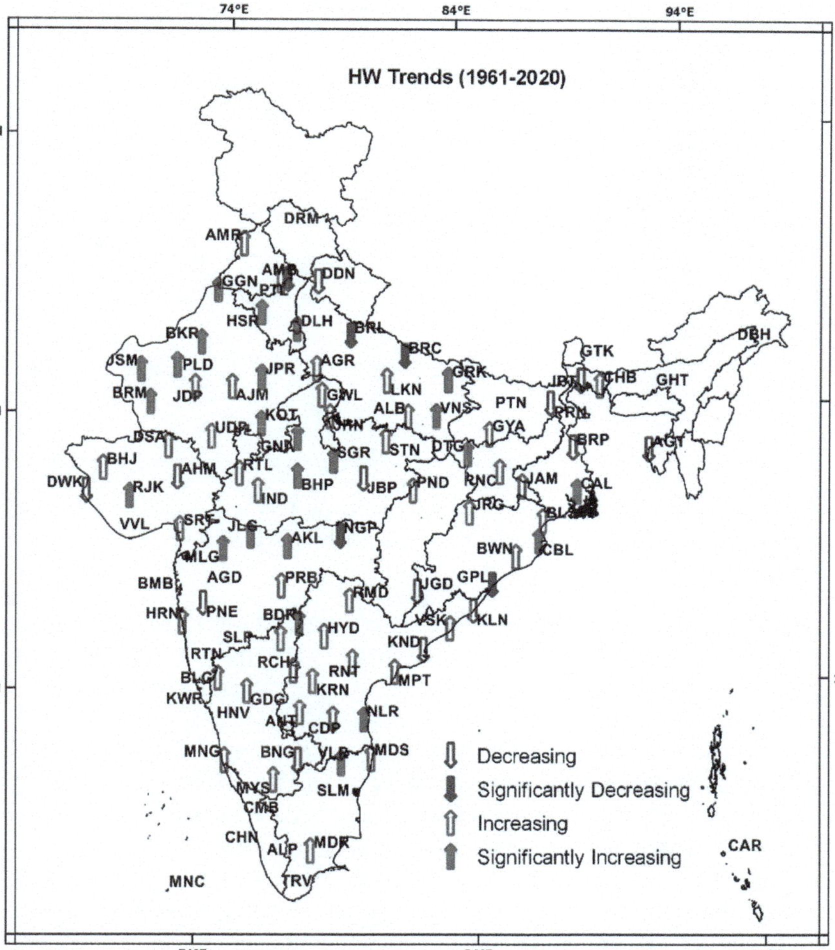

Fig. 5 Trends in HW days of 103 stations during the AMJ season for the period 1961–2020 nonparametric Mann–Kendall test were used to test the significance of the trends. Red rising (blue falling) arrows represent the increasing (decreasing) trends. Filled arrows represent the trends significant at 5% level

the inclusion of Jammu and Kashmir and the exclusion of coastal Andhra Pradesh. It is shown in Fig. 6a that most areas of the country except northeast India and some parts of southern Peninsula have experienced on an average ≥2 CW days per season. Most part of north and northwest of CCZ experienced ≥6 CW days per season. The average SCWs of 1–4 days were mainly experienced over isolated areas of northwestern CCZ (Fig. 6b).

Month-wise distribution of CW days (Fig. 7a–c) during December shows 2 CW days or more were mainly experienced by most parts of Rajasthan and some parts of west J and K, west Madhya Pradesh, Punjab, and Gujarat. Among the 3 months

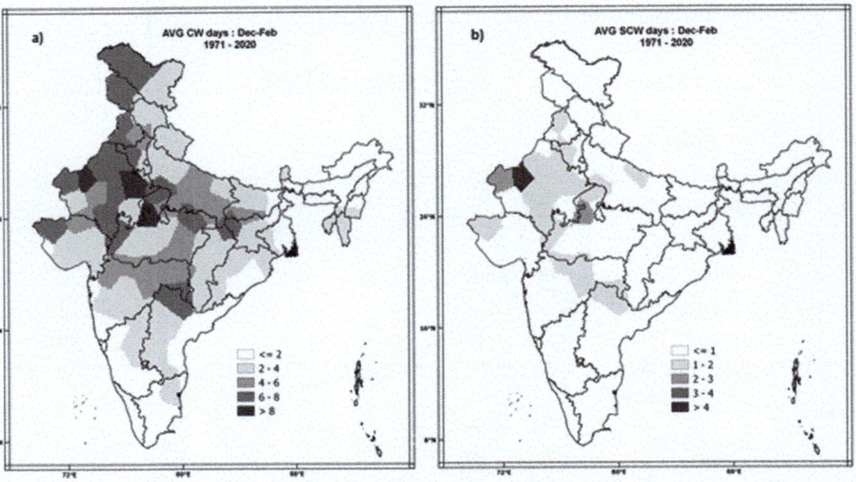

Fig. 6 **a** Average number of CW days over India during the DJF season computed using the CW information for the period of 1971–2020. **b** Same as (**a**), but for SCW days

of the season, average number of CW days experienced was relatively more during the month of January as compared to December and February months. Month-wise, SCW days were experienced mainly during January and February (about 1–2 days on an average and mostly over northwest India).

The decade-wise spatial distribution of season average of CW days (Fig. 8a–e) shows overall decrease in the average CW days over the country going from 1971–1980 to 1991–2000 with systematic and noticeable decrease over north, northwest, and northeast India and slight increase over some southern parts of Central India. However, during the recent decades (2001–2010) and (2011–2020) there is slight increase in the spatial coverage and frequency of CW days compared to the previous decade (1991–2000). It is also noticed that between the recent two decades, namely,

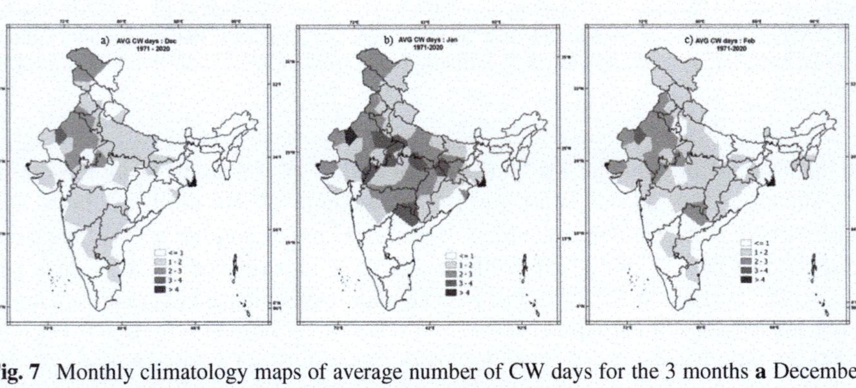

Fig. 7 Monthly climatology maps of average number of CW days for the 3 months **a** December, **b** January, and **c** February over India for the period 1971–2020

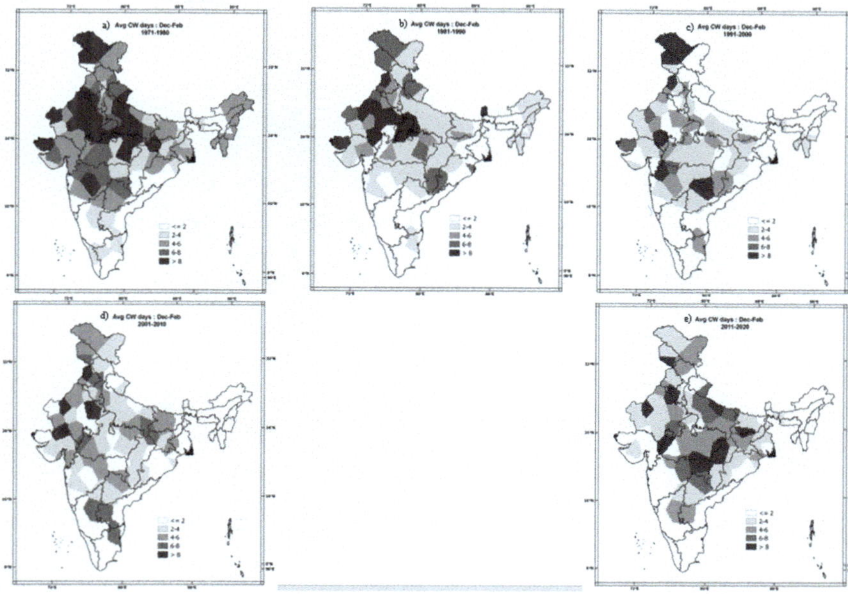

Fig. 8 Mean number of CW days over India during the cold weather season (DJF) for the five decades, viz., **a** 1971–80, **b** 1981–90, **c** 1991–2000, **d** 2001–2010, **e** 2011–2020

2001–2010 and 2011–2020 the frequency of ≥6 CW days has increased over parts of north along the plains of Himalayas, east and east-central India.

During the winter months (DJF), there is incursion of cold winds into northwest and central India as a result of passage of western disturbances (WDs) over the region. WDs are transient disturbances in the mid-latitude westerlies and are followed by occurrence of cold waves mostly over the areas north of 20° N and rarely in areas south of this latitude.

Persistency of CW/SCW Conditions

The duration and intensity of CW events determine the amount of thermal stress on the human body. The duration of the most frequent CW/SCW spells is about 1–2 days in most of the stations though some individual CW/SCW spells lasted for more than 10/5 days continuous in some stations.

5.2 *CW Days and ENSO*

The composite spatial maps of mean CW and SCW days over India for the El- Nino and La Nina are given in Fig. 9a, b. It can be observed that the areas of 8 CW days or more on an average were maximum during La-Nina years and minimum during the El-Nino years. In the case of SCW also [figures not given], increase (decrease) in the

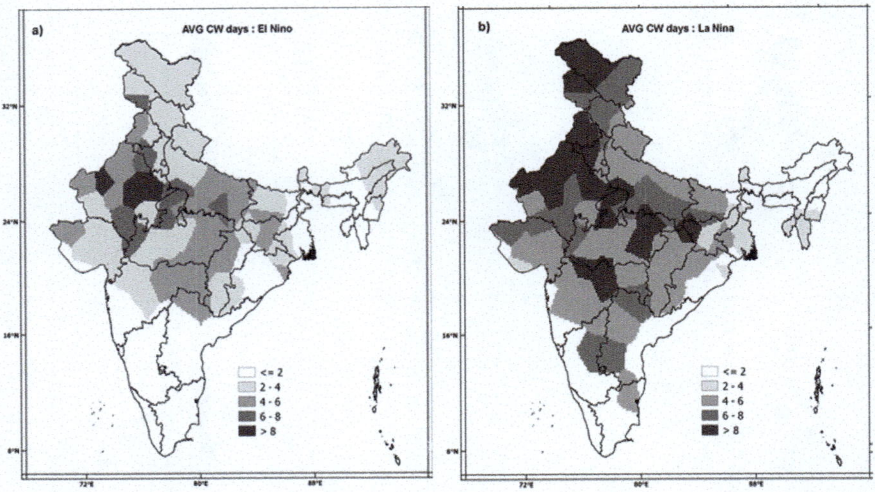

Fig. 9 Composite of average number of CW days during the DJF season for **a** 18 El-Nino years, **b** 19 La-Nina years

frequency of the SCW days was observed over central parts of the country during La-Nina (El-Nino) case compared to the climatology map. This is in tune with the study by Ratnam et al. (2016a, b) that the cold waves during the La-Nina years are more dominant ones as compared to those formed during the El-Nino years.

5.3 *Trends in the Station-Wise Seasonal CW/SCW Days*

In Fig. 10, red rising arrows (blue falling arrows) are used to represent stations with increasing (decreasing) trend in the CW days with filled arrows showing trends significant at the 95% confidence level. Decreasing trends were observed at most of the stations (Fig. 10) north of about 18°N with significant trends observed in 28 stations. On the other hand, three stations from Central India, one from north India and two from east, and one from south peninsula (Pendra, Ratlam, Nagpur, Gorakhpur, Balasore, Ranchi, and Ananthapur) showed significant increasing trends. In case of SCW (figure not shown), decreasing trends were observed at many stations of CCZ with significant trends observed at nine stations from northwestern part and four stations from central part of CCZ. The decreasing trends in the CW/SCW days were also corroborated by the increasing trends (figure not shown) in the season averaged (DJF) minimum temperature anomalies. This is in tune with the findings of Kothawale and Rupa Kumar (2005) that the recent accelerated warming over India is manifested equally in day-time and night-time temperatures.

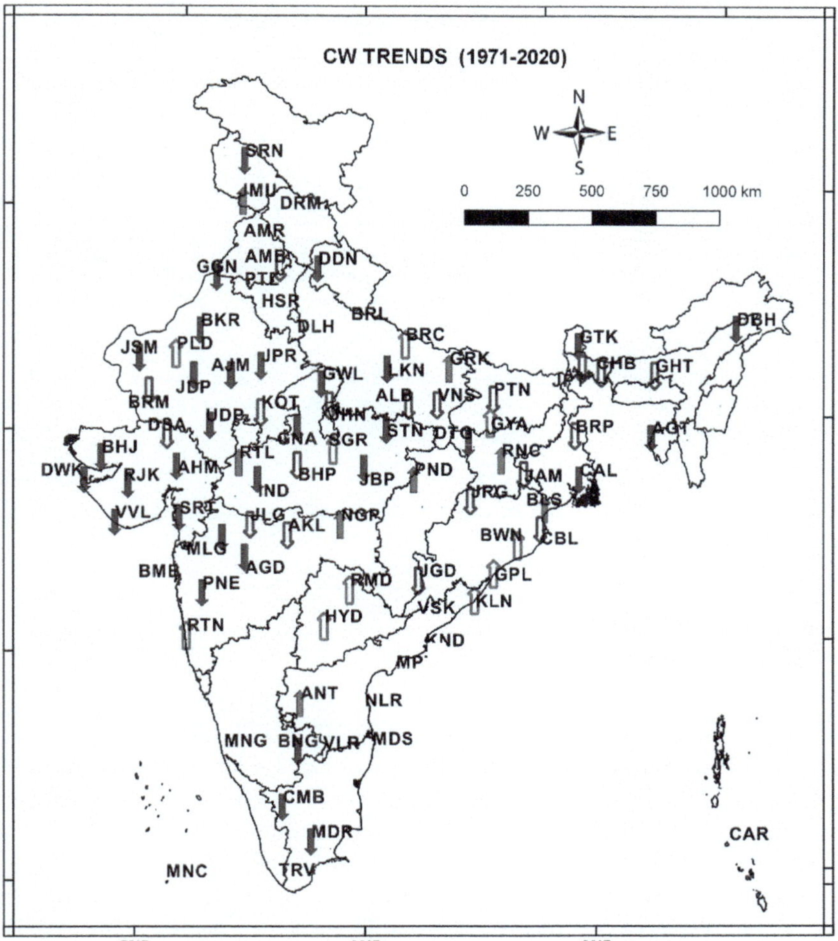

Fig. 10 Trends in CW days of 86 stations during the DJF season for the period 1971–2020 nonparametric Mann–Kendall test were used to test the significance of the trends. Red rising (blue falling) arrows represent the increasing (decreasing) trends. Filled arrows represent the trends significant at 5% level

6 Human Mortality Associated with the Extreme Temperature Events

Figure 11 shows the loss of human lives due to hot/cold waves in India during the period 1971–2020. Highest number (2081) of deaths related to HWs was reported in 2015 followed by 1998. Highest number (1061) of deaths related to CWs was reported in 1985 followed by 2003. As seen in this figure, no noticeable trend is visible in the time series of both HW and CW deaths. However, the decade (2011–2020) registered the highest number of deaths due to heat wave events and the decade (1981–1990)

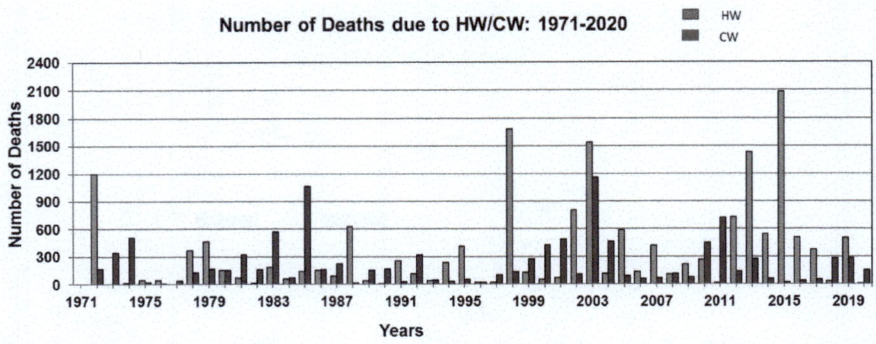

Fig. 11 Number of deaths in India annually due to HW/CWs for the period 1971–2020 obtained from media reports and IMD's annual disaster weather reports. The red (blue) bars correspond to deaths due to HW/SHWs (CW/SCWs)

registered the highest number of deaths due to cold wave events compared to the other four decades.

7 Conclusions

From the examination of various features of the HW (SHW) during the study period (1961–2020) from 103 stations uniformly distributed over the country and that of CW (SCW) during the study period (1971–2020) from 86 stations uniformly distributed over the country, we can make the following conclusions.

During the hot weather season (AMJ), many stations from north, northwest, central, east, and northeast Peninsula India (together called CHZ) experienced highest number of HW/SHW with relatively highest frequency during May. The islands are not affected by heat waves. An increase in the frequency, spatial coverage, and persistency of the HW/SHW days was observed during recent three decades (1991–2000, 2001–2010, 2011–2020) compared to previous three decades. On the other hand, during the cold weather season (DJF), stations from CCZ that are nearly same as CHZ but include Jammu and Kashmir and Ladakh and exclude coastal Andhra Pradesh are more prone to CW/SCW with relatively highest frequency during January. It is noticed that in the recent decades (2001–2010) and the decade 2011–2020, there is a slight increase in the spatial coverage and frequency of CW days compared to the previous decade (1991–2000). It is also noticed that in the recent two decades, namely, 2001–2010 and 2011–2020 the frequency of $\geq$6 CW days has increased over parts of north along the plains of Himalayas, east, and east-central India.

An association between the occurrence of the extreme temperature events (HW/SHW and CW/SCW) and ENSO phases was observed. Noticeable increase (decrease) in the frequency and spatial coverage of HW/SHW days compared to their climatological values were observed during the El-Nino (La-Nina) years over the regions that generally experience these extreme events. Similarly, noticeable decrease (increase) in the frequency and spatial coverage of CW/SCW days compared to their climatological values were observed during the El-Nino (La-Nina) years over the regions that generally experience these extreme events.

Significant increasing trends were observed in the frequency and spatial coverage of the HW/SHW days over CHZ. This was caused by the significant increasing trends in the HW/SHW days of majority of the stations from the region which in turn was caused by the significant increasing trends in the season averaged maximum temperatures of these stations. Decreasing trends are observed in the frequency and spatial coverage of the CW/SCW days over many stations in CCZ. This in turn was caused by the significant increasing trends in the season averaged minimum temperatures of these stations.

For issuing heat/cold wave forecasts, inputs such as surface and upper air charts, change chart, T-phigram, radar and satellite products, and guidance from various global and regional Numerical Weather Prediction (NWP) models are used. The models used by IMD include IMD Global Forecast System (GFS), NCEP GFS, NCMRWF Unified Model (NCUM), Global Ensemble Forecast System (GEFS), NCMRWF Ensemble Prediction System (NEPS), ECMW, Regional Weather Research and Forecasting (WRF) Model of IMD and NCMRWF regional (NCUAR) model. Real-time extended range predictions are made using the IMD GFS model by the India Meteorological Department (Mandal et al. 2019).

At present, the heat/cold wave bulletins (color coded and impact based) in text as well as graphic format for next 5 days are updated four times in a day (based on 05:30, 08:30, 14:30, and 17:30 h IST) in All India Weather Forecast Bulletin (https://mausam.imd.gov.in/imd_latest/contents/all_india_forcast_bulletin.php) by NWFC, IMD, New Delhi. Heat and Cold Wave Monitoring and Warning Services 165 in the morning and evening (at 08:00 and 16:00 h IST) special heat wave guidance bulletins (http://internal.imd.gov.in/pages/heatwave_mausam.php) are also issued by NWFC. The district-wise heat/cold wave warnings are issued by Meteorological Centers/Regional Meteorological Centers of IMD. Thereafter, these warnings are shared with concerned State Government Authority, National Disaster Management Authority, Media, and other stakeholders like Indian Railway, Health departments, Power Sector, etc. In addition, regional centers also provide temperature forecast for all the major cities of respective state as per the state disaster management requirement. In adding to above, warnings are also communicated by all digital modes of communications like over phone (to senior disaster managers), WhatsApp, SMS, e-mail, Facebook, Twitter, Instagram, and weekly weather in the form of audio–video model and uploaded in YouTube (https://mausam.imd.gov.in/imd_latest/contents/pdf/forecasting_sop.pdf).

References

Annual Climate Statement-2020 (2021) https://imdpune.gov.in/Statement%20of%20Climate%20of%20India-2020.pdf

Attri SD, Tyagi A (2010) Climate profile of India. Met. Monograph, Environment Meteorology No. 01/2010, India

Bedekar VC, Dekate MV, Banerajee AK (1974) Heat and cold waves in India forecasting manual-Part IV-6. India Meteorological Department

Chaudhury SK, Gore JM, Sinha Ray KC (2000) Impact of heat waves over India. Curr Sci 79(2):153–155

Coumou D, Rahmstorf S (2012) A decade of weather extremes. Nat Clim Change 2:491–496. https://doi.org/10.1038/nclimate1452

De US (2001) Climate change impact: regional scenario. Mausam 52:201–212

De US, Dube RK, Prakasa Rao GS (2005) Extreme weather events over India in the last 100 years. J Ind Geophys Union 9(3):173–187

Gilbert RO (1987) Statistical methods for environmental pollution monitoring. Van Nostrand Reinhold, New York

IMD (2002) Recommendation regarding the revised criteria for declaring heat wave/cold wave, DDGM (WF). UOI. No. W-969/1304 to 1365 dated February 2002, India Meteorological Department

IPCC (2019) Contributors to the IPCC special report on climate change and land. In: Shukla PR, Skea J, Calvo Buendia E, Masson-Delmotte V, Pörtner H-O, Roberts DC, Zhai P, Slade R, Connors S, van Diemen R, Ferrat M, Haughey E, Luz S, Neogi S, Pathak M, Petzold J, Portugal Pereira J, Vyas P, Huntley E, Kissick K, Belkacemi M, Malley J (eds) Climate change and land: an IPCC special report on climate change, desertification, land degradation, sustainable land management, food security, and greenhouse gas fluxes in terrestrial ecosystems

Kothawale DR, Rupa Kumar K (2005) On the recent changes in surface temperature trends over India. Geophys Res Lett 32. https://doi.org/10.1029/2005GL023528

Mandal R, Joseph S, Sahai AK, Phani R, Dey A, Chattopadhyay R, Pattanaik DR (2019) Real time extended range prediction of heat waves over India. Sci Rep 9(1):1–11

Meehl G, Tebaldi S (2004) More intense, more frequent, and longer lasting heat waves in the 21st century. Science 305:994–997. https://doi.org/10.1126/science.1098704

Natarajan KK (1964) A note on the hot days of Madras (1875–1963). Indian J Meteorol Geophys 14:431–436

Pai DS, Thapliyal V, Kokate PD (2004) Decadal variation in the heat and cold waves over India during 1971–2000. Mausam 55(2):281–292

Pai DS, Nair SA, Ramanathan AN (2013) Long term climatology and trends of heat waves over India during the recent 50 years (1961–2010). Mausam 64(4):585–604

Raghavan K (1966) A climatological study of severe heat waves in India. Indian J Meteorol Geophys 17:581–588

Raghavan K (1967) Climatology of severe cold waves in India. Indian J Meteorol Geophys 18:191–196

Rathore LS, Attri SD, Jaswal AK (2013) State level climate change trends in India. Meteorological Monograph, No. ESSO/IMD/EMRC/02/2013

Ratnam JV, Behera SK, Ratna SB, Rajeevan M, Yamagata T (2016a) Anatomy of Indian heatwaves. Sci Rep 6(1):1–11

Ratnam JV et al (2016b) ENSO's far reaching connection to Indian cold waves. Sci Rep 6:37657. https://doi.org/10.1038/srep37657

Revadekar JV, Kothawale DR, Rupa Kumar K (2009) Role of El Nino/La Nina in temperature extremes over India. Int J Climatol 29:2121–2129

Rohini P et al (2016) On the variability and increasing trends of heat waves over India. Sci Rep 6:26153. https://doi.org/10.1038/srep26153

Rupa Kumar K, Krishankumar, Pant GB (1994) Diurnal asymmetry of surface temperature trends over India. Geophys Res Lett 21(8):677–680
Subbaramayya I, Surya Rao DA (1976) Heat wave and cold wave days in different states of India. Indian J Meteorol Hydrol Geophys 27:436–440
Trenberth KE et al (2007) Observations: surface and atmospheric climate change. In: Solomon S et al (eds) Climate change 2007: the physical science basis. Cambridge University Press, pp 235–336

Economic Impacts of Air Pollution and Fog in India and Prediction Efforts

Sachin D. Ghude, Sandip Nivdange, D. M. Chate, and N. R. Karmalkar

Abstract Increasing concentrations of air pollutant are directly affecting human health and crop production and shown to have negative impact on economic sector of India. Similarly, wintertime fog in India severely hinders the flight operations and causes significant financial losses to the aviation industry. It is therefore important to undertake a quantitative to the estimated losses to the economy due to both air pollution and fog in India. This chapter provides overview of the estimated losses due to air pollution impact on health and crop production in India and estimates for the economic loss to the aviation sector due to interruption to flight operation due to dense fog at Delhi International Airport. The present-day premature mortalities due to $PM_{2.5}$ and O_3 exposure caused economic cost of approximately 640 billion USD, which is a factor of 10 higher than total expenditure on health by public and private expenditures in India. Similarly, the damage to crop due to O_3 exposure caused economic cost of approximately 1.29 ± 0.47 billion USD2005 annually which is sufficient to feed approximately 94 million poor people living below poverty line in India under the provision of the National Food Security Ordinance. On the other hand, the wintertime fog in northern India hinders the flight operations led to a total economic cost of approximately 3.9 million USD (248 million Indian rupees) due to flights affected by heavy fog spells at Delhi International Airport over 5 years.

S. D. Ghude (✉)
Indian Institute of Tropical Meteorology, Ministry of Earth Sciences, Pune, India
e-mail: sachinghude@tropmet.res.in

S. Nivdange · N. R. Karmalkar
Savitribai Phule Pune University, Pune 411007, India

D. M. Chate
Centre for Development of Advanced Computing, Pune, India

V. K. Gahalaut and M. Rajeevan (eds.), *Social and Economic Impact of Earth Sciences*,
https://doi.org/10.1007/978-981-19-6929-4_10

1 Introduction

Anthropogenic emissions and air quality are inexorably linked from their emission sources to their impacts on human health, agriculture, and ecosystems. Anthropogenic emissions of pollutants into the atmosphere have caused extreme changes in the Earth system, altering air quality, climate, and nutrient flow in every ecosystem. Increasing economic growth and energy demand have led to a significant increase in emissions of air pollutants in India during the past few decades. Due to rapid urbanization and industrialization, India is experiencing environmental issues and related health effects. The health impacts concomitant with air pollution are not just a matter of environmental science research and epidemiology but have become an important social science issue in India. The onset of winter brings frequent harmful pollution over North India and causes cardiovascular and respiratory diseases. It also brings a thick fog and haze layer that clouds visibility and creates an unhealthy situation for public health, and hazardous situation for aviation, road transport, rail transport, economic losses to airlines, and thoroughly disrupts human activities (~450 million people) in the northern region of India. Although air pollution and fog are two separate phenomena, they negatively impact human activity and have economic consequences. Air pollution and fog in an urban environment can bring down productivity and prove costly by rising health costs, reducing crop yield, and impacting ecosystems and biodiversity.

2 Health Impact of Air Pollution

It is anticipated that the increasing concentrations of particulate matter and ozone will lead to substantial effects on health and the environment. The environmental risk caused by exposure to these pollutants has increased in many parts of the world, including India, due to human activity (Lelieveld et al. 2015). The number of premature mortalities due to exposure to outdoor air pollution in 2010 amounted to around 3 million people globally, while they are projected to be 6–9 million in 2060 considering a liner–nonlinear and linear concentration–response function, respectively. The number of premature mortalities is unequally distributed across the globe, with the highest number of deaths projected in China and India (OECD Policy Highlights). In a review of 178 research articles, Lu (2020) systematically examines the psychological (affective, cognitive, behavioral), economic, and social effects of air pollution beyond its physiological and environmental impact. This study found evidence that air pollution causes decrease in happiness and life satisfaction and increases annoyance, anxiety, mental disorders, self-harm, and suicide. Cognitively, it impairs functioning and decision-making. Socially, it exacerbates criminal behavior. And economically, it hurts work productivity. The market impacts of outdoor air pollution are projected to gradually rise from 0.3% in 2015 to 1% of global GDP by 2060. Costs related to additional health expenditures and labor productivity losses dominate in the long run. The annual global welfare (non-market) costs of premature motilities due to exposure to outdoor air pollution, calculated using estimates of the individual willingness-to-pay to reduce the risk of premature death, are projected to rise to USD 18–25 trillion in 2060. In addition, the costs of pain and suffering from illness are estimated at about USD 2.2 trillion by 2060. Pollution has a profound economic impact on gross domestic product (GDP) (OECD Policy Highlights). Pollution is responsible for 7% of annual healthcare spending in middle-income countries that are heavily polluted and rapidly developing. A World Bank study released in 2016 revealed that India lost more than 8.5% of its GDP in 2013 due to the cost of increased welfare and lost labor due to air pollution. At its current size of $2.6 trillion, the loss equals about $221 billion. A study by the Indian Institute of Technology, Bombay, found that air pollution cost of Mumbai and Delhi is $10.66 billion (approximately Rs. 70,000 crore) in 2015, or about 0.71% of the country's GDP. The costs to society due to air pollution (part of which is direct productivity loss) in the largest Indian cities are estimated as high as nearly one-tenth of the income generated from these cities from all economic activities. This indicates that India suffers from a disproportionally heavy health burden of urban air pollution by international comparison (Lvovsky 1998).

Poor air quality is an important societal issue, specifically for developing countries like India. Figure 1 shows the model simulated mean surface O_3 and $PM_{2.5}$ distribution over India on 36 km spatial resolution for the winter (December–February) and summer (March–May) season (Ghude et al. 2012). It can be seen that the levels of higher O_3 and $PM_{2.5}$ concentration vary strongly between regions and seasons. $PM_{2.5}$ levels are significantly high over Indo-Gangetic Region (IGP) during the winter and summer seasons. On the other hand, O_3 levels along the IGP are high during summer and low during winter. Higher $PM_{2.5}$ and O_3 levels over the IGP region during the summer season suggest the greatest exposure to outdoor air pollution to the population in this region compared to other parts of India. The rapid expansion of industrial, urban, and traffic emissions has significantly increased air pollution, especially over the last two decades (Ghude et al. 2008, 2009, 2013, 2016; Jena et al. 2015). Therefore, the potential health risk is higher for populations in these regions. As indicated by the 2011 census of India's statistics, around 32% (~0.4 billion people) of India's 1.2 billion population lives in urban areas. About 78% of the total 141 cities in the country exceed the $PM_{2.5}$ standard, 90 cities have critical levels, and 26 have the most critical levels, surpassing the PM standard by over three times (CPCB 2014). The theoretical minimum-risk concentrations for daily mean $PM_{2.5}$ exposure has a range of 5.8–8.0 μgm^{-3}, and for ozone, it is 36.6 μgm^{-3}. In India, as per the national ambient air quality standard, the safe limit for daily mean $PM_{2.5}$ exposure is 60 μgm^{-3}, and for ozone, the permissible limit is 100 μgm^{-3} for daytime 8-h average. A recent study by Ghude (2014) assesses the impact of only outdoor air pollution (Fig. 2) on premature mortalities linked to CEV, COPD, IHD, LC, and ALRI, and O_3-related mortalities by COPD in 2011, respectively, using a three-dimensional chemistry transport model. The estimated premature mortality is widespread in India. The Indo-Gangetic Plain (IGP) shows the highest estimated premature mortalities due to both ground-level $PM_{2.5}$, which is about 42% for $PM_{2.5}$ exposure and 45% for O_3 exposure compared to all India numbers.

The state with the highest premature mortalities due to $PM_{2.5}$ exposure is Uttar Pradesh, which accounts for about 15% (about 86,000 excess cases) of all premature mortalities in India during 2011, followed by Maharashtra (10%), West Bengal (9%), and Bihar (8%). Other states with high premature mortalities due to $PM_{2.5}$ are Andhra Pradesh, Tamil Nadu, Gujarat, Karnataka, Madhya Pradesh, Orrisa, and Rajasthan, which collectively make up for 32% of the countrywide premature mortalities. Estimated nationwide premature mortalities related to $PM_{2.5}$ exposure in 2011 were about 570,000 people. For adults (age >25 years), premature mortalities linked

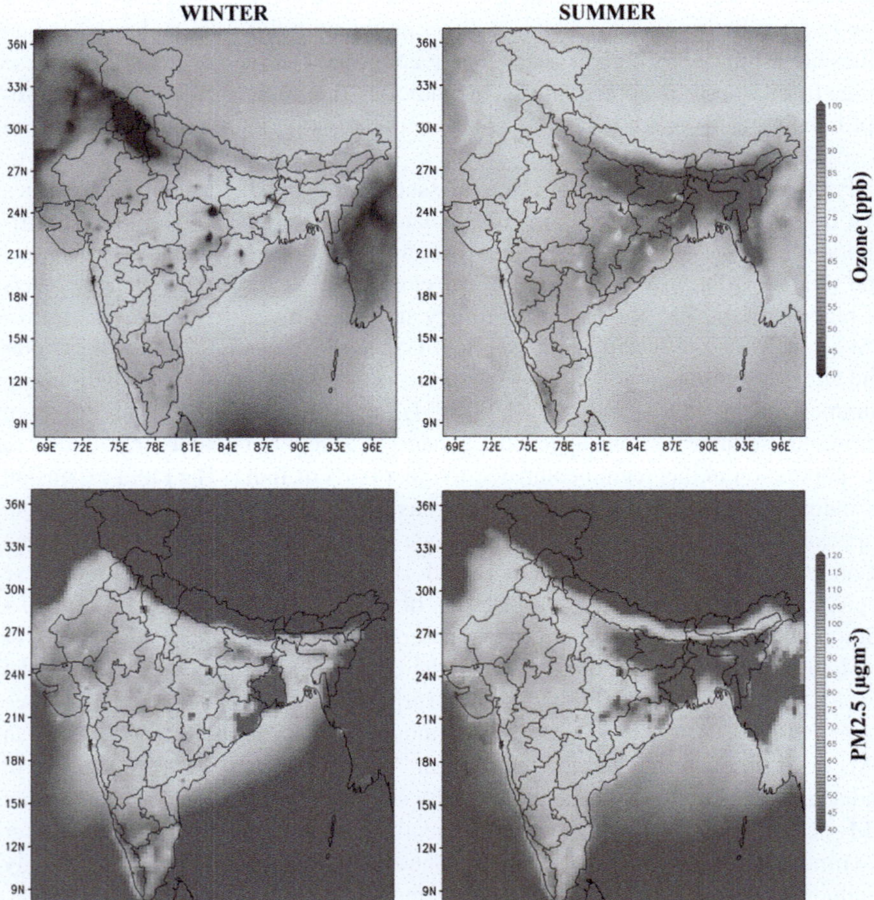

Fig. 1 Model simulated spatial distribution of **(top)** daytime 8-h mean surface ozone (in ppb) and **(bottom)** 24-h mean $PM_{2.5}$ (in $\mu g/m^3$) concentration during summer and winter season over India in 2011

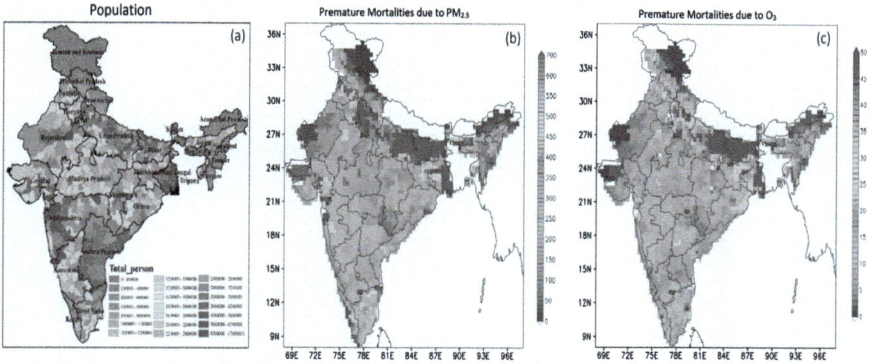

Fig. 2 Distribution of **a** district-wise India's population and total premature mortalities due to **b** $PM_{2.5}$ and **c** O_3 exposure in 2011 (Unit: premature mortalities per grid box)

to IHD, CEV (stroke), COPD, and LC are estimated to be about 250,000, 190,000, 120,000, and 2700 people, respectively. The estimated health burden in terms of life years lost due to $PM_{2.5}$ exposure was about 3.4 ± 1.1 years for India. Regionally, the lowest lost life expectancy is noticed in Jammu and Kashmir (~0.6 ± 0.2 years), followed by Himachal Pradesh and Sikkim (~1.2 ± 0.4 years). Delhi suffered the greatest loss of life expectancy of about 6.3 ± 2.0 years, followed by West Bengal (6.1 ± 1.9 years) and Bihar (5.7 ± 1.9 years), which are also the top three polluted states in India. The economic cost of estimated premature mortalities associated with $PM_{2.5}$ and O_3 exposure was about 640 (350–800) billion USD in 2011, a factor of 10 higher than total expenditure on health by public and private expenditure. The state-wise economic cost due to premature mortality of $PM_{2.5}$ and O_3 exposure is largest in Uttar Pradesh (98 billion USD), followed by Maharashtra (62 billion USD), West Bengal (57 billion USD), and Bihar (53 billion USD) (Fig. 3).

However, it should be noted that the estimated mortalities are subject to several sources of uncertainty including uncertainties in the model parameterizations (e.g., emissions and physical/chemical parameterizations) and uncertainties in the exposure response functions. Ghude et al. (2016) estimated overall uncertainty of 21% in premature mortality due to combination of above input parameters translate into about.

3 Impact of Air Pollution on Agriculture

Long-term exposure to the high concentration of surface O_3 damages vegetation with a substantial reduction in crop yields and crop quality (Krupa et al. 1998; Morgan et al. 2006). This is a primary concern for developing countries like India. Agriculture in India is demographically the broadest economic sector, ranking worldwide second in farm output. It is the principal source of livelihood for more than 58% of the

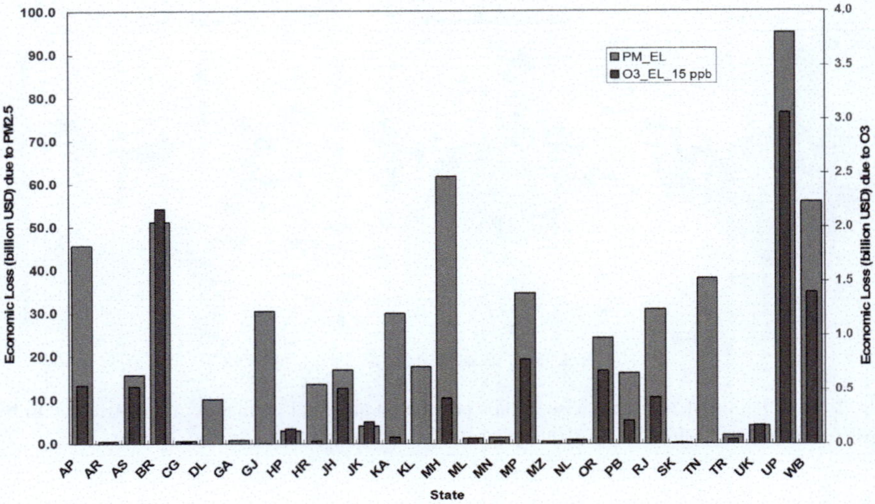

Fig. 3 Estimates of the state-wise **a** economic loss due to total mortalities by $PM_{2.5}$ and O_3 during 2011

population and hence plays an important role in the overall socio-economic fabric of India. Wheat and rice are the most important crops for ensuring food security for India's poor population, and cotton and soybean are known to be commercial crops. A recent study over India (Ghude 2014) showed substantial areas of crop production experience sufficient damage to vegetation and crop yield loss (see Sect. 2 for details of these calculations) due to ozone (O_3) exposure (Fig. 4). For the top 10 wheat- and rice-producing states, O_3-induced fractional loss of wheat is greatest in Maharashtra (~17%) followed by Madhya Pradesh (~8%), Gujarat (~8%), West Bengal (~6%), and Uttaranchal (~5%). In terms of weight, greatest loss of wheat is noticed in Uttar Pradesh (~0.6 million tons (Mt)) and Madhya Pradesh (~0.5 Mt), which accounts for about 3.2% of total wheat lost in India during 2005. Although Punjab and Haryana are the second and third largest wheat-producing states in India, wheat loss is significantly less (<1%) due to exposure to low-ozone values (<30 ppb) during wheat-growing winter seasons. For major rice-producing states like West Bengal, Orissa, Andhra Pradesh, and Uttar Pradesh, O_3-induced fractional loss of rice is between 1 and 3% (0.1–0.3 Mt), whereas it is greatest in the state of Punjab (0.8 Mt, more than 8%). On national scale, cotton suffered the highest fractional loss of 5.3 ± 3.1% followed by wheat (5.0 ± 1.2%), soybean (2.7 ± 1.9%), and rice (2.1 ± 0.9%). In terms of weight, wheat is most affected crop amounting losses of the order of 3.5 ± 0.8 Mt, double than that of the averaged wheat exported (1.8 ± 2.0 Mt) during this decade. National aggregated yield loss of wheat and rice of 5.6 Mt in 2005 is roughly about 9.2% of the cereals required every year (61.2 Mt) under the provision of the food security bill, or sufficient to feed approximately 94 million poor people (~35%) living below poverty line in India.

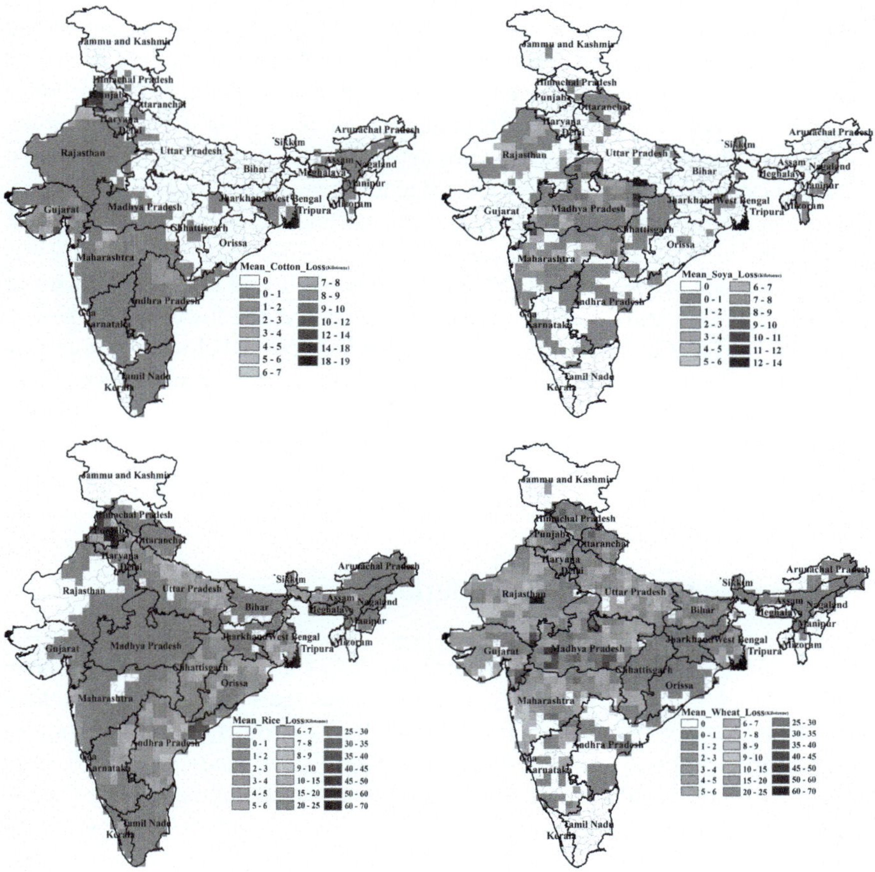

Fig. 4 Average O_3-induced crop production loss from AOT40 metrics for cotton, soybean, rice, and wheat during 2005. The production loss numbers are given in kilotons/grid box (Ghude et al. 2013)

The estimated economic damage due to crop damage caused by O_3 exposure is about 1.29 ± 0.47 billion USD in 2005. Among the four crop types, most of the economic losses in terms of absolute damage occur for wheat of about 620 ± 150 million USD in 2005. For rice, the estimated economic damage is about 540 ± 230 million USD in 2005. This is a point of concern because India ranks second worldwide in farm outputs, and a significant share of the world's rice (~22%) and wheat (12%) production comes from India. Flooding due to climate variability is a significant problem for rice farming, especially in the lowlands of South and Southeast Asia. Flooding already affected about 10–15 million hectares of rice fields in South and Southeast Asia, causing an estimated 1 billion USD in yield losses per year (Bates et al. 2008). The economic loss (0.54 ± 0.23 Mt) due to O_3-induced rice damage in India is half of the estimated annual loss of 1 billion USD due to flooding in South and Southeast Asia. Warming since 1981 declined global wheat production by 5.5%, resulting in ~19 Mt/yr (about 2.6 billion USD) wheat loss and roughly 40 Mt (equivalent to 5 billion USD) loss combined for wheat, maize, and barley globally due to global warming (Lobell et al. 2011; Lobell and Field 2007). Economic loss due to O_3-induced wheat damage alone in India is one-third of the estimated 5 billion USD per year, the present-day losses due to crop globally, and half of the estimated 2.6 billion USD loss due to wheat caused by global warming. In a similar study, Sharma et al. (2019) estimated the nationwide economic cost due to ozone-induced yield loss of about ~5 billion USD and ~1.5 billion USD for wheat and rice, respectively, in 2014–15. States of UP and MP sustain losses of more than 1 billion USD each for wheat, while for rice, the losses are highest in UP and Punjab at ~0.3 billion USD each. Global and regional climate change may also directly affect crop production through changes in monsoonal rainfall patterns in India, temperature, atmospheric conditions, soil moisture, land-use change, and local conditions. However, little is known about the combined effects of ozone pollution and climate change on agriculture, and this requires further research.

4 Economic Impact of Air Pollution

India's north and northeast plains, especially the Indo-Gangetic plains (IGP) region, are well known for dense fog throughout the winter season. Fog in the northern region of India is predominantly formed during the peak winter season over various temporal and spatial scales under a variety of meteorological conditions (Ghude et al. 2017; Singh and Kant 2006). The other unique characteristic of the fog layer in this region is that, once it is developed, it sometimes persists from a few days to weeks over vast areas, with only a partial lifting in the late afternoon. A stable atmosphere, lower surface temperatures, and calm winds persisting for most of the day or night as well as an ample moisture supply favor the development and persistence of fog in Delhi and the airport area. Low-visibility conditions arising from the fog often create problems for the aviation sector in India. The occurrence of dense fog often affects flight operations at airports in the northern region of India in terms of flight delays, cancellations, and diversions, causing passengers to be severely affected and airlines to bear severe economic losses (Airport Charged Revised 2009; Singh and Kant 2006). Recent studies on fog in India have highlighted considerable socio-economic concerns due to the alarming increase in fog and pollution and the persistence of fog in winter over the IGP (Press Clippings 2016; Jenamani and Tyagi 2011). At present, India is the 9th largest aviation market in the world. Domestic air traffic has quadrupled (from 13 million passengers to 52 million passengers). An almost similar increase in international traffic has also been seen during the same period (CAPA India Ground Handling 2014). The Indira Gandhi International (IGI) Airport of Delhi is one of the busiest airports in the country in terms of both passengers and cargo traffic. A survey conducted by the National Council of Applied Economic Research suggests that IGI Airport is one of the highest contributors of the Delhi Gross Development Product (GDP), contributing approximately 0.45% to the national GDP and 13.53% to the Delhi GDP (in 2009–2010 FY) (Jenamani 2007). Dense fog affects aviation severely at IGI Airport (Jenamani and Tyagi 2011), and flight operations are often hindered due to delays, diversions, or cancellations, causing significant economic losses. Dense fog in the early morning or evening hours creates more loss to the airlines than if it occurs at midnight or in the late morning hours. Recent studies show that there has been an increase in the frequency and intensity of fog over the northern parts of the country during the winter season, and morning poor visibility days have increased significantly during the last five decades (DGCA 2012; Kumar Jenamani 2012). Figure 5 shows the trend in number of dense and general fog days between the years 1981 and 2016 at Delhi. It can be seen that number of fog days have doubled in last four decades. An increase in dense fog hours at an airport will affect more flight operations and can create more economic losses to the airlines due to the increased diversion, delay, and cancellation of flights. When fog occurs, (1) air traffic on the ground slows down due to a reduced taxiing speed and (2) runway occupancy time (ROT) increases. Air traffic controllers (ATCs) may have to increase the spacing between each aircraft landing (Kulkarni et al. 2019) and takeoff to ensure safe operations, which can reduce airport capacity by 40% compared to its normal capacity. Additionally, ATCs must protect landing system signals for an aircraft arrival at two nautical miles from touchdown. For this to occur, previous arrivals and departures must be cleared from these areas. Because of this, a scheduled flight may be held at its origin, diverted back to the other airport, or in the worst-case scenario it may be cancelled.

A recent study (Kulkarni et al. 2019) estimated the economic cost incurred to Indian airlines during 2011–2015 due to fog for numbers of delayed, diverted, and

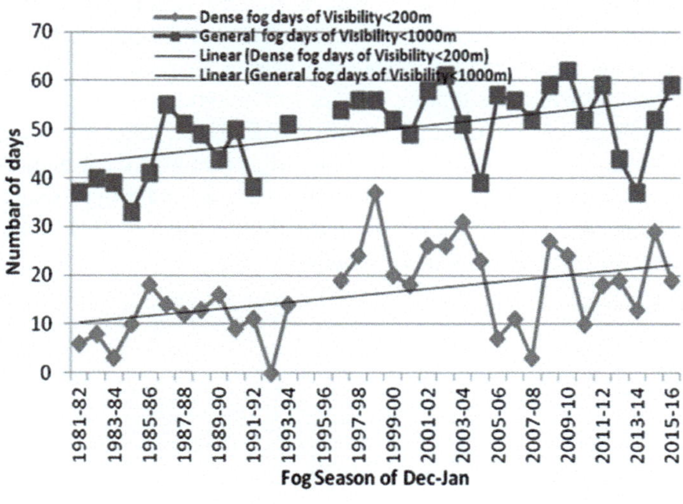

Fig. 5 Temporal evaluation of a number of fog days (December–January) from 1981 to 2016

cancelled flights from domestic and international sectors at IGI Airport, New Delhi. Based on the breakdown of charges as per the Center for Asia Pacific Aviation (CAPA) (CAPA India Ground Handling 2014) and reported values, the estimated costs of delays, diversions, and cancellations are approximately 2681.35 USD, 7456.85 USD, and 8171.64 USD for one domestic flight, and 5665.96 USD, 21,934.53 USD, and 8171.64 USD for one international flight, respectively. The total number of flights cancelled, diverted, and delayed due to dense fog during 2011–2015 was 351, 398, and 567, respectively. This led to a total economic cost of approximately 3.9 million USD (248 million Indian rupees) due to flights affected by heavy fog spells at IGI Airport over 5 years. In 2013–2014, dense fog hours were approximately 160 h, and most of the events occurred at midnight or in the early morning hours and continued until the next day (DGCA 2012). This resulted in the largest economic loss of approximately 1.78 million USD (120 million INR) due to the heavy fog spells in the winter of 2013–2014. To avoid incidents of high disruption and to bring flight diversions to zero, experts from DGCA-IMD-AAI prepared strict guidelines (DGCA 2012) for all flight landings at IGI Airport during night–morning operations when the airport is vulnerable to the likely occurrence of fog events as per the IMD meteorological forecast. As a result, there was a drop of approximately 88%, 55%, and 36% in delays, diversions, and cancellations of flights, respectively, irrespective of a greater number of dense fog hours (~10% higher than the preceding winter season) in the winter of 2014–2015 (Fig. 6). Consequently, a significant reduction in economic losses to the airlines due to dense fog was observed during the winter of 2014–2015 (0.675 million USD). This is approximately a 65% reduction in economic losses compared to the earlier winter season (2013–2014). It shows that the economic cost of fog is directly related to the number of dense fog hours, but that the implementation of the DGCA-IMD-AAI guidelines during 2014–2015 has reduced significantly the cost of fog to the airlines.

In conclusion, the implementation of strict policies that reduce pollution levels in the urban environment will certainly address and reduce the psychological, social, and environmental effects and the economic costs of air pollution. These policies can include incentivizing or adopting cleaner technologies for energy combustion, implementing air quality standards, emission standards for automobile sectors, fuel quality standards, taxes on emissions, etc. Similarly, promoting the use of the landing training system and the effective air traffic guidance, accurate fog forecasts of the onset, and dissipation of dense fog of various visibility ranges by IMD are important to minimize the economic losses to the aviation sector.

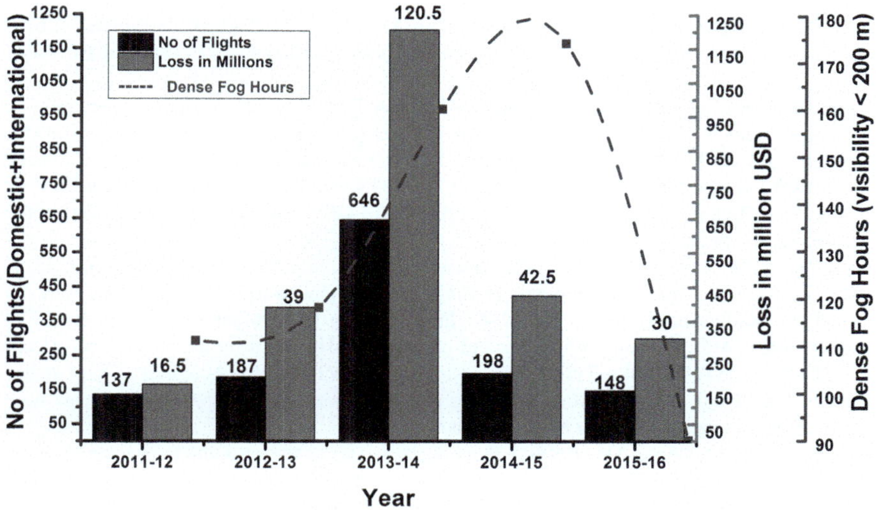

Fig. 6 Total economic losses due to dense fog at IGI Airport between 2011 and 2016. The black bar shows the total number of flights affected, the red bar indicates the economic loss (in millions USD), and the blue line shows the total number of dense fog hours for visibility less than 200 m

References

Airport Charged Revised (2009) Airport authority of India "Airport Charged_Revised" 2009 Eng_Report_2011.Pdf

Bates B, Wu S, Zbigniew Kundzewicz AW, Palutikof J (2008) Intergovernmental panel on climate change WMO UNEP climate change and water

CAPA India Ground Handling (2014) Centre for Asia Pacific Aviation India Private Limited (CAPA) India Ground Handling Report 2014. [online] https://docplayer.net/21483481-Capa-india-ground-handling-report-2014-table-of-contents-sample-extracts.html. Accessed 22 Sept 2020

CPCB (2014) Annual report 2014–15

DGCA (2012) DGCA report: Government of India Ministry of Civil Aviation report of working group on civil aviation for formulation of twelfth five year plan

Ghude SD (2014) Reductions in India's crop yield due to ozone. Geophys Res Lett (April):6413–6419. https://doi.org/10.1002/2014GL061184

Ghude SD, Fadnavis S, Beig G, Polade SD, van der ARJ (2008) Detection of surface emission hot spots, trends, and seasonal cycle from satellite-retrieved NO_2 over India. J Geophys Res 113(D20):D20305. https://doi.org/10.1029/2007JD009615

Ghude SD, Van der ARJ, Beig G, Fadnavis S, Polade SD (2009) Satellite derived trends in NO_2 over the major global hotspot regions during the past decade and their inter-comparison. Environ Pollut 157(6):1873–1878. https://doi.org/10.1016/j.envpol.2009.01.013

Ghude SD, Pfister GG, Jena CK, Emmons LK, Kumar R, van der ARJ (2012) Satellite constraints of nitrogen oxide (NO_X) emissions from India based on OMI observations and WRF-Chem simulations. Geophys Res Lett 40(x):423–428. https://doi.org/10.1029/2012gl053926

Ghude SD, Kulkarni SH, Jena C, Pfister GG, Beig G, Fadnavis S, Van Der RJ (2013) Application of satellite observations for identifying regions of dominant sources of nitrogen oxides over the Indian subcontinent. J Geophys Res Atmos 118(2):1075–1089. https://doi.org/10.1029/2012JD017811

Ghude SD, Chate DM, Jena C, Beig G, Kumar R, Barth MC, Pfister GG, Fadnavis S, Pithani P (2016) Premature mortality in India due to PM2.5and ozone exposure, Geophys Res Lett 43:4650–4658. https://doi.org/10.1002/2016GL06894

Ghude SD, Bhat GS, Prabhakaran T, Jenamani RK, Chate DM, Safai PD, Karipot AK, Konwar M, Pithani P, Sinha V, Rao PSP, Dixit SA, Tiwari S, Todekar K, Varpe S, Srivastava AK, Bisht DS, Murugavel P, Ali K, Mina U, Dharua M, Rao YJ, Padmakumari B, Hazra A, Nigam N, Shende U, Lal DM, Chandra BP, Mishra AK, Kumar A, Hakkim H, Pawar H, Acharja P, Kulkarni R, Subharthi C, Balaji B, Varghese M, Bera S, Rajeevan M (2017) Winter fog experiment over the Indo-Gangetic plains of India. 112(4)

Jena C, Ghude SD, Pfister GG, Chate DM, Kumar R, Beig G, Surendran DE, Fadnavis S, Lal DM (2015) Influence of springtime biomass burning in South Asia on regional ozone (O_3): a model based case study. Atmos Environ 100:37–47. https://doi.org/10.1016/j.atmosenv.2014.10.027

Jenamani RK (2007) Alarming rise in fog and pollution causing a fall in maximum temperature over Delhi on JSTOR. [online] https://www.jstor.org/stable/24099461. Accessed 22 Sept 2020

Jenamani RK, Tyagi A (2011) Monitoring fog at IGI Airport and analysis of its runway-wise spatio-temporal variations using Meso-RVR network

Krupa SV, Nosal M, Legge AH (1998) A numerical analysis of the combined open-top chamber data from the USA and Europe on ambient ozone and negative crop responses. Environ Pollut 101(1):157–160. https://doi.org/10.1016/S0269-7491(98)00019-0

Kulkarni R, Jenamani RK, Pithani P, Konwar M, Nigam N, Ghude SD (2019) Loss to aviation economy due to winter fog in New Delhi during the winter of 2011–2016. Atmosphere (Basel) 10(4):1–10. https://doi.org/10.3390/ATMOS10040198

Kumar Jenamani R (2012) Development of intensity based fog climatological information system (daily and hourly) at IGI airport, New Delhi for use in fog forecasting and aviation

Lelieveld J, Evans JS, Fnais M, Giannadaki D, Pozzer A (2015) The contribution of outdoor air pollution sources to premature mortality on a global scale. Nature 525(7569):367–371. https://doi.org/10.1038/nature15371

Lobell DB, Field CB (2007) Global scale climate-crop yield relationships and the impacts of recent warming. Environ Res Lett 2(1):14002–14009. https://doi.org/10.1088/1748-9326/2/1/014002

Lobell DB, Schlenker W, Costa-Roberts J (2011) Climate trends and global crop production since 1980. Science (80-) 333(6042):616–620. https://doi.org/10.1126/science.1204531

Lu JG (2020) Air pollution: a systematic review of its psychological, economic, and social effects. Curr Opin Psychol 32:52–65. https://doi.org/10.1016/j.copsyc.2019.06.024

Lvovsky K (1998) Economic costs of air pollution with special reference to India, South Asia Environ. Unit, World Bank, pp 1–26. [online] http://siteresources.worldbank.org/PAKISTANEXTN/Resources/UrbanAir/Economic+costs+of+air+pollution+KL.pdf

Morgan PB, Mies TA, Bollero GA, Nelson RL, Long SP (2006) Season-long elevation of ozone concentration to projected 2050 levels under fully open-air conditions substantially decreases the growth and production of soybean. New Phytol 170(2):333–343. https://doi.org/10.1111/j.1469-8137.2006.01679.x

OECD (n.d.) The economic consequences of outdoor air pollution policy highlights

Press Clippings of Airport Authority of India—Public Relations Department (2016) Financial Express, New Delhi, India

Singh J, Kant S (2006) Radiation fog over north India during winter from 1989–2004

Advances in Ocean State Forecasting and Marine Fishery Advisory Services for the Indian Ocean Region

T. M. Balakrishnan Nair, K. Srinivas, M. Nagarajakumar, R. Harikumar, Kumar Nimit, P. G. Remya, P. A. Francis, and K. G. Sandhya

Abstract India is adept in implementing the latest technological advancements and research findings with its fully operational ocean information and advisory services for the Indian Ocean. The operational services viz. Ocean State Forecast (OSF) and Marine Fishery Advisory Services (MFAS) support not only the stakeholders in India, but also those in other neighbouring countries in the region, for safe navigation and cost-effective fishing. The OSF and MFAS programs of the Indian National Centre for Ocean Information Services (INCOIS) have set high standards, with accurate and timely delivery of services. The backbone of this system are, robust in-situ and satellite observation systems, state-of-the-art computational facilities running multi-model simulations and importantly, adoption of the latest Information and Communication Technology (ICT) tools for effective dissemination to various stakeholders. The numerical models include wave models, Ocean General Circulation Models (OGCM), tidal models and atmospheric models. State-of-the-art data assimilation schemes are also incorporated in these operational models for providing accurate forecasts. During recent times, INCOIS has made great strides from providing general forecasts to custom-made and impact-based ocean state forecasts, general Potential Fishing Zone (PFZ) advisories to species-specific advisories, based on the user feedback. The user-customised, tailor-made products have been generated and supplied to the fishermen, defence authorities, maritime boards, port authorities, and offshore industries during the last few years. The latest among these are the satellite-based dissemination services for fishing vessels operating hundreds of kilometers away from the shore through **NAV**igation with **I**ndian **C**onstellation (**NAVIC**) and **G**agan **E**nabled **M**ariner's **I**nstrument for **N**avigation and **I**nformation (GEMINI) systems. In addition to direct dissemination from INCOIS, the reach of these services to end users is also catered by NGO partners and their downstream dissemination systems.

T. M. Balakrishnan Nair (✉) · K. Srinivas · M. Nagarajakumar · R. Harikumar · K. Nimit · P. G. Remya · P. A. Francis · K. G. Sandhya
Indian National Centre for Ocean Information Services (INCOIS), Ministry of Earth Sciences, Ocean Valley, Pragathi Nagar (BO), Nizampet (SO), Hyderabad 500090, India
e-mail: bala@incois.gov.in

K. Nimit
e-mail: nimitkumar.j@incois.gov.in

V. K. Gahalaut and M. Rajeevan (eds.), *Social and Economic Impact of Earth Sciences*,
https://doi.org/10.1007/978-981-19-6929-4_11

Keywords Ocean State Forecast · Marine Fishery Advisory, Yellowfin Tuna, Potential Fishing Zone (PFZ), · Cyclones · Early Warning · *Kallakadal* · Advisory Services

1 Introduction

The oceans are strong pillars of the economy of any maritime country, contributing to the "Blue Economy" of that country. Because of the depleting resources on the land, countries world over are looking at the exploitation of resources from the sea for food, minerals, energy, etc. India also looks forward to exploiting the blue economy resources for its benefit, ensuring a balanced approach between conservation and development. With the Arabian Sea to the west and the Bay of Bengal to the east and the vast expanse of the Indian Ocean to the south, a wide range of marine activities dominates the seas around India, contributing to the economic growth of the country (Nayak 2020). Prior information on the state of the seas surrounding the Indian subcontinent is vital for the smooth operational activities, not only for those who are venturing out into the sea but also for those at the sea shore (Balakrishnan Nair et al. 2013, 2014, 2020; Francis et al. 2013, Harikumar et al. 2015; Francis et al. 2020a, b; Sandhya et al. 2016, 2018; Remya et al. 2020). Major cities of India are also located along the coastline, some of which keep witnessing the brunt of oceanogenic natural hazards, every year. For example, cyclones keep devastating relatively small stretches of the coast every year, causing loss of life and damages to coastal infrastructure. The swell waves as well as the perigean spring tides have also caused impacts, the former one causing major issues such as erosion and damage to the boats as well as rough sea conditions nearshore and inundation while the latter mostly causing nuisance flooding, at low lying areas (Remya et al. 2016). In some rare cases, a combination of factors like swell waves along with the perigean spring tides have increased the impacts on the coastline. Sustained high waves such as that witnessed during the southwest monsoon (June–September) season, have caused operational difficulties to the user communities and large scale erosion at the near shore regions.

The sustained and focused R&D efforts, ocean modeling and real-time observations of INCOIS continuously enhance the user experience and they can now make informed decisions based on the forecasted sea state conditions and advisories, alerts and warnings, thereby supporting livelihoods and saving lives. Marine activities in the seas around India are very diverse, viz. artisanal/mechanised fishing, port and shipping activities, coastal tourism, oil-natural gas-mineral explorations, defence interests and marine research. INCOIS recently operationalized an impact-based forecast viz. Small Vessel Advisory and Forecast Services System (SVAS) in an attempt to forewarn the sea farers (particularly fishermen) about the impending dangerous sea states (Aditya et al. 2020). The data collected from a suite of observational systems is also very important for validating the numerical model outputs, both in real-time and in hindcast mode. For example, statistical bias correction to the forecasts is applied to the direct ocean model forecasts using real-time observations from different areas of the Indian Ocean (Harikumar et al. 2016). This procedure is especially important during the cyclones, when there could be large errors (mostly under-estimation) in the forcing wind fields, resulting in under-estimated wave fields. The forecasts have also improved greatly with the implementation of the data assimilation schemes in numerical models, which make use of the real-time observations (in-situ and remotely sensed data).

Productivity of oceans depends on nutrient availability in sun-lit upper waters known as, euphotic zone (Platt and Sathyendranath 1988). Oceanographic phenomena such as upwelling help contribute to much of this requirement, by entraining nutrients into the mixed layer depth and in turn, allowing phytoplankton to sustain the food-web with the help of photosynthesis. Stronger the upwelling, more nutrients will be supplied to the mixed layer. Upwelled colder waters result in lowered Sea Surface Temperature (SST), providing a handy signature in detecting upwelling zones with the help of remotely-sensed data. Productive waters may initially attract only planktivorous fishes but eventually attract the preying bigger fishes. This is the very reason how SST was harnessed as a tool towards the first of fishery resource predictions (Choudhury et al. 2007). Commercially-important species such as Tuna, are believed to have temperature specificity (Zainuddin et al. 2004). This has been reflected in studies that show a positive correlation of specific temperature ranges with better hooking rates (Kumari et al. 2009). Ocean-color missions have provided more than one dimension to our understanding of ecosystem-level interactions in the ocean. Cooler SST signature with a higher concentration of chlorophyll, relative to surrounding waters, indicates upwelling-induced productivity (Tummala et al. 2008) and has been correlated with higher Catch per Unit Effort (Friedland et al. 2012). This article gives an overview of progress in Ocean State Forecasting and Marine Fishery Advisory Services developed and disseminated from INCOIS.

2 Ocean State Forecast (OSF) System

The short term advisories based on the ocean state forecasts (5–7 days), are sought by many users who venture out into the sea for their varied operations. For providing good quality forecasts seamlessly, requires a robust ocean observation network in real-time, state-of-the-art models and computing facilities, an advanced dissemination system and importantly a dedicated team of experts in relevant fields, for the smooth operational functioning of the system. A simple Process Chain chart depicting the OSF generation is presented in Fig. 1. Most of the forcing data for operational models is received between 1130 and1230 h daily from the European Centre for Medium Range Weather forecasts (ECMWF)/National Centre for Medium Range Weather Forecast (NCMRWF), New Delhi. After that, the pre-processing of the model inputs, running of the numerical models on high performance computers (HPC) and then post-processing of the outputs take place. The end-to-end operations (generation, quality check, value addition and dissemination) of ocean state forecast services takes about 4 h daily. In this process, the Forecast Assessment and

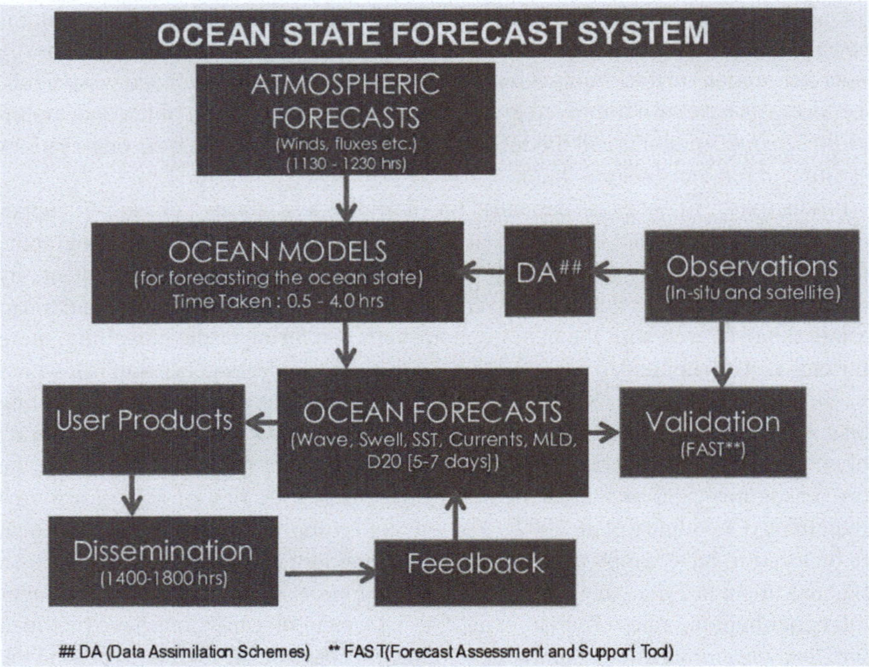

Fig. 1 The process chain of generation of OSF advisories

Support Tool (FAST, an in-house developed tool, which acts as a decision support system) plays a major role, as it is used for comparing the multi-model forecasts vis-a-vis the observations from different platforms (wave rider buoys, met-ocean buoys, Automatic Weather Stations, satellite data) and only then disseminated to the stakeholders, after applying the bias-corrections (to take care of the over-estimation or under-estimation of the models), if any.

The forecasts are generated by a suite of state-of-the-art numerical models (Table 1), which are setup and fine-tuned to simulate and predict the Indian Ocean features accurately. The wave models used are Spectral Wave Model (Sirisha et al. 2019), WAVEWATCH III (MWW3, Remya et al. 2020) and **S**imulating **WA**ves **N**earshore (SWAN, Aditya et al. 2020). The Ocean General Circulation models being used are **R**egional **O**cean **M**odeling **S**ystem (ROMS, Francis et al. 2020a, b; Effy et al. 2020) and **G**lobal **O**cean **D**ata **A**ssimilation **S**ystem (GODAS, Ravichandran et al. 2013). For oil spills, the **G**eneral **NOAA** **O**il **M**odeling **E**nvironment (GNOME, Prasad et al. 2014) model is used whereas the Leeway model is used for the **S**earch **A**nd **R**escue **A**id **T**ool (SARAT, Breivik and Allen 2008). TASK-2000 (Srinivas and Dinesh Kumar 2002) is used for tidal analysis and prediction. Wave models are forced by wind parameters, whereas Ocean General Circulation models are forced by atmospheric parameters like surface wind speed, specific humidity, surface air temperature, precipitation and short wave and long wave radiation, along with tracer and

Table 1 The wave models, OGCModels, oil spill models, SARAT and tides that are run operationally

Numerical models	Spatial resolution (degrees)
Ocean wave models	
Spectral wave model	1.000°–0.070°
MWW3 (Multi-grid WAVEWATCH III)	1.000°–0.050°
SWAN (**S**imulating **WA**ves **N**earshore)	0.003° × 0.005°
Ocean general circulation models	
ROMS (Regional Ocean Modeling System)	0.125° × 0.125°
GODAS (Global Ocean Data Assimilation System)	Up to maximum 0.25°
HYCOM (HYbrid Coordinate Ocean Model)	Up to maximum 0.06°
Online Oil Spill Advisory (OOSA)	
GNOME (General NOAA Oil Modeling Environment)	–
Search And Rescue Aid Tool (SARAT)	
Leeway Model	–
Tide	
TASK—2000 (Tidal Analysis Software Kit)	–

velocity boundary conditions at the lateral boundaries (for regional models) extracted from INCOIS-GODAS (Francis et al. 2020b). Atmospheric forecast products from different meteorological forecasting agencies (NCMRWF, NCEP and ECMWF) are used for forcing these wave and ocean general circulation models in the forecast mode, for ensuring redundancy and ensembling. Global and regional forecasts, differ mainly in their model domains, spatial and temporal resolutions, extent of validations carried out, etc. The models are set up using the concept of 'multiple grid' with very fine resolution for the specified coastal area, providing more accurate data at the coast and coarse resolution in the open ocean region. Further, data assimilation schemes are implemented in these operational models aimed at the improvement of forecast quality (Details of the data, which are assimilated in the models are given in Table 2).

3 Category of Ocean Services

The OSF system at INCOIS is capable of predicting surface and sub-surface features of the Indian Ocean, 10 days in advance at three-hourly intervals for parameters such as (1) Height, direction and period (of wind waves and swell waves) (2) Wind speed and direction (3) Surface and sub-surface currents (4) Surface and sub-surface temperature (5) Mixed Layer Depth (the depth of the well mixed upper layer of the sea) (6) Depth of the 20° isotherm (a measure of the depth of the thermocline) (7)

Table 2 Details of the satellite/in-situ observations assimilated into the operational models at INCOIS

Model	Assimilated parameter	Satellite	In-situ	Forecast parameters
Wave and swell surge forecasting system (WAVEWATCH III, SWAN, ADCIRC)	Significant wave height	Altimeter–Jason2, Jason3, Saral-AltiKa	Nil	Height, direction and period (of both wind waves and swell waves), swell surge (arrival time and extent of inundation)
Regional ocean forecast (ROMS) system	SST Temperature and salinity profiles	GHRSST L2 track data	Argo, XBT and moored buoys including RAMA moored buoys	Sea surface currents, sea surface temperature, mixed layer depth, depth of the 20 °C isotherm, temperature and salinity profiles
Global ocean analysis (INCOIS-GODAS) system	Temperature and salinity profiles (SST is relaxed with 5 day time scale from OI SST)	NOAA OI (for daily SST relaxation)	Argo, XBT and moored buoys, RAMA	Sea surface currents, sea surface temperature, mixed layer depth, depth of the 20 °C isotherm, temperature and salinity profiles
Basin wide ocean forecast (HYCOM)	Temperature and salinity profiles, SST, SLA	Jason3, Saral-AltiKa GHRSST	Argo, moored buoys, RAMA	Sea surface currents, sea surface temperature, mixed layer depth, depth of the 20 °C isotherm, temperature and salinity profiles
HWRF-HYCOM coupled model	Temperature and salinity profiles, SST, SLA	Jason3, Saral-AltiKa GHRSST	Argo, moored buoys, RAMA	Cyclone intensity and track forecast

ADCIRC—The ADvanced CIRCulation model; RAMA—Research Moored Array for African-Asian-Australian Monsoon Analysis and Prediction; GHRSST—Group for High-Resolution Sea Surface Temperature; NOAA OI—NOAA Optimum Interpolation (OI) Sea Surface Temperature

Astronomical tides (8) Oil Spill trajectory, during spill occurrences (9) Location of the lost persons or objects in the sea.

The OSF services have been categorised as regular services, early warning services and user-customised services (Table 3). The forecasts are available separately for the following regions: Arabian Sea, Bay of Bengal, Northern Indian Ocean, Southern Indian Ocean, Red Sea, Persian Gulf and South China Sea. Further, INCOIS provides more detailed forecast information for specific locations such as fish landing centres (for more than 1500 Indian locations), small fishing harbours, major and minor ports, etc. as well as for the coastal waters of the maritime states, union territories and island regions of India. These forecasts are provided for India and neighbouring countries. The forecasts for all the nine coastal states of India are updated daily along with that for the island territories of Andaman, Nicobar and Lakshadweep. Forecasts for the countries Sri Lanka, Maldives, Seychelles, Comoros, Madagascar and Mozambique are also updated daily.

Table 3 Categories of ocean state forecast services/products

Serial number	Regular forecast services	Early warning services	Customised products for blue economy sector
1	Coastal forecast	High wave/Swell alert/Warning	Search And Rescue Aid Tool (SARAT)**
2	Location specific forecast	INCOIS-IMD joint bulletins during depressions/cyclones	Online Oil Spill Advisory system (OOSA)**
3	Tropical cyclone heat potential	Kallakadal warning (coastal flooding due to swell surge)	Port and harbour information system
4	Predicted astronomical tide	Kondalkattu warning (location specific forecast on high wind waves)	Forecast along ship–routes
5	Regional forecast	Tidal flood advisory (Perigean/Proxigean)	Webmap services
6	Forecast for the Islands	Small vessel advisory	Navy specific forecasts
7	Forecast for neighbouring countries (Sri Lanka, Seychelles, Maldives, Mozambique, Madagascar and Comoros)		IVL demarcation and IVL forecast for maritime boards
8	Global forecast products		Oil industry specific forecast and analysis products

** These also come under emergency services

4 Verification and Reliability of Services

For testing the accuracy and reliability of any forecast parameters, INCOIS conducts detailed validations based on in-situ (near shore wave rider buoys, deep sea buoys, shipborne wave height meters, Automatic Weather Stations and other deep sea met-ocean buoys) and satellite measurements (Table 4). As far as the Indian Ocean is concerned, high waves occur mainly due to sustained southwest monsoon winds, cyclones and Southern Indian Ocean meteorological conditions. Hence the validation exercises of the wave models are carried out at different conditions to ensure the reliability of wave forecasts (Sandhya et al. 2018; Sirisha et al. 2019; Remya et al. 2020). An example of such validation of the wave models at some west coast locations (Ratnagiri, Karwar and Kozhikode) during 2016 is shown in Fig. 2. There is a very good agreement between the observed and first day forecast wave heights at the three locations. Besides, there is also a remarkable similarity in the annual march of the Significant Wave Height (SWH), at the three locations. However, there was an extended period at Ratnagiri, when the buoy data was not available, but the trend of the forecast suggests good agreement with the reality (on comparing with the nearest station Karwar). Two different observational data sets viz. buoys and altimeters, have been used for the evaluation of the model skill, in their study. Further, the reliability of the wave model was tested separately for deep and for coastal waters, and reliable performance (correlation >0.8 and Scatter Index <30%) was found for both. The year 2019 was unusual, with many cyclones forming in the Arabian Sea. Significant Wave Height during Very Severe Cyclonic Storm—***Vaayu*** (June 10–17, 2019) and the actual track are shown in Fig. 3, just before the landfall. Cyclone impacts were less frequent along the Indian west coast as compared to the east coast, with a frequency of one for every four events along the east coast, but this trend seems to be changing during recent times (Murakami et al. 2017).

High waves, which are of long periods, without any sign in the local winds, sometimes cause severe flooding events along the south-west coast of India (especially low lying Kerala coast), locally known as the Kallakkadal events and cause major societal problems along the coasts (Remya et al. 2016). These are mainly caused by southern ocean swells generated due to the high winds in the southern ocean (Sabique et al. 2012). The high swells generated by southern ocean wind systems can sometimes even co-exist with a cyclone, thus contributing to the cyclone-generated wave heights (Sandhya et al. 2016). The link between North Indian Ocean (NIO) high swell events and the meteorological conditions over the Southern Indian Ocean (SIO) has been explored by Remya et al. (2016), using a combination of in-situ measurements and model simulations for the year 2005 (when a major swell wave event occurred along Kerala coast). Their study shows that such natural hazards along the NIO coasts can be predicted at least 2 days in advance if the meteorological conditions of the SIO are properly monitored (Fig. 4). Their study confirms that these events are caused by the swells propagating from south of 30° S. In all cases, 3–5 days prior to the high swell events in NIO, they observed a severe low pressure system, called the Cut-Off Low (COL) in the Southern Ocean. These COLs are quasi-stationary in nature, providing

Table 4 Results of statistical analysis of significant wave height (observed versus buoy) at different buoy locations (Sandhya et al. 2018)

Buoy ID	Seasons (no. of entries)	R	Bias (m)	RMSE (m)	SI	SDB (m)	SDM (m)	Buoy mean (m)	Model mean (m)
Gopalpur	Pre (712)	0.86	0.13	0.17	0.23	0.22	0.20	0.75	0.88
	Mon (753)	0.67	−0.16	0.27	0.20	0.29	0.17	1.36	1.20
	Post (864)	0.96	0.08	0.25	0.35	0.56	0.35	0.72	0.80
Visakhapatnam	Pre (465)	0.66	0.21	0.25	0.36	0.17	0.15	0.69	0.90
	Mon (569)	0.81	−0.15	0.26	0.17	0.34	0.18	1.53	1.38
	Post (769)	0.93	0.13	0.25	0.29	0.50	0.37	0.86	0.99
Puducherry	Pre (511)	0.86	0.05	0.12	0.20	0.20	0.12	0.60	0.65
	Mon (970)	0.64	−0.07	0.15	0.22	0.17	0.12	0.69	0.61
	Post (862)	0.91	−0.11	0.19	0.20	0.34	0.24	0.95	0.83
Tuticorin	Pre	–	–	–	–	–	–	–	–
	Mon (483)	0.65	0.02	0.16	0.17	0.18	0.20	0.97	0.99
	Post (426)	0.78	−0.07	0.13	0.18	0.17	0.18	0.75	0.68

Pre—Pre-monsoon, Mon—Summer monsoon, Post—Post-monsoon, R—Correlation Coefficient, RMSE—Root Mean Square Error, SI—Scatter Index, SDB—Standard deviation of buoy data, SDM—Standard Deviation of model data

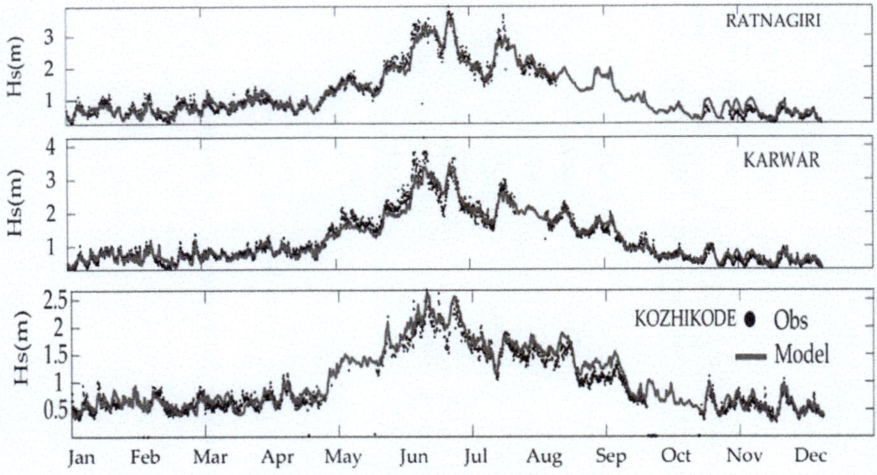

Fig. 2 Validation of the Significant Wave Height forecast of 24 h lead time at west coast locations during the year—2016 (Remya et al. 2020)

strong (~25 ms^{-1}) and long duration (~3 days) surface winds over a large fetch; essential conditions for the generation of long-period swells. The intense equatorward winds associated with COLs in the SIO trigger the generation of high waves, which propagate to NIO as swells. These swells cause high wave activity and sometimes Kallakkadal events along the NIO coastal regions, further depending on the local topography, angle of incidence, and tidal conditions. INCOIS now successfully forecasts such events using a multi-model based operational wave forecasting system. Forecast of the swell surge event of 21–22 April, 2018 along the west coast of India and Lakshadweep, with swell heights between 2 and 3 m and periods between 17 and 22 s was well appreciated by the users. This was a very major one, with impacts right from Kerala to Maharashtra coast ("Swell waves pound coastal belt in Ernakulam". *The New Indian Express, April 23, 2018*). The buoy deployed at Seychelles by INCOIS is primarily meant to monitor the propagation of such swell events to the northern Indian Ocean. The distance between Kollam and Seychelles buoy is 2750 km, and it takes almost a day for such swells from Seychelles to reach the tip of peninsular India.

The Ocean General Circulation system has a very good skill in predicting the thermal structure of the deep and coastal ocean, even with 3–5-day lead time (Francis et al. 2020b). The correlation between the observed and the predicted temperature were very high (0.7–0.9) and the RMSEs were less than 1 °C in the top 50 m of the ocean at most of the buoy locations. Even sudden SST drop during the passage of tropical storms was accurately predicted by the system even with 3-day lead time.

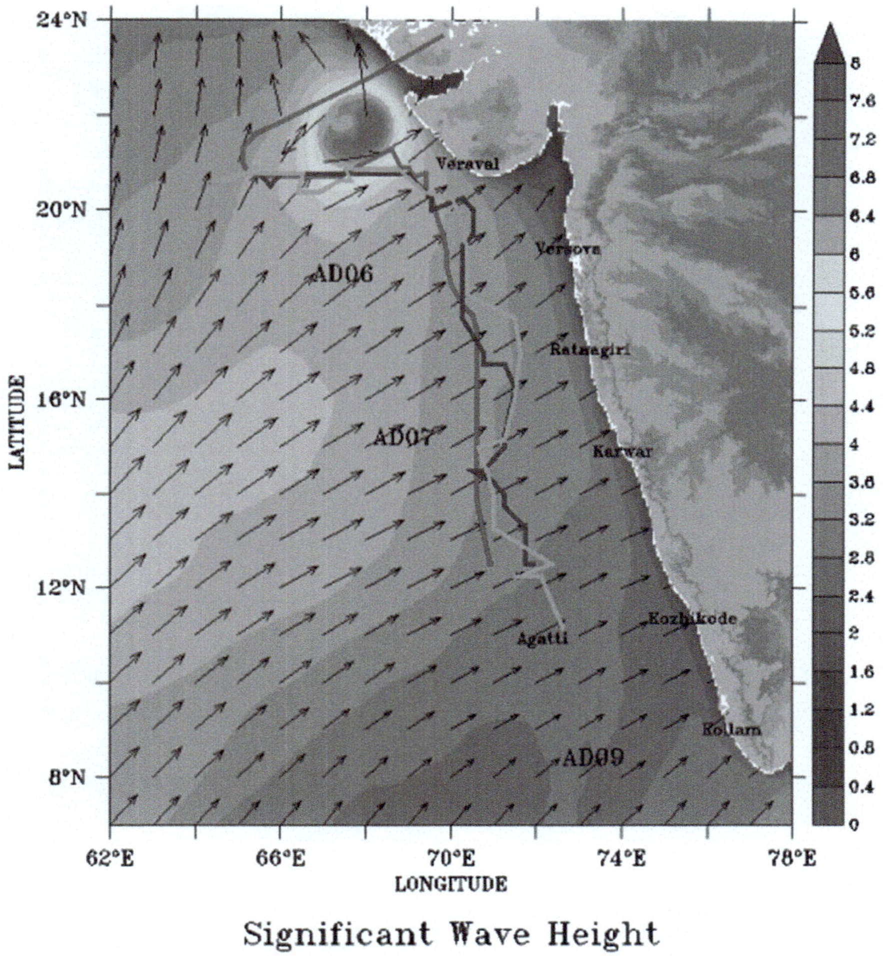

Fig. 3 Significant Wave Height (m) and direction during Very Severe Cyclonic Storm—*Vaayu* (June 10–17, 2019), along with the actual track of the cyclone (red). The locations where the met-ocean buoys and WRBs are deployed are also indicated

5 Early Warning Services

Early Warning Service is a very important service to the user community, for minimizing the damages to life and property in the sea and also at the shoreline, by providing advance information on the event, and the possible impacts. INCOIS has been issuing Early Warnings for high waves, tropical cyclones (Joint Bulletins with IMD), swell surges, Kondalkattu and tidal flood events, that occur along the Indian coastlines.

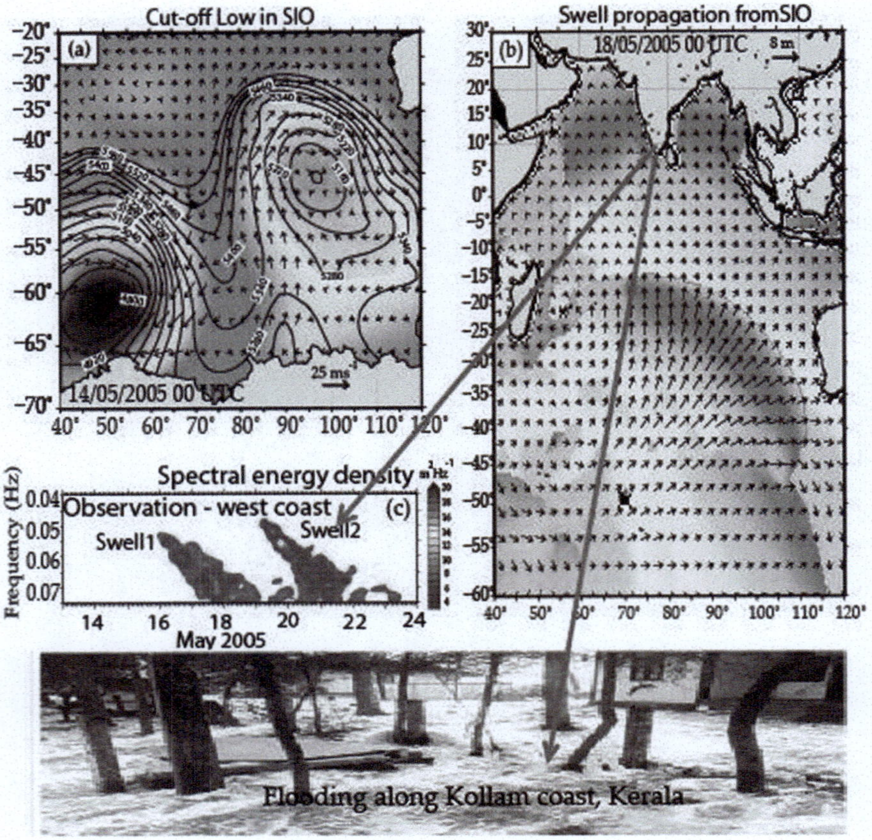

Fig. 4 Model and observations supporting the swell surge related damages, along Kerala coast during May, 2005 (Remya et al. 2016)

5.1 High Wave Warning

High wave alert is issued when the waves (Total Significant Wave Height) are expected to be between 3.0 and 3.5 m or when the swell waves are expected to be between 2.5 and 3.0 m. High wave warning is issued when the wind waves are forecasted to be above 3.5 m or when the swell waves are forecasted to be above 3.0 m. These high waves can be expected during sustained high winds, typically during the southwest monsoon months, as well as during the cyclones, that occur and make landfall on the coastline of the Indian subcontinent. However, we have to bear in mind that a high swell wave dominated field can be far more damaging than the local wind generated wave.

5.2 INCOIS-IMD Joint Bulletin During Tropical Cyclones

INCOIS also issues three-hourly Joint Bulletins with IMD during the periods of tropical cyclones, providing information on the Significant Wave Heights, swell heights, and surface currents, for each coastal district of the state, along with track and intensity of the system and the warnings from IMD. It also updates the observed/predicted data on the same from a suite of met-ocean buoys and WAve Monitoring Along Nearshore (WAMAN) Buoys deployed in the seas around India, and in the vicinity of the path of the system.

5.3 Swell Surge (Kallakkadal) Warning

The swell surge forecast system at INCOIS is an innovative system designed to predict *Kallakkadal*/Swell Surge that occurs along the Indian coast, particularly the west coast. The system can predict *Kallakkadal* and warnings will be issued at least 2–3 days in advance, which will help the local authorities for contingency plans and to reduce damage. If the forecast peak wave period is 15–18 s, a rough sea alert is issued, while a swell surge warning is issued if the forecast peak wave period is more than 18 s. Kerala, along the southwest coast of India, is particularly prone to damages due to this event, being a low lying land, that too with high population density.

5.4 Kondal Kattu Alert

"Kondal Kattu" is a gusty wind which occurs near Rameswaram coast (southern Tamil Nadu) during April–May i.e. pre-monsoon season (Sirisha et al. 2019). Near Rameswaram coast, winds are generally below 10 m/s during the pre-monsoon season. However, during this event, wind suddenly increases to nearly 25 m/s, sustaining for nearly 2–3 h, causing damage to the boats which are moored at the coast. The low pressure system that forms over Salem or the west coast of Sri Lanka region causes these gusty winds. To forecast these gusty winds and associated sea waves, a high resolution WRF model and spectral wave model have been set up at INCOIS for Rameswaram coast (southern tip of India).

5.5 Perigean Spring Tide Alerts

The possibility of flooding in the low lying coastal areas, especially during high tide timings throughout the perigean spring tide dates, is predicted, three days in advance. Sustained strong onshore winds, high nearshore waves/swells and heavy rainfall and associated discharge can cause nuisance flooding (public inconveniences) especially at high tide times along low lying coastal areas. The commercial capital of India, Mumbai witnesses these impacts every year, especially during the southwest monsoon season. These events are indicative of locations, where the impact can be far higher in the future, due to the projected sea level rise scenarios.

5.6 Small Vessel Advisory and Forecast Services System

Small Vessel Advisory and Forecast Services System (SVAS) is an innovative impact-based forecasting and advisory system implemented at INCOIS recently, for issuing timely advisories to small vessels operating in the Indian coastal waters (Aditya et al. 2020). The objective of SVAS is to reduce the number of accidents caused by capsizing of vessels at sea, by warning the users regarding potential zones and times at which the risk of vessel overturning can take place, three days in advance. This system is based on a newly coined index viz., 'Boat Safety Index' (BSI) derived from wave model forecast outputs such as significant wave height, wave steepness, directional spread and the rapid development of wind sea. To make the advisories boat-specific, the warning is issued with the region of danger in conjunction with the category/size of the boat in an inclusive manner, by using the significant wave height information from the wave model. Thus SVAS is categorized according to the beam width of the vessel. Advisories are valid for small vessels of beam width up to 7 m. This limit covers the entire range of beam widths of the fishing vessels used in all the 9 coastal states and union territories of India. The advisory system is also validated for accidents that occurred in the past and found to be useful in identifying the potentially dangerous areas at sea.

6 Customised Products

The three most important customised products are SARAT, OOSA and forecast along ship-routes: SARAT and OOSA are very important during extreme weather and untoward incidents in the sea and are therefore basically emergency services, as a quick remedial action by concerned is required in both the cases. Customised forecast for the port and shipping sector also comes in this category. Another useful service is to provide Inland Vessel Limits (IVL) demarcation, which is used by ports and harbours and shipping sector. A little more description on these services is presented below.

6.1 Search And Rescue Aid Tool (SARAT)

Search and rescue operations in the sea are extremely challenging tasks throughout the world, and can be compared to the proverbial searching for a needle in the haystack. This is more challenging in the Indian context, where the fishermen venture far out into the open sea in smaller boats, without using modern communication gadgets, often not being aware of the developing weather systems. Even the conventional communication gadgets do have spatial reception limitations and cannot reach out to the fishermen far out in the sea. In foul weather, when the winds and waves can be high, the fishermen can be easily off boarded, the boats may get capsized, and boats may even drift with engine failure. In these events, more lucky fishermen may stay alive, as it is merely a chance. It is, therefore, extremely important to trace the person/boat quickly to save the life/property. SARAT system helps to define the most probable search area for missing persons/objects at the sea, to the search responders *viz.* Indian Coast Guard/Indian Navy.

The backend tool is the Leeway model (Breivik and Allen 2008), which is a Monte Carlo-based stochastic ensemble trajectory model that calculates the motion of objects on the sea surface under the influence of wind and surface currents. The SARAT system has been developed mainly to support the Indian Coast Guard, Indian Navy, Coastal Security Police and disaster management authorities in their operations to minimize loss of life and property. All the requests and responses are provided in local languages of all coastal states, enabling the local fishermen to use it immediately to search for their fellow fishermen/boats in distress. The accuracy of SARAT has been extensively verified using the drifting buoys and other objects deployed at the sea, and found to have reasonable operational accuracy.

6.2 Online Oil Spill Advisory (OOSA)

INCOIS has been issuing oil spill advisories to the coastal communities during oil spill incidents since 2011. The advisory on oil drift and spread is disseminated to Indian Coast Guard (oil spill responders), port authorities, Maritime Boards, and Pollution Control Boards (PCBs). The INCOIS oil spill trajectory prediction system includes an oil spill trajectory prediction model in diagnostic mode, forced with high resolution met-ocean forcings. The Operational Online Oil Spill Advisory (OOSA) system was operationalised during May 2014, the oil spill responders can get the advisory on the drift pattern of the spilled pollutant instantly. The back end of the OOSA is the General NOAA Oil Modeling Environment (GNOME), a standard Eulerian/Lagrangian spill trajectory model designed to meet the needs of the users through three different user modes: Standard, GIS Output, and Diagnostic (Prasad et al. 2014).

6.3 Services for Ports and Shipping Sector

Dotted along the long Indian coastline are numerous ports (both major and minor) which have contributed greatly to the economic growth of the country. Taking cognizance of the importance of ports and inland Shipping, *Sagar Mala*, a port linked industrialization project was introduced in the year 2017 by the Ministry of Ports, Shipping and Waterways, Government of India with an objective to reduce logistics cost for export-import as well as domestic trade with minimal infrastructure investment. For the efficient operations involving the entry, exit and loading or unloading of these vessels (be it a small fishing boat or large oil tanker) at these ports, one critical factor is advance information on sea state parameters in the vicinity, like locally generated wind waves, remotely forced swell waves, currents, winds and tides, along with warnings on cyclones and other extreme weather events. Since many ports cannot handle huge oil tankers, offshore port activities like Single Point Mooring (SPM) operations become important. This is a loading buoy anchored offshore, that serves as a mooring point and interconnect for large tankers loading or unloading gas or liquid products to the shore based facilities requiring accurate sea state information. INCOIS is presently issuing forecasts for such locations, for facilitating the discharge of oil, the main advantages are being able to handle extra large vessels (with large drafts), save fuel and time, and importantly large quantity of cargo can be easily handled.

INCOIS also provides daily OSF updates along with meteorological data and warnings, on way points, for the standard routes like Chennai-Port Blair and Kolkata-Port Blair, as these are the regular routes used by passenger ships. Unlike the fishing operations, which are suspended for brief periods along the east and west coast of India, shipping and port activities take place throughout the year, even during an extended period of foul weather (southwest monsoon season—June–September). Safe transit during such times is also very important. Demarcation of Inland Vessel Limits (IVL) for various coastal states and island territories of India greatly benefits the coastal shipping/inland navigation sectors. Apart from the climatologically-derived monthly averaged trend of the IVLs, 7-day forecasts of IVL are also provided online in the form of dynamic IVLs. Apart from the climatologically-derived monthly averaged trend of the IVLs, 7-day forecasts of IVL are also provided online in the form of dynamic IVLs.

7 Marine Fishery Advisory Services (MFAS)

The short term advisories on the location of fish availability are popularly known as Potential Fishing Zones (PFZs). Advisories are provided on a daily basis using satellite data on different parameters like Sea Surface Temperature (SST) and Chlorophyll. SST, a parameter that indicates the conducive environment for fish, retrieved from NOAA-AVHRR and Chlorophyll content in the sea water, indicating the availability of food for fish, retrieved from the Indian Remote Sensing Satellite, Oceansat-2 and/or NASA's MODIS-Aqua satellites are used for the operational generation of PFZ advisories (Fig. 5). The application of satellite derived sea surface temperature (SST) and chlorophyll-a for the demarcation of PFZs has been well demonstrated in the Indian waters (Nayak et al. 2007; Dwivedi et al. 2005; Choudhury et al. 2007; Deshpande et al. 2011). In 2002, INCOIS created a digital database for coastline, bathymetry, major landing centres, and lighthouses along the coastal states from the NHO maps. With the availability of digital data and the Geographic Information System (GIS) tools, complete data processing and advisories delineation has been made into software (ERDAS Imagine and ArcGIS) based digital processing system. Subsequently, automatic frontal detection and generation of advisories were introduced and thus reduced the process time from 3 h to 30 min. One of the main limitations in the generation of PFZ advisories is cloud cover which obstructs the satellites in observing the radiation values that are used for retrieval of geophysical parameters like SST or Chlorophyll. To address this challenge, INCOIS started using the SST and Chl-a data from multiple satellites and optimally interpolated SST data from Global High Resolution Sea Surface Temperature (GHRSST) project enhancing the frequency of PFZ advisories from weekly thrice to daily advisories. The process chain in generation of PFZ advisories is in Fig. 8. A detailed description on the PFZ advisories and validation are given below.

INCOIS is providing PFZ advisories to the fishermen community, on a daily basis, utilizing remotely sensed SST and chl-*a* (Mohanty et al., 2017). In the present

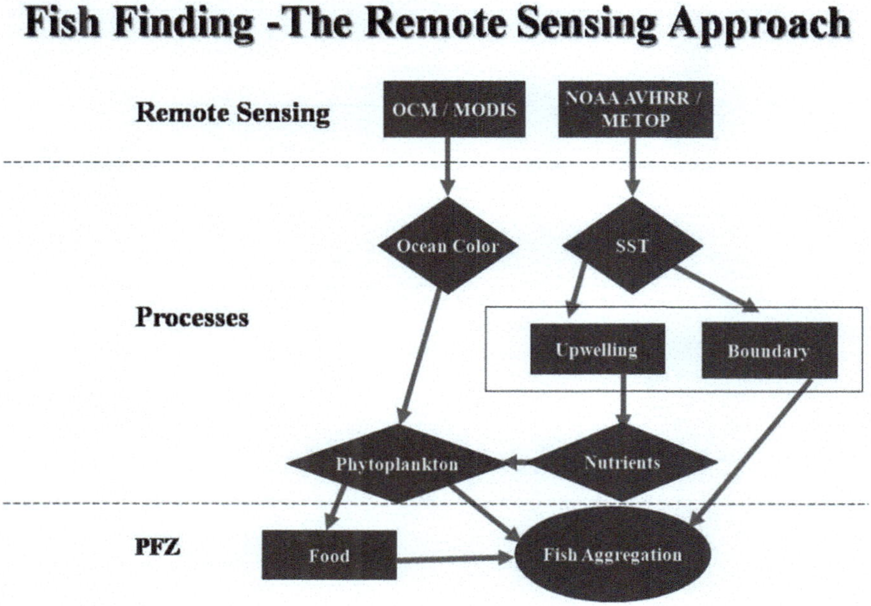

Fig. 5 Schematic of remote sensing approach of PFZ

setup, SST is acquired from NOAA-AVHRR, having spatial resolution of 1.1 km, and chl-*a* from either Oceansat-2 OCM (spatial resolution: 360 m) or MODIS-Aqua (spatial resolution: 1 km). The PFZs are identified as regions of strong SST gradients coinciding with higher chl-*a* concentration. In the existing operational mechanism, the thermal fronts are identified, as a first step, using the algorithm prescribed by Cayula and Cornillon (1992). The algorithm employs an overlapping window of 32 × 32 pixels over the entire region to detect fronts. The ratio of variance between the two populations to the variance within the populations is assigned as 0.76. The mean temperature difference between the two populations is kept as 0.45 °C for identifying the strong thermal fronts. The potential fronts from satellite chl-*a* is detected manually using Aeronautical Reconnaissance Coverage Geographic Information System (ArcGIS) software. Subsequently, the frontal vectors delineated from satellite SST and chl-*a* are superimposed to identify the common fronts. A tolerance limit of 15 km has been fixed to identify common fronts derived from both SST and chl-*a* (Nammalwar et al. 2013). In the absence of chl-*a* data, the strong persistent thermal fronts from SST data are identified as PFZs (Fig. 6).

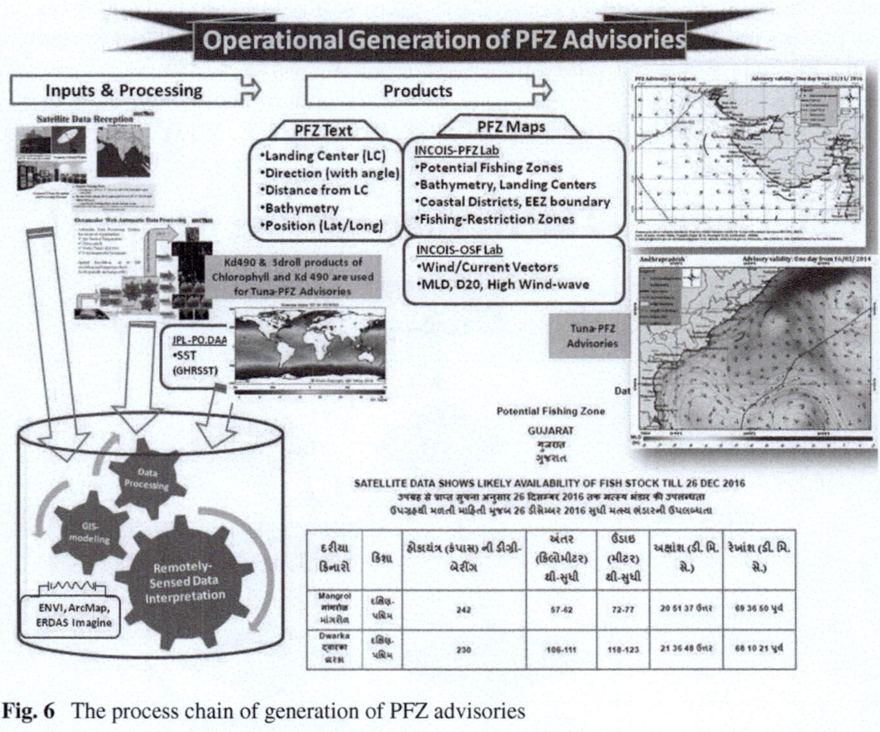

Fig. 6 The process chain of generation of PFZ advisories

7.1 Species Specific PFZ Advisories for Yellowfin Tuna

INCOIS started experimental tuna advisories with 3-day rolling satellite images of SST and chlorophyll with the aim to identify the matured productive fronts that have aggregated forage fish which in turn, will attract tuna to these regions. Additionally, tuna being a visual predator, one more parameter of water clarity, as perceived from satellite data of first optical depth (Kd 490) was also taken into consideration to avoid delineating tuna PFZs in turbid waters.

It has been argued that temperature plays a dominant role in defining the migration of tuna (Brill 1994). Although tuna movement often exhibits oscillatory patterns to balance energy conservation and foraging, it prefers staying mostly close to the base of mixed layers (Block et al. 1997) to avoid deeper cold water. Upwelling processes can be detected with the help of satellite derived SST values. Divergence translates in lowering Sea Surface Height, which can be observed through satellite altimetry (negative SSH-anomaly (SSHa)) and is strongly correlated with oxycline depth (Prakash et al. 2013). In order to understand this, INCOIS conducted a controlled experiment, wherein yellowfin tuna were tagged using recovery-independent pop-up satellite archival tags (PSATs) in the Indian Ocean. Analysis of tagged yellowfin tuna movements with reference to SSHa showed that the fishes tend to spend more time (~70%) in the periphery of divergence zones (between ± 6 cm SSHa) with frequency peak slightly towards negative side (Mean $= -1.2$, Median $= -1.5$). This supports the SST-based observations that show frontal adherence and also that yellowfin tuna is attracted to an extent, with the presence of food induced by divergence and upwelling (Nimit et al. 2019). However, the fact that the fishes avoid going further towards the core of divergence, suggests that there could be driving forces other than food, which might control yellowfin tuna horizontal movements.

For operational purposes, apart from standard OGDR, IGDR and GDR altimeter products, French Portal AVISO provides a range of global altimeter products such as MSLA (Maps of Sea Level Anomaly), MADT (Maps of Absolute Dynamic Topography) and absolute geostrophic velocities, along-track SLA and ADT etc. However, non availability of real-time data made it an unviable option for operational purposes. As an alternative, upon request, the SSHa data from CCAR (Colorado, USA) is being made available to INCOIS via dedicated FTP access since 2013 (Nimit et al. 2015). Additionally, oxycline depth derived from SSHa using conversion proposed by Prakash et al. (2013) helps in the generation of 3D advisories, which informs fishermen of the maximum depth of fishing (Fig. 7).

7.2 PFZ Validation

Towards estimating the accuracy level and to get the region-specific economic benefit of PFZ advisories, several researchers have conducted independent validation experiments in Indian seas. Specific validation experiments engaging two identical boats,

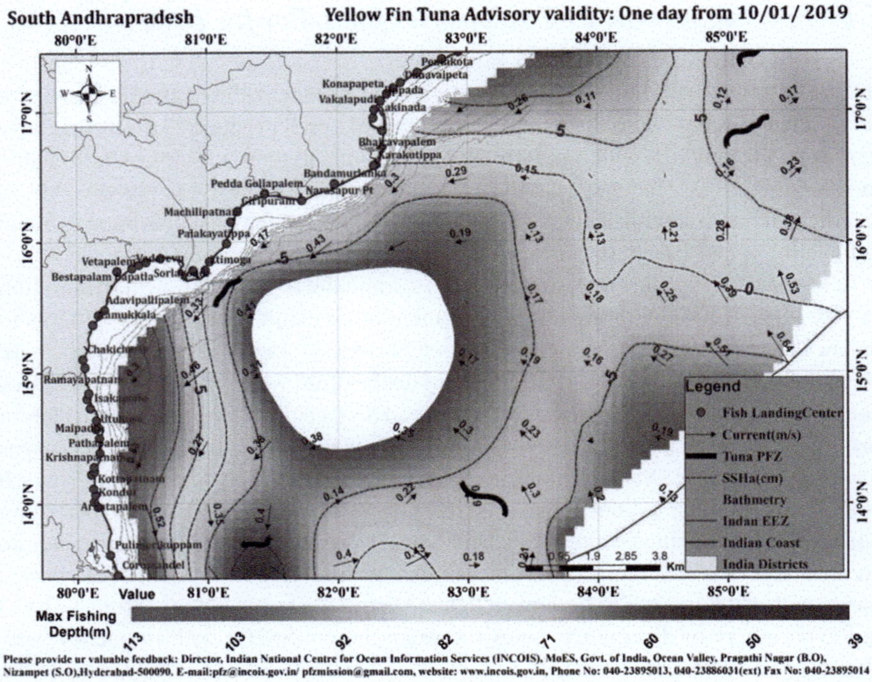

Fig. 7 An example of a Yellowfin Tuna PFZ advisory map with Maximum Fishing Depth (MFD) information

one for fishing in the notified area and the other for fishing away from the notified area have consistently shown that the net profits increase by 2–5 times due to the marked increase in Catch Per Unit Effort (CPUE) by 2–4 times (Nayak et al. 2007; Tummala et al. 2008). A major portion of the profit always comes from the savings on the cost of fuel due to the avoidance of search for potential fishing grounds. On an average, the utilization of PFZ advisories for fishing reduced the time spent on fishing by 30–70%.

The controlled validation experiment off West Bengal showed that the CPUE in the PFZ notified area was almost two folds higher than the mean CPUE obtained from the fishing operations carried away from the notified areas. The experiment was conducted with two identical boats, one operated in the notified area and another operated away from the notified area (Fig. 8). Both the fishing boats had gill netting facility and the boat that operated in the PFZ notified area did fishing for 24 h (08 hauls casted) and was able to fetch total fish catch of 417 kg having selling cost of Rs. 87,000/-. The fishing boat that operated away from the PFZ area and did fishing for 48 h (16 hauls casted) and was able to fetch 203 kg having a selling cost of Rs. 40,500/-. The boat that operated in the PFZ area had a 3.5 fold increase in profit, as compared to the boat that operated away from the notified area (Table 5).

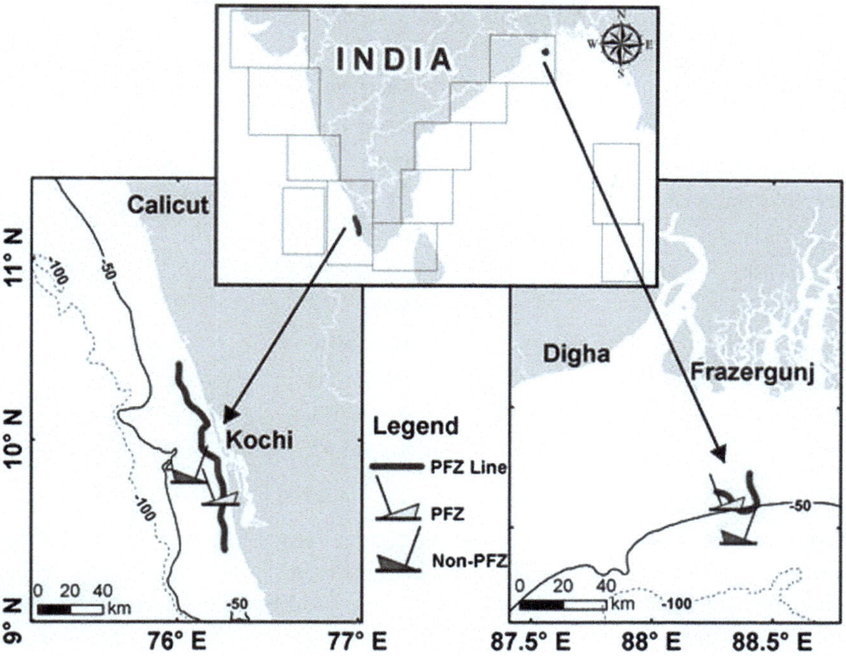

Fig. 8 Schematic representation of controlled validation experiment performed off Kerala and West Bengal

Table 5 Details of simultaneous fishing operations including economics (West Bengal)

Details	PFZ	Non-PFZ
Name of the boat	Bhawani Ma	Bramhoni Ma
Type of net	Gill net	Gill net
Duration of total trip	45 h (including dead run)	72 h (including dead run)
Number of fishing hours	24	48
Number of hauls	8	16
Number of fishermen engaged	10	10
Total catch (kg)	417	203
Major species caught	*Pampus argenteus, Tenualosa ilisha*	Indo-pacific king mackerel, *Tenualosa ilisha*, Cat fish
Approximate cost of total catch	87,000	40,500
Total expenditure in fishing operation (Rs.)	35,000	26,000
Net profit (Rs.)	52,000	14,500

Table 6 Details of simultaneous fishing operations including economics (Kerala)

Details	PFZ	Non-PFZ
Name of the boat	MRR-29	MRR-27
Type of boat	Mech. ring seine	Mech. ring seine
Duration of total trip	09 h	08 h
Number of fishing hours	01	01
Number of hauls	01	01
Number of fishermen engaged	20	20
CPUE in (kg)	8000	550
Major species caught	Oil sardine	Oil sardine
Approximate cost of total catch (Rs.) @16/kg	Rs. 200,000	Rs. 13,750
Total expenditure in fishing operation (Rs.)	Rs. 36,075	Rs. 9,475
Net profit	Rs. 163,925	Rs. 4,275

A similar experiment was carried out off Kerala coast using two identical Ring Seiners boats on November 18, 2009. One boat operated in the notified area, while the other one operated away from the notified area and both the fishing boats did fishing for one hour. The boat in the notified area fetched 8000 kg of Oil Sardine, with selling cost of Rs. 200,000/- and the other was able to catch 550 kg of Oil Sardines, with selling cost of Rs. 13,750/-. The experiment showed more than 38.5 fold increase in profit for the boat operated in the PFZ notified area as compared to the boat that operated away from the notified area (Table 6).

8 The Dissemination of Service products

Even though the primary dissemination mode for OSF/PFZ is the INCOIS website, it also provides these services through other modes like e-mail, mobile phones (in the form of SMSs and mobile apps), TV, Radio and Electronic Display Boards to all the Stakeholders, in multilingual modes too (Fig. 9). The Information and Communication Technology tools have been exploited to the maximum for providing in-time quality forecasts and advisories to the users. Further, there is strong collaboration with NGOs and coastal research centres as well as universities, for further dissemination. One significant and noteworthy achievement during recent times is the transmission of warning messages to the fishers far out in the sea (1500–6000 km from Indian the coast) through satellite communications, by collaborating with Indian Space Research Organisation (ISRO) for NAVIC and Airport Authority of India (AAI) for Gagan Enabled Mariner's Instrument for Navigation and Information (GEMINI), both through satellite communications.

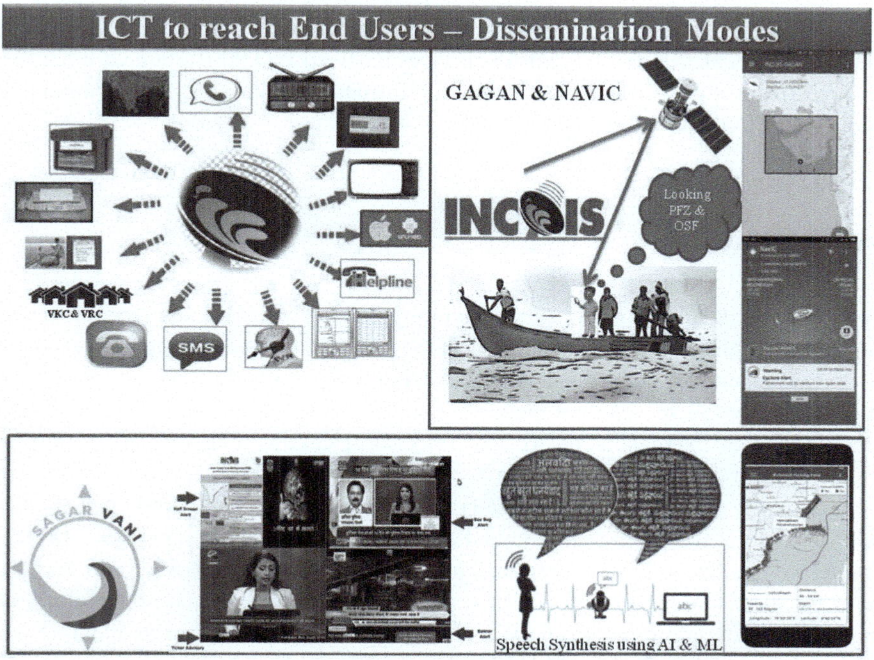

Fig. 9 The multi-modal dissemination system used by OSF and MFAS

9 The Users and Feedback

The users of INCOIS products and services include fisher and coastal populations, Maritime Boards, Indian Navy, Indian Coast Guard, shipping sector, energy sector, hydrocarbon industries, port authorities, Pollution Control Boards, Disaster Management Agencies, NGOs and many research organisations. However, fishers dominate the user profile and every attempt is made to provide the best services to them, given their socio-economic background. There are 7 lakh registered users, largely comprising of fishermen utilizing these services, for their operational requirements and safety (Balakrishnan Nair et al. 2020). The user survey indicated that the OSF and MFAS information is reaching to 16 lakh users on daily basis through various modes of dissemination. User feedback and delayed mode evaluation suggest not only that the ocean state forecasts and Marine Fishery Advisory forecasts are more than 80% accurate, but also that the forecasts/information reach the maximum end users on time, which is also equally crucial for saving life and property, and for smooth regular operations of these users. Recent feedback on the Ocean State Forecast Services and the Marine Fisheries Advisory Services of INCOIS has been presented in Harikumar et al. (2020). The overall cumulative economic benefits due to OSF service realised by Indian Navy, Indian Coast Guard, oil and gas exploration etc. as per National Center for Applied Economic Research (NCAER), computations exceed Rs. 3.7 trillion (Venkatesan et al. 2015). OSF/MFAS services received ISO 9001:2008 certification from the Ministry of Communications and Information Technology, Government of India on 29 December 2014.

10 Future Plans

A general feedback from the user community in the context of sea state forecasts indicates that there is a realization of what the weather or sea state might be like, but there is frequently a lack of proper understanding of what could be the impacts of the weather or the sea state on the users (WMO 2015). Accordingly, as per WMO (No. 1150 in 2015) manual, all weather forecasting centres of its member countries are advised to upgrade their weather forecast and warning services to multi-hazard impact-based, rather than a generalised one. The latest update of WMO manual (No. 558 in WMO 2018) puts forth a guideline that warning should be provided for the dangerous sea states. INCOIS captures upcoming sea state conditions through short term forecasts and issues specialised warnings/alerts to the coastal population and seafarers. In this context, INCOIS is in the process of transforming and updating the forecasts, warnings, advisories and alerts into multi-hazard impact-based forecast and warning services, for the wide user communities. For example, the alerts/warnings on rough sea conditions due to the high period swell waves are presented along with the category of user community likely to be affected, how it would be affected, and what kind of precautions to be taken. Importantly, the development of such a system can only be achieved by a sustained and close interaction with the end users along with feedback collection, and the incorporation of the same.

At present, MFAS provides PFZ information only for the productive fronts. In order to extend the scope of the service, fish biomass availability is to be linked with environmental parameters. This can be achieved with the help of conventional fishing operations along with observations from advanced fishery acoustics, environmental DNA (eDNA) and egg sampling (e.g. Continuous Underway Fish-Egg Sampler or CUFES) techniques (Holmes et al., 2021; Nimit K., 2021; Reddem et al., 2021) . Such an approach also helps take into account ecosystem-based sustainable fisheries. Another step in the same direction would be to make use of primary and secondary productivity information in addition to sea surface temperature and chlorophyll (Nimit et al., 2016; Chakraborty et al., 2019). Bycatch reduction and anti-poaching aspects may be focused on to aid the conservation efforts and make the service further environmentally responsible. Doorstep delivery of service by dissemination at sea (onboard) can help fishermen immensely by paving the way for nowcasting services. Similarly, integration of information on hazards to fishery (harmful algal blooms, oil spill, jelly-fish swarms, micro plastics accumulation, etc.) would be readily useful to the fishermen (Pattiaratchi et al., 2022). Fully coupled and developed biogeochemical numerical models can help in forecasting important parameters for one or more of these aspects. In addition to the capture fishery, providing adequate focus towards services for mariculture would help our country meet its nutritional, medicinal and bio-fuel requirements to a great extent. All of these thrust areas will play very crucial roles in coming decades for realizing India's blue revolution goals.

Acknowledgements The authors are very much thankful to the Director, INCOIS for the encouragement and support. Ocean State Forecast Services and Marine Fisheries Advisory Services are flagship programs of INCOIS and we thank the Secretary, Ministry of Earth Sciences, and Government of India for encouragement and continued financial support. This is INCOIS contribution number 425.

References

Aditya ND, Sandhya KG, Harikumar R, Balakrishnan Nair TM (2020) Development of small vessel advisory and forecast services system for safe navigation and operations at sea. J Oper Oceanogr 1–16. https://doi.org/10.1080/1755876X.2020.1846267

Balakrishnan Nair TM, Sirisha P, Sandhya KG, Srinivas K, Sabique L, Nherakkol A, Krishna Prasad B, Kumari R, Jeyakumar C, Kaviyazhahu K, Ramesh Kumar M, Harikumar R, Shenoi SSC, Nayak S (2013) Performance of the ocean state forecast system at Indian National Centre for Ocean Information Services. Curr Sci 105(2):175–181

Balakrishnan Nair TM, Remya PG, Harikumar R, Sandhya KG, Sirisha P, Srinivas K, Nagaraju C, Nherakkol A, Krishna Prasad B, Jeyakumar C, Kaviyazhahu K, Hithin NK, Kumari R, Sanil Kumar V, Ramesh Kumar M, Shenoi SSC, Nayak S (2014) Wave forecasting and monitoring during very severe cyclone Phailin in the Bay of Bengal. Curr Sci 106(8):1121–1125

Balakrishnan Nair TM, Harikumar R, Srinivas K, Anuradha M, Kaviyazhahu K et al (2020) Ocean state forecast services for the maritime community. Geogr You 20(6–7):18:25

Block BA, Keen JE, Castillo B, Dewar H, Freund EV, Marcinek DJ, Brill RW, Farwell C (1997) Environmental preferences of yellowfin tuna (Thunnus albacares) at the northern extent of its range. Mar Biol 130(1):119–132

Breivik Ø, Allen AA (2008) An operational search and rescue model for the Norwegian Sea and the North Sea. J Mar Syst 69(1–2):99–113

Brill RW (1994) A review of temperature and oxygen tolerance studies of tunas pertinent to fisheries oceanography, movement models and stock assessments. Fish Oceanogr 3:204–216

Cayula JF, Cornillon P (1992) Edge detection algorithm for SST images. J Atmos Oceanic Tech 9(1):67–80

Choudhury SB, Jena B, Rao MV, Rao KH, Somvanshi VS, Gulati DK, Sahu SK (2007) Validation of integrated Marine Fishery Advisory Services (IPFZ) forecast using satellite based chlorophyll and sea surface temperature along the east coast of India. Int J Remote Sens 28(12):2683–2693

Chakraborty K, Kumar N, Girishkumar MS, Gupta GVM, Ghosh J, Udaya Bhaskar TVS, Thangaprakash VP (2019) Assessment of the impact of spatial resolution on ROMS simulated upper-ocean biogeochemistry of the Arabian Sea from an operational perspective. J Oper Oceanogr 12(2):116–142. https://doi.org/10.1080/1755876X.2019.1588697

Deshpande SP, Radhakrishnan KV, Bhat UG (2011) Direct and indirect validation of Marine Fishery Advisory Services advisory off the coast of Uttara Kannada. Karnataka, JSIRS 39(4):547–554

Dwivedi RM, Solanki HU, Nayak S, Gulati D, Somvanshi VS (2005) Exploration of fishery resources through integration of ocean colour with sea surface temperature: Indian experience. Indian J Mar Sci 34(4):430–440

Effy BJ, Francis PA, Ramakrishna SSVS, Mukherjee A (2020) Anomalous warming of the western equatorial Indian Ocean in 2007: role of ocean dynamics. Ocean Model 101542. https://doi.org/10.1016/j.ocemod.2019.101542

Francis PA, Vinayachandran PN, Shenoi SSC (2013) The Indian Ocean forecast system. Curr Sci 104(10):1354–1368

Francis PA, Abraham J, Chatterjee A, Mukharjee A, Shankar D, Vinayachandran PN (2020a) Structure and dynamics of under currents in the south-east coast of India. Ocean Dyn 01340-9

Francis PA, Jithin AK, John EB, Chatterjee A, Chakraborty K, Paul A, Balaji B, Shenoi SSC, Biswamoy P, Singh AMP, Deepsankar B, Siva Reddy S, Vinayachandran PN, Girish Kumar MS, Udaya Bhaskar TVS, Ravichandran M, Unnikrishnan AS, Shankar D, Prakash A, Aparna SG, Harikumar R, Kaviyazhaku K, Suprit K, Venkat Shesu R, Kiran Kumar N, Srinivasa Rao N, Annapurnaiah K, Venkatesan R, Suryachandra Rao A, Rajagopal EN, Prasad VS, Das Gupta M, Balakrishnan Nair TM, Pattabhi Rama Rao E, Satyanarayana BV (2020b) High-resolution operational ocean forecast and reanalysis system for the Indian Ocean. Bull Am Meteor Soc. https://doi.org/10.1175/BAMS-D-19-0083.1

Friedland KD, Stock C, Drinkwater KF, Link JS, Leaf RT et al (2012) Pathways between primary production and fisheries yields of large marine ecosystems. PLoS ONE 7(1):e28945. https://doi.org/10.1371/journal.pone.0028945

Harikumar R, Hithin NK, Balakrishnan Nair TM, Sirisha P, Krishna Prasad B, Jeyakumar C, Nayak S, Shenoi SSC (2015) Ocean state forecast along ship-routes: evaluation using ESSO-INCOIS real-time ship-mounted wave height meter and satellite observations. J Atmos Oceanic Technol. https://doi.org/10.1175/JTECH-D-15-0047.1

Harikumar R, Balakrishnan Nair TM, Rao BM, Rajendra Prasad P, Ramakrishna Phani P, Nagaraju C, Ramesh Kumar M, Jeyakumar C, Shenoi SSC, Nayak S (2016) Ground zero met-ocean observations and attenuation of wind energy during cyclonic storm Hudhud. Curr Sci 110(12). https://doi.org/10.18520/cs/v110/i12/2245-2252

Harikumar R, Nagarajakumar M, Nair TMB (2020) Empowering seafarers of India. Geogr You 20(6–7):76–79

Holmes EE, BR S, Nimit K, Maity S, Checkley JrDM, Wells ML, Trainer VL (2021) Fish Oceanogr 30(6):623–642. https://doi.org/10.1111/fog.12541

Kumari B, Raman M, Mali K (2009) Locating tuna forage ground through satellite remote sensing. Int J Remote Sens 30(22):5977–5988

Nimit K (2021) Ideas and perspectives: ushering the indian ocean into the UN Decade of Ocean Science for Sustainable Development (UNDOSSD) through marine ecosystem research and operational services–an early career's take. Bio geo sci 18(12):3631–3635 https://doi.org/10.5194/bg-18-3631-2021

Murakami H, Vecchi GA, Underwood S (2017) Increasing frequency of extremely severe cyclonic storms over the Arabian Sea. Nat Clim Change 7:885–889

Mohanty PC, Mahendra RS, Nayak RK, Kumar N, Kumar TS, Dwivedi RM (2017) Persistence of productive surface thermal fronts in the northeast Arabian Sea. Reg Stud Mar Sci 16216–224. S2352485516303073. https://doi.org/10.1016/j.rsma.2017.09.010

Nammalwar P, Satheesh S, Ramesh R (2013) Applications of remote sensing in the validations of Potential Fishing Zones (PFZ) along the coast of North Tamil Nadu, India. Indian J Geo-Mar Sci 42(3):283–292

Nayak S (2020) Towards blue economy: a perspective. Indian J Geosci 74(3):191–196

Nayak S, Srinivaskumar T, Nagarajakumar M (2007) Satellite based fishery service in India. Full Pict 256–257

Nimit K, Nagaraja Kumar M, Swetha N, Nayak J, Rose PB, Srinivasa Kumar T (2015) Utility of Sea Surface Height anomaly (SSHa) in determination of potential fishing zones

Nimit K, Lotlikar A, Srinivasa Kumar T (2015) (2016) Validation of MERIS sensor's coast colour algorithm for waters off the west coast of India. Int J Remote Sens 37(9):2066–2076. https://doi.org/10.1080/01431161.2015.1129564

Nimit K, Nagaraja Kumar M, Berger AM, Bright RP, Prakash S, Uday Bhaskar TVS, Srinivasa Kumar T, Rohit P, Tiburtius A, Ghosh S, Varghese SP (2019) Oceanographic preferences of Yellowfin tuna (*Thunnus albacares*) in warm stratified oceans: a remote sensing approach. Int J Remote Sens 41(15):5785–5805. https://doi.org/10.1080/01431161.2019.1707903

Pattiaratchi C, van der Mheen M, Schlundt C, Narayanaswamy BE, Sura A, Hajbane S, White R, Kumar N, Fernandes M, Wijeratne S (2022) Plastics in the Indian Ocean–sources transport distribution and impacts. Ocean Sci 18(1):1–28. https://doi.org/10.5194/os-18-1-2022

Platt T, Sathyendranath S (1988) Oceanic primary production: estimation by remote sensing at local and regional scales. Science 241:1613–1620

Prakash S, Prakash P, Ravichandran M (2013) Can oxycline depth be estimated using sea level anomaly (SLA) in the northern Indian Ocean? Remote Sens Lett 4:1097–1106

Prasad SJ, Balakrishnan Nair TM, Francis PA, Vijayalaksmi T (2014) Hindcasting and validation of Mumbai oil spills using GNOME. Int Res J Environ Sci 3(12):18–27

Ravichandran M, Behringer D, Sivareddy S, Girishkumar MS, Chacko N, Harikumar R (2013) Evaluation of the global ocean data assimilation system at INCOIS: the tropical Indian Ocean. Ocean Model 69:123–135. https://doi.org/10.1016/j.ocemod.2013.05.003

Remya PG, Vishnu S, Praveen Kumar B, Balakrishnan Nair TM, Rohith B (2016) Teleconnection between the North Indian Ocean high swell events and meteorological conditions over the Southern Indian Ocean. J Geophys Res Oceans 121(10):7476–7494

Remya PG, Rabi Ranjan T, Sirisha P, Harikumar R, Balakrishnan Nair TM (2020) Indian Ocean wave forecasting system for wind waves: development and its validation. J Oper Oceanogr 1–16. https://doi.org/10.1080/1755876X.2020.1771811

Reddem VS, Muthalagu R, Bekkam VR, Eluri PRR, Jampana V, Nimit K (2021) Ocean fronts detection over the Bay of Bengal using changepoint algorithms–A non-parametric approach. Oceanologia 63(4):438–447. S0078323421000439. https://doi.org/10.1016/j.oceano.2021.05.003

Sabique L, Annapurnaiah K, Balakrishnan Nair TM, Srinivas K (2012) Contribution of Southern Indian Ocean swells on the wave heights in the Northern Indian Ocean—a modeling study. Ocean Eng 43:113–120

Sandhya KG, Remya PG, Balakrishnan Nair TM, Arun N (2016) On the co-existence of high-energy low-frequency waves and locally-generated cyclone waves off the Indian east coast. Ocean Eng 111:148–154

Sandhya KG, Murty PLN, Aditya Deshmukh N, Balakrishnan Nair TM, Shenoi SSC (2018) An operational wave forecasting system for the east coast of India. Estuar Coast Shelf Sci 202:114–124

Sirisha P, Remya PG, Modi A, Tripathy RR, Balakrishnan Nair TM, Venkateswara Rao B (2019) Evaluation of the impact of high-resolution winds on the coastal waves. J Earth Syst Sci 128(8), art. no. 226

Srinivas K, Dinesh Kumar PK (2002) Tidal and non-tidal sea level variations at two adjacent ports on the southwest coast of India. Indian J Mar Sci 31(4):271–282

Tummala SK, Masuluri NK, Nayak S (2008) Benefits derived by the fisherman using Marine Fishery Advisory Services (PFZ) advisories. Proc SPIE Int Soc Opt Eng 7150:71500N

Venkatesan R, Munjal P, Sharma A, Meattle C (2015) Economic benefits of dynamic weather and ocean information and advisory services in India and cost and pricing of customized products and services of ESSO-NCMRWF & ESSO-INCOIS (Phase III), 171 pp

WMO (2015) WMO guidelines on multi-hazard impact-based forecast and warning services (WMO. No. 1150), 34 pp

WMO (2018) Manual on marine meteorological services, Volume I Global aspects (WMO. No. 558), 73 pp

Zainuddin M, Saitoh SI, Saitoh K (2004) Detection of potential fishing ground for albacore tuna using synoptic measurements of ocean color and thermal remote sensing in the northwestern North Pacific. Geophys Res Lett 31(20)

Satellite-Based Marine Ecological Services for the Indian Ocean Region

Sanjiba K. Baliarsingh, Alakes Samanta, Aneesh A. Lotliker, Prakash C. Mohanty, R. S. Mahendra, and T. M. Balakrishnan Nair

Abstract India has a vast coastline of ~7,500 km encompassing the west coast (eastern Arabian Sea), east coast (western Bay of Bengal), and islands. The coastal water of India is rich in a wide variety of marine biotic and abiotic resources that support the livelihood of millions through fishery, recreational activities, tourism, marine commerce, maritime logistics, education-research, etc. However, frequent phytoplankton blooms, jellyfish swarming (and beach stranding), and bleaching of coral reefs are exerting adverse impacts on the coastal water quality and ecological services. Indian National Centre for Ocean Information Services (INCOIS), an autonomous body under the Indian Ministry of Earth Sciences, has taken the initiative to monitor and disseminate information/warnings on the above-mentioned perturbations. INCOIS has operationalized satellite-based Algal Bloom Information Service (ABIS) and Coral Bleaching Alert System (CBAS) for monitoring phytoplankton blooms and coral bleaching, respectively. For monitoring jellyfish swarming, environmental triggers have been identified based on favorable conditions and a conceptual framework has been prepared for the possible generation of Jellyfish Aggregation Advisory Service (JAAS). Apart from assisting policy-makers/researchers in developing sustainable ocean management strategies, demarcating marine protected areas, and conducting scientific research, these services also complement other satellite-based ecosystem services.

Keywords INCOIS · India · Satellite · Algal Bloom · Coral bleaching · Jellyfish swarming

---

S. K. Baliarsingh · A. Samanta · A. A. Lotliker (✉) · P. C. Mohanty · R. S. Mahendra · T. M. B. Nair
Indian National Centre for Ocean Information Services (INCOIS), Ministry of Earth Sciences, Hyderabad 500090, India
e-mail: aneesh@incois.gov.in

V. K. Gahalaut and M. Rajeevan (eds.), *Social and Economic Impact of Earth Sciences*,
https://doi.org/10.1007/978-981-19-6929-4_12

1 Introduction

The resources from India's marine environment provide a livelihood to more than 3.5 million people through recreation, fishing, and other economic activities (Nayak 2017). The productive coastal waters of India support large fisheries. Phytoplankton are the major primary producers and form the base of the marine food chain, and they are ubiquitous in the Indian waters, including the estuaries, lagoons, coastal waters, and open oceans where adequate sunlight reaches. When growth-promoting nutrients are available in adequate quantity, phytoplankton can overgrow and generate blooms. The problem becomes more severe when the bloom is caused by toxic species. This results in biomagnifications of toxins in the food chain, which finally reaches humans through fish. There are many instances of human death due to phytotoxin. Many of the phytoplankton bloom events have resulted in the disruption of ambient water quality, making inhabitable conditions for other biotas, and these events are categorized as Harmful Algal Bloom (HAB). The recurrent phytoplankton blooms pose a threat to the marine ecosystems, and the associated ecology and socio-economy (Oyeku and Mandal 2021). There is a demand for systematic high-frequency monitoring to identify the bloom regions, to study the bloom causative factors, and to understand the consequences.

The coastal water also caters to tourism in terms of swimming, snorkeling, diving, and beach recreational activities. However, the jellyfish swarming in coastal waters and beach stranding events make them unfit for tourism. Jellyfish toxic stinging to humans can cause several medical complications such as respiratory arrest, abdominal cramps, nausea, vomiting, headache, anxiety, and hypertension (Lucas et al. 2014). The jellyfish swarming as well as beach stranding events are on the rise in both east and west coast of India, resulting in environmental deterioration (Baliarsingh et al. 2020).

As like phytoplankton and jellyfish, coral reefs are also abundant in Indian waters and play a pivotal role in providing a home for a wide variety of marine species, including fish, mollusks, worms, crustaceans, echinoderms, sponges, tunicates, and other cnidarians. Coral reefs are huge biological structures in coastal oceans that have evolved over millennia to their current state. The major reef formations in India are restricted to the Gulf of Mannar, Palk Bay, Gulf of Kutch, Andaman, and Nicobar Islands, and the Lakshadweep islands. In India, coral reefs are being damaged and destroyed at an alarming rate. They are under a great deal of stress as a result of anthropogenic pressures and interventions. However, we cannot be precise about how much and where because of the special difficulties of monitoring underwater. However, over the last decade, global coral reef ecosystems have deteriorated significantly, posing a serious threat to the essential coral ecosystems. One of the major reasons to the rising deterioration of reef health has been coral bleaching. Under certain environmental pressures, the symbiotic link between algae (zooxanthellae) and their host corals breaks off, resulting in coral bleaching. As a result, the zooxanthellae are expelled from the host. The corals display their white underlying calcium carbonate coral skeleton in the absence of symbiotic algae, and the damaged coral colony turns pale in hue. Coral bleaching can be triggered and sustained by a variety of environmental factors. Massive coral bleaching has been recorded worldwide, especially in Indian waters, as a result of unusually warm waters in recent years. This demands that coral bleaching occurrences must be better understood, monitored, and predicted.

Indian National Centre for Ocean Information Services (INCOIS), an autonomous body under the Ministry of Earth Sciences, Government of India has taken the initiative to monitor and disseminate information/warnings on the above-mentioned perturbations. A detailed account on the backdrop and generation of different marine ecological services for monitoring of algal bloom, coral bleaching, and jellyfish aggregation is provided in the following sections.

2 Algal Blooms and Their Harmful Effects

The algal blooms may also exert adverse effects on the fisheries. In addition, the introduction of non-native species to coastal waters through ballast water exchange, aquaculture practices, climate change, and cultural eutrophication are also potential causes of increment in frequency as well as the spatial extent of bloom events. The algal blooms are regarded as HAB events when exerting a deleterious effect on the environment (Anderson 1994; Hallegraeff 2010; Trainer et al. 2020; Sahu et al. 2020; Baliarsingh et al. 2016, 2020). HABs include both toxic and non-toxic phytoplankton species. The toxins released from bloom-forming toxic algal species primarily belonging to diatom, dinoflagellates, and cyanobacteria have the potential to be biomagnified, reaching higher trophic levels.

Human ailments linked to the intake of seafood and exposure to contaminated aerosols have been linked to phytoplankton species that produce strong toxins. Phytotoxins have been linked to the periodic deaths of marine mammals, birds, and other animals dependent on the marine food web, in addition to their negative effects on human health. Non-toxic algal blooms, on the other hand, have a significant impact on ecosystems due to resultant events of dissolved oxygen concentration decline, reduced food web efficiency, pathogenic bacteria stimulation, and other ecological implications.

In Indian waters, algal bloom studies indicate a higher number of reports along the west coast in comparison to the east coast. A total of 101 algal bloom events have been reported in Indian waters, with 68 occurrences on the west coast and 33 on the east coast during the period of 1908–2009 (D'Silva et al. 2012). A recent study has updated the reported occurrences to 154 during the period of 1908 to 2017 (Oyeku and Mandal 2021). In the last decade (2008 and 2017), a total number of 50 phytoplankton bloom incidents have been reported in the Indian waters, which drives scientific attention (Oyeku and Mandal 2021). Blooms of phytoplankton groups such as diatoms, dinoflagellates, cyanobacteria, raphidophytes, and haptophytes have been reported from coastal waters of the eastern Arabian Sea, while blooms of diatoms, dinoflagellates, and cyanobacteria were reported from coastal waters of the western

Bay of Bengal. Heterotrophic dinoflagellate *Noctiluca scintillans* and diazotrophic cyanobacteria *Trichodesmium erythraeum* were the most frequently occurring bloom species on the west coast of India. In addition to *N. scintillans* and *T. erythraeum*, recurring blooms of diatomic species *Asterionellopsis glacialis* are regular algal bloom scenarios of the east coast of India. In regards to the spatial distribution of algal blooms along the west coast, the highest prevalence of bloom incidents were observed in coastal waters off Kerala, Mangalore, and Goa, while along the east coast, bloom incidents were observed from coastal waters off Tamil Nadu and Odisha (D'Silva et al. 2012).

3 Algal Bloom Detection and Monitoring Service

In terms of large spatio-temporal monitoring of phytoplankton, ocean colour remote sensing provides a cost-effective opportunity. Although in situ observations are the most accurate, remotely sensed measurements are cost-effective and able to detect quickly and provide synoptic scenarios on wide spatio-temporal scale (Baliarsingh et al. 2017; Lotliker et al. 2018). Specific bio-optical algorithms are required to retrieve algal bloom information from ocean colour satellites, and several such schemes have been developed. Ocean colour satellite missions have provided more than one dimension to our understanding of ecosystem-level interactions in ocean.

INCOIS has taken an initiative on "Algal Bloom Information Service (ABIS)" using ocean colour remote sensing technology with a suite of appropriate bio-optical algorithms. An Automatic Data Processing Chain (ADPC) is established for acquisition, processing and dissemination of ocean colour data products from Moderate Resolution Imaging Spectroradiometer onboard Aqua satellite (MODIS-Aqua) at near real-time. The ADPC acquires level 1A data from NASA-GSFC (https://oceancolor.gsfc.nasa.gov/) and processes to level 3 using SeaWiFS Data Analysis System (SeaDAS) at 1 and 4 km spatial resolution. A schematic workflow of the ADPC is shown in Fig. 1.

The standard products generated using ADPC, such as chlorophyll concentration (CHL), Sea Surface Temperature (SST) and Diffuse Attenuation Coefficient at 490 nm (KD_490) are utilized to generate Potential Fishing Zone (PFZ) advisory. In addition, the value-added products such as Algal Bloom Index (Shanmugam 2011), Phytoplankton Class (Dwivedi et al. 2015), Bloom Indices (Ahn and Shanmugam 2006), 30 days rolling chlorophyll and SST anomaly are utilized for detection and monitoring of algal blooms.

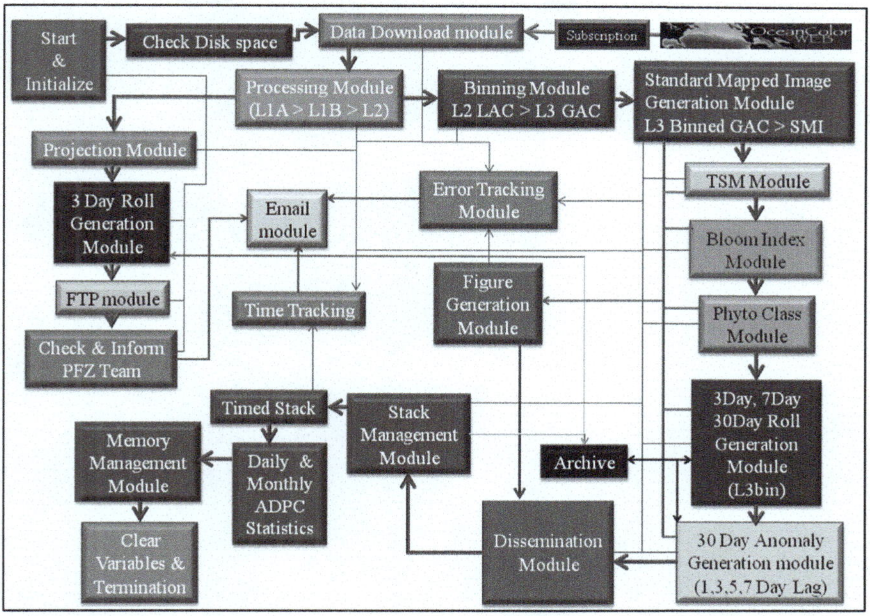

Fig. 1 Schematic workflow of currently operational INCOIS-ADPC

4 Algal Bloom Monitoring Hotspots

INCOIS has developed an algal bloom detection and monitoring service to provide information on the bloom status in four selected bloom hotspot regions such as the northeastern Arabian sea, Coast of Kerala, Gulf of Mannar, and Coast of Odisha. Figure 2 shows the extent of the above-mentioned regions on the map.

The bloom rank order in the northern Arabian Sea was diatom, cyanobacteria, and dinoflagellates, according to studies on phytoplankton dynamics. Contrary to this pattern, the phytoplankton population has been dominated by green-colored *N. scintillans* since the 1990s (Matondkar et al. 2004; Parab et al. 2006; Gomes et al. 2008). Especially, the open ocean waters of the northeastern Arabian Sea (north of 15° N) infests with green *N. scintillans* during the winter months. In general, detrimental consequences of large-scale *N. scintillans* bloom results in mortality of aquatic fauna attributed to the accumulation of toxic levels of ammonia in the ambient medium. In addition, the senescence phase of the bloom results in decline of dissolved oxygen levels in the water column. The spatial distribution of algal blooms along the western continental shelf has been reported with the highest prevalence from the Kerala coast. The nutrient influx due to strong upwelling during the southwest monsoon serves as a conducive factor for phytoplankton bloom. There were 23 dinoflagellate species that were responsible for blooms along the Kerala coast. The majority of raphidophyte (*Chattonella marina*) blooms have been documented along the Kerala coast (Subrahmanyan 1954; Jugnu and Kripa 2009). *Microcystis aeruginosa* cyanobacterial blooms

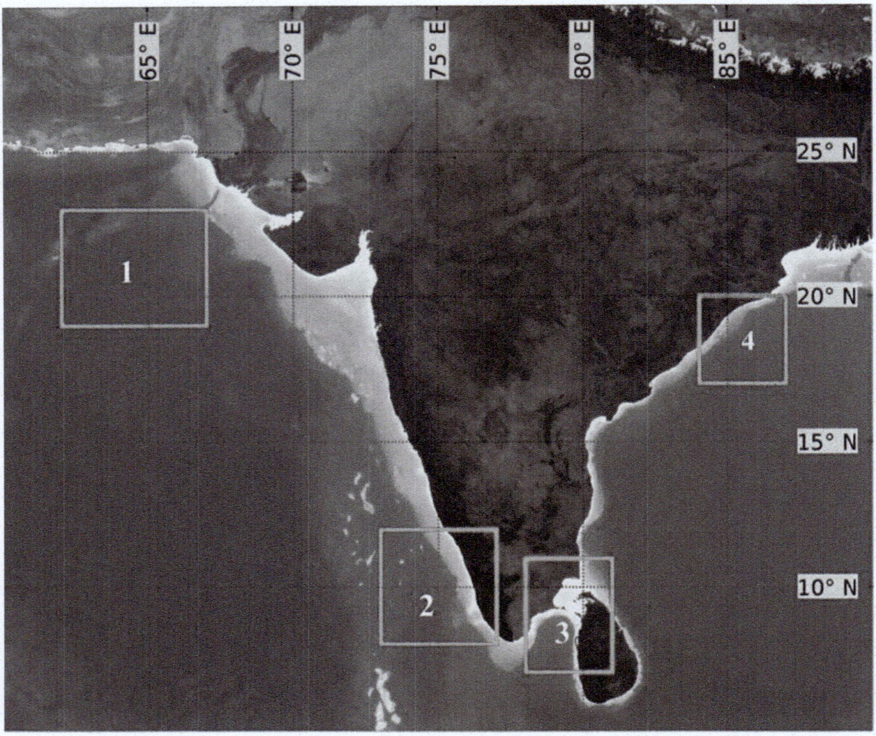

Fig. 2 Extents of algal bloom detection and monitoring regions of interest i.e. northeastern Arabian Sea, Coast of Kerala, Gulf of Mannar and Coast of Odisha

have appeared in Kerala's inshore seas (Padmakumar et al. 2008). The Gulf of Mannar also experiences the intense bloom of green *N. scintillans*. The bloom's occurrence and persistence along a narrow swath of the coast from Kilakarai to Pamban was attributed to favourable environmental conditions such as higher sea surface temperature and salinity, low pH, lack of water currents, high nutrient concentrations, low precipitation, and a favourable wind along the shore (Gopakumar et al. 2009). The southern coast of Odisha experiences recurring bloom of heterotrophic red *N. scintillans* off Rushikulya estuarine region. This site has its importance for periodic stay of the migratory sea turtles (*Lepidochelys olivacea*) for mass-mating and nesting. The higher phase of bloom occurred during a period of high abundance of jellyfish, which may have aided bloom formation by grazing on other phytoplankton grazers, especially copepods. In the falling phase of bloom, ammonium concentrations were found to be higher and dissolved oxygen concentration dropped (Baliarsingh et al. 2016).

5 Operational Algal Bloom Information Service Generation

Geophysical products selected for this service are SST, chl-*a* (O'Reilly et al. 1998; Hu et al. 2012), chl_abi (Shanmugam 2011), BI: Bloom Index (Ahn and Shanmugam 2006), Phytoplankton ID (Dwivedi et al. 2015), Phytoplankton size-fraction (Sahay et al. 2017. The current ADPC uses a standard atmospheric correction scheme implemented by NASA OBPG in L2GEN executable of SeaDAS. L2GEN is used to generate L2 LAC files, using L1B LAC and corresponding GEO files created by modis_L1B.py and modis_GEO.py of SeaDAS, respectively from L1A LAC files fetched from the NASA-GSFC portal. The L2 LAC files are then mosaicked together and projected to cover the region of interest. All the algorithms other than the standard ones are implemented using Graph Processing Toolkit (GPT) of SeaDAS and uses the above-mentioned projected and mosaicked files as input. The standard products are fetched from the INCOIS-ADPC and are cropped as per requirement. After generating data files of required derived variables processed by the respective modules, images are prepared using "WriteImage" operator and are used for dissemination. The data files are then used in the statistics and decision-making module to take a decision on whether there exists a bloom condition or not and if exists what is the extent of bloom. The statistics and the figures are then disseminated in the INCOIS-ABIS webpage, marking status of hotspot as green for normal, yellow for watch and red for warning condition using a automated statistical decision-making process. This completes the web dissemination process, and the data files and figures are archived for any future references. In addition to the mapped images of geophysical products, a table of quantitative measures of the bloom status is added for ready reference and comprehension. Some of such parameters are Average Biomass concentration, Biomass Standard deviation, Average SST, SST Standard deviation, Average Bloom Indices, BI Standard deviation, Dominant Species/Class, Spread of Noctiluca if present, Spread of Different species, Dominant Phytoplankton Size Class, etc. A schematic workflow diagram of the ABIS developed in INCOIS is shown in Fig. 3. INCOIS is disseminating near real-time (NRT) mode ABIS since February 2020 on daily basis. A screenshot of INCOIS ABIS dissemination web portal is illustrated as Fig. 4.

6 ABIS: Future Scope and Directions

Blooms' implications have been debated from both positive and negative perspectives. For instance, diatom blooms have the potential to be beneficial to fisheries, but they can also be harmful to fisheries and human health. Therefore, INCOIS has initiated ABIS and the phase-wise implementation of the ABIS is given in Table 1.

Researchers, ecologists, and environmentalists are prospective users of the service to investigate the impact of algal blooms on the environment and, as a result, human

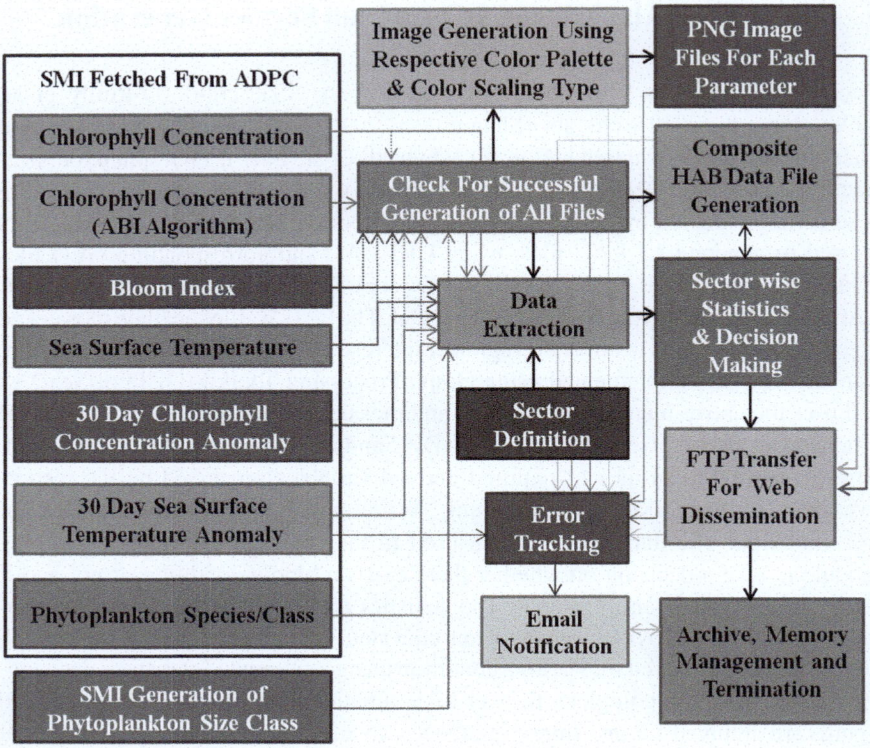

Fig. 3 Schematic workflow of Algal Bloom Information Service (ABIS)

health. ABIS aims at assisting policymakers in developing sustainable ocean management and decision-making strategies, as well as demarcating marine protected areas and conducting scientific research programmes to learn more about the blooms. It will also complement existing ecosystem-based services such as the PFZ, in indicating a safer fishing zone.

7 Jellyfish Swarming and Their Harmful Effects

Jellyfish are a large section of marine fauna that can be found in both coastal and open ocean environments. These gelatinous animals represent in two forms, meroplankton (transient plankton phase of some invertebrates) and holoplankton. Some jellyfish species form dense aggregations during favorable environmental conditions. Jellyfish growth and reproduction rate vary extensively with changes in the environmental conditions, which leads to remarkable patchy distributions, dense aggregations, and subsequent collapse (Purcell 2005). The jellyfish aggregation events in the coastal

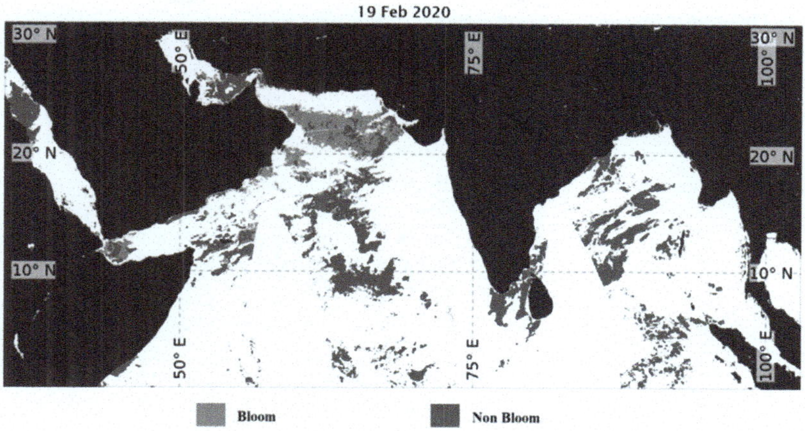

29,680 SqKm Spread of green Noctiluca in North Eastern Arabian Sea(NEAS), No Information for green N

No Information for red Noctiluca in North Eastern Arabian Sea(NEAS), No Information for red Noctiluca ir

90,640 SqKm Spread of Diatoms in North Eastern Arabian Sea(NEAS), No information Spread of Diatoms

Hotspot Statistics

Parameter Description (unit)	North Eastern Arabian Sea (NEAS)	Kerala Coast (KRL)	Gulf of Mannar (GoM)	Gopalpur Coast (GPLPR)
Average Biomass Concentration (mg/m3)	5.663	0.1105	0.5688	0.2483
Standard Deviation of Biomass Concentration (mg/m3)	9.254	0.04266	0.6849	0.09404
Average Sea Surface Temperature (°C)	24.73	28.83	28.23	26.58
Standard Deviation of Sea Surface Temperature (°C)	0.7545	0.9924	1.074	0.7565
Average Bloom Index	-0.01062	-0.6519	-0.7401	-0.7105
Standard Deviation of Bloom Index	0.3815	0.1143	0.1074	0.05468
Average Biomass Concentration Anomaly (mg/m3)	1.239	-0.05760	-0.07930	-0.1029
Average Sea Surface Temperature Anomaly (°C)	0.3590	-0.3554	0.1684	0.4625
Spread of Green Noctiluca (Sq-Km)	29680	-999	112	-999
Spread of Red Noctiluca (Sq-Km)	-999	-999	-999	-999
Spread of Diatoms (Sq-Km)	90640	-999	496	16
Picophytoplankton Abandance (%)	23.33	46.14	40.83	44.13
Nanophytoplankton Abandance (%)	17.00	34.36	30.25	32.79
Microphytoplankton Abandance (%)	59.67	19.51	28.92	23.07
Status				

Status Color

Normal (Bloom Pixels < 50%) Watch (Bloom Pixels > 50% & <75%)

Warning (Bloom Pixels > 75%) No Data

Fig. 4 ABIS dissemination webpage showing algal bloom events in northeastern Arabian Sea

Table 1 Phase-wise implementation of ABIS

Phase	Service type	Remarks
Phase-I	Information service (completed and operational)	• Based on satellite data • Now-casting/Near real-time • Spatial and temporal extent of the bloom
Phase-II	Advisory service	• Based on satellite data • Now-casting/Near real-time • Spatial and temporal extent of the bloom • Probable impact on ecosystem (fisheries, water quality etc.)
Phase-III	Predictive advisory service	• Based on satellite and model data • Forecasting • Spatial and temporal extent of the bloom • Probable impact on ecosystem (fisheries, water quality etc.)

region in form of swarming and beach stranding often result in water quality deterioration, trophic malfunction, and socio-economic losses. In addition, the jellyfish outbreaks exert a significant impact on tourism, recreational activities, and human health through toxic stinging. The episodes of jellyfish swarming have received large attention in recent times. However, because of broad classifications and limited observation tools, studying these gelatinous animals is highly subjective (Baliarsingh et al. 2020). Jellyfish swarming and beach stranding occurrences have been observed sporadically in the peninsular Indian waters and coastal districts (Fig. 5).

Jellyfish swarming has had a negative influence on tourism, fishing, and the water intake of coastal power plants. Jellyfish swarms have also been recorded obstructing fishing activities in Indian coastal seas. Jellyfish swarms clogged fishing nets in estuaries and coastal seas off the coast of Kerala. The presence of a significant number of jellyfish in the coastal waters of the northwestern Bay of Bengal has had an impact on plankton ecology (Baliarsingh et al. 2016). India has a vast coastline of ~7,500 km encompassing the west coast, east coast and islands. The Indian coast's geographic structure varies in distinct pockets, which can act as a trapping medium for aggregating wind/current-driven jellyfish assemblages. Most notably, an increase in the amount of contaminants in coastal water can degrade water quality, allowing jellyfish to proliferate and multiply. In this regard, rapidly expanding aquaculture, mariculture, and cage culture activities pose a risk of acting as a stimulant for jellyfish swarming by supplying substrate for the jellyfish's benthic stages and degrading coastal water quality (Baliarsingh et al. 2020). In consideration to the aforementioned adverse effects of jellyfish swarming and after receiving request from several quarters for jellyfish swarming information, INCOIS has taken initiative towards the development of Jellyfish Aggregation Advisory Service (JAAS).

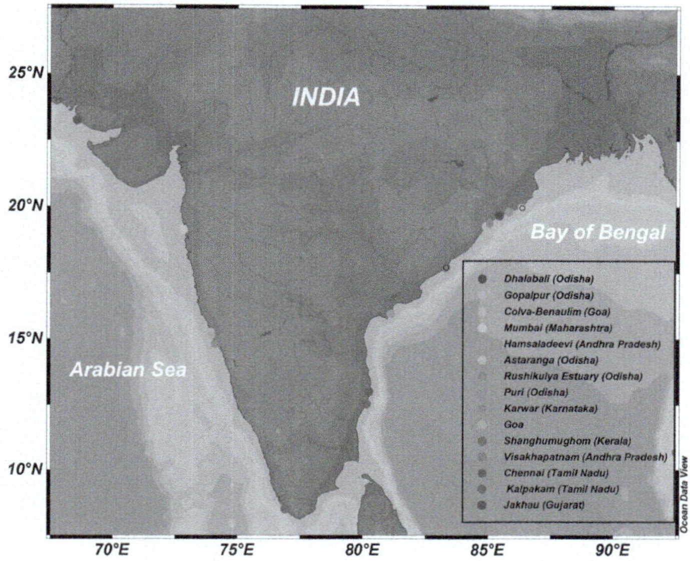

Fig. 5 Jellyfish swarming and beach stranding events reported along the Indian coast during the period from 1995 to 2019

8 Conceptual Framework for Detection and Monitoring of Jellyfish Swarming

As an initial step towards the development of JAAS, the jellyfish swarming and beach stranding locations in Indian waters have been demarcated (Fig. 5). Subsequently, the environmental parameters setting conducive conditions for jellyfish swarming have been identified based on a thorough literature survey on jellyfish ecology (Fig. 6). Based on the availability of potential data sources (Table 2), a conceptual framework has been prepared for the development of JAAS (Fig. 7).

9 JAAS: Future Scope and Directions

INCOIS has envisaged the importance of JAAS and initiated activities towards its development. The future plans based on the JAAS can be categorized as (i) coastal water quality monitoring and implementation of management strategies, (ii) timely warning to coastal power plants to avoid seawater intake screen damage, (iii) advisory for fishermen on the swarming zones to avoid clogging of nets, (iv) early-warning to safeguard coastal tourism and recreational activities, and (v) scientific research scope on studying spatio-temporal jellyfish aggregation pattern. A schematic showing the road map of JAAS is illustrated in Fig. 8.

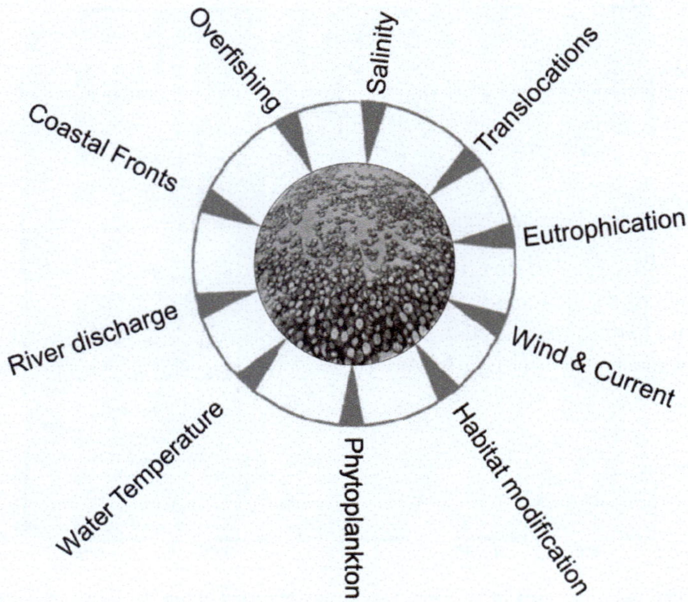

Fig. 6 Potential environmental parameters and conditions that plays important role in causing jellyfish swarming and beach stranding events

Table 2 Potential data source of environmental parameters for generation of jellyfish aggregation monitoring advisory using the conceptual framework

Parameters	Potential data source
Sea surface current	Satellite, coastal radar, ocean general circulation model
Wind	Satellite, automatic weather station, ocean general circulation model
Water temperature	Satellite, argo, autonomous data buoy, ocean general circulation model
Salinity	Argo, autonomous data buoy
Dissolved oxygen	Argo, autonomous data buoy, physical–biogeochemical model
Nutrient	Argo, autonomous data buoy, physical–biogeochemical model
Phytoplankton	Satellite, autonomous data buoy, physical–biogeochemical model
Algal bloom	Satellite, autonomous data buoy
Zooplankton	Physical–biogeochemical model
Fishing	Responsible authorities (e.g. fisheries department, fish landing centre, fishing harbor, fishermen community)
Habitat	Satellite, responsible authorities (e.g. revenue department)
River discharge	Water gauge, responsible authorities (e.g. water resource department)

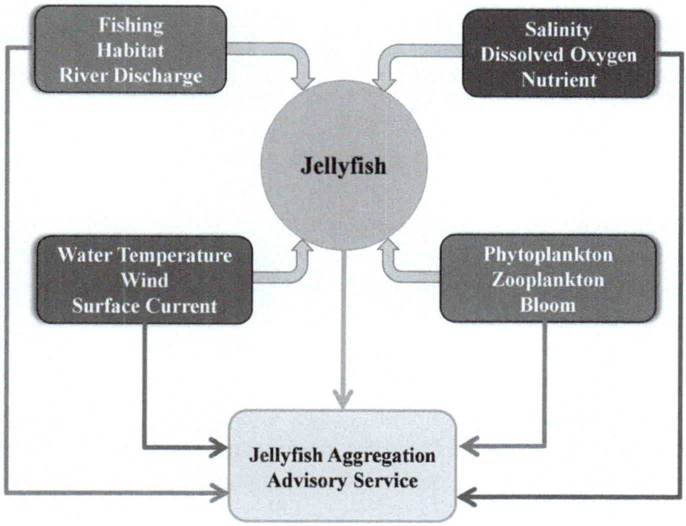

Fig. 7 Conceptual framework towards the development of JAAS

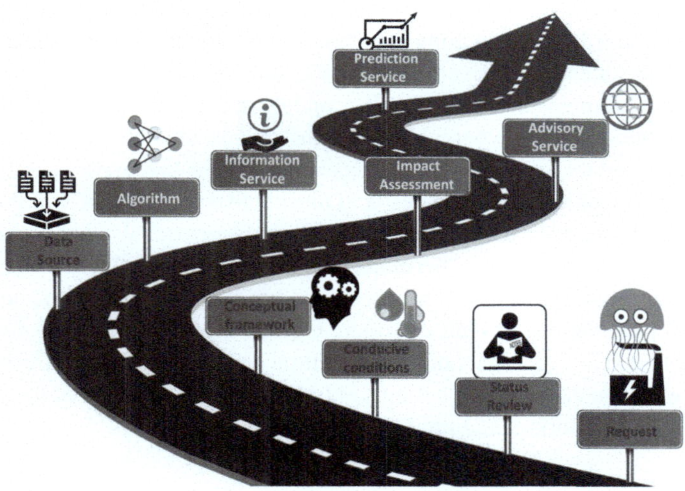

Fig. 8 Road map of INCOIS-JAAS

10 Coral Bleaching and Their Harmful Effects

Corals are marine organisms that belong to the phylum Cnidaria and can be found in tropical waters at depths of up to 200 m. Scleractinians, which are responsible for reef construction, make up the majority of coral reefs. Corals produce the reef by symbiotically living with microalgae (zooxanthellae) and secreting a calcium carbonate exoskeleton. It is a vital ecosystem in the shallow marine environment, providing a home for a variety of organisms. The biodiversity of the coral habitat is abundant, with a high density of species per unit area. According to estimates, coral reef ecosystems host about 0.5 million species worldwide (Spalding et al. 2001).

Under specific environmental pressures, the symbiotic link between algae (zooxanthellae) and their host corals breaks off, resulting in coral bleaching. As a result, the zooxanthellae are expelled from the host. The corals display their white underlying calcium carbonate coral skeleton in the absence of symbiotic algae, and the damaged coral colony turns pale in hue. Coral bleaching can be triggered and sustained by a variety of environmental factors. Massive coral bleaching has been recorded worldwide in recent years as a result of unusually warm waters. Coral bleaching can occur even 1–2 °C increase in ambient water temperature during the summer months (Berkelmans and Willis 1999; Reaser et al. 2000). Coral death can occur because of prolonged partial/total bleaching events in coral habitats. Long-term ecological and social consequences of severe bleaching events include the extinction of reef-building corals, changes in benthic habitat, and, in some circumstances, changes in fish populations. Even if favourable conditions remain after the incident, severely bleached reefs can take several years to recover.

Coral reefs are one of India's oldest and most active ecosystems. Coral reefs not only provide a haven for a diverse range of marine species but also help to safeguard the coastline from erosion. Furthermore, people living along India's ~7,500 km-long coastline rely on coral reefs for their livelihood. In Indian waters, coral reefs are present in the regions of Gulf of Kutch, Gulf of Mannar, Andaman & Nicobar, Lakshadweep Islands, and Malvan. As the natural and anthropogenic stressors exert an adverse effect on the coral reef systems of India, there is a strong need for improved understanding, monitoring, and prediction of coral bleaching.

11 Coral Bleaching Detection and Monitoring

The use of satellite oceanography to map and monitor coral reefs is a practical and cost-effective technology. Satellite remote sensing is a key technique for providing near-real-time synoptic views of the global oceans and the capacity to monitor global coral reef areas. As a result, the nocturnal Sea Surface Temperature (SST) is an important metric to assess the thermal conditions and degree of coral bleaching in order to provide early warnings. There are various remote sensing satellites circling the globe that can offer SST data on a regular basis, both during the day and at night.

This makes it possible to create a coral reef bleaching warning system that generates early-warning alerts and bulletins in real-time. In February 2011, INCOIS launched the Coral Bleaching Alert System (CBAS). With the support of satellite-derived SST, the CBAS applies a model that measures the thermal stress accumulated in the coral surroundings.

12 Coral Bleaching Alert System

CBAS uses the Multi-Channel Sea Surface Temperature (MCSST) data from Advanced Very High Resolution Radiometer (AVHRR) onboard National Oceanic and Atmospheric Administration (NOAA) series satellites to assess the thermal stress on the coral reef environments. The thermal stress on coral reefs is calculated using bi-weekly composite AVHRR SST data (Mohanty et al. 2013). In each pixel, the Maximum Monthly Mean (MMM) is extracted, which indicates that up to MMM threshold, coral can survive without thermal stress. The intensity of thermal stress analysis is carried out using HotSpot (HS) analysis methods. The HotSpot is the degree of SST increases beyond the thermal tolerance level of the coral reef. Therefore, HS is equal to subtracting the value of MMM from current bi-weekly data, which results from the different levels of stress on coral. The HS value less than zero is categorized as "no Stress", the HS value within the range of 0 to 1 °C is categorized as "watch," and values above 1 °C is a threshold for thermal stress as a warning leading to coral bleaching. To confirm the coral bleaching due to thermal stress, the Degree of Heating Weeks (DHWs) is carried out to determine the intensity of bleaching. DHWs is calculated as the cumulative sum of 24 bi-weekly HS (only >= 1 °C) and categorized into different levels of warning. The value of DHWs within the range of 0–4 °C is marked as a warning, the range of 4–8 °C is marked as significant bleaching (Alert Level-I), and above 8 °C is considered as widespread bleaching (Alert Level-II). A flow chart of the methodology is shown in Fig. 9. A case study on coral bleaching advisory issued by INCOIS and field photographs of the Andaman coral reef region during April 2016 is shown in Figs. 10 and 11, respectively.

The intensity of coral bleaching can also be measured using the persistence of Marine Heat Waves (MHW) on spatio-temporal scale and is used to improve the present INCOIS CBAS. MHW is a discrete prolonged anomalously warm water event in a particular location and time period. It is induced by a coupled atmospheric and oceanic factors that have a significant impact on the structure and function of marine ecosystems. The intensity, duration, and spatial extent of the MHW should be described in relation to a baseline period (climatology) and a specific time of year. That means that an MHW isn't just for the summer, but for some biological applications, heat waves in the winter are also important to consider. As a result, it is required to compute and forecast MHW on a spatio-temporal scale in relation to climatology, as well as develop a link with sensitivity studies on marine habitat. The Intensity of Marine Heat Wave (IMHW) is calculated using AVHRR SST data to address the impact of MHW on coral reef, based on the following method (Fig. 12).

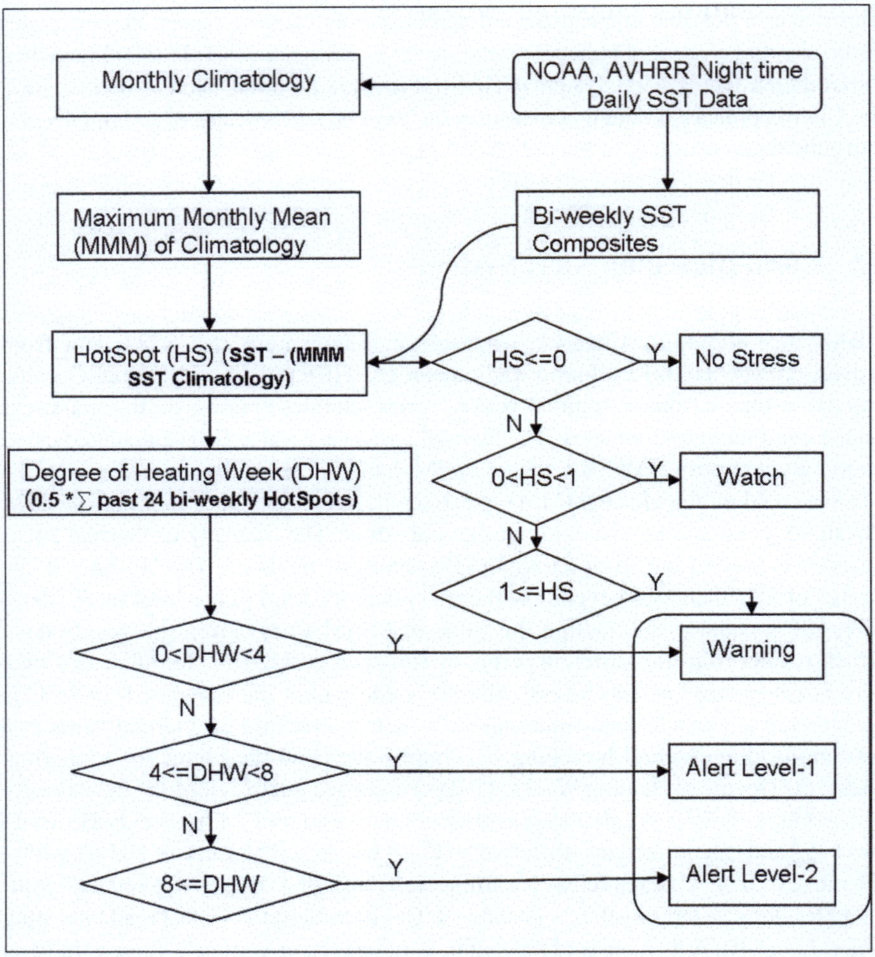

Fig. 9 Flow chart depicting the methodology and the decision criteria for CBAS

A linear least square method has established the relationship between IMHW with in situ observations at different (% of bleaching) coral bleaching conditions. The coefficient value of resulted regression fitted Eq. 1 is used to calculate the percentage of coral bleaching as a dependent parameter with the IMHW taken as an independent parameter. Figure 13 shows a scatter plot of IMHW against in situ observed PCB. As shown in the linear regression model, both parameters are significantly correlated ($R^2 = 0.59$) with each other. These derived coefficients are used to generate the PCB in spatio-temporal scale and is depicted in Fig. 14.

$$\mathrm{PCB}(\%) = 7.767 * (\mathrm{IMHW}) - 141.7 \quad (1)$$

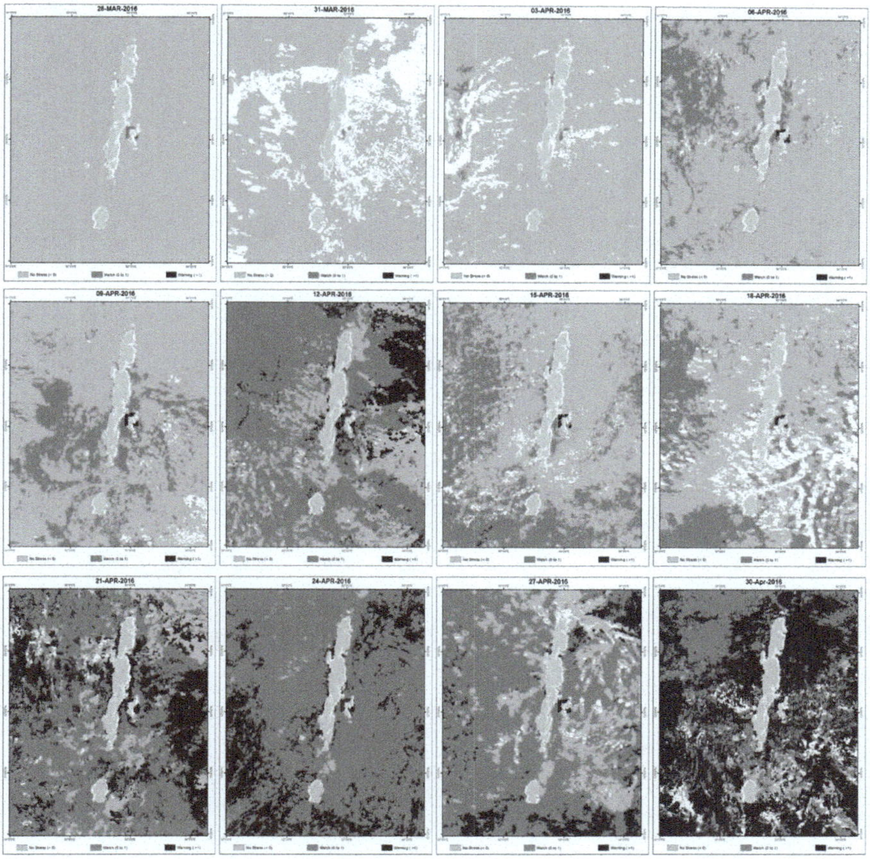

Fig. 10 Time-series showing bi-weekly HotSpot for the Andaman Islands (*Source* ESSO-INCOIS)

13 CBAS: Future Scope and Directions

At present, INCOIS is disseminating a web advisory on CBAS up to different alert/warning levels of bleaching qualitatively based on thermal stress. However, for the improvement of this service, the percentage of coral bleaching at the particular time window can be quantified based on IMHW. In addition to this, assessing the health of coral reef resilience-resistance to bleaching due to climate change and ocean acidification is envisioned to be carried out based on water quality parameters (pH, dissolved inorganic carbon, carbonate, bicarbonate, and CO_2).

Fig. 11 Underwater photographs of coral reef showing primary signs of coral bleaching (*Source* ESSO-NIOT)

14 Conclusion

Ocean-related environmental concerns have grown in importance during the twentieth century. Meeting the United Nation's directives (SDG-14), which aims to conserve and sustainably use the ocean and its resources, requires addressing these challenges. It's needless to mention the dependence of the countries on the ocean and marine resources for socio-economy. On the other hand, anthropogenic activities are exerting a negative impact on marine water quality and harming the ocean biota. In the context of Indian waters, the increasing prevalence of algal blooms, jellyfish

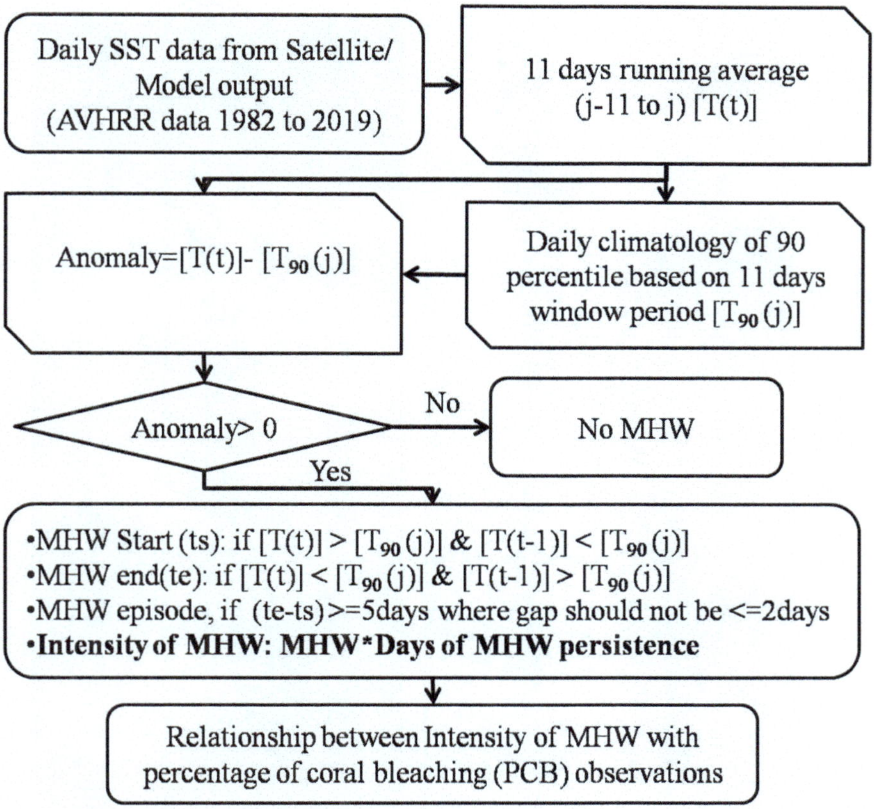

Fig. 12 Flowchart depicting the methodology adopted for the study

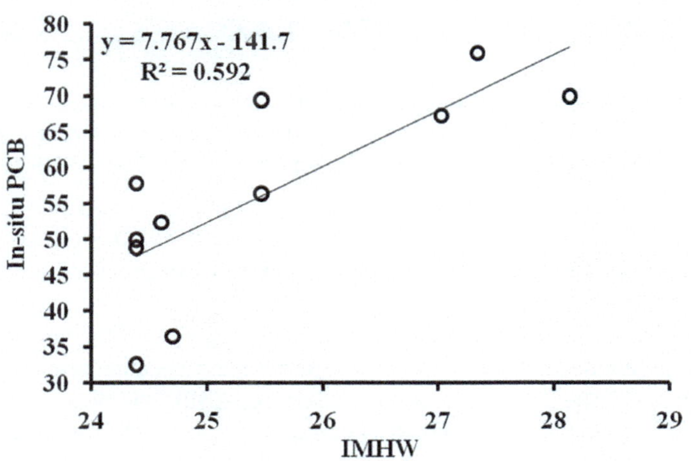

Fig. 13 Scatter plot showing the Intensity of Marine Heat Wave (IMHW) plotted against observed percentage of Coral Bleaching (PCB). (Image reprinted with permission from Springer Nature, Mohanty et al. 2021)

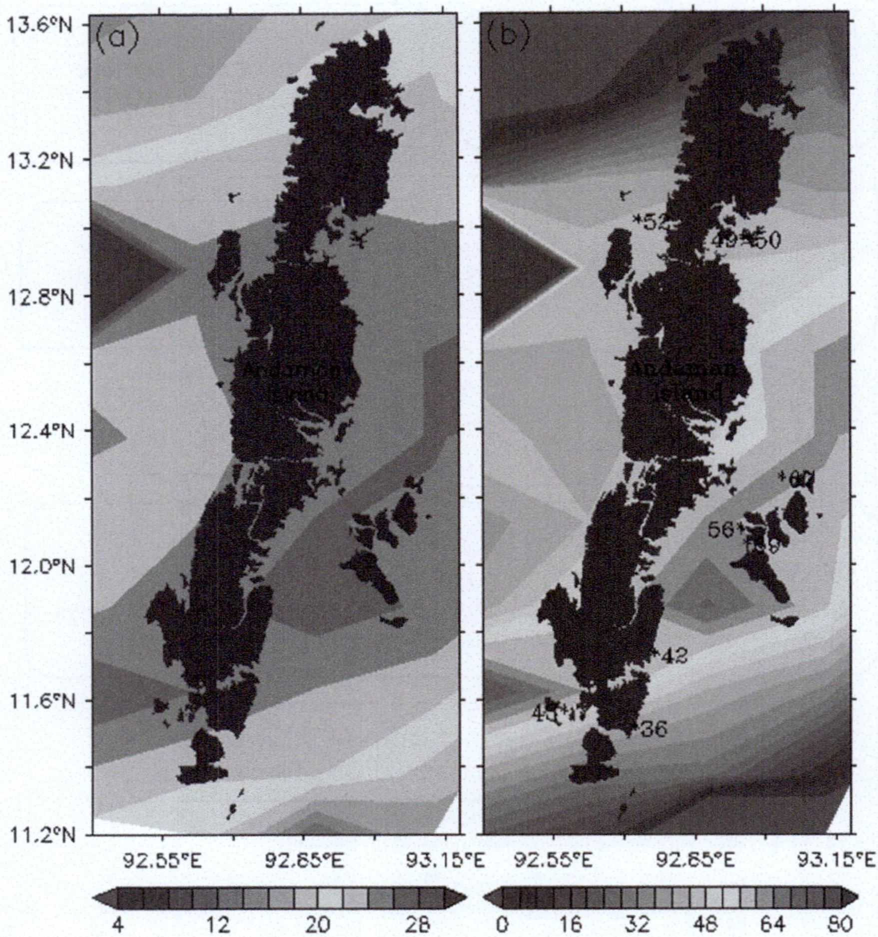

Fig. 14 **a)** Spatial patterns of Intensity of Marine Heat Wave (IMHW) during May 2010 and **b**) PCB computed empirically based on Eq. 1 along with in-situ observed PCB value at selected points (marked in value points). (Image reprinted with permission from Springer Nature, Mohanty et al. 2021)

swarming/beach stranding occurrences and coral reef bleaching are all causes of concern.

INCOIS has taken the call being the responsible agency of the Government of India and operationalized marine ecological services for satellite-based monitoring of algal blooms (Algal Bloom Information Service: ABIS) and bleaching of corals (Coral Bleaching Alert System: CBAS). For monitoring jellyfish swarming, environmental triggers have been identified based on conducive conditions, and a conceptual framework has been prepared for the possible generation of Jellyfish Aggregation Advisory Service (JAAS).

The operational services of INCOIS are being upgraded, and the conceptual services are being completed. The ABIS will be enhanced to include (i) nowcasting/near real-time mode to provide the most up to date feasible information, (ii) estimation of the bloom's spatial and temporal extent and (iii) information on the bloom's likely impact on the environment (fisheries, water quality, etc.). Similarly, INCOIS is currently disseminating a service on CBAS up to different alert/warning levels of coral bleaching qualitatively as per the thermal stress. However, using IMHW, the proportion of coral bleaching at a specific time window is planned to be computed to improve the existing CBAS. In addition to this, assessing the health of coral reef resilience-resistance to bleaching due to climate change and ocean acidification is envisioned to be carried out based on water quality parameters. As jellyfish swarming and beach stranding events are on the rise, the conceptual framework of JAAS is being developed into a real model that will be able to provide (i) timely warning to coastal power plants to avoid seawater intake screen damage, (ii) advisory for fishermen on swarming zones to avoid clogging of nets, and (iii) early-warning to protect coastal tourism and recreational activities. INCOIS' marine ecological services are/will be used for many purposes by researchers, ecologists, environmentalists, ocean resource managers, pollution monitoring agencies, tourist agencies, fishers, and policymakers.

Acknowledgements The authors are thankful to the Director, Indian National Centre for Ocean Information Services (INCOIS), Ministry of Earth Sciences, Hyderabad, India for the encouragement. This is INCOIS contribution no. 476.

References

Ahn YH, Shanmugam P (2006) Detecting the red tide algal blooms from satellite ocean color observations in optically complex Northeast-Asia Coastal waters. Remote Sens Environ 103(4):419–437

Anderson DM (1994) Red tides. Sci Am 271:52–58

Baliarsingh SK, Lotliker AA, Trainer VL, Wells ML, Parida C, Sahu BK, Srichandan S, Sahoo S, Sahu KC, Kumar TS (2016) Environmental dynamics of red *Noctiluca scintillans* bloom in tropical coastal waters. Mar Pollut Bull 111(1–2):277–286

Baliarsingh SK, Dwivedi RM, Lotliker AA, Sahu KC, Kumar TS, Shenoi SSC (2017) An optical remote sensing approach for ecological monitoring of red and green *Noctiluca scintillans*. Environ Monit Assess 189(7):1–10

Baliarsingh SK, Lotliker AA, Srichandan S, Samanta A, Kumar N, Nair TB (2020) A review of jellyfish aggregations, focusing on India's coastal waters. Ecol Process 9(1):1–9

Berkelmans R, Willis BL (1999) Seasonal and local spatial patterns in the upper thermal limits of corals on the inshore Central Great Barrier Reef. Coral Reefs 18(3):219–228

D'Silva MS, Ani AC, Naik RK, D'Costa PM (2012) Algal blooms: a perspective from the coasts of India. Nat Hazards 63:1225–1253

Dwivedi R, Rafeeq M, Smitha BR, Padmakumar KB, Thomas LC, Sanjeevan VN, Prakash P, Raman M (2015) Species identification of mixed algal bloom in the Northern Arabian Sea using remote sensing techniques. Environ Monit Assess 187(2):51

Gokul EA, Shanmugam P (2016) An optical system for detecting and describing major algal blooms in coastal and oceanic waters around India. J Geophys Res: Oceans 121(6):4097–4127

Gopakumar G, Sulochanan B, Venkatesan V (2009) Bloom of *Noctiluca scintillans* (Macartney) in Gulf of Mannar, southeast coast of India. J Mar B Assoc India 51(1):75–80

Gomes H, Matondkar SGP, Parab SG, Goes JI, Pednekar S, Al-Azri ARN, Thopill PG (2008) Unusual blooms of green *Noctiluca miliaris* (Dinophyceae) in the Arabian Sea during the winter monsoon. Geophys Monogr Ser 185:347–363

Hallegraeff GM (2010) Ocean climate change, phytoplankton community responses, and harmful algal blooms: a formidable predictive challenge 1. J Phycol 46(2):220–235

Hu C, Lee Z, Franz B (2012) Chlorophyll a algorithms for oligotrophic oceans: a novel approach based on three-band reflectance difference. J Geophys Res 117(C1)

Jugnu R, Kripa V (2009) Effect of *Chattonella marina* [(Subrahmanyan) Hara etChihara 1982] bloom on the coastal fishery resources along Kerala coast, India. Indian J Geo Mar Sci 38(1):77–88

Lotliker AA, Baliarsingh SK, Trainer VL, Wells ML, Wilson C, Bhaskar TU, Samanta A, Shahimol SR (2018) Characterization of oceanic Noctiluca blooms not associated with hypoxia in the Northeastern Arabian Sea. Harmful Algae 74:46–57

Lucas CH, Gelcich S, Uye SI (2014) Living with jellyfish: management and adaptation strategies. In: Pitt K, Lucas C (eds) Jellyfish blooms. Springer, Dordrecht. https://doi.org/10.1007/978-94-007-7015-7_6

Matondkar SGP, Bhat SR, Dwivedi RM, Nayak SR (2004) Indian satellite IRS-P4 (OCEANSAT). Monitoring algal blooms in the Arabian Sea. Harmful Algae News 26:4–5

Mohanty PC, Mahendra RS, Bisoyi H, Srinivasa KT, George G, Nayak S, Sahu BK (2013) Assessment of the coral bleaching during 2005 to decipher the thermal stress in the coral environs of the Andaman Islands using remote sensing. Eur J Remote Sens 46:417–430. https://doi.org/10.5721/EuJRS20134624

Mohanty PC, Kushabaha A, Mahendra RS, Nayak RK, Sahu BK, Rao E, Kumar TS (2021) Persistence of marine heat waves for coral bleaching and their spectral characteristics around Andaman coral reef. Environ Monit Assess 193(8):1–9

Nayak S (2017) Coastal zone management in India—present status and future needs. Geo-Spat Inf Sci 20(2):174–183

O'Reilly JE, Maritorena S, Mitchell BG, Siegel DA, Carder KL, Garver SA, Kahru M, McClain CR (1998) Ocean color chlorophyll algorithms for SeaWiFS. J Geophys Res 103:24937–24953

Oyeku OG, Mandal SK (2021) Historical occurrences of marine microalgal blooms in Indian peninsula: probable causes and implications. Oceanologia 63(1):51–70

Padmakumar KB, Sanilkumar MG, Saramma AV, Sanjeevan VN, Menon NR (2008) Green tide of Noctiluca miliaris in the Northern Arabian Sea. Harmful Algae News 36:12

Parab SG, Matondkar SGP, Gomes HDR, Goes JI (2006) Monsoon driven changes in phytoplankton populations in the eastern Arabian Sea as revealed by microscopy and HPLC pigment analysis. Cont Shelf Res 26:2538–2558

Purcell JE (2005) Climate effects on formation of jellyfish and ctenophore blooms: a review. J Mar Biol Assoc UK 85(3):461–476

Reaser JK, Pomerance R, Thomas PO (2000) Coral bleaching and global climate change: scientific findings and policy recommendations. Conserv Biol 14(5):1500–1511

Sahu BK, Baliarsingh SK, Samanta A, Srichandan S, Singh S (2020) Mass beach stranding of blue button jellies (Porpitaporpita, Linnaeus, 1758) along Odisha coast during summer season. Indian J Geo-Mar Sci 49(6):1093–1096

Sahay A, Ali SM, Gupta A, Goes JI (2017) Ocean color satellite determinations of phytoplankton size class in the Arabian Sea during the winter monsoon. Remote Sens Environ 198:286–296

Shanmugam P (2011) A new bio-optical algorithm for the remote sensing of algal blooms in complex ocean waters. J Geophys Res: Oceans 116(C4)

Shanmugam P (2012) CAAS: an atmospheric correction algorithm for the remote sensing of complex waters. Ann Geophys (09927689) 30(1)

Shanmugam P, Suresh M, Sundarabalan B (2013) OSABT: an innovative algorithm to detect and characterize ocean surface algal blooms. IEEE J Sel Top Appl Earth Observ Remote Sens 6(4):1879–1892

Spalding M, Spalding MD, Ravilious C, Green EP (2001) World atlas of coral reefs. Univ of California Press

Subrahmanyan R (1954) On the life-history and ecology of Hornellia marina gen. et sp. nov., (Chloromonadineae), causing green discoloration of the sea and mortality among marine organisms off the Malabar Coast. Indian J Fish 1(1&2):182–203

Trainer VL, Moore SK, Hallegraeff G, Kudela RM, Clement A, Mardones JI, Cochlan WP (2020) Pelagic harmful algal blooms and climate change: lessons from nature's experiments with extremes. Harmful Algae 91:101591

Varunan T, Shanmugam P (2015) A model for estimating size-fractioned phytoplankton absorption coefficients in coastal and oceanic waters from satellite data. Remote Sens Environ 158:235–254

Augmentation of Water—Can Oceans Help?

Purnima Jalihal

Abstract Water stress is an important issue throughout India today, and ways to augment water are needed urgently. One of the methods is the desalination of sea and brackish water. There are various desalination techniques, and both thermal and membrane systems have been addressed in this chapter. The negative impact of the desalination system is discussed. National Institute of Ocean Technology (NIOT) under Ministry of Earth Sciences (MoES) has developed a technology called Low Temperature Thermal Desalination (LTTD), which has been successfully operating in the Lakshadweep islands for several years with no visible environmental impact. An offshore barge-mounted plant and one in a power plant using the condenser reject heat have also been demonstrated successfully. To reduce the usage of fossil fuels for powering desalination systems, it may be prudent to use renewable energies. Lastly, the ecological and environmental costs should be considered to arrive at the cost of any technology. MoES through its institute NIOT thus has made a tremendous societal impact through this technology developed for the first time in the world and implemented in Indian waters.

1 Introduction

Life on earth is possible due to the existence of water. Three important factors that have sustained the earth in its present form are the atmospheric cover, the oceans and the plate tectonics or movements in its crust. There is a delicate balance of pressures and temperatures which maintain the various physico-chemical properties and sustain life while sustaining the water bodies. The water bodies on the globe are affected by the hydrological cycle consisting of precipitation which changes from place to place and various environmental parameters like temperature and humidity. On land, water bodies like lakes and rivers tend to dry or flood depending upon the climatic situation. In the sea, however, the levels are dependent on oceanographic parameters such as bathymetry and geological features and changing tidal levels

P. Jalihal (✉)
National Institute of Ocean Technology, Chennai 600100, India
e-mail: purnima@niot.res.in

V. K. Gahalaut and M. Rajeevan (eds.), *Social and Economic Impact of Earth Sciences*,
https://doi.org/10.1007/978-981-19-6929-4_13

on a daily basis. Today due to satellite measurements, new modelling methods and computational simulations for oceanic and atmospheric parameters, some anomalies are being observed. Sea-level rise and temperature increase are being observed today which have been directly linked to climate change and which has become a concern across the globe. One of the areas that is impacted by it is water. Rising temperatures have disrupted hydrological cycles and leading to changing rainfall patterns and consequently increasing dry areas. Thus water scarcity is the direct fallout.

2 Water Stress

There are many other reasons which contribute to the water stress today. They are rising population in developing countries as also poor or non-implementation of conservation methods and pollution control. Today all over the world, many regions are water-stressed due to various reasons. In India, the reasons range from a mammoth population, pollution of water bodies, changing climatic conditions, poor water management and wasteful practices. Technology Information, Forecasting and Assessment Council (TIFAC) under Department of Science & Technology (DST) took up the task of understanding several sectors and the effect of climate change on them, and one of those is water (GTWG 2019). The report classifies the water-related issues into (a) Augmentation and (b) Conservation and Water Management. It is well known that India receives sufficient rainfall, but water capture is among the lowest in the world. Due to this, the groundwater has been excessively tapped leading to the depletion of groundwater resources. Additionally, water bodies are severely polluted due to the non-implementation of environmental norms in a controlled manner.

Water pollution can be of three distinct types viz. untreated domestic wastewater, untreated industrial effluents let out in open sea or land or river, as also existing ground contaminants. All these need to be addressed in a different manner. When such water is treated to improve its quality, it becomes available for use and thus existing water sees an increase in availability. Water conservation especially in the domestic sector has started getting importance with policies slowly falling in place for its implementation.

However, the growing needs of the increasing population for domestic and industrial uses make the demand much greater than what is made available through treatment and conservation.

There is clearly a need to augment existing water sources by other means. One possible solution for augmentation is desalination which is applicable for brackish water in land as well as for seawater at the coast. Today, desalination has become a technologically viable solution for water-stressed areas.

3 Desalination Methods

Desalination refers to a process wherein good and pure water is separated from the sea or brackish water using externally applied energy. The Middle East has been a leader in desalination. Saudi Arabia, United Arab Emirates, Kuwait and Israel rely heavily on desalination as a source of clean water. These countries have hardly any fresh groundwater, so desalination is a necessity. The UN predicts that by 2025, 14% of the world will have to depend on desalination (Werft 2016).

India has a long coastline of nearly 7500 km with a very large Exclusive Economic Zone (EEZ) as shown in Fig. 1. This makes it important to consider desalination as an important solution to the water stress in the coastal areas where nearly 30% of the total population resides.

There are various methods employed for desalination today, but most can be classified either under thermal or membrane types. Figure 2 shows how various technologies can be classified.

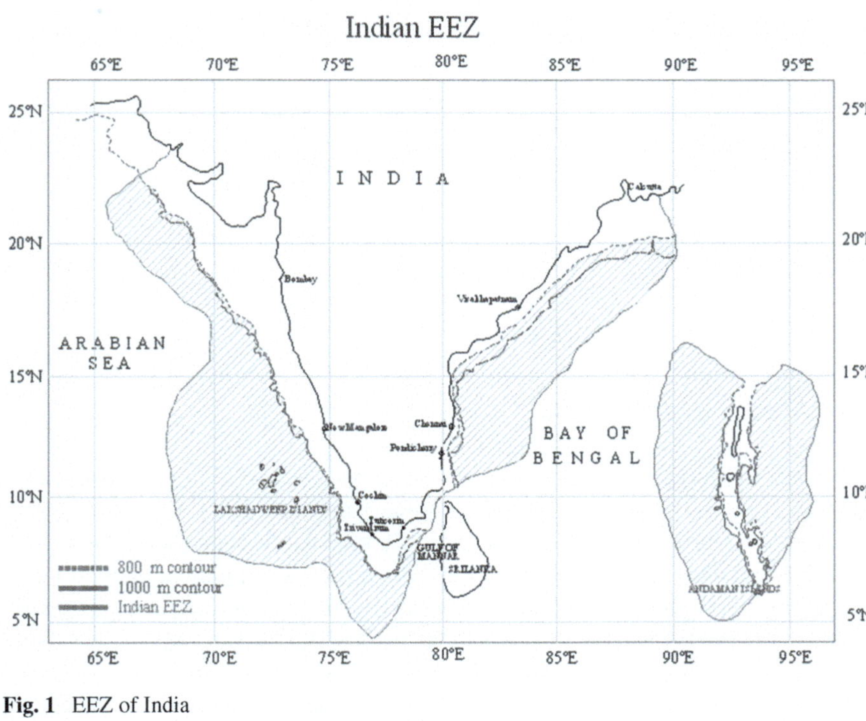

Fig. 1 EEZ of India

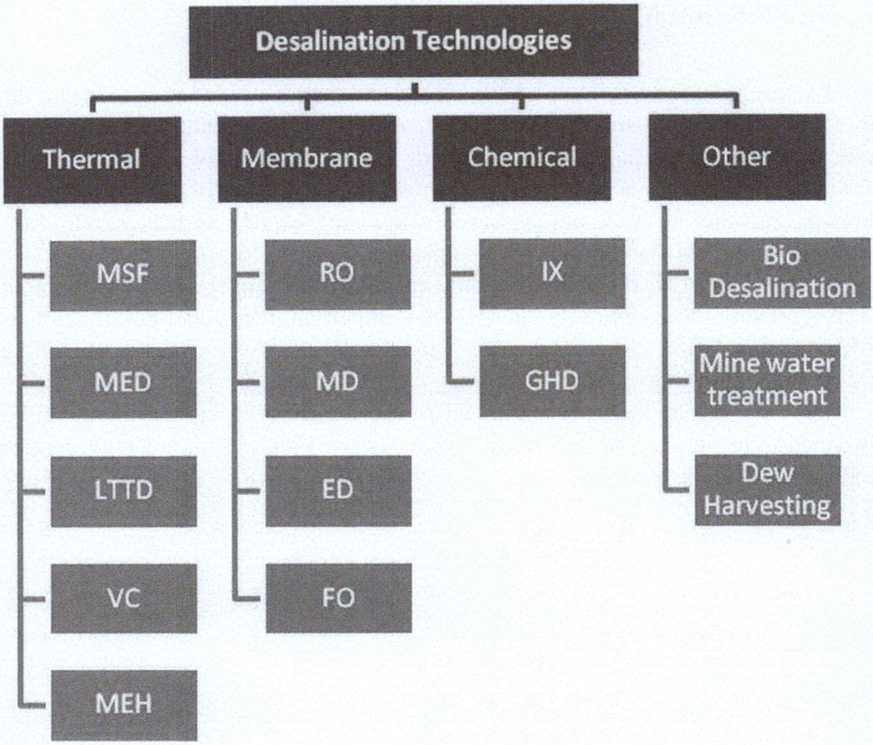

Fig. 2 Classification of desalination technologies

3.1 Thermal Technologies

3.1.1 Multi-stage Flash (MSF)

Multi-stage Flash (MSF) desalination process plays a vital role in the provision of fresh water in many areas of the world, particularly in the Middle East countries, and it accounts for approximately 34% of the world's desalination capacity (El-Ghonemy 2017). It is the most reliable and mature desalting process with more than 40 years of experience in its design, operation, material selection and maintenance (Darwish and Alsairafi 2004). It needs high temperature and pressure steam and several stages. It is very capital-intensive; however, the systems are robust and proven. In MSF, seawater is heated by steam and then fed into a series of vessels (effects) where reduced pressure leads to immediate boiling (flashing), and the steam generated is condensed in a sequence of stages, producing high-quality fresh water.

3.1.2 Multi-effect Desalination (MED)

In the multi-effect distillation (MED) process, the saline water is desalinated by means of evaporation and subsequent condensation though here the vapour passes inside the tubes, which is the reverse of the MSF process. MED is getting more attention among thermal desalination technologies due to its major advantages such as low energy consumption compared to MSF, higher overall heat transfer coefficients, less specific area as compared to MSF, low operating steam temperature and also that other low-grade heat sources can be used to power it. The multi-effect distillation process can be found in various industries, like sugar, paper and pulp, dairy, textiles, acids and desalination.

3.2 Membrane Methods

3.2.1 Reverse Osmosis (RO)

The Reverse Osmosis membrane technology has developed over the past 40 years to a 44% share in world desalting production capacity, and an 80% share in the total number of desalination plants installed worldwide (Greenlee et al. 2009). Among the different available techniques, Reverse Osmosis (RO) has proved to be the most reliable, cost-effective and energy-efficient in producing fresh water. Pressure is used to drive water molecules across the membranes, and the energy needed to drive water molecules across the membranes is directly related to the salt concentration. Therefore, RO has been most often used for brackish waters that are lower in salt concentrations (Buros 2000) compared to seawater since the salt content is so high in the sea.

The operating pressure for brackish water systems ranges 15–25 bar, and for seawater systems from 54 to 80 bar (the osmotic pressure of seawater is about 25 bar). Since the pressure required to recover additional water increases as the brine stream is concentrated, the water recovery rate of RO systems tends to be low. A typical recovery value for a seawater RO system is only 40% (Spiegler et al. 1994).

Because all RO membranes and devices are susceptible to fouling, the RO process usually cannot be applied without pre-treatment. RO feed streams must be compatible with the membrane and other materials of construction used in the devices. If the feed stream contains incompatible compounds, these must be removed in pre-treatment, or some other compatible device and/or membrane must be considered (Younos and Tulou 2005).

The heart of the reverse osmosis process is the membrane. The membranes can be made of natural polymers like cellulose acetate or synthetic polymers such as aromatic polyamides (Jalihal and Prabhakar 2019). Membranes require strength to withstand the operating pressure, a chemical surface that would attract water and pores to transport water. Thin film composite membranes are being extensively used in RO and nano filtration (NF) applications. RO is the most popular method used today with energy requirements having reduced due to the introduction of energy recovery systems.

Hollow fibre modules used for seawater desalination consist of bundles of hollow fibres in a pressure vessel. Different membrane configurations have their own drawbacks and strengths. The major components constituting a RO system are

1. Intake and pre-treatment systems;
2. High-pressure pumps;
3. Appropriate membranes and module design;
4. Post-treatment of product water;
5. Membrane cleaning and chemicals for all stages.

3.2.2 Forward Osmosis (FO)

Osmosis, or as it is currently referred to as forward osmosis, has new applications in separation processes for wastewater treatment, food processing and seawater/brackish water desalination. Forward Osmosis (FO) is an emerging low-energy desalination technology with several merits over the other conventional pressure-based reverse osmosis (RO) desalination technologies. Unlike the pressure-driven RO desalination process, the driving force in the FO process is the osmotic pressure produced naturally by the concentrated draw solution (Cath et al. 2006).

3.3 Electrodialysis (ED)

The ED process is effective with salt removal from feedwater in which the water cost is relatively high but lower than that of conventional distillation processes of the same capacity. Electrodialysis (ED) has been used for many years; this is an electrochemical process for the separation of ions across charged membranes from one solution to another under the influence of an electrical potential difference used as a driving force. This process has been widely used for the production of drinking and to process water from brackish water and seawater, treatment of industrial effluents, and recovery of useful materials from effluents and salt production (Oztekin and Altin 2016).

3.4 Membrane Distillation (MD)

Membrane distillation (MD) is a thermal, membrane-based separation process (Lawson and Lloyd 1997; Lei et al. 2005). The driving force for the MD process is quite different from other membrane processes, being the vapour pressure difference across the membrane rather than an applied absolute pressure difference, a concentration gradient or an electrical potential gradient, which drives mass transfer through a membrane (Lawson and Lloyd 1997; Schneider and van Gassel 1984). High energy consumption and brine disposal problem are faced in the RO process due to the limited recovery of water. These problems may be overcome by membrane distillation. MD is an emerging technology for brackish water desalination and is not yet fully implemented in industry.

4 Demerits and Impacts of Membrane Desalination Systems

As can be seen from the earlier section, there are several different technologies, each of which has different processes and components involved.

It is important to understand the impacts of these desalination technologies. It is understood that when the natural state of seawater, which has 35,000–40,000 ppm levels of dissolved solids, undergoes a process of salt removal, the portion left over after fresh water is obtained may have properties significantly different from the seawater that entered. Membrane methods like Reverse Osmosis remove salts, and fresh water is the output. However, thick salt solution or brine gets left behind. In thermal systems, there is no brine formation, since only vapour is condensed. The flowchart in Fig. 3 depicts a desalination system.

The common units of any desalination system as shown in Fig. 3 are the intake and outfall systems. The intake systems essentially are engineering structures and need careful design and proper installation techniques. However, they can also cause harm by impingement and entrainment, if badly designed. Plants and animals, including

Fig. 3 Components of a desalination system

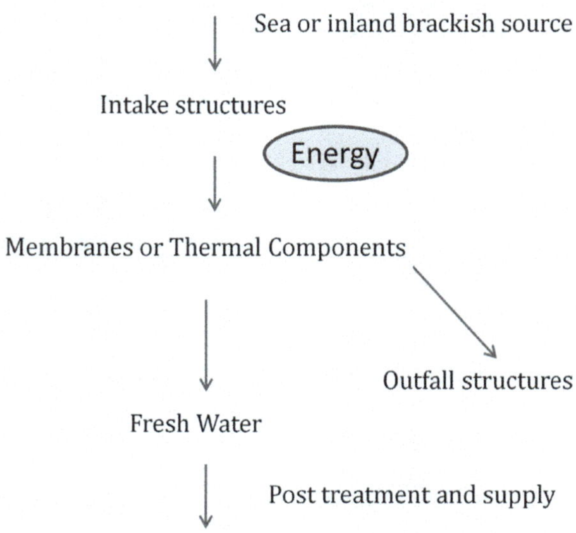

Fig. 4 Outfall Pipeline

eggs and juveniles, can be trapped and killed. The outfalls seen in Fig. 4 on the other hand are important structures because.

(i) the effluent after desalination is let out through them.
(ii) the quantity and quality of effluent is extremely important from the environmental perspective in terms of polluting potential.
(iii) the location and design can affect shipping routes, fishing vessels and marine animals by affecting salinity, temperature and localized flow velocities.

Membrane systems can have adverse impacts as enumerated below.

1. Chemicals used for pre-processing of the intake water or the desalination process:

   Popular processes like RO require chemicals for pre-processing of the intake water itself. These chemicals are chlorine, carbon dioxide, hydrochloric acid, etc. and eventually are let out in the discharge system and can lead to polluting the coastal sea. This can lead to a serious environmental impact on biological systems.
2. Effluents from the systems other than fresh water

Membrane processes leave behind very heavily concentrated solutions called brine. If this is let out to sea, the salinity levels and, in turn, overall seawater quality can deteriorate leading to various issues. Today, many computational studies are being taken up to understand the speed, direction and quantity of the spread or dispersal when the effluent is let out. It may be recalled that computational models can be run for short time frames and with idealistic equations and assumptions regarding boundary conditions. The actual effects for the long term are yet to be predicted realistically. Additionally, several inputs like bathymetry, wave and current directions, etc. are needed for the models which may not be available, thus making them inaccurate. The negative impacts can have cascading effects in terms of destruction not only of flora and fauna but increase in salt water intrusion. Soil quality can be impacted on land, and groundwater can get affected and this, in combination with other climatic changes, can have serious consequences. The super-saturated salt water also decreases oxygen levels in the water causing animals and plants to suffocate.

The organisms most commonly affected by the brine and chemical discharge are the planktons, especially phytoplanktons, which being the base of the food chain form the basis of marine life.

3. Disposal of components of the system

Membranes used in RO systems in India today are mainly imported especially for seawater desalination. The pressures in these systems are high since osmosis is being used to force salt particles out of the seawater or brackish water. Hence, the membranes need to be extremely strong structurally to be able to withstand large pressures. Thus, once the membranes have outlived their utility, their destruction or disposal is a challenge. In the near future, it is likely that huge piles of membranes from large-scale desalination plants may become an environmental hazard.

On the other hand, desalination systems using MSF, MED and other thermal methods face none of the above challenges and hence are eminently suitable to be used not only as augmentation methods but also in the context of climate change. However, all these thermal processes need steam at high pressure and high temperature. The use of steam is expensive as it needs components to be designed to be robust and fail-proof. Thermal desalination is also energy-intensive and requires higher energy than membrane methods, which can lead to huge CO_2 emissions when using conventional energy.

National Institute of Ocean Technology (NIOT) under Ministry of Earth Sciences (MoES) has pioneered a thermal desalination process for regions near the coast and islands wherein no steam is required and has been found to be ecologically safe, and the next section discusses this.

5 Ocean-Based Low Temperature Thermal Desalination (LTTD)

The ocean has an interesting feature where the surface waters are warm, and as we go into deeper waters the temperature drops. In tropical countries like India, the surface waters are generally warm. Figure 5 shows a general temperature profile in the ocean. This temperature gradient can be used to generate power and/or water.

The Low Temperature Thermal Desalination (LTTD) process utilizes the temperature gradient between two water bodies to evaporate the warmer water at low pressures and condense the resultant vapour using the colder water to obtain high-quality fresh water. The thermal gradient between different layers of the ocean water column provides huge reservoirs of warm and cold water that can effectively be utilized for power generation and desalination (Rognoni et al. 2008, Kathiroli and Jalihal 2008); Fig. 6 shows a schematic diagram of the LTTD unit.

The main components that are required for a LTTD plant are an evaporation chamber, condenser, pumps and pipelines to draw warm and cold water, and a vacuum

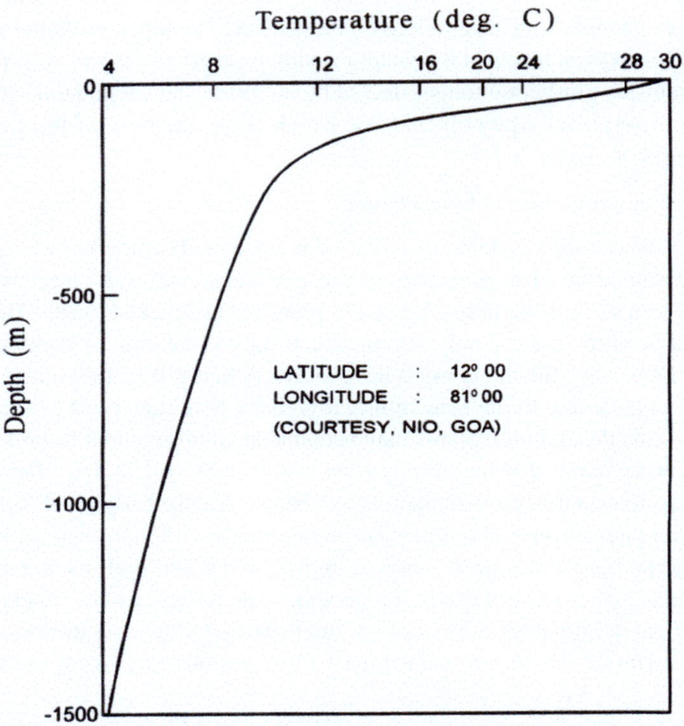

Fig. 5 Temperature Profile in the Ocean

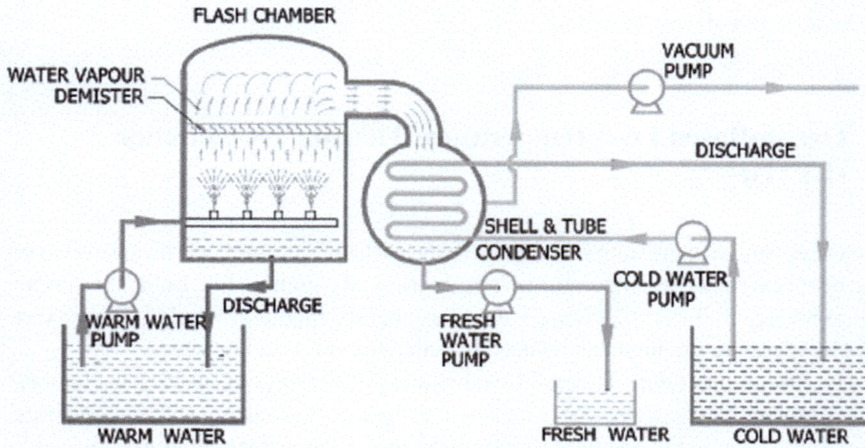

Fig. 6 Schematic of LTTD unit

pump to maintain the sub-atmospheric pressures. One of the advantages of the process is that it can be implemented even with a low-temperature gradient of about 8–10 °C between the two water bodies.

Thus, it can be used in the ocean scenario with surface water and deep sea cold water or in a power plant using surface seawater as the cold water and the condenser reject hot water as the warm water.

This section discusses the successes and projects commissioned using this technology.

Starting from a 0.5 m^3/day experimental laboratory set up in 2003, NIOT has been working on the LTTD process to develop the technology from the laboratory to field levels.

A 100 m^3/day land-based plant was commissioned at Kavaratti island in Lakshadweep in 2005 (Fig. 7) for the first time in the world. This plant has been continuously generating fresh water for the past 14 years to meet the drinking water needs of the island community. The water is of excellent quality as well as meeting the design quantity. The plant is housed in a structure on the shore. The bathymetry at the island is such that water at 10–12 °C is available at a depth of 350 m at a distance of around 300–400 m from the shore which is the source of cold water. The cold water is brought to the surface through a 600 m long High Density PolyEthylene pipe (HDPE). Figure 8 shows the cross-section of the plant and the unusual configuration of the cold water pipe in an invented catenary configuration. The temperature of seawater at the surface is 28–30 °C, which is the source of warm water. This first-ever plant has become the main source of drinking water for the islanders, and the health of the people has improved considerably. The water-related diseases like diarrhoea, hypertension, etc. have reduced, and the societal impact has been tremendous. Figure 9 shows the local islanders using this water, and their first reaction was that even the tea tastes better!

The summary of the impact on islanders and NIOT is as below.

For islanders:

- The plant has become the main source of good drinking water.
- The health of islanders improved considerably according to medical surveys.
- Local islanders are trained to run and maintain the plant leading to self-reliance.

For NIOT:

- This is the first-ever plant in the world from concept to commissioning using naturally occurring temperature differences fully designed and implemented by the institute.
- Expertise in complex civil construction and installation of pipelines was achieved.
- Engineering has been translated into true societal benefit.

Subsequently, a scaled-up offshore barge-mounted plant was taken up in 2007 with requirements of the mainland as the target. The plant was of 1000 m^3/day capacity, mounted on a barge and moored in over 500 m water depth around 40 km off Chennai. The plant was successfully commissioned and it produced water of potable quality.

Fig. 7 Desalination Plant at Kavaratti

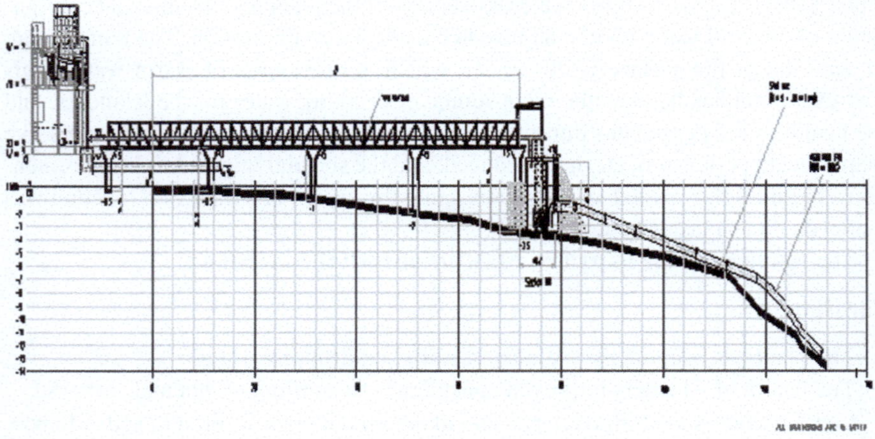

Fig. 8 Cross-section of Kavaratti Plant

Even though all the plant components had the same configuration as those of the land-based island plant, the configuration of the cold water pipe was completely different because the pipe had to be connected vertically below the floating barge that houses the plant. A long 1 m diameter High Density Poly Ethylene (HDPE) pipe was assembled near the shore on the mainland and towed to deep waters. The barge was moored using a single-point mooring and then the HDPE pipe was connected to the barge. The plant was operated for a few weeks during its sea trial in 2007 (Fig. 10) and subsequently decommissioned.

Fig. 9 Islanders using the water from the LTTD plant

Fig. 10 Barge-mounted desalination Plant

Fig. 11 Desalination Plant at Minicoy and Agatti Islands

Two more island plants in Agatti and Minicoy were commissioned successfully in 2011 (Fig. 11).

It is interesting to note that the same technology can be applied to any scenario where two bodies of water at different temperatures are available. Coastal thermal power plants discharge huge amounts of condenser reject water into the nearby ocean. The temperature of condenser reject water is 8°–10 °C above ambient temperature which causes environmental and thermal pollution. This available thermal gradient between the condenser reject water and nearby surface seawater can be utilized in the LTTD process to evaporate the warmer condenser reject water at low pressure and to condense the resultant vapour with the colder surface seawater to obtain fresh water (Venkatesan et al. 2015).

With this concept, a land-based 150 m^3/day capacity desalination plant was established in 2009 in the North Chennai Thermal Power Station (NCTPS) to demonstrate the utility of the process for any coastal thermal power plant that discharges huge amounts of condenser reject water into the nearby sea (Fig. 12). The 600 MW NCTPS plant discharges about 100,000 m^3/hr of condenser reject water at about 38–43 °C. In order to reduce the thermal pollution issues arising out of mixing this water with the nearby seawater at 26–30 °C, NCTPS lets the water run through a long open channel where the water is brought to about 33–37 °C. The NIOT plant just tapped the intake to the condenser and outlet from the condenser to demonstrate the technology. The water discharged after the process is cooler than the water entering the desalination system.

Large quantities discharged at lower temperatures can truly prove to be beneficial to the environment. Thus, if this system is adopted when power plants are in the early stages of planning, the LTTD plant can play the role of cooling towers, wherein energy is being pumped today to cool the outlet water. Thus, LTTD using condenser reject heat can truly play a positive role in the framework of water stress as well as climate change.

In general the positives of LTTD desalination plants are the following:

1. Utilization of vast renewable energy in the sea for generating fresh water.
2. Installation and operational simplicity.
3. Less maintenance issues and thus sustainability.

Fig. 12 NCTPS desalination Plant

4. Almost nil environmental hazard due to non-discharge of chemicals or waste or brine as occurs in membrane-based desalination plants.
5. Reduction in power plant cooling water discharge temperature to prevent thermal pollution.
6. No steam is required and no components need to be replaced or destroyed periodically.

This indigenous technology designed in-house in NIOT has received international acclaim and is a path-breaking technology which has benefitted the island community greatly, and studies are now on, to scale up for an offshore platform for mainland needs.

6 Desalination Plants and Energy

Desalination is considered to be an energy-intensive process. The electrical energy used in the process is usually derived from fossil fuels, the burning of which emits carbon dioxide resulting in the pollution of the environment. Thus, a measure of the viability of a desalination process is the amount of energy being consumed, since it is a major component of the O&M cost. Electrical energy for desalination is utilized from coal, thermal or nuclear grids. In islands, diesel generators are needed to supply power. To make the desalination process green, it is prudent to consider the use of renewable energies. To this end, Solar-PV systems combined with RO are being tried. While this can certainly reduce the pollution arising from the energy consumed, the inherent issues with RO remain.

A few years ago, the Department of Science & Technology funded a project for an industry with technical support from NIOT for a Solar Multi-effect Distillation plant at Ramanathapuram in Tamil Nadu. This plant is considered green due to the usage of solar energy and is also safe environmentally, since it is a thermal desalination system. However, the fluctuating and short availability of insolation makes its viability at this point in time questionable. The footprint for solar systems on land is also high, which is not desirable in this age of high land costs.

In the year 2003, NIOT also demonstrated a wave energy-powered RO plant. This was the first plant, wherein energy from the sea was being used to desalinate seawater. This plant, however, was a demonstration unit and hence was decommissioned after the local fishing community used the water for around 3 years.

The NIOT is now therefore embarking on the first of its kind system at Kavaratti island, where the LTTD plant will be powered by Ocean Thermal Energy Conversion. This means the plant will be self-powered and not dependent on diesel generators. The design caters to generating power just sufficient to power the desalination process. No extra power will be pumped to the grid. This hybrid system has the merits of clean and green energy, and environmentally safe desalination, and since energy is produced from the existing ocean thermal gradient and not from diesel the operational and maintenance costs reduce drastically.

The success of this plant can pave the way to making fossil fuel-free desalination plants, especially in islands thus creating fresh water for pristine locations with no contribution to climate change.

7 The Socio-Economic Angle

Having mentioned the environmental impacts, it is also essential to talk about costs. Today due to energy recovery systems, RO plants consume less energy than thermal systems thus making RO the preferred technology across the globe. In the context of the LTTD energy consumption, in the islands it is around Rs.7–8 kwh/m^3 of water produced. In the case of RO, it is around 4 kwh/m^3 of water. However for the islands, the cost of transporting the pre- and post-processing chemicals needs to be taken its account. Also during the monsoons, ships and cargo vessels do not ply, hence if chemicals are not available in the islands, the RO plant may not produce fresh water. RO also produces a large quantum of brine which is detrimental for islandic areas.

However, it is important to take the environmental factors into account for understanding the true cost. Apart from the capital cost and operating cost models, it is necessary to consider the environmental and ecological cost per litre of desalinated water.

Environmental cost per litre of desalinated water is arrived at on the basis of additional energy consumed per litre of desalinated water over the technology option with the least specific energy consumption. In the Indian context, one megawatt hour (MWh) energy consumption is assumed to imply a tonne of carbon dioxide emission.

If a process involves the reduction of specific energy consumption by one MWh, it is assumed to have earned one certified emission reduction (CER).

Ecological cost per litre of desalinated water is arrived at as the change in GDP per litre of desalinated water in the 'project catchment area' due to the introduction of a particular technology.

Concepts like the Leontief inverse can be used in mathematical terms to factor in the above costs so that the true cost of a desalination system can be arrived at (Venkatesan 2014). By carrying out this study for the Kavaratti plant and waste heat recovery plant showed that the ecological costs of RO Plant are extremely high and this needs serious consideration while making choices of desalination methods. Currently, efforts are on to scale up the technology in power plants and to implement OTEC-powered desalination in Kavaratti, which can further reduce its costs since we are using green energy.

8 Conclusions

The chapter addresses how water stress can be mitigated using desalination. While membrane systems have several inherent drawbacks, the choice of desalination methods in site-specific. The NIOT under MoES has developed the technology and is harnessing fresh water using the ocean temperature difference and has successfully implemented it in islands, offshore and in power plants. Costing should consider ecological as well as environmental costs so that the ill effects of the desalination method get factored in to give more accurate water costs. The LTTD technology scores high when these aspects are considered. The technology, if implemented in coastal power plants will greatly reduce thermal pollution. The societal impact of the installed plants has been tremendous and has put India at the forefront of developing new techniques in desalination.

References

Buros OK (2000) The ABCs of desalting (p 30). Topsfield, MA: International Desalination Association

Cath TY, Childress AE, Elimelech M (2006) Forward osmosis: principles, applications, and recent developments. J Membr Sci 281(1–2):70–87

Darwish MA, Alsairafi A (2004) Technical comparison between TVC/MEB and MSF. Desalination 170(3):223–239

El-Ghonemy AMK (2017) Performance test of a sea water multi-stage flash distillation plant: Case study. Alexandria Eng J

Global Technology Watch Group (2019) "Water", Department of Science & Technology, TIFAC

Greenlee LF, Lawler DF, Freeman BD, Marrot B, Moulin P (2009) Reverse osmosis desalination: water sources, technology, and today's challenges. Water Res 43(9):2317–2348

Kathiroli S, Jalihal P (2008) "Up from the Deep", Civil Engineering, Magazine of American Society for Civil Engineers, Vol. 78

Lawson KW, Lloyd DR (1997) Membrane distillation. Journal of membrane. Science 124(1):1–25
Lei Z, Chen B, Ding Z (2005) Membrane distillation. In: Lei Z, Chen B, Ding Z (Eds) Special distillation processes. Elsevier Science: Amsterdam, the Netherlands, pp 241–319
Oztekin E, Altin S (2016) Wastewater treatment by electrodialysis system and fouling problems. Turkish Online J Sci Technol 6(1)
Purnima J, Prabhakar S (2019) Desalination technologies. Water Futures of India, INSA, pp 362–400
Rognoni M, Kathiroli S, Jalihal P (2008) Low temperature thermal desalination (LTD). New Sustain J Nuclear Desalination 3(1):69–78
Schneider K, van Gassel TJ (1984) Membrandestillation. Chemieingenieurtechnik 56(7):514–521
Spiegler KS, El-Sayed Y, Primer AD (1994) Balaban desalination Publications. Santa Maria Imbaro
Venkatesan G, Iniyan S, Jalihal P (2015) A Desalination method utilizing low-grade waste heat energy. Desalination Waste Treatment 56180:2037–2045
Venkatesan R (2014) Comparison between LTTD and RO process of sea-water desalination: an integrated economic, environmental and ecological framework. Curr Sci, 378–386
Werft M (2016) Is Desalination the answer to water shortages. Global Citizen
Younos T, Tulou KE (2005) Overview of desalination techniques. J Contemp Water Res Educ 132(1):3–10

Emerging Blue Economy Paradigm and Technological Developments in India

N. Vedachalam, G. A. Ramadass, and M. A. Atmanand

Abstract With the land-based resources depleting faster, effective exploration and sustained harvesting of the Ocean resources, taking into account the economic growth, social needs and the health of the Ocean environment is essential. India with a 7600 km long coastline and an exclusive economic zone of 2.3 million km^2 and spear heading the blue economic growth and security of all the countries in the Indian Ocean region have initiated policies for leveraging the growth of the Indian economy through the blue economy. The paper discusses the strategic technology developments undertaken in India for the harvesting of the living and non-living blue economic resources, and for protecting the ocean ecosystems.

1 Introduction

The global population is expected to exceed 8 billion by 2030 would require 30% more water, 40% more energy and 50% more food than at present. More than 40% of the world population live within 100 km from the coastline and it continues to be in the uptrend (David et al. 2017). With the depleting land-based resources, the oceans, with an estimated asset value of more than US$ 24 trillion are a promising strategic frontier for economic growth, water and food security. The present ocean economy, which is ~US$ 2 trillion and 7th largest economy in the world, is expected to exceed US$ 3 trillion and support more than 40 million jobs by 2030 (United Nations World Bank Group 2017).

Based on the 1982 United Nations Convention on the Law of the Sea (UNCLOS), the countries have exclusive rights and jurisdiction over the resources within their Exclusive Economic Zone (EEZ) that extend up to 200 nautical miles (Fig. 1). The UNCLOS articles 136, 137.2, and 145 cover the common heritage of mankind, resources and the protection of the marine environment, respectively (United Nations Convention on the Law of the Sea n.d.). The total global EEZ area belonging to various nations is ~137 million km^2. An EEZ of ~12, 11.3, 8.5 and 2.3 million

N. Vedachalam (✉) · G. A. Ramadass · M. A. Atmanand
Ministry of Earth Sciences, National Institute of Ocean Technology, Chennai, India
e-mail: veda1973@gmail.com

V. K. Gahalaut and M. Rajeevan (eds.), *Social and Economic Impact of Earth Sciences*,
https://doi.org/10.1007/978-981-19-6929-4_14

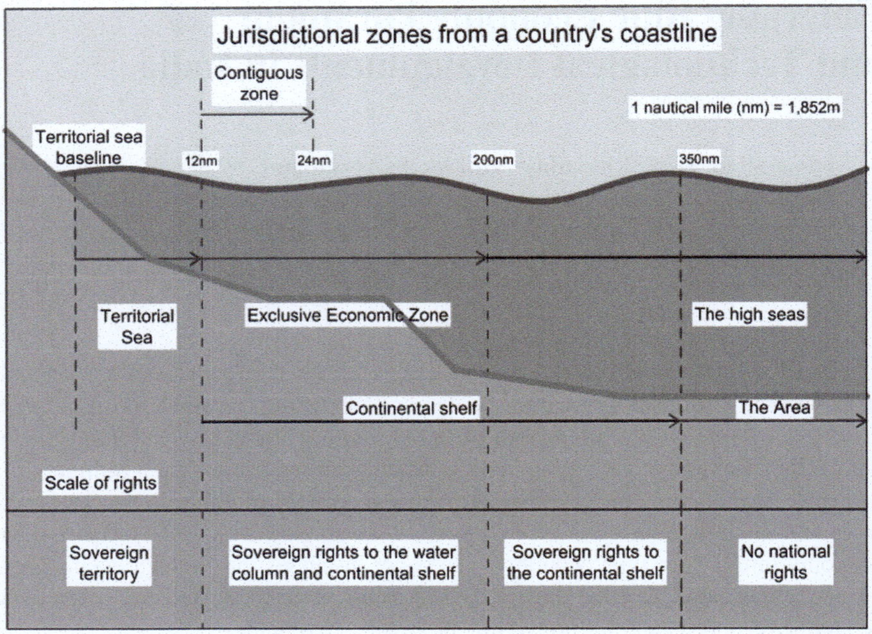

Fig. 1 Jurisdictional zones from a country's coastline (United Nations Convention on the Law of the Sea n.d.)

km^2 belongs to France, US, Australia and India. The EEZ of India includes 1.64 million km^2 near the Indian mainland and Lakshadweep, and 0.66 million km^2 in the Andaman and Nicobar Island area (United Nations Convention on the Law of the Sea n.d.).

Based on the foundations of the 2012 Rio + 20 UN Conference on sustainable development, the global community announced its commitment to the Sustainable Development Goals (SDG) 2030. The SDG Goal 14 is related to the sustainable development of the ocean resources by which the economic activity is in balance with the long-term capacity of the ocean ecosystems so that oceans remain resilient and healthy (Life Below Water Why It Matters 2017). Subsequently, the UN has proclaimed a "Decade of Ocean Science for Sustainable Development (2021–2030)" to support efforts to reverse the cycle of decline in ocean health and gather ocean stakeholders worldwide behind a common framework that will ensure ocean science can fully support sustainable ocean developments (United Nations Decade of Ocean Science for Sustainable Development n.d.).

2 Emerging Blue Economy Paradigm

The concept of the blue economy stems from the Gunter Pauli's book "The Blue economy: 10 years, 100 innovations and 100 million jobs" published in 2010, which expresses that the blue economy shall shift the society from scarcity to abundance (Pauli 2010). The blue element insists on the innovative approaches required to conserve the oceans, while reaping their benefits in an equitable and sustainable way (Fig. 2). The blue economy is thus based on resilient systems, persistent innovation and advances in achieving integrated ecological, economic and social well-being. The pillars essential for transforming the traditional "Ocean and marine economy" to a "Blue" or "Sustainable" economy require appropriate governance in the sustained utilization of the ocean, coastal and marine economies, vision, technology, management, monitoring and time-bound regulatory reforms (Patil et al. 2016). Thus the fast emerging blue economy paradigm (Table 1) requires proper estimation of the size of the opportunity, nature of risks involved, identification of sustainable ocean asset investment, investment framework and scaling up the capital investments of the blue industries.

With the blue economies of the United States, China and the European Union estimated to be ~US$1.5, 0.1 and 0.5 trillion, respectively, the countries are moving from individual sector-based silo models towards implementation of integrated multi-sector policy frameworks for ocean management (Blue Economy Vision 2025). The approach will help the countries to reconcile the huge ocean potential, at the same time, address the increasing pressures on the marine environment such as over-exploitation, declining biodiversity, pollution and climate change.

Considering its strategic interest, India is seeking to extend its EEZ to 350 miles, by which the total EEZ area will equal its land area. India is a key member in the Indian Ocean Rim Association (IORA), an inter-governmental organization comprising of 21 member states and 9 dialogue nations aimed at strengthening regional cooperation and sustainable development within the Indian Ocean region. The IORA blue economy dialogue held at Goa in August 2015 passed the Goa declaration

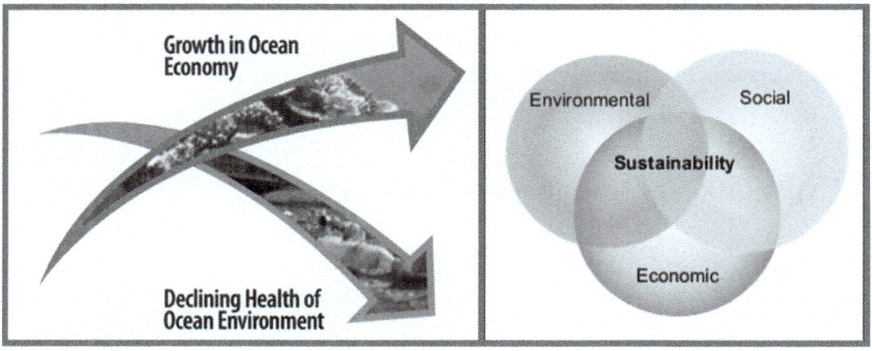

Fig. 2 Concept of sustainable blue economy (Life Below Water Why It Matters 2017)

Table 1 Components of the Blue Economy (United Nations World Bank Group 2017)

Type of activity	Ocean service	Industry
Extraction of non-living resources	Minerals	Seabed Mining
	Energy	Oil and natural gas
		Renewables
	Fresh water	Desalination
Harvesting of living resources	Seafood	Fisheries
		Aquaculture
	Marine biotechnology	Pharmaceuticals, Chemicals
Ocean commerce	Transport and trade	Shipping
		Port infrastructure and services
	Tourism and recreation	Tourism
		Coastal Development
Response to ocean health challenges	Ocean monitoring and surveillance	Technology and R&D
	Carbon sequestration	Blue Carbon
	Coastal protection	Habitat protection and restoration
	Waste disposal	Assimilation of nutrients and wastes

stressing the need to identify the thrust areas of the blue economy (Prospects of the Blue Economy in India n.d.). Since 2018, National Institution of Transforming India (NITI), the Indian Government's apex policy think-tank was conducting discussions with all the stakeholders to identify the potential areas of blue economy that are to be placed on the national strategic focus, and to formulate policies for blue growth (HamantMaini and LipiBudhraja n.d.). Multiple ministries under the Government of India (GoI), including the Ministry of Earth Sciences (MoES) have undertaken various programs for fostering the strategic blue growth of India.

3 Exploitation of Non-living Resources

3.1 Sea Bed Mineral Mining

The developments of offshore mineral resources are complimentary for the industrial and economic growth. The blue mineral resources include the seafloor poly-metallic sulfides around the hydro-thermal vents, cobalt-rich crusts on the seamounts and the poly-metallic manganese nodules on abyssal plains. The poly-metallic nodules comprise manganese (28%), nickel (1.3%), copper (1.1%), cobalt (0.2%), molybdenum (0.059%) and rare earth metals (0.081%). They also contain traces of other elements of commercial interest, including platinum and tellurium required for manufacturing photovoltaic cells and catalytic technology. The seafloor sulfides are rich in copper, gold, zinc, lead, barium and silver. The cobalt-rich crusts contain manganese, iron and a wide array of trace metals including cobalt, copper, nickel and platinum. Cobalt is used for manufacturing super-alloys and in the modern battery technology (Hein et al. 2013).

In the Central Indian Ocean Basin (CIOB), economically interesting contiguous nodule fields occur in a region covering ~700,000 km^2 in water depths ranging 3000–6000 m in between 10°–16°30′S and 72°–80°E. Within this region, a nodule-rich area called the Indian Ocean Nodule field covering 300,000 km^2, hosts ~1400MT of manganese nodules with an average abundance of 4.5 kg/m^2. About 21.84MT of nickel, copper and cobalt resources are located in the hydrothermal sulfides in the southern Indian Ocean and cobalt crusts in the Afanasy Nikitin sea mount area (Fig. 3) (Sharma 2017). The economics of exploitation depends on the grade, tonnage, oceanographic conditions and the harvesting technology (Miller et al. 2018). The International Seabed Authority (ISA) has issued 30 contracts covering a global area of >1.4 million km^2 to various countries for mineral exploration, and in the process of formulating regulation for commercial mining (Exploration Contracts n.d.).

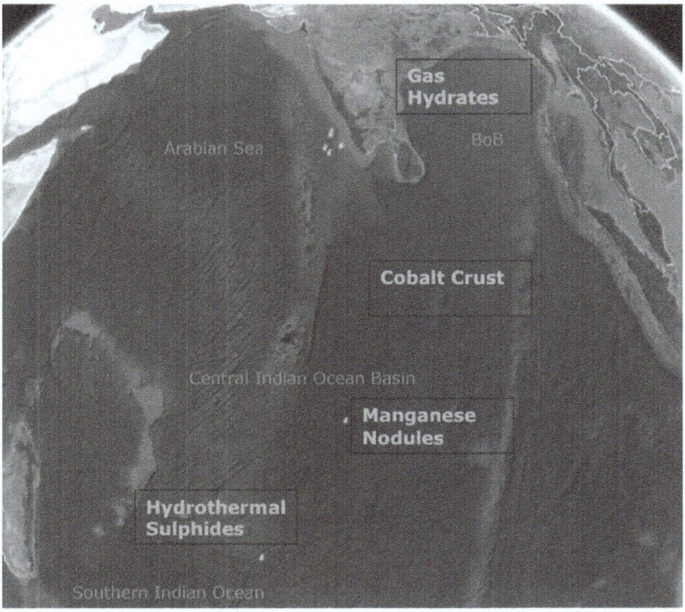

Fig. 3 Mineral and energy resources in the Indian Ocean (Sharma 2017)

Based on extensive surveys conducted by the CSIR-National Institute of Oceanography (NIO) over four decades, India has been allocated an area of 75,000 km^2 in the poly-metallic nodules location and 10,000 km^2 in the hydrothermal sulfide area for carrying out exploratory activities, that includes testing of mining equipment. After carrying out field demonstration of a crawler-based mining machine at ~500 m water depth (Fig. 4), the MoES-National Institute of Ocean Technology (NIOT) has undertaken the development of a 6000 m depth-rated demonstrative poly-metallic nodule mining machine (Fig. 4) (Estimation of Reliability of Underwater Poly-metallic Nodule Mining Machine 2015). For carrying out in-situ shear and bearing strength measurements of the deep sea bed, a 6000 m depth-rated in-situ soil tester is developed and qualified at 5462 m water depth in the CIOB by NIOT. Deep-ocean moorings are developed by NIOT and CSIR-NIO to generate base line oceanographic data and to understand the potential impacts of the deep-ocean mining. The development of the metallurgical processes for extraction of metals is being developed by the CSIR-Institute for Mineral and Material technology (IMMT) and CSIR-National Metallurgical Laboratory (NML) (Vedachalam et al. 2018).

For enabling deep-ocean mineral exploration, a 6000 m depth-rated electric work-class Remotely Operated Vehicle (ROSUB 6000) with the capability to webcast sea bed visual in real time to the shore is developed. It was used in the polymetallic nodule site in the CIOB during 2010 and in the hydrothermal sulfides site at the Rodriguez Triple Junction in the Central Indian Ridge during 2013 (Ramadass et al. 2010, 2015). In order to enable deep ocean manned missions, development of a 6000 m depth-rated human-occupied scientific submersible with an operational endurance of 12 h and emergency endurance of 96 h is undertaken by NIOT (Ramadass et al.

Fig. 4 Polymetallic module mining under qualification (Vedachalam et al. 2018)

Fig. 5 ROSUB6000 onboard TDV Sagar Nidhi prior to deployment (Vedachalam et al. 2018)

2020). Other strategic technologies required for the long-term deep-ocean mineral surveys, such as autonomous underwater vehicle homing and docking, subsea wireless optical communication and wireless power transfer are developed and demonstrated by NIOT (Vandavasi et al. 2018; Analysis of Subsea Inductive Power Transfer Performances Using Planar Coils 2016) (Fig. 5).

3.2 Offshore Hydrocarbons

Globally ~36% of the natural gas resources are located offshore (Offshore Energy Outlook 2019). In India, about 26 offshore sedimentary basins covering ~3 million km^2 host ~28 billion tons (BT) of conventional hydrocarbons. About 67% of these resources are located offshore. The Mumbai High hosts about 9.2BT, the sedimentary basin with an area of ~1.3 million km^2 in the east and west coasts of India spanning from 400 m water depth up to the EEZ hosts about 7 BT, and the rest of the resources are concentrated in the Krishna-Godavari (KG), Cauvery and Kerala Konkan basins. The KG basin hydrocarbon production accounts to ~40% of the India's in-house production (Annual Report n.d.). Production wells have been established in the KG basin at 2483 m water depths to foster the natural gas production from deep waters. For effective exploitation of the ultra-deep water hydrocarbon resources, Indian government along with private participation has planned to invest ~US$ 5 billion in the deep water projects in the KG basin.

Natural gas hydrates are crystalline clathrates composed of low-molecular weight gases, mostly methane, engaged in a lattice of hydrogen-bonded water molecules. In the marine settings, gas hydrates occurs below the sea floor at water depths ranging between 800 and 3000 m. Depending on the formation mechanism and the reservoir characteristics, changes in the pressure and temperature conditions of the reservoir shall result in the dissociation of gas hydrates producing methane gas and water. One 1 m^3 of gas hydrates produces ~164 m^3 of methane gas and 0.9 m^3 of water at standard temperature and pressure conditions (Sloan and Koh 2008). Considering the strategic importance of the natural gas hydrates located in the continental margins of India, the National Gas Hydrate Program (NGHP) led by the Directorate General of Hydrocarbons (DGH) and supported by Oil and Natural Gas Corporation (ONGC), major oil companies and national scientific organizations including NIOT, the National Geophysical Research Institute (NGRI) and the National Institute of Oceanography (NIO) have performed two detailed drilling expeditions NGHP-01 and 02 during 2006 and 2015, respectively in the KG, Mahanadi, and Andaman convergent margin (Vedachalam et al. 2015). The down-hole logging and coring done using the research vessel JOIDES resolution and Japan's deep drilling vessel Chickyu has confirmed the presence of large, highly saturated gas hydrate accumulations in the coarse-grained sand-rich depositional systems in the KG basin (Fig. 6). Scientific expeditions conducted in the KG basin using NIOT's deep water work class ROV ROSUB 6000 and NIO vessel Sindhu Sadhana brought out chemosynthetic habitat abundance which confirms the presence of gas hydrates (Ramadass et al. 2010; Dewangan et al. n.d.). Efforts are undertaken to establish a pilot scale well in the KG basin for long-term production capability assessment (Expedition n.d.).

The NIO and NGRI are pursuing activities towards delineation and resource estimation. For ground truth validation and spatial quantification of the gas hydrate resource, NIOT is developing a 3000 m-depth rated sea bed based wire-line autonomous coring system (ACS) (Fig. 7) capable of taking in-situ gas hydrate core samples below the sea floor (Ramesh et al. n.d.).

3.3 *Offshore Renewable Energy*

Sustainable marine energy plays a vital role in the economic development and climate adaptation. The global offshore wind energy capacity has quintupled from ~3.2 GW in 2010 to ~20 GW in 2019. With the policy support and the technology advances the global offshore wind energy installed capacity is expected to reach 160 GW in 2040, with US, China, India, UK, Netherlands and France targeting to realize ~50, 40, 18, 15,12 and 9 GW, respectively (Global Energy Transformation n.d.; REN21 2018) (Table 2).

For fostering the growth of the offshore wind energy in India, the National Offshore Wind Energy Authority has been established to carry out resource assessment in the EEZ. The offshore wind resource assessment by the MoES-Indian

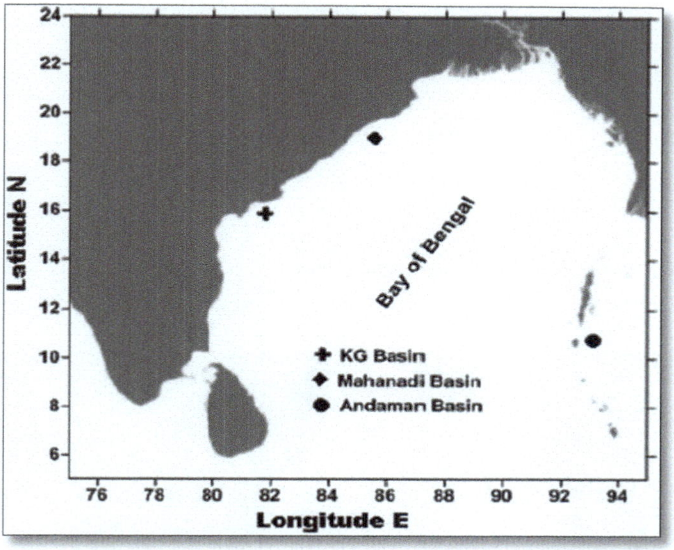

Fig. 6 Location of major gas hydrate provinces of India (Vedachalam et al. 2015)

Fig. 7 Autonomous Coring System on-board vessel prior to deployment (Ramesh et al. n.d.)

National Center for Ocean Information Systems (INCOIS) based on the long-term satellite wind data (Fig. 8) indicates high wind energy potential in off-Kanyakumari, Gujarat offshore, Rameshwaram and Jakhau (Vedachalam et al. 2018).

Table 2 Offshore wind energy potential (Global Energy Transformation n.d.; REN21 2018)

	EU	USA	China	India
ATP (GW)	2700	2085	500	350
Water depth (m)	1000	1000	5–50 m	
Distance from shore (nm)	5 nm	200 nm	50 nm	EEZ
Wind speed (m/s)	>8	>7	>8	>8

ATP: Assessed Technical Potential; nm: Nautical mile

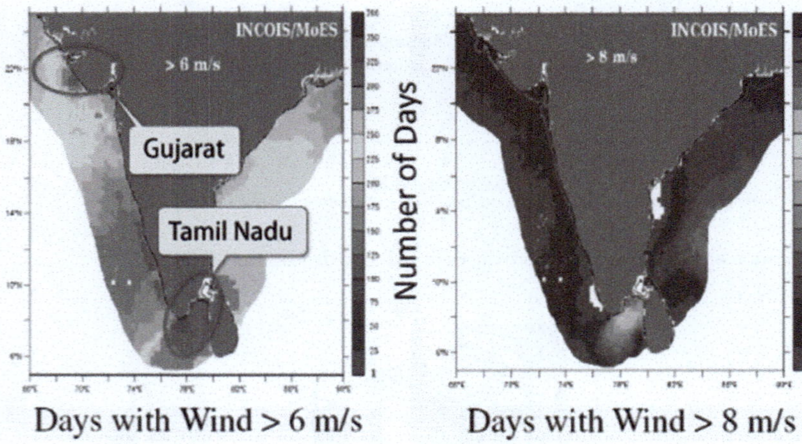

Fig. 8 Indian offshore wind energy resource assessment (Vedachalam et al. 2018)

For carrying out wind speed, direction, temperature and humidity measurements at 20 m elevation from the sea level, LIDAR-based data collection platforms are installed by NIOT and MNRE-National Institute of Wind Energy (NIWE) in the Gulf of Kutch and Gulf of Khambhat (Fig. 9). Geotechnical investigations, structural studies and commercial viability analysis are undertaken (Raju et al. 2020).

Other ocean energy technologies in various stages of development include marine current, wave power and ocean thermal energy conversion (OTEC) face stiff competition from other onshore options. India is a member of the International Energy Agency-Ocean Energy Systems (IEA-OES) Group and has taken up activities to accelerate the growth of ocean energy systems in the Asian region. The underwater current measurements carried out in Kadamtala in the Andaman Islands revealed the continuous water current potential of about 2.5 m/s, and suitable hydrokinetic turbines are being developed by NIOT (Biren et al. 2020).

Fig. 9 LIDAR-based based platform in Gulf of Kutch installed by MNRE and NIOT (Vedachalam et al. 2018; Raju et al. 2020)

3.4 Ocean Desalination

Securing adequate quantities of clean water to meet the needs of the growing population is a major challenge.. Coastal communities are increasingly turning to the sea to meet their drinking water needs, while in inland there is a tendency for groundwater to become increasingly brackish over time. Around 60% of the desalination capacity treats seawater, with the remainder treating brackish and less saline feed water (Water Supply Diversification Desalination n.d.).

Membrane-based desalination technologies based on reverse osmosis and electro dialysis are employed in ~93% of the plants, whereas thermal-based technologies, including multi-stage flash evaporation, multi-effect distillation and vapour compression are used in the rest. NIOT has designed and implemented Low Temperature Thermal Desalination (LTTD) plants with a capacity of 100 m^3/day in three Islands in the Union Territory of Lakshadweep, where a long cold water is used to draw water from water depth around 300 m using a high density polyethylene pipe (Fig. 10). The

Fig. 10 Desalination plant operating in Agatti Island of India (Ramadass et al. 2010)

plants that are operating over a decade have proved multiple socio-economic advantages including health of the island community. Efforts are underway for installing LTTD-based desalination plants in six more islands of India (Ramadass et al. 2010; Venkatesan et al. 2020).

4 Harvesting of Living Resources

Effective and sustained marine bio-prospecting is essential to pursue human health, offering sustainable supply of high quality food, developing sustainable sources of energy alternates to the conventional hydrocarbons, new industrial products and processes with low greenhouse gas emissions.

4.1 Seafood-Fisheries and Aquaculture

Seafood is an important source for protein, especially micro-nutrients and omega-3 fatty acids and sustained production provides long lasting benefits (Food and Agriculture Organization of the United Nations 2020; Thompson et al. December 2017) and efforts are underway to increase the production (PFZ n.d.) With more than 2.4

lakh fishing crafts, 6 major fishing harbors, 62 minor fishing harbors, 3432 fishing villages, 1511 landing centers and ~4 million fishing population, India stands 4rd in the fish production. More than 50 different types of fish and shellfish products are exported to 75 countries. The sector contributes 1.1% of the national GDP and creates annual export earnings of about US$ 5 billion. Out of total fish production of5.5 MT (~25% of the global fish production), marine contributes 3.7MT and production from ~73,000 km^2 of inland water bodies is ~1.8MT (Thompson et al. 2017).

It is identified that the potential fish production in India could be increased up to 8.4MT, ~30% increase from the current production. The integrated development and management of fisheries provides subsidies for the conversion of fishing trawlers into deep sea fishing vessels to promote deep sea fishing. With the aid of the Indian satellite-based oceanography, INCOIS provides reliable and timely advisories on the potential fish aggregation zones with specific references to more than 500 fish-landing centers along the Indian coast. They also provide information on the algal blooms for near real-time assessment of the primary productivity (MEDAS n.d.). Programs are undertaken by MoES-Center for Marine Living Resources and Ecology (CMLRE) to address the bio-geo-chemistry of the eastern Arabian Sea and biological responses including fishery resources in the eastern Arabian and taxonomically-resolved marine species records from the northern Indian Ocean through systematic surveys (Culture and Biotechnology n.d.).

In order to identify effective fish aggregation methods, multi-point moored open sea cages made of high density polyethylene of diameters >9 m capable of withstanding turbulent sea states were developed and demonstrated by NIOT. Culturing commercially important marine fin-fishes such as the Asian Seabass, Cobia, Pompano, Milkfish, Parrot fish and the Giant Travelly were demonstrated with encouraging results in the North Bay in the Andaman Islands, Olaikuda in Tamilnadu and Kothachathram in Andhra Pradesh (Patil et al. 2016) (Fig. 11).

4.2 Marine Biotechnology-Pharmaceuticals and Chemicals

Marine biodiversity is a major untapped resource especially in countries with high endemic biodiversity. With more than 1600 new marine species discovered every year, the number of predicted species is ~10 times higher than the 200,000 catalogued species. By 2025, the global market for marine-derived pharmaceuticals and marine technologies for industrial applications are estimated to reach ~US$ 5billion and US$6.4 billion, respectively (Protocol n.d.; Department of Bio Technology n.d.) (Table 3).

India's Department of Biotechnology through various research centres is carrying out activities including fish genomics and transcriptomics, fish and shellfish diseases, immune-stimulants and antimicrobial peptides. Development of the bioactive molecules, biomaterials, bio-surfactants, marine action-bacteria,

Fig. 11 Open sea fish cages in Andaman Islands deployed by NIOT (Patil et al. 2016)

Table 3 Strategic applications of marine biotechnology (Nagoya Protocol n.d.)

Marine conservation	Marine organism production
Monitoring environmental change	Organism cultivation and Collection
Pollution prevention and control	Disease control and monitoring
Biodiversity conservation and ecosystem recovery	Marine bio-safety and Mass production e.g. seaweeds
New Materials	Organism-based Technology
Drug discovery	Bioprospecting
Industrial Materials	Marine genome sequence and bioinformatics
Health supplements, nutraceuticals	Metagenomics and other omics technologies
Biofuels and bioenergy	
Bio refining	

DNA barcoding and molecular taxonomy, cell lines and diagnostics were also pursued through adoption of molecular tools and techniques. Studies on the development of quorum sensing inhibitors and anti-infective from marine micro-organisms for the control/prevention of pathogenic vibrios in shrimp aquaculture are also implemented. A project for molecular characterization of the bio-film produced by coral associated bacteria isolated from Andaman Sea is carried out (Deepsea Sampling and Microbial Culture Facility n.d.).

The marine-derived enzymes and algae, which are key for the food, nutritional, cosmetic, pharma and bio-fuel industries, are studied at NIOT. Large-scale cultivation systems such as bubble column photo-bioreactor, tubular photo-bioreactor and raceway pond systems are developed and suitable culture conditions were optimized for the mass culture of marine microalgae, and the extracted lipids from the mass culture ventures are trans-esterified into biodiesel (Fig. 12). Economically viable methodologies are being studied for extraction of multiple by-products sequentially and translate it to commercial scale (Vedachalam et al. 2018).

For fostering the studies on the applications of the deep sea piezophilic micro-organisms in the health and medical sectors, a deep-ocean microbial sampling and incubation system capable of bringing the deep-ocean micro bio-resources to the surface and incubating them by maintaining their ambient pressure (Fig. 13) is established at NIOT (Ganeshkumar et al. 2020; Marine Biotechnology Strategic Research and Innovation Roadmap n.d.). The Indian Government has taken initiatives to establish the Institute of Ocean Biology in partnership with the globally reputed research labs to explore the scope of marine biology and marine bio-technology in India (Developments in international seaborne trade, Review of maritime transport. 2018).

Fig. 12 Marine micro-algae related work at NIOT (Vedachalam et al. 2018)

Fig. 13 Deep sea sampling and microbial culture facility at NIOT (Ganeshkumar et al. 2020)

5 Ocean Commerce

5.1 Trade, Transport and Tourism

A robust technology-driven maritime sector with effective port infrastructure is essential to achieve sustained growth (Beyer 2018), In India, during 2018, the major ports with a capacity of 1.45 BT handled 0.7 BT and non-major ports handled ~0.5 BT. With the total vessel fleet of 1400 with gross registered tonnage (GRT) of ~10 million tons, ~451 vessels with a capacity of 88% are used for overseas trade and ~938 coastal vessels are used for ferrying goods within the country. The inland waterway transport is an economical means of domestic transport. The port performance statistics of the global major economies are shown in Table 4.

In India, the national maritime development programs worth US$ 11.8 billion are being implemented to accelerate the ocean commerce and transport. The coastline

Table 4 Port performance statistics of major economies (Ministry of Shipping 2016; Travel Tourism Economic Impact 2018)

Criteria	India	China	US
Contribution of water ways in domestic transport	<1%	24%	6%
Total turnaround time in days	4.5	1.0	1.2
Port capacity stock in % of GDP	1	3	10
Number of shipyards	7	70	45
Container traffic in million TEU	11	185	44

of India covering 13 states and Union Territories are located on the international trade routes and has about 14,500 km of navigable waterways. Water-borne transport accounts only 6% of the total freight transport movement in India computed in tonne-km basis. Considering the need for efficient, eco-friendly and economical domestic freight services utilizing the coastal and inland waterways, a national perspective plan, Sagarmala, aimed in the port modernization, effective port connectivity, port-led industrialization was formulated in 2016 by the Ministry of Shipping (Fig. 14). Under the Sagarmala Programme, the government has envisioned a total of 189 projects for modernization of ports involving an investment of Rs 1.42 trillion (US$ 22 billion) by the year 2035. The successful realization of the program, involving an infrastructure mobilization of US$ 60 billion, aspires to reduce the logistics costs for EXIM and domestic cargo by US$ 6 billion annually, double the share of waterways, boost exports by US$ 110 billion, create 4 million direct jobs, 6 million indirect jobs, and increase the commercial vessel fleet to 1600 by 2025. With the present 143 existing ports and another 76 proposed ports shall add more fillip to the blue commerce (Tourism Council 2018).

India's 43% of coastline with sandy beaches, 11% with rocky land and 31 mangrove areas could be developed as potential tourist hubs. In India, tourism industry contributed 2% to the total GDP and is projected to reach 6.4% by 2022. In 2017, ~10.4 million foreign tourists visited India, mostly along the coastal places earning $27 billion. The government has undertaken 17 coastal development projects across the country in the past four years (India Tourism Statistics 2018; Ravichandran 2015).

6 Response to Ocean Health Challenges

The ocean and coastal ecosystem produce 50% of the oxygen that we breathe, absorb 30% of the global GHG emissions and absorb more than 90% of the added heat. Hence, healthy oceans and coasts help to mitigate climate change and its impacts. Coral reefs and mangrove provide coastal protection, and oceans contain immense genetic resources. The stress on the oceans are constantly increasing with declining biodiversity, habitat degradation, increasing ocean acidity, accumulating plastics and other wastes. So, these natural assets are to be managed to support human development in the future. Similarly, tropical oceans have to be monitored for events such as cyclones and tsunamis to provide advance warning for protecting life and coastal property, and the Polar oceans are to be monitored from the climate change and sea level rise perspectives.

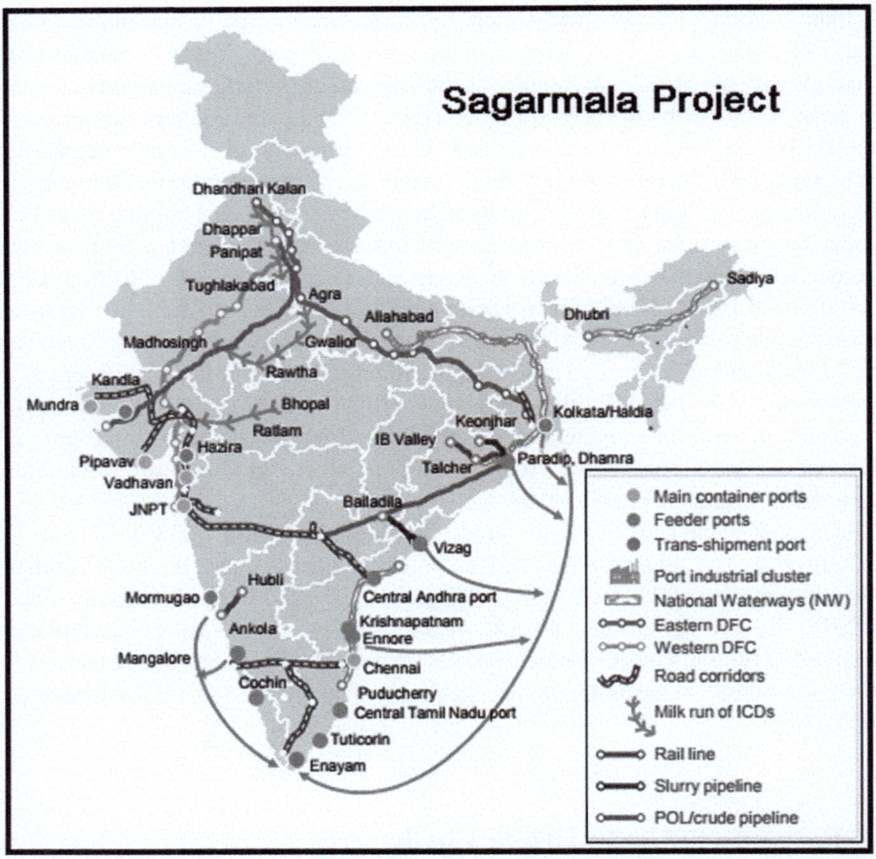

Fig. 14 Efficient evacuating network using inland waterways under Sagarmala (India Tourism statistics 2018)

6.1 Ocean Hazard and Health Monitoring

The Indian maritime zone is dominated by a range of economic activities including port operations, conventional fishing, hydrocarbon exploration, transportation, marine research and defense activities. Sustained real-time ocean observations are vital to understand the ocean dynamics and variability required for improving weather prediction, ocean state forecast, climate change studies and other oceanographic services. The Indian ocean observational networks are configured for real-time and delayed-mode coastal and offshore observations using offshore and coastal-located moored surface buoys (including Research Moored Array for African-Asian-Australian Monsoon Analysis and Prediction mooring network (RAMA) and the Ocean Moored buoy Network for northern Indian Ocean (OMNI) buoys networks), acoustic Doppler water current profile moorings, deep-ocean wave buoys, coastal wave rider buoys, equatorial current meter moorings, and tsunami buoys for deep-sea water level measurements and subsurface floats (ARGOs) (McPhaden et al. n.d.; Reliability Metrics from Two Decades of Indian Ocean Moored Buoy Observation Network 2018).

The incidence of tropical cyclones is higher in the northern Indian Ocean, specifically in the Bay of Bengal because of the unique geography. Although, it is not possible to completely avoid natural disasters, their effects can be minimized by taking some long-term, short-term mitigation measures and improved response mechanism. Hence continuous ocean observation and timely dissemination of information to various end users is essential.

The OMNI mooring measures the surface meteorological variables and temperature and salinity profiles from the surface down to 500 m depth and the. Over the past two decades, the moored data buoys are deployed (Fig. 15) ranging from the coastal waters to the deep oceans spanning between 63°E to 93°E and 6°N to 20°N for collecting the meteorological, water surface and subsurface parameters; as well as the tsunami water level data. The collected data is transmitted to the NIOT Mission Control Center located at NIOT in Chennai and to the Indian Tsunami Early Warning Center (ITEWC) in INCOIS, Hyderabad (Venkatesan et al. 2016a; Srinivasa Kumar et al. 2016).

During a tsunami event, the sea level inputs from the tsunami buoys serve as critical inputs to the ITEWC's pre-run scenario database, which computes the tsunami travel times (Fig. 16) and wave run-up wave heights, is essential for the timely generation and dissemination of the tsunami advisories. The ITEWC has been serving as the primary source of the tsunami advisory for India and as a tsunami service provider for the entire Indian Ocean region (PRATYUSH n.d.).

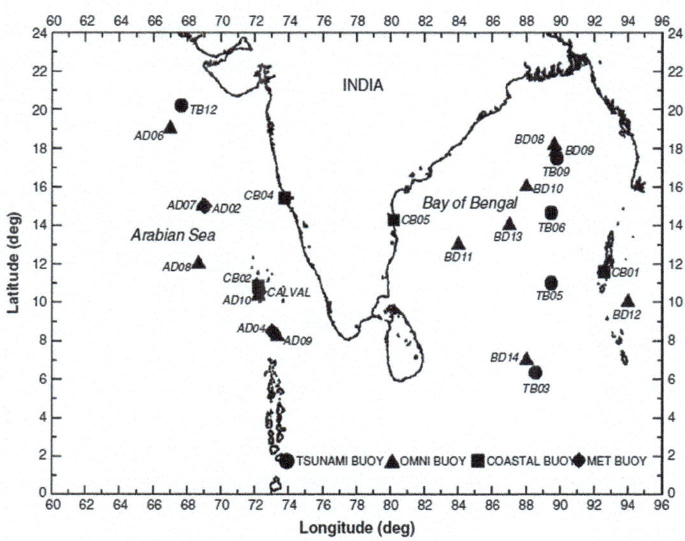

Fig. 15 Moored surface buoys deployed in Indian waters (Venkatesan et al. 2016a)

Fig. 16 Strategic placement of the tsunami buoys

During the past two decades, the Indian moored buoy networks have detected more than 41 cyclones and 11 water level change events used to monitor tsunami waves. The four subsurface ADCP moorings along the equator and one near the coast of Java are used to observe current profiles in the upper ocean, and three deep current-meter moorings with ADCPs in the central and eastern equatorial regions. At present, there are 769 active Argo floats in the Indian Ocean with 491 floats to the north of 40°S (McPhaden et al. n.d.). The data acquired during various events serve as important inputs to various agencies including the Indian Meteorological Department (IMD) and for understanding the Indian Ocean dynamics essential for improved modeling of the evolution of the seasonal monsoons and cyclones. The supercomputer Pratyush established at the Indian Institute of Tropical Meteorology (IITM) in Pune and the National Center for Medium Range Weather Forecasting (NCMRWF) in Noida augments India's capability in hazard mitigation and improving the weather and climate forecasting services (Ramesh et al. 2016).

To understand the influence of cryosphere on the bio-geochemical cycles, biological productivity and increase in sea level on the global oceans, the MoES-National Center for Polar and Ocean Research (NPOR) is pursuing integrated and multi-disciplinary approach in analyzing ice shelf dynamics, paleo-climatic, long-term mass balance, energy balance and hydrolological balance. The NIOT deployed the PROVe in the new Indian ice shelf barrier region in Antarctica for understanding the shelf ice thickness (Fig. 17) (Venkatesan et al. 2016b). The first Indian multi-sensor moored observatory, IndArc, for enabling long-term in-situ data collection comprising physio-chemical and oceanographic sensor suite such as salinity, temperature, dissolved oxygen, photo-synthetically active radiation, fluorescence, turbidity, nitrate, water currents and ambient noise for enabling understanding fresh water inputs to the global oceans was deployed by NIOT and NPOR in Kongsfjorden in Arctic waters (Ramana Murthy et al. n.d.) (Fig. 17).

Fig. 17 a NIOT's PROVe deployed in Eastern Antarctica and b NIOT-NPOR IndArc mooring in Arctic

6.2 Protection of Coast and Its Ecosystems

Coastal ecosystems include beaches, fish stocks, coral reefs and mangrove forests. The total economic values of the coral reefs are estimated between US$ 0.1–0.6 million/km^2.

Coasts undergo constant changes, as rivers, near-shore currents and waves move sediments inside, outside and within the near-shore zone. The morphological evolution tends to accelerate due to increase in sea level and under extreme events, such as tsunamis, storms and tropical cyclones. The anthropogenic activities also leave a strong footprint in deterioration of the coastal environment.

The Indian shorelines which are sensitive to erosion are being monitored and measures have been initiated. Coastline-specific solutions based on the sedimentation process and littoral drift were implemented in the Kadalur-Periyakuppam village near Chennai and the coastline of Puducherry for effective erosion control and resilience (Ramesh et al. n.d.).

The corals of the Andaman Islands that have the highest coral diversity among the Indian reef were victims thermal stress due to increase in temperatures. Coral reef surveys were conducted at the North Bay, Chidiyatapu, Jolly Buoy, Red skin and Grub Islands of the south Andaman district using the shallow water/Polar remotely operated vehicle (PROVe) (Fig. 18) which was developed by NIOT under the unmanned underwater vehicle program. The observations revealed that most of the coral ecosystems are in the resilient stage after major events such as the 2004 devastating tsunami and the beaching events during 2005 and 2010 (NCCR n.d.).

The National Center for Coastal Research (NCCR) carries out health monitoring of estuarine, coastal waters and water quality measurements and ecosystem modeling of regions under stress and coastal circulation and sediment transport modeling along the Indian Coast (Coastal and Marine Environmental Monitoring n.d.). The National

Fig. 18 NIOT- PROVe in South Andaman coral reef expedition (NCCR n.d.)

Center for sustainable coastal management (NCSCM) under the Ministry of Environment and Forests (MoEF) is active in coastal protection, conservation, rehabilitation, management and policy design of the coasts and promotes integrated and sustainable management of coastal and marine areas in India (Sheps et al. 2009).

6.3 Carbon Dioxide Sequestration

As a part of Carbon storage and sequestration (CCS), options are being investigated to store carbon by increasing the oceanic inventory by biological, chemical and physical methods. Deep ocean waters are highly unsaturated and its excess solubility and stability due to the formation of hydrates enable oceans to provide a sink for CO_2over a centennial time scale (CCS n.d.). Monitoring programs for the assurance of offshore geological storage by understanding the variability and heterogeneity of marine carbonate chemistry are underway. In India, the possibilities of sequestrating 105, 360 and 200 $GtCo_2$ in deep saline aquifers, deep water and basalt formations are identified (Fig. 19).

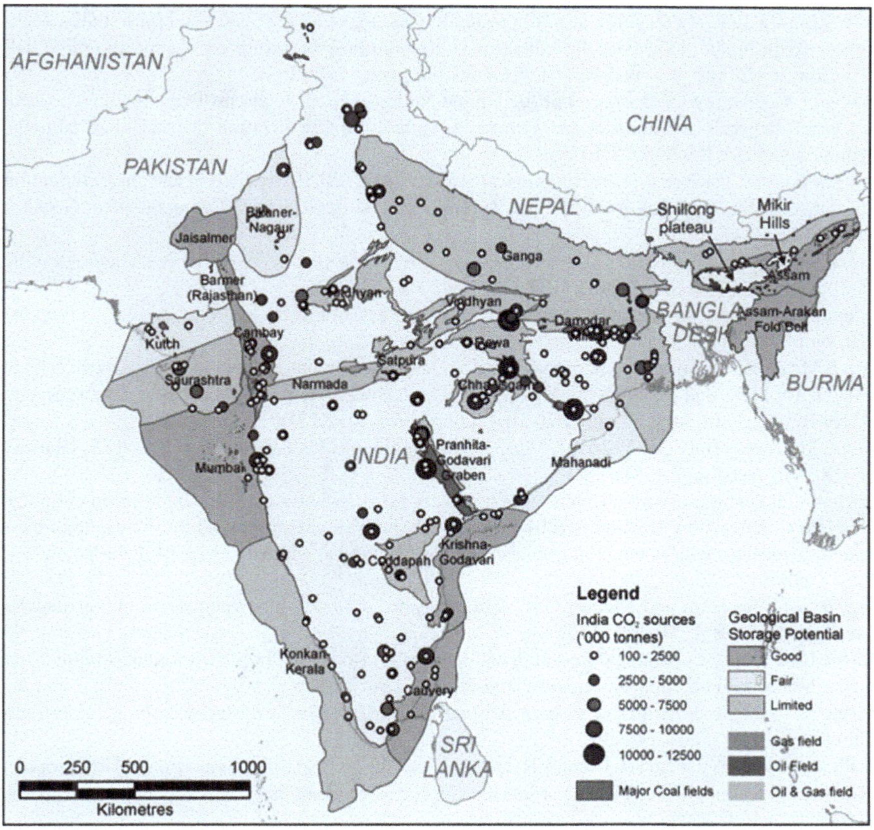

Fig. 19 Identified CCS storage locations in India (CCS n.d.)

References

Alluri SK, Gujjula D, Dhinesh G, Kumar SV, Murthy MV (2020) Offshore wind projects for the Indian coast-Experiences and challenges for its realization. Curr Sci 118(11):1774–1777

Annual Report, 2017–18. Department of Bio technology, Ministry of Science & Technology Government of India

Beyer A (2018) Inland waterways, transport corridors and urban waterfronts

Bijla DL, Bogaarta PW, Dekkera SC, Stehfestb E, de Vriesa BJM, van Vuurena DP (2017) A physically-based model of long-term food demand. Sci Direct Global Environ Change 45:47–62

Blue Economy Vision 2025. http://ficci.in/spdocument/20896/Blue-Economy-Vision-2025.pdf

CCS deployment in the context of regional developments in meeting long term climate change objectives, IEAGHG Technical Report, 2017–07

Coastal and marine environmental monitoring. http://ncscm.res.in/cms/cia/about.php

Deepsea sampling and microbial culture facility http://www.niot.res.in/index.php/node/index/177/

Developments in international seaborne trade, Review of maritime transport. UN (2018)

Dewangan P, Sriram G, Kumar A, Mazumdar A,. Pekati A, Mahale V, Reddy SSC, Babu A, Widespread occurrence of methane seeps in deep water regions of Krishna-Godavari basin in Bay of Bengal. Marine Petrol Geol. https://doi.org/10.1016/j.marpetgeo.2020.104783

Expedition 03 aims at carrying out pilot production testing of at least one site in the Indian deep-water environment. However, the execution of Expedition-03 depends on the success of NGHP Expedition-02. http://www.dghindia.org/index.php/page?pageId=39

Food and Agriculture Organization of the United Nations (2020) Fisheries Department. The state of world fisheries and aquaculture. Food and Agriculture Organization of the United Nations, Rome 2020. ISBN 978-92-5-132692-3

Ganeshkumar A, Dharani G, Kirubagaran R, Atmanand MA (2020) Production and characterization of antimicrobial peptides form Bacillus subtilis isolated from deep-sea core samples. Curr Sci 118(11):1725–1730

Global Energy Transformation, Irena. https://www.irena.org/-/media/Files/IRENA/Agency/Publication/2018/Apr/IRENA_Report_GET_2018.pdf

HamantMaini, LipiBudhraja, Ocean based blue economy: an insight into the SAGAR as the Last Growth Frontier

Hein J, Mizell K, Koschinsky A, Conrad T (2013) Deep-ocean mineral deposits as a source of critical metals for high- and green-technology applications: comparison with land-based resources. Ore Geol Rev 51:1–14. https://doi.org/10.1016/j.oregeorev.2012.12.001

India Tourism statistics (2018). http://tourism.gov.in/sites/default/files/Other/ITS_Glance_2018_Eng_Version_for_Mail.pdf

International Seabed authority, Exploration contracts. https://www.isa.org.jm/exploration-contracts

Life Below Water Why It Matters (2017) Sustainable development goals. http://www.un.org/sustainabledevelopment/wpcontent/uploads/2016/08/14_Why-it-Matters_Goal-14_Life-BelowWater_3p.pdf

Ministry of Petroleum and Natural Gas, Annual report 2018–19. http://petroleum.nic.in/sites/default/files/APR_E_1718.pdf

Marine Biotechnology Strategic Research and Innovation Roadmap. http://www.marinebiotech.eu/launch-marine-biotechnology-researchand-innovation-roadmap

Marine Ecosystem Dynamics of eastern Arabian Sea (MEDAS). http://www.cmlre.gov.in/programms.html

McPhaden MJ, Busalacchi AJ, Cheney R, Donguy JR, Gage KS, Halpern D et al (1998) The tropical ocean global atmosphere (TOGA) observing system: a decade of progress. J Geophys Res Oceans 103(C7):14169–240

Miller KA, Thompson KF, Johnston P, Santillo D (2018) An overview of seabed mining including the current state of development, environmental impacts, and knowledge gaps. Front Mar Sci 4:418

Ministry of Shipping, Indian Ports Association, Final report for Sagarmala (Vol. III), November 2016. http://www.chennaiport.gov.in/pdf/frsc.pdf

Murthy MV, Ravichandran V, Vendhan M, Kiran AS, Raju SK, Avula AK, Varthini S, Abhishek T (2020) Shore protection measures along Indian Coast–Design to implementation based on two case studies. Curr Sci 118(11):1768–1773

NCCR weblink. http://www.icmam.gov.in/

Nagoya Protocol. https://www.cbd.int/abs/about/

Open Sea Cage Culture, Marine Biotechnology, National Institute of Ocean Technology. https://www.niot.res.in/index.php/node/index/177/

Offshore Energy Outlook, 2019. World Energy Outlook Series. https://www.iea.org/

Pattanaik B, Vishwanath A, Jalihal P, Rao YVN, Karthikeyan A, Sajeev KS, Shipin VP (2020) Performance evaluation of power module during demonstration of wave-powered navigational buoy. Curr Sci 118(11):1712–1717

Patil, Pawan G, Virdin J, Diez SM, Roberts J, Singh A (2016) Toward a blue economy: a promise for sustainable growth in the Caribbean. World Bank

Pauli, Gunter A (2010) The blue economy: 10 years, 100 innovations, 100 million jobs. Paradigm publications

PRATYUSH 4PF HPC System. http://pratyush.tropmet.res.in/

Prospects of the blue economy in India. IORA Blue Economy Dialogue, August 2015. http://www.ris.org.in/sites/default/files/Goa%20Declaration.pdf

Potential Fishing Zone (PFZ) Advisory. http://www.incois.gov.in/MarineFisheries/PfzAdvisory

Ramadass GA et al (2010) Deep-ocean exploration using remotely operated vehicle at gas hydrate site in Krishna-Godavari basin, Bay of Bengal. Curr Sci 99(6):809–815

Ramadass GA, Ramesh S, Subramanian AN, Sathianarayanan D, Ramesh R, Harikrishnan G, Pranesh SB, DossPrakash V, Atmanand MA (2015) Deep ocean mineral exploration in the indian ocean using remotely operated vehicle (ROSUB 6000). In: Underwater technology (UT), IEEE, pp 1–8. IEEE

Ramadass GA et al (2020) Challenges in developing deep water human occupied vehicles. Curr Sci 118(11):1687–1693

Ramesh S, Sathianarayanan D, Ramesh R, Harikrishnan G, Vadivelan A, Ramadass GA (2016) Qualification of polar remotely operated vehicle at East Antarctica. IEEE Oceans, USA

Ramesh S, Ramadass GA, Prakash VD, Sandhya CS, Ramesh R, Sathianarayanan D, Vinithkumar NV (2017) Application of indigenously developed remotely operated vehicle for the study of driving parameters of coral reef habitat of South Andaman Islands, India. Curr Sci J 113(12):2353–2359

Ramesh S, Murthy KNVV, Hussain SM, Ramasamy S, Ramadass GA, Occurrence of Nummulitic coralline limestone from the Palar Basin deep sea, East coast of India, Current Science. https://doi.org/10.1080/1064119X.2018.1551447

Ravichandran M (2015) Indian Ocean is no more under-observed, ocean digest. Q Newslett Ocean Soc India 2(3):2–5

Renewables 2018 global status report—REN21. http://www.ren21.net/wp-content/uploads/2018/06/17-8652_GSR2018_FullReport_web_final_.pdf

Sharma R (2017) Deep Sea Mining: Resource potential, technical and environmental considerations. Springer. ISBN: 978-3-319-52556-3

Sheps KM, Max MD, Osegovic JP, Tatro SR, Brazel LA (2009) A case for deep-ocean CO_2 sequestration. Energy Procedia 1(1):4961–4968. ISSN: 1876-6102. https://doi.org/10.1016/j.egypro.2009.02.328

Sloan ED, Koh CA (2008) Clathrate hydrates of natural gases, 3rd edn. CRC Press, Florida

Srinivasa Kumar T, Venkatesan R, Vedachalam N, Padmanabhan S, Sundar R (2016) Assessment of the reliability of the Indian Tsunami early warning system. Marine Technol Soc 50(3):92–108

Thompson CC, Kruger RH, Thompson FL (2017) Unlocking marine biotechnology in the developing world. Trends Biotechnol Cellpress 35(12):1119–1121. https://doi.org/10.1016/j.tibtech.2017.08.005

Travel & tourism economic impact 2018, World Travel and Tourism council. https://www.wttc.org/-/media/files/reports/economic-impact-research/regions-2018/world2018.pdf

United nations convention on the law of the sea. http://www.un.org/depts/los/convention_agreements/texts/unclos/closindx

United Nations Decade of Ocean Science for Sustainable Development (2021–2030). https://en.unesco.org/ocean-decade

United Nations World Bank Group (2017) The potential of the blue economy increasing long term benefits of the sustainable use of marine resources for small island developing states and coastal least developed countries. The World Bank Group, USA

Vandavasi BN, Arunachalam U, Narayanaswamy V, Raju R, Vittal DP, Muthiah R, Gidugu AR (2018) Concept and testing of an electromagnetic homing guidance system for autonomous underwater vehicles. Appl Ocean Res 73:149–159

Vedachalam N, Srinivasalu S, Rajendran G, Ramadass GA, Atmanand MA (2015) Review of unconventional hydrocarbon resources in major energy consuming countries and efforts in realizing natural gas hydrates as a future source of energy. J Nat Gas Sci Eng 26:163–175

Vedachalam N, Ravindran M, Atmanand MA (2018) Technology developments for the strategic Indian blue economy. Marine Georesour Geotechnol, 1–17

Venkatesan G, Prakash kumar LSS, Jalihal P (2020) "Influence of non-condensable gases on performance of the low temperature thermal desalination plants. Curr Sci 118(11):1718–1724

Venkatesan R, Sundar R, Vedachalam N, Jossia Joseph K(2016a) India's ocean observation network: relevance to society. Marine Technol Soc J 50(3):34–46

Venkatesan R, Krishnan KP, Arul Muthiah M, Kesavakumar B, Divya DT, Atmanand MA, Rajan S, Ravichandran M (2016b) Indian moored observatory in the Arctic for long-term in situ data collection. Int J Ocean Climate Syst 7(2):55–61

Vittal DP, Arunachalam U, Narayanaswamy V, Arumugam V, Raju R, Ramadass GA, Atmanand MA (2016) Analysis of subsea inductive power transfer performances using planar coils. Marine Technol Soc J 50(1):17–26

Varshney N, Rajesh S, Aarthi AP, Ramesh NR, Vedachalam N, Ramadass GA, Atmanand MA (2015) Estimation of reliability of underwater polymetallic nodule mining machine. Marine Technol Soc J 49(1):131–147

Venkatesan R, Vedachalam N, Muthiah MA, Sundar R, Kesavakumar B, Ramasundaram S, Jossia Joseph K (2018)Reliability metrics from two decades of Indian Ocean moored buoy observation network. Marine Technol Soc J 52(3):71–90

Water Supply Diversification Desalination, Water Research Foundation. http://www.waterrf.org/knowledge/WaterSupplyDiversification/FactSheets/SupplyDiversification_Desal_FactSheet.pdf

Coastal Research—Beach Restoration and Protection

M. V. Ramana Murthy, Vijaya Ravichandran, Mullai Vendhan, Satya Kiran Raju Alluri, and J. Ram Kumar

Abstract Coastal erosion and flooding are serious threats across many coastal areas. They become globally more severe due to human-induced changes and accelerated sea-level rise. A significant proportion of beaches are undergoing long-term chronic erosion globally and on the Indian coast that leads to the loss of land and degradation of habitats. Several approaches have been attempted to control erosion and protect the coastal areas by adopting a hard coastal defence structure in response to these threats. Ministry of Earth Sciences through National Centre for Coastal Research (NCCR) and National Institute of Ocean Technology have devised nature-based solutions for the protection and restoration of beach. Here, we describe two recent beach protection and restoration schemes that use non-traditional soft approaches to prevent erosion and restore beaches. A detailed description of the environmental conditions at the site, implemented schemes and the morphological evolution are discussed. Both sites show significant growth in beach width and volume post implementation, contributing to socioeconomic development. Similar strategies can be adopted and implemented at other sites that are undergoing chronic erosion.

1 Introduction

Coastal environments are one of the most favourable and attractive destinations for modern settlements. Recent decades have seen rapid development in many coastal regions around the world, resulting in the fact that today 15 out of the top 20 mega cities in the world are located in coastal zones (Small and Nicholls 2003). Coastal zones have acted as a driving force of economic growth for East and Southeast Asian

M. V. Ramana Murthy (✉) · S. K. R. Alluri
National Centre for Coastal Research, Pallikaranai, Chennai, India
e-mail: mvr@nccr.gov.in

V. Ravichandran · M. Vendhan · J. Ram Kumar
National Institute of Ocean Technology, Pallikaranai, Chennai, India

V. K. Gahalaut and M. Rajeevan (eds.), *Social and Economic Impact of Earth Sciences*,
https://doi.org/10.1007/978-981-19-6929-4_15

countries (Han and Yan 1999). The Indian mainland, with a coast of more than 5700 km is well placed to harness and benefit from the many advantages that coastal zones endow.

In general, the coast comprises many different coastal landforms like dunes, beaches, cliffs, spits, tidal flats, coastal vegetation, estuaries, etc. Amongst them, sandy beaches are the most prominent, accounting for more than 43% of India's coastline (Kankara et al. 2018). Sandy beaches are very beneficial to the coastal communities, providing many different amenities; being an integral part of the life and livelihood of these communities. However, coastal erosion has become one of the most common problems that coastal communities encounter globally. Based on the observations from satellite images, it is estimated that nearly 28,000 km^2 of the land area was lost due to erosion globally during the period 1984–2015 (Mentaschi et al. 2018). A recent study estimated that 24% of the world's sandy beaches are undergoing erosion. The study also pointed out that the majority of marine protected areas are eroding gradually, placing India amongst the global hotspots under the threat of erosion (Luijendijk et al. 2018).

1.1 The Indian Scenario

The National Centre for Coastal Research has recently published a report titled "National Assessment of Shoreline Changes along Indian Coast" which studied the changes along the Indian coast for the period 1990–2016 based on satellite images and field observations (Kankara et al. 2018). The report states that 34% of the coastline in mainland India is undergoing erosion. The situation is more aggravated in some states like West Bengal (63%), Puducherry (57%), Kerala (45%), Tamil Nadu (41%), etc., which experience higher rates of erosion when compared to other parts of the country. All over India, several important and famous sandy beaches are undergoing erosion (Puri, Visakhapatnam, Mahabalipuram, Varkala, Alapuzha, etc.). The causes of such chronic erosion may vary from place to place and can be either natural or man-made. Sea level rise and the impacts of climate change are expected to further worsen the problem of erosion in the coming decades (Zhang et al. 2004; Nicholls et al. 1995; Ranasinghe et al. 2012; Reguero et al. 2019). By considering together with the predicted increase in the coastal population (Neumann et al. 2015) and increased focus on the development of port and coastal infrastructure, it is clear that there is an urgent need to develop sustainable solutions to tackle the problem of beach erosion.

2 Coastal Protection Techniques

Traditional methods of coastal protection like groins and seawalls, while proving capable of protecting inland infrastructure in the short term, end up shifting the problem to adjacent coasts. Even after the construction of these structures, erosion will continue to take place, leading to steepening of beach slopes, scouring, and subsidence of the structure (Schoonees et al. 2019; Rangel-Buitrago et al. 2018). Hence, building such hard structures can cause a severe impact on natural processes and hinders, perennially, the accumulation of sediments. However, there is a huge and urgent public demand for such structures when the inland infrastructure is threatened severely, leaving administrators with little choice or time. Such demands lead to a significant stretch of the coast being armoured with seawalls, adversely affecting the natural processes. For example, in Kerala, a state with 560 km long coastline, almost 350 km, i.e. more than 60% of the coast, is already covered with seawalls (Sundar and Murali 2007).

Different measures have been undertaken in many parts of the world to sustainably overcome the problem of coastal erosion. There is a huge impetus to move towards nature-based solutions and soft protection measures (Morris et al. 2018). Countries like the Netherlands have policies that primarily focus on beach nourishment to protect sandy coasts against erosion (Roeland and Piet 1995). A wide beach offers protection against coastal flooding during extreme events by reducing the energy of the incoming waves (Hanley et al. 2014). This aspect of the beach gains prominence as there is an increasing risk of coastal flooding due to cyclonic events (storm surges) and sea-level rise (Kirezci et al. 2020; Vitousek et al. 2017). However, beach nourishment is a costly process, and it's not affordable everywhere. It becomes essential that the methods we develop are optimal from an ecological as well as economic perspective for India. Coastal engineers have been increasingly turning towards nature to develop such methods. For example, the coral islands of Lakshadweep have lagoons on the western side that are protected by reef crest as shown in Fig. 1. The reefs form a barrier such that the waves break on this reef, reducing the wave energy reaching the lagoon beaches and making them stable. This kind of natural phenomena have inspired similar engineered solutions. Two recent examples of such nature-based concepts to protect coastal stretches from erosion are presented here. The two sites, Puducherry and Kadalur Periyakuppam, present different sets of challenges as Puducherry is a coast with a long history of hard coastal structures (harbour breakwaters, seawalls, and groins). At the same time, at Kadalur Periyakuppam there was no coastal intervention prior to the present project. While the solutions vary from case to case, the broad principles like holistic understanding of the system, least interference in the natural process, and flexibility of the proposed solution remain the same. The proposed solutions are also ecologically advantageous as they provide hard substrata that can host and attract new species, increasing biodiversity (An ecological perspective on the deployment and design of low-crested and other hard coastal defence structures 2005).

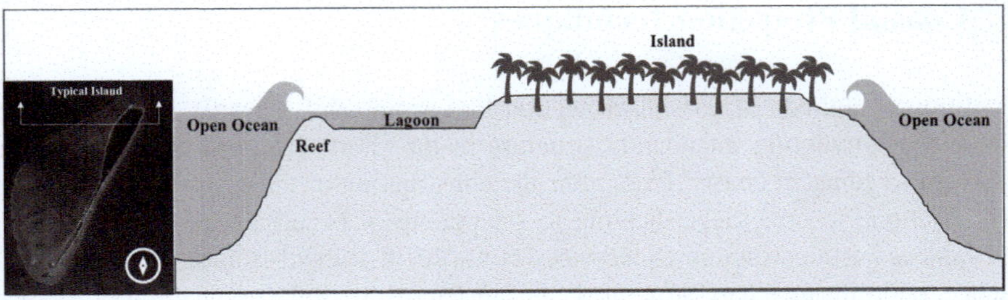

Fig. 1 Cross-section of typical reef structure in islands of Lakshadweep

3 Case Studies

3.1 Site Description

Puducherry and Kadalur Periyakuppam are two sites on the east coast of India that have been undergoing severe coastal erosion due to anthropogenic and natural causes. At Puducherry, the erosion has been so severe and chronic, the beach was considerably eroded over the past decades and came to be replaced with rubble mound seawalls (Muthusankar et al. 2017). The construction of harbour breakwaters to the south of the city impeded the longshore sediment transport resulting in intensified erosion north of the harbour. Several measures that are implemented to prevent erosion like groins and seawalls, complicated the natural nearshore processes and shifted the problem to the adjacent coast further north (Mohamed Rajab and Thiruvenkatasamy 2016). While the seawalls protected the inland infrastructure, in a tourism hotspot like Puducherry, it is not ideal to focus only on hard protection without accounting for the enormous value a beach can offer which led to demands for restoration of the beach. At Kadalur Periyakuppam, a fishing hamlet, there has been a steady erosion of beach leading to a loss in beach width to the extent of more than 60 m–90 m over the past decades, with only a very narrow beach remaining at the time of the study (Ramana Murthy et al. 2020). Cyclones such as Thane (2011), Nilam (2012), Madi (2013), HudHud (2014), and Vardah (2016) have caused severe damage to the Kadalur Periyakuppam coast resulting in the loss of over 40 m of beach width. The fishing community living here is highly dependent on the space provided by the beach. The complete erosion of beach or replacement with seawalls will severely impact their life and livelihood. Hence, the restoration of the beach is of utmost importance to improve the socioeconomic well-being of the people.

3.2 Concept Evolution

The two sites are in different stages in terms of the evolution of the coastal profiles due to erosion. Whereas in Puducherry city, the beach has been completely replaced by rubble mound seawalls, at Kadalur Periyakuppam, there exists still a narrow beach. While protecting the inland infrastructure, the seawalls at Puducherry did not solve the root causes of erosion. Hence, the seabed continued to erode and the beach profile steepened. This process has important implications for the design of beach protection or beach restoration strategy. The nearshore profiles at Puducherry and Kadalur Periyakuppam and the theoretical equilibrium beach profile are shown in Fig. 2. It can be seen from the figure that the profile at Puducherry is much steeper than the equilibrium profile, whereas the profile at Kadalur Periyakuppam is very similar to the expected equilibrium profile. The strategies adopted for beach protection or restoration must take into account these varied realities to evolve appropriate solutions.

By analysing the variation between the equilibrium profile and the existing profile at Puducherry, it is observed that close to 800 m^3 of sand per m of the coastline has been eroded over the past decades. To recreate beach to the extent of 2 km as in the present case, 1,600,000 m^3 of sand will be required. A solely structural solution will not be enough to overcome such a huge deficiency in the available sediment quantity. At Puducherry, sediments dredged from the harbour mouth as part of regular maintenance provide a readily available source of sediment supply. However, sand nourishment without a mechanism to prevent the nourished sand from being freely transported out of the protected coastal stretch will require frequent re-nourishments and hence uneconomical. The proposed solution at Puducherry involves two structural mechanisms to restrict sediment movement with sand nourishment in-between them. It involves a submerged wedge-shaped reef on the northern side and an emerged reef on the southern side with nourishment using dredged sand from the harbour mouth in-between the structures. The northern reef is submerged, which allows sediment

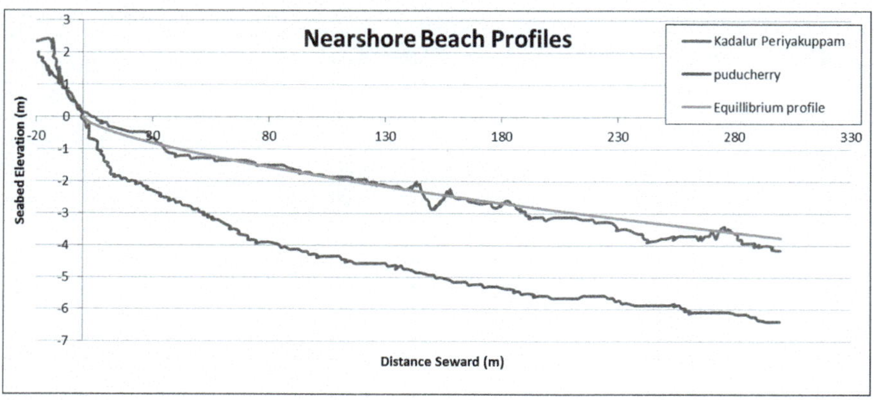

Fig. 2 Nearshore Seabed profiles at Puducherry and Kadalur with equilibrium beach profile

bypassing and avoids downdrift erosion. In contrast, the southern reef is an emerged structure that impedes sand movement towards the southern side, preventing siltation issues in the harbour.

The aim at Kadalur Periyakuppam is to prevent further erosion and increase the beach width to achieve a new, more favourable dynamic equilibrium. While traditional solutions like groins may be able to achieve the objective, the downdrift erosion will worsen, and hence it is not desirable from a holistic perspective. Submerged breakwaters, shore parallel structures placed at a certain distance seawards from the shoreline, provide an alternative. They usually consist of multiple segments covering the vulnerable coast with small gaps in-between individual segments. Their function is to decrease the wave energy, as shown in Fig. 3, reaching the vulnerable coastal stretch by breaking large waves offshore while letting smaller waves pass through to decrease downdrift erosion. This reduction in wave energy brings about a gradient in the wave energy and longshore sediment transport in the alongshore direction, leading to the accretion of sediments. Geo-tubes are preferred for the construction of the submerged breakwaters since it doesn't forestall any future interventions and gives coastal engineers and planners flexibility in the future.

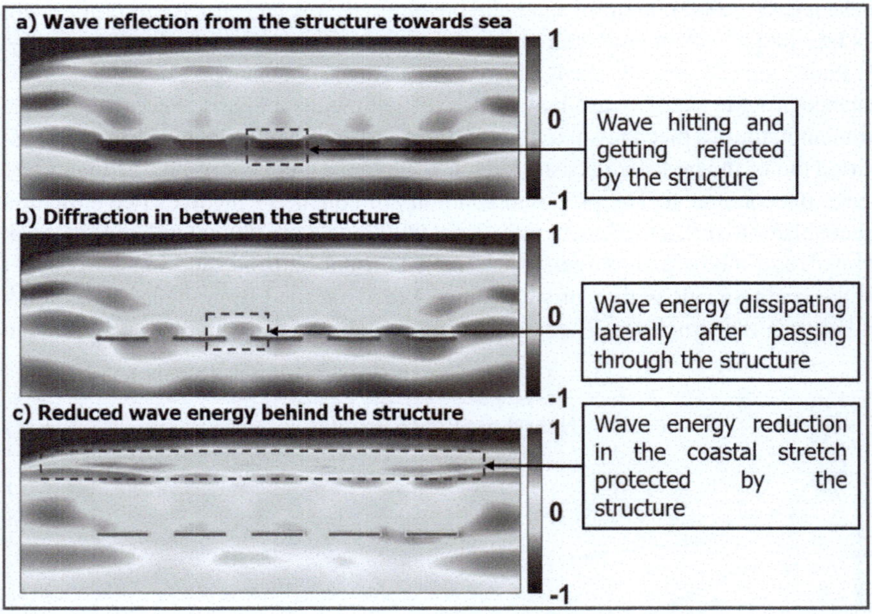

Fig. 3 Numerical simulation of wave interaction with the submerged breakwaters. **a** The increased wave heights seaward of the structure due to wave reflection **b** wave diffraction through the gaps in the breakwater system **c** Reduction in wave heights and wave energy at the shore

3.3 Environmental Conditions

An accurate estimate of the environmental conditions is a vital factor for testing different coastal protection schemes using numerical models. Extensive field measurements were conducted to observe waves, tides, currents, sediment characteristics, etc., at the respective sites. Both sites are micro-tidal coasts with a tidal range of about 1.2 m. The wave rose diagram at the two sites is given in Fig. 4. The waves approach the coast from the southeast during eight months of the year (March–October) and from the northeast during the other four months (November–February), coinciding with the northeast monsoon. Though waves from the northeast occur for only four months in a year, these months are characterized by higher significant wave heights. Swells are seen mainly during the southwest monsoon, while predominant low-period wind waves are noticed during the northeast monsoon season. The distribution plots of wave parameters in Fig. 4 show that the occurrence frequency of significant wave heights of more than 0.5 m is higher at Kadalur Periyakuppam than at Puducherry. Also, Kadalur Periyakuppam experiences more long-period waves than Puducherry with $T > 8$ s for about 35% of the time, whereas for Puducherry it stands at 27% of the time. The orientation of the coast plays an important role in longshore sediment transport. It varies between the two sites with Puducherry coastline oriented at 11° to the north and Kadalur Periyakuppam coast oriented at around 33° to the north. At both these locations, the net longshore sediment transport is towards the north. The average annual net sediment transport rate estimated at Puducherry and Kadalur Periyakuppam are 72,000 m^3/year and 76,000 m^3/year, respectively.

3.4 Numerical Modelling

Spectral wave and hydrodynamic models are constructed and calibrated with field observations and then used for hindcasting the nearshore wave-current climate. The detailed design after numerical testing can allow studying the response of the coast to different configurations. Different schemes are evaluated based on the width and volume of beach gained, profile evolution, long-term beach stability, etc. (Fig. 5).

3.4.1 Beach Protection Schemes

For Puducherry, a submerged North reef of triangular steel wedge of dimensions 60 × 50 m with a maximum height of 2.5 m at the base and projecting into the sea for a length of 170 m with a width of 110 m at the base tapering towards the sea is used. The reef was made of steel and weighs around 900 tons. It was constructed on the working platform and launched into the sea using airbags, floated to the desired location, and positioned by ballasting. The reef is designed to naturally bypass the sand as the crest of the reef is maintained at the lowest low water level. Simulation of the event corresponding to the southwest monsoon indicates that the nearshore reef

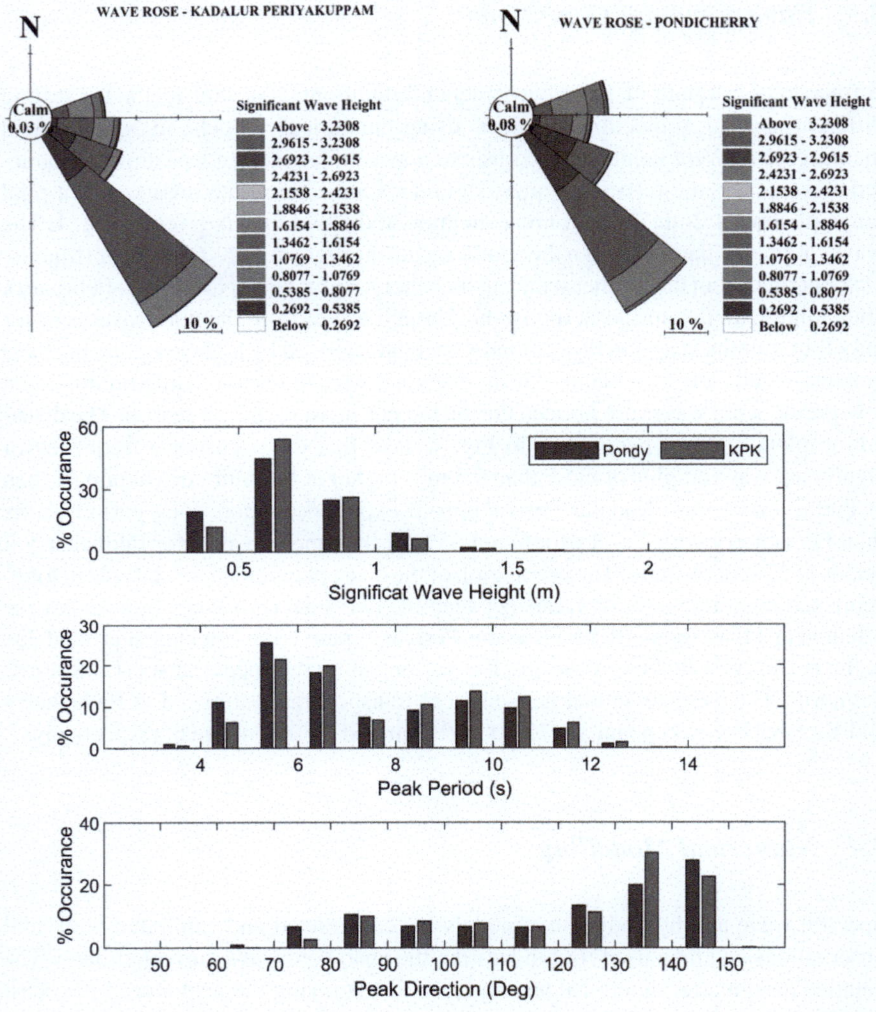

Fig. 4 Wave rose and wave parameter distribution for Puducherry and Kadalur Periyakuppam

structure rotates the waves, which helps in reducing the longshore current towards the north. This reduces the longshore transport capacity and leads to accretion at the south of the reef structure. On the southern side, an emerged reef is proposed to prevent the loss of sand towards the south during the northeast monsoon season, and the beach was nourished to the extent of 3 lakh m^3 (Fig. 6).

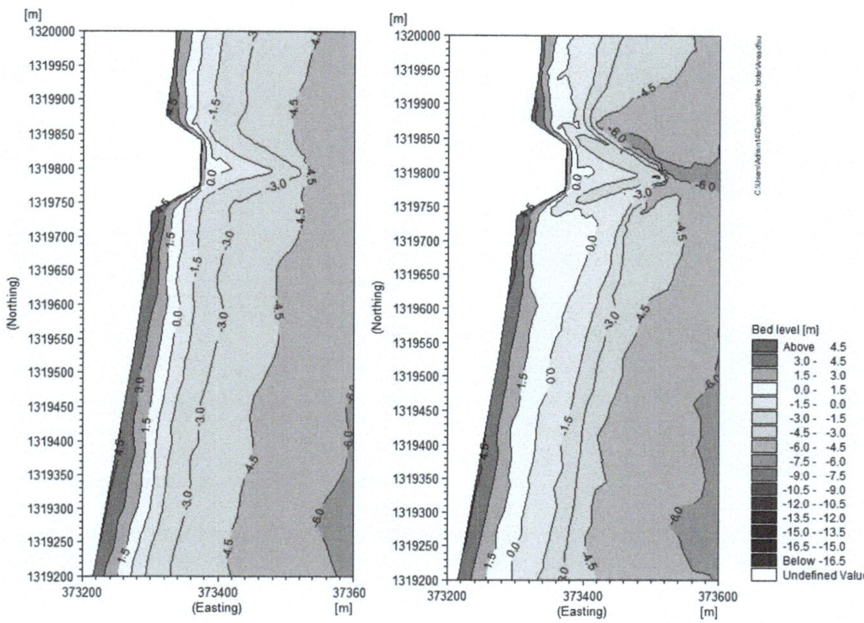

Fig. 5 Morphodynamic modelling for design of coastal protection scheme. Left panel shows the seabed contours at time, t = 0 and the right panel shows the morphological evolution of the bed at t = 3 years

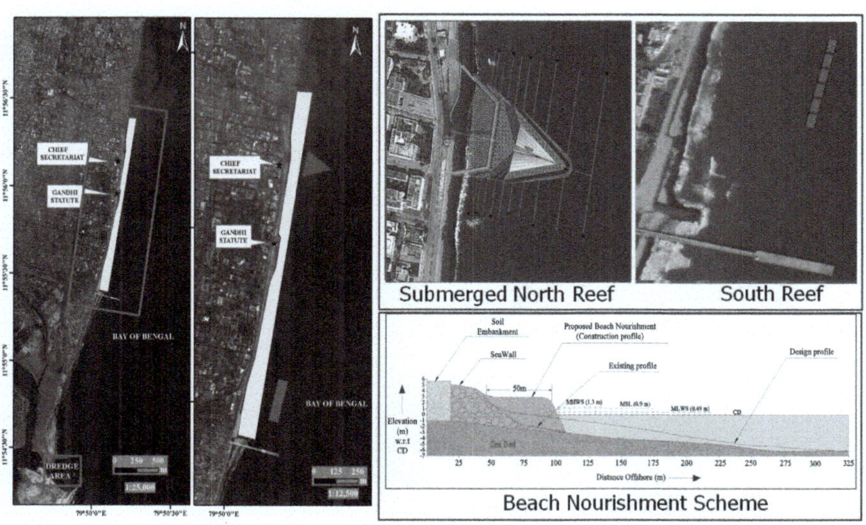

Fig. 6 Beach restoration scheme at Puducherry. The left panel shows the overall layout of the scheme with background of satellite imagery. The top right panel shows the configuration of the reef structures. The bottom right panel gives the profile view of the existing profile with the seawall and the placement of the nourishment sand over the profile

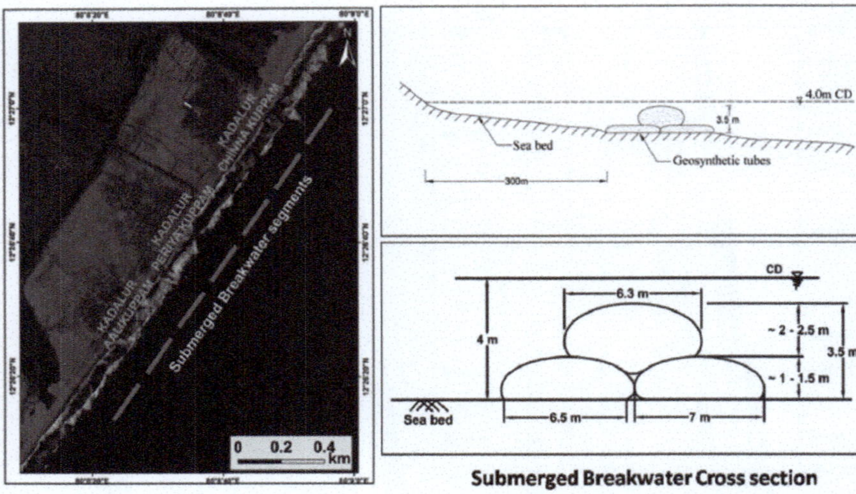

Fig. 7 Submerged breakwaters to prevent erosion at Kadalur Periyakuppam. The left panel shows the overall layout. The top right panel shows the cross shore profile, and the bottom right panel shows the typical placement of the geotubes

For Kadalur Periyakuppam, various layouts of submerged breakwaters were tested using numerical models by varying different parameters like the number of segments, length of each segment, gap length between segments, submerged height of breakwaters, etc. After careful analysis of the response of the shoreline to different layouts, a layout of 7 segments with each segment of 200 m length placed with a gap of 60 m in-between them was finalised. The structure is 3.5 m high, placed at 4 m (with respect to C.D) contour. The cross-section is made up of three geosynthetic tubes, two at the base and one on top, each 15 m in circumference and 25 m in length (Fig. 7).

3.5 Construction Methodology

The nearshore submerged reef at Puducherry includes three components: (1) Work area, (2) Rock base, and (3) Steel wedge are executed in that order. The work area walls and rock base are constructed using rock boulders as per the design and filled with sand. These together help in stabilising the base for the construction and launching of the reef. The triangular reef is fabricated in the work area using 25 mm steel plates. The total weight of the wedge is about 900 t and it has a floating draft of 0.50 m. To launch the reef into the sea, airbags are placed under the reef and inflated, gradually shifting the weight of the reef onto the airbags. When the reef was fully resting on the airbags, it was pulled into the sea using a network of 3 winches, 2 winches offshore on a boat for pulling into the sea, and 1 winch onshore for controlling the descent into the sea. The reef was floated to the required location, positioned, and then ballasted by pumping water inside. The sand was filled into its inner chambers to increase the in-place stability of the reef further.

The submerged reef at Kadalur Periyakuppam consists of geosynthetic tubes 15 m in circumference and 25 m long, filled using locally available sand from the seabed. The sand for filling the geosynthetic tubes is converted to slurry by mixing with water in a hopper using a mobile crane and pumped through HDPE floating pipeline for ease of filling. A pump was used to transfer slurry over a pipeline length of 400 m. The pipeline is connected to a pontoon anchored over the submerged position of the geosynthetic tube. From the pontoon, a flexible hose carries the sand slurry from the HDPE pipeline to the inlet of geosynthetic tube. A jack up platform with a crane was used for positioning, preparatory works, and storage of geosynthetic tubes. The geosynthetic tubes used are permeable and have an apparent opening size of 250 microns. The sand used for filling had a mean size of 500 microns. The geosynthetic tube material was chosen such that it retains the sand and allows quick drainage of water. The tubes are placed in position using anchor blocks and are filled with sand slurry pumped through floating pipelines (Fig. 8).

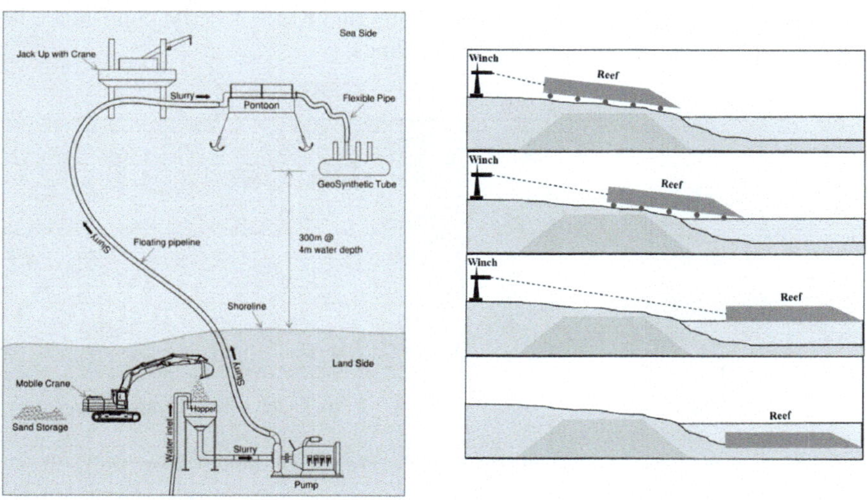

Kadalur Periyakuppam Puducherry

Fig. 8 Construction methodology of proposed solutions at Kadalur Periyakuppam (left) and Puducherry (right). The left panel shows the layout of different instruments and the mechanism of filling up of the geotubes. The right panel shows the stages involved in the launching of the steel reef and positioning at Puducherry

3.6 Evaluation of Project Post Implementation

The coastal changes resulting from the implemented interventions are monitored at periodic intervals and analysed. A sandy beach of significant width has been restored in Puducherry, the first time Puducherry city having a beach in recent decades. At Kadalur Periyakuppam, the submerged structures have led to stabilisation and growth of beach width. The beach formation in both sites is shown in Fig. 9. The introduction of these structures also led to marine growth, attracting various species to the structure, as shown in Fig. 10 with underwater images taken close to the structures. The hard substrate provided by the structures hosts species that require such substrate to survive, leading to increased bio-diversity in the vicinity of the structures.

As part of a study on new and innovative coastal resilience approaches for India and Bangladesh, the impact of various coastal interventions in increasing coastal resilience was evaluated (Marchand 2020). Many implemented projects to improve coastal resilience were analysed by reviewing various parameters mainly under three heads: (1) Effectiveness (Flood risk, coastal erosion, and wind risk) (2) Sustainability (Environmental, Social, and Economic), and (3) Implementations (Success/failure factors, Time–space, applied innovations, and know-how). It is found that the projects that adopt a holistic approach like in Puducherry and Kadalur Periyakuppam are far better than projects that adopt traditional methods.

a) Puducherry b) Kadalur Periyakuppam

Fig. 9 Aerial view of beach formation after implementation of the project at two demonstration sites

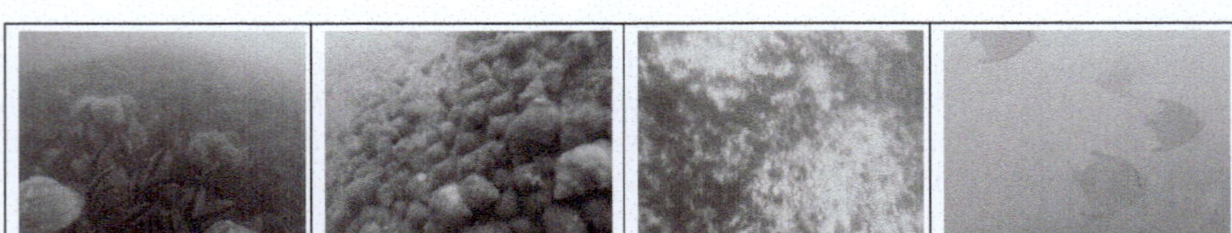

Kadalur Periyakuppam Puducherry

Fig. 10 Underwater pictures of marine growth and ecology near the Reefs

4 Conclusions

Natural and anthropogenic activities together have rendered our coasts vulnerable to coastal erosion. Many coastal stretches are ravaged by erosion, putting at risk the lives and livelihoods of coastal communities. With the expected worsening of the problem due to various factors like sea-level rise, climate change, and coastal infrastructure development, it is of critical importance to develop and implement sustainable solutions to overcome coastal erosion. The two case studies presented here are examples of such solutions. Though these sites are located close by and experience similar environmental conditions, they require very different solutions as they are in different phases of coastal erosion. Puducherry is a heavily armoured coast, whereas Kadalur Periyakuppam is pristine with no coastal intervention. It is important to understand that every coast is different. Thus, a thorough understanding is necessary before planning for any intervention. The problem has been approached holistically where the focus is more on beach protection and recreation than the narrow focus on inland infrastructure protection as is the case traditionally. The implemented strategies produced desirable results by restoring and protecting the beach as expected. The beach restoration at Puducherry received appreciation from all key stakeholders. It led to a spike in people, locals as well as tourists, flocking to the seafront. The submerged breakwaters at Kadalur Periyakuppam have prevented further erosion and led to increased beach width, safeguarding the beach that has important socioeconomic implications for the local community. The artificial reef structures also provide a substratum that attracts various species increasing the bio-diversity and leading to the reef sites becoming a hotspot for small-scale fishing activities. With looming uncertainty about the future climate and sea-level rise, coastal interventions must be flexible enough to allow for modifications in a changing environment. Studies related to engineering interventions should be carried out to understand the thresholds. The approaches presented here are broadly applicable to the coasts where the erosion process is undergoing. Ministry of Earth Sciences proposed to work on similar solutions at various locations along the Indian coast, as shown in Fig. 11.

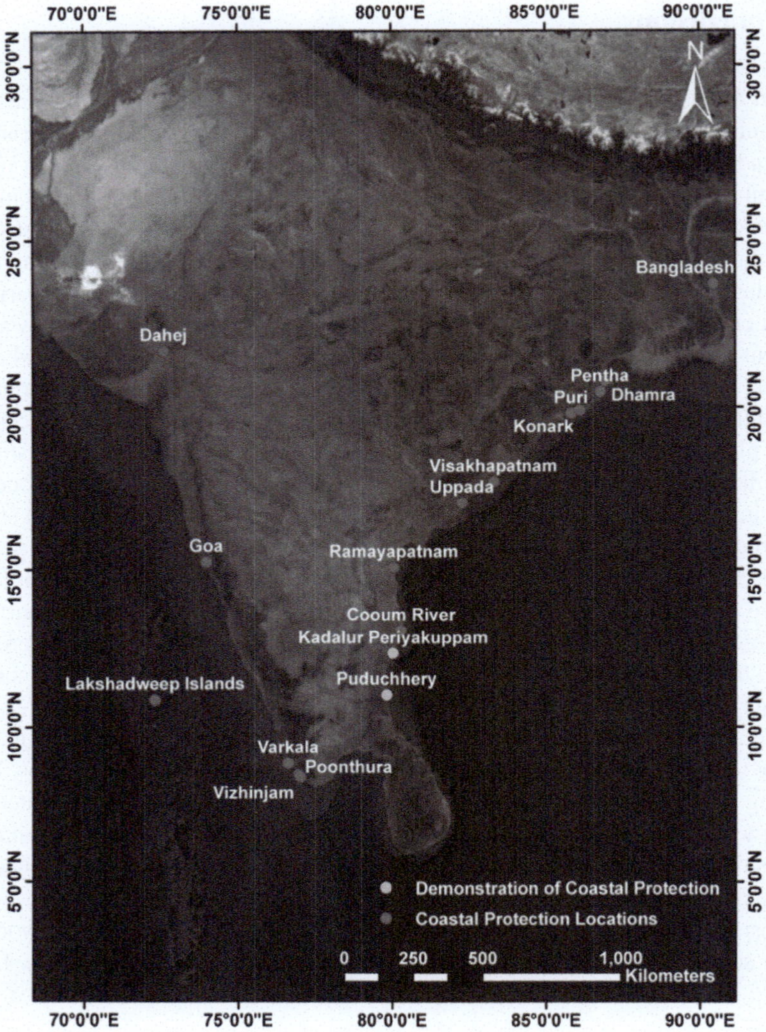

Fig. 11 Location of coastal protections and beach restoration projects under progress by the Ministry of Earth Sciences

Acknowledgements The authors wish to express their sincere thanks to former directors Dr. M. A. Atmanand and Dr. Satish Shenoi for their continued support and valuable guidance throughout the execution of these projects.

References

An ecological perspective on the deployment and design of low-crested and other hard coastal defence structures (2005) Coast Eng 52(10–11):1073–1087
Han SS, Yan Z (1999) China's coastal cities: development, planning and challenges. Habitat Int 23(2):217–229
Hanley ME et al (2014) Shifting sands? Coastal protection by sand banks, beaches and dunes. Coast Eng 87:136–146
Kankara R, Murthy MV, Rajeevan M (2018) National assessment of shoreline changes along Indian coast
Kirezci E et al (2020) Projections of global-scale extreme sea levels and resulting episodic coastal flooding over the 21st Century. Sci Rep 10(1):1–12
Luijendijk A, Hagenaars G, Ranasinghe R, Baart F, Donchyts G, Aarninkhof S (2018) The state of the World's beaches. Nature Publishing Group
Marchand M (2020) Phase 1 report of component 1: improving empirical evidence and analytical support on investments in coastal resilience. Deltares and Associates
Mentaschi L, Vousdoukas MI, Pekel JF, Voukouvalas E, Feyen L (2018) Global long-term observations of coastal erosion and accretion. Sci Rep 8(1):12876
Mohamed Rajab P, Thiruvenkatasamy K (2016) Shoreline change studies due to construction of breakwaters at Ariyankuppam river mouth in Puducherry—a union territory of India, South India. Indian J Sci Technol 9(45)
Morris RL, Konlechner TM, Ghisalberti M, Swearer SE (2018) From grey to green: efficacy of eco-engineering solutions for nature-based coastal defence. Glob Chang Biol 24(5):1827–1842
Muthusankar G, Jonathan MP, Lakshumanan C, Roy PD, Srinivasa-Raju K (2017) Coastal erosion vs man-made protective structures: evaluating a two-decade history from southeastern India. Nat Hazards 85(1):637–647
Neumann B, Vafeidis AT, Zimmermann J, Nicholls RJ (2015) Future coastal population growth and exposure to sea-level rise and coastal flooding—a global assessment. PLoS One 10(3)
Nicholls RJ, Leatherman SP, Dennis KC, Volonté CR (1995) Impacts and responses to sea-level rise: qualitative and quantitative assessments. J Coast Res. Coastal Education & Research Foundation, Inc., pp 26–43
Ramana Murthy MV et al (2020) Shore protection measures along Indian Coast-Design to implementation based on two case studies. Curr Sci 118(11)
Ranasinghe R, Callaghan D, Stive MJF (2012) Estimating coastal recession due to sea level rise: beyond the Bruun rule. Clim Change 110(3–4):561–574
Rangel-Buitrago N, Williams AT, Anfuso G (2018) Hard protection structures as a principal coastal erosion management strategy along the Caribbean coast of Colombia. A chronicle of pitfalls. Ocean Coast Manag 156:58–75
Reguero BG, Losada IJ, Méndez FJ (2019) A recent increase in global wave power as a consequence of oceanic warming. Nat Commun
Roeland H, Piet R (1995) Dynamic preservation of the coastline in the Netherlands. J Coast Conserv 1(1):17–28
Schoonees T et al (2019) Hard structures for coastal protection, towards greener designs. Estuaries Coasts 42(7):1709–1729
Small C, Nicholls RJ (2003) A global analysis of human settlement in coastal zones. J Coast Res 19(3):584–599
Sundar V, Murali K (2007) Planning of coastal protection measures along Kerala Coast. Indian Institute of Technology, Madras
Vitousek S, Barnard PL, Fletcher CH, Frazer N, Erikson L, Storlazzi CD (2017) Doubling of coastal flooding frequency within decades due to sea-level rise. Sci Rep 7(1):1–9
Zhang K, Douglas BC, Leatherman SP (2004) Global warming and coastal erosion. Clim Change 64(1–2):41–58

Developing Ocean Technology

M. A. Atmanand, N. Vedachalam, and G. A. Ramadass

Abstract Ocean technology developments in India commenced quite late compared to many developed countries. This has resulted in India struggling in the last two decades to make up for the time delay. The activities taken up in India in this area are enumerated in this article. The non-living resources and their method of utilization, along with the challenges are briefly indicated. The details of ocean energy, desalination, minerals which form the main resource are explained. The technology developments on living resources are vast and only a brief on cage culturing is touched upon here. Various platforms including ships and other ocean observation systems to provide data for weather prediction, climate change studies and tsunami warning are briefly touched upon.

1 Background

The world oceans cover about 2/3 of the earth's surface, extending over 360 million square kilometers. They dominate the surface of our planet. Even today their features and potential are not fully explored and are poorly understood. We also lack an in-depth understanding of the critical role that our oceans play in influencing the earth and climate systems. Though humans ventured into the sea and started long voyages as early as 4500 B.C., their main purpose was to discover new colonies and to find sea routes for improving trade and connectivity with other human settlements. The size and endurance of ships limited how far these ships could go. As technology for building and powering ships has evolved the reach and spread of human activities expanded significantly. Though today we have a good understanding of the extent and surface behavior of the oceans of the world and their connections we still know very little of what lies under the waters. The first scientific voyages were taken up

---

M. A. Atmanand (✉)
Indian Institute of Technology, Chennai, India
e-mail: atmanandma@gmail.com

N. Vedachalam · G. A. Ramadass
National Institute of Ocean Technology, Chennai, India

V. K. Gahalaut and M. Rajeevan (eds.), *Social and Economic Impact of Earth Sciences*,
https://doi.org/10.1007/978-981-19-6929-4_16

during 1831–36 by the H.M.S. Beagle which also carried Charles Darwin. This was later, followed by the H.M.S. Challenger that set out to understand the features of oceans and the ocean bed during 1872–76.

With the gradual development of ocean-going research ships, humans began to realize the great potential for resources in the deep sea as well as those located below the sea bed. The Ocean exploration missions conducted during the nineteenth and twentieth centuries were the first attempts to document the depth of the oceans, charting of key bathymetric features and identification and study of marine life.

Despite the important discoveries made during these missions, we still have only a cursory understanding of the deep ocean. About 95 percent of the ocean floor remains unexplored, much of it located in harsh environments such as the polar latitudes and the Southern Oceans. On virtually every expedition, oceanographers are discovering fascinating new lifeforms that shed new light on our understanding of evolution and their associated ecosystems.

The ocean covers nearly 70% of the earth. Thus 70% of the earth resources are still untapped. If tapped judiciously, these resources, some of which are non-renewable, can last for thousands of years to come. A list of non-living resources occurring at different depths is given in Table 1.

Preliminary evidence indicates that immense new energy sources exist in the deep sea. The amount of carbon bound in frozen gas hydrates on the seafloor is conservatively estimated to be twice the total amount of carbon existing in all the other known fossil fuels on Earth.

Advances in deep-sea technologies have also made it easier to locate shipwrecks and historical artifacts lost in the ocean depths. The continued exploration of marine archaeological sites will help us to better understand human history and our global cultural heritage.

Ocean exploration also offers an unprecedented opportunity to engage the general public in marine science and conservation. Exploration missions to the depths of the ocean provide images of ancient human artefacts, amazing creatures, and never-before-seen ecosystems. This kind of popular excitement and support can be an enormous asset in sustaining exploration projects over the long term. Given the importance of the ocean in human history and in regulating climate change, guaranteeing food security, providing energy resources, and enabling worldwide commerce, it is astounding that we still know so little about it. This is due primarily to the lack of a long-term, large-scale national commitment to ocean exploration.

While there are offshore oil and gas harvesting platforms and coastal mineral harvesting installations in India, the full extent of all our ocean resources is not understood and documented.

Table 1 Non-living resources in the oceans

Resource (Non-Living)	Depth (in meters)	Technology	
		Exploration	Exploitation
Energy	Up to 1000	Observations	Wave, Tide, Ocean Thermal Energy Conversion (OTEC)
Fresh water	Surface and <1000		Reverse Osmosis, Low Temperature Thermal Desalination (LTTD)
Placer deposits	On shore and estuaries	Geological Survey	Excavation
Oil & natural gas	Up to 3000	Seismic Survey	Drilling and pumping
Gas hydrates	1000–3000	Seismic Survey	Depressurization, Thermal stimulation etc
Cobalt crust		Sonar, deep Tows, Underwater Vehicles and samplers	Underwater Mining Machines with appropriate cutting devices and risers
Poly metallic sulfides	2000–5000	Sonar, deep Tows, Underwater Vehicles and samplers	Underwater Mining Machines with appropriate cutting devices and risers
Poly metallic nodules	>4500	Sonar, deep Tows, Underwater Vehicles and samplers	Underwater Mining Machines with appropriate cutting devices and risers

2 Resources in Oceans Around India and Challenges in Their Utilization

Even though Oceans are so vast and India can boast of a coast line of about 7516 km (Source, NCAOR), it is a paradox that we know more about surface of Mars than the Oceans. The Exclusive Economic Zone (EEZ) of India is 2.37 million km^2 (Source: World EEZ-Vol 4). The resources within this area are available to India for use. A proposal for extending the continental shelf by about 0.6 million square km, put up by India is pending with the International Sea Bed Authority (ISBA), of the United Nations.

Apart from the above, India has been allotted an area of 150,000 km^2 in the Central Indian Ocean Basin (CIOB) for developing technology for mining of polymetallic nodules by the International Sea Bed Authority, United Nations. India also has lodged a claim an area near the Rodrigues Triple Junction (RTJ) for the mining of polymetallic sulfides.

The EEZ is almost equal to 80% of the land area of our country. Even though, the fisheries resources in shallow waters are harvested to a reasonable extent, fishing resources that are located in deeper waters are not fully exploited as yet.

Similarly, oil and gas resources are being harvested by oil companies from the offshore platforms in Bombay high and Godavari basins, but mainly in shallow waters of 100 m depth. Very few of them operate at depths of 1000 m.

Except for this limited knowledge on fish and oil and gas reserves close to the coastline the rest of India's EEZ remains largely unexplored. Neither the physical properties of the ocean water column nor the resources on and below the seabed have been mapped.

3 Ships and Other Platforms

Normally, mobile platforms with mechanical/electrical systems to reach deep waters are necessary to cover the vast areas of the oceans. These mobile platforms are the research vessels having different laboratories to acquire and analyze data. Apart from the Research Vessel (R.V.) Gaveshni, Ocean Research Vessel (ORV) Sagar Kanya and Fisheries Ocean Research Vessel (FORV) Sagar Sampada, the Government of India was hiring research vessels from aboard for some specific programs of ocean-related data collection. With the creation of the Department of Ocean Development in 1982, the Government of India started a major initiative in Antarctica for polar science, hiring ice breakers from abroad. The few research vessels of Government of India could not contribute substantially to the major data collection in all the ocean areas around India. With the establishment of the National Institute of Ocean Technology (NIOT) at Chennai in 1993 and that of the Indian National Centre for Ocean Information and Services (INCOIS) at Hyderabad during 1997 and the National Centre for Antarctic and Ocean Research (NCAOR) in Goa in 1999, there was a quantum jump in the ocean science and technology related activities in India. A dedicated program to deploy and operate moored buoys for ocean and atmospheric data collection was started by NIOT along with a Vessel Management Cell (VMC) in 1997. This cell acquired two coastal research vessels, Sagar Paschmi and Sagar Purvi in 1998 and an ocean-going vessel Sagar Manjusha in 2005. A unique sophisticated technology demonstration vessel 'Sagar Nidhi' was designed for deep sea operations and was built in Italy in 1997. Sagar Paschimi and Purvi has been decommissioned and recently replaced with Sagar Tara and Sagar Anveshika for coastal research. These four vessels of NIOT and the earlier three vessels were always in great demand for ship board data collection and for deployment/servicing of moored data buoys in areas mostly under our EEZ. But these seven ships are grossly insufficient to collect the entire data from ocean water column in the Indian waters and so more ships are essential to be procured (Fig. 1).

Fig. 1 Indian Ocean Research Vessel Sagar Nidhi

4 Underwater Vehicles

Apart from fixed and mobile platforms on the surface of the sea, a variety of underwater vehicles have been developed for the purposes of underwater surveys, data collection, water and soil sampling, search and rescue/recovery. Some of the seabed vehicles can be used for sea bed sampling and for mining demonstrations. The type of vehicles could be.

(I) Autonomous under vehicles (AUVs) for few hours of underwater operation
(II) Remotely operated vehicles (ROVs) connected by a tether to the ship for power and data transmissions
(III) Sea bed crawler, also with tethers to the ship.
(IV) Human submersible for taking humans to a depth of 6000 m water depth.

The components of the above vehicles have to be designed for Deep Sea Survival. Hence, they are of complicated design and expensive.

5 Harnessing of Resources

The estimated quantity of resources is given in the Table 2.

Table 2 Estimated quantity of resources

S. no	Details	With Present EEZ
1	Minerals (a) Beach placer deposits (b) Deep sea nodules in the pioneer area in CIOB	632 Million Ton Co: 1.82 Million Ton Ni: 10.47 Million Ton Cu: 9.5 Million Ton Mn: 247 Million Ton
2	Non-renewable energy (a) Hydro carbon (oil & gas) (b) Gas hydrates	350 Million Ton (50% of national reserves) 1 TCM natural gas (67% of national reserve) 1,894 trillion m^3
3	Renewable ocean Energy (OTEC, Wave & tidal)	>Several Terra Watts
4	Food (fisheries)	3.934 Million ton
5	Fresh water	Unlimited

5.1 Renewable Energy and Fresh Water

Twenty major cities are located along Indian coast line and the water requirement for all these cities in 2008 stood at 6,267 million liters per day (MLD). The coastal cities experiencing tremendous growth are Mumbai, Chennai, Surat, Kolkata and Vizag. The projected water requirement for all coastal cities in 2026 is estimated to be around 23,607 MLD, a four-fold increase from 2008. Many large scale desalination units are being established in the country with international support. In spite of this, regions like Tamilnadu, Sourashtra, Kutch and Rajasthan face severe drinking water shortage. Among the different available processes existing in the world market, Sea water Reverse Osmosis (SWRO) enjoys the highest share of 53% followed by Multi-Stage Flashing (MSF) with 25%. MED contributes to 8% and ED to 3%, whereas other types of desalination processes occupy a total of 11%. Hence, development of a suitable desalination technology is an urgent need. Establishing higher capacity desalination plants by utilizing the waste energy and ocean thermal energy is found to be feasible for meeting drinking water need of the mainland.

India is blessed with abundant solar radiation. The average daily solar radiation in India is 4–7 kW-h/m^2, compared with the global average of 2.5 kW-h/m^2. Hence, the use of solar energy in thermal desalination processes is one of the most promising applications in coastal locations. Although everybody recognizes the strong potential of solar thermal energy to seawater desalination, the process is not yet developed at the commercial level. More studies are needed to increase the confidence levels for sustainability and commercialization. Solar desalination system consists of three major subsystems, viz. solar energy collectors, storage and desalination system. Improvements in each subsystem can augment the yield of the solar desalination plant.

A variety of different technologies are currently under development throughout the world to harness this energy in all its forms including waves. Deployment is currently limited but the sector has the potential to grow, fueling economic growth, reduction of carbon footprint and creating jobs not only along the coasts but also in the inland along its supply chains. Given the long-term energy need through this abundant source, action needs to be taken now on the R&D front in order to ensure that the ocean energy sector can play a meaningful part in achieving our objectives in the coming decades. Most types of technologies are currently in the demonstration stage or the initial stage of commercialization. Also, sensitivity of the performance of the energy devices to ocean parameters and their variability and non-availability of specific data at desired locations lead to a strong need for resource assessment with a prime focus on the energy extraction. The knowledge of resources available within the territorial waters will give a good understanding and would lead to more efficient designs of energy devices suitable to the site-specific conditions. This can be done by carrying out detailed resource mapping along the Indian coasts with a focus on requirements of energy devices for Indian conditions.

Commercial viability studies of offshore wind energy showed promising results for the states of Tamilnadu and Gujarat and these studies can be extended to other parts of the coast by measuring the wind profile with LIDAR. The substructure that supports offshore wind turbine in ocean accounts almost for 40% of the project. Hence, to improve the economics of offshore wind projects, it is essential to develop reliable and innovative substructure concept along with an indigenized installation methodology. This can be achieved through continued R&D efforts modeling and demonstration. The demonstration projects aimed at the analysis of the behavior of fixed/floating substructure concepts, which can lead to demonstration of these concepts in the field, so that industry can take up these concepts for development of the offshore wind project.

Marine organisms like algae can be a potential bioenergy feedstock for the developing nation. The high growth rate and photosynthetic capacity of marine algae can produce much higher amount of bioenergy when compared to terrestrial plants and considered as a better alternative than bioethanol from sugarcane. Generation of biodiesel from marine organisms is one of promising areas of research.

5.2 Minerals

Sea bed has varying soil conditions starting from sand to clay and rock, just as land. Also, as in land, the topography varies depending on many parameters. For the exploration of the sea bed, there are many tools in use. The first one is the sub-bottom profiler, which impinges acoustic waves into the sea bed and measures the reflections from the sea bed from which the properties below sea bed are determined. In order to do ground trothing, it is necessary to measure the bearing strength and shear strength in-situ. Soil tester has been developed and tested at 5289 m water depth for soil property measurement in the Central Indian Ocean Basin (CIOB).

5.3 Challenges

The deep-sea environment is a very difficult space to work with. The design, erection, operation and maintenance of a device/system in deep sea are associated with many problems because of the following factors:

The average density of sea water is around 1030 kg/m^3. This value is about 1000 times of that of the density of air. In deep waters of 5000 m to 10,000 m depth, the ambient water pressure exerted on the submerged body varies from 500 to 1000 bar. The mechanical design of all submerged bodies has to withstand such large ambient pressures. These factors of density, buoyancy and pressure in deep sea environment make it extremely complicated and expensive to design and operate systems in deep sea.

Corrosion of materials in sea water possesses a great danger to the safety of the underwater equipment. Large offshore oil and gas platforms have failed and taken many lives due to the corrosion attacks on the materials of construction, especially those of pipelines. Since the impact of corrosion on carbon steel components (carbon steel is the most commonly used in offshore industry) is high, a variety of alternate materials like stainless steel and nickel alloys have been developed. But use of these material increases the cost of the project by a factor of 3 to 10.

5.4 Living Resources

5.4.1 Offshore Cage culture—An Effort Towards Realizing the Blue Economy

We are in an era, where every coastal nation is seriously framing their strategy for the "Blue Economy" due to the limitation in land-based resources. India, being a seafront nation, shares 2.14% of land and 5.59% of oceanic waters of the globe and has a greater potential for mainstreaming its sea-based economy as its primary source for development. The growth of the Indian economy though projected to accelerate at a speed of 7.4% (valued at US $2.1 trillion), the contribution of fisheries sector including the freshwater resources (around 38%) of the country is only 1.3%. Hence, "Farming the Blue" initiative of the Ministry of Earth Sciences is expected to propel India into its future, to meet its raising demand for proteins. Understanding the increasing demand for fish protein and the sea farming potential of the country, NIOT has taken a few initiatives to enhance its fish production capabilities using the marine waters. NIOT is the first organization in India to develop sea cages, pens for lobster and mud crab fattening. The institute has designed floating collar gravity cages for marine finfish farming that can withstand turbulent sea conditions of Indian Seas and has also come out with an innovative concept of rearing post-larval fishes (as small as 1 g) in specially designed nursery cages in the sea. To support the livelihood security of the traditional and island fishers, the institute has designed and deployed

Artificial Reef structures along Odisha coast and oceanic Fish Aggregation Devices in appropriate sites off Andaman and Lakshadweep Islands. To assess the potential of meritic waters of the nation, a geospatial analysis of indian seas has been executed, which revealed the presence of vast pristine sea space combined with tropical climate, preferable for fish farming within the cost-effective mariculture zone.

Following the various experimental culture trials and testing of sea cages using a variety of marine finfish species in different environmental conditions, NIOT has worked out the economics of the cage farming at various culture intensities and feed inputs. By accepting the ground realities and the knowledge gaps that obstruct the expansion of offshore fish farming in the country, NIOT with diversified the combination of field experimental results that has charted out a comprehensive plan for the production of initially 1 million metric ton, which can later be expanded to a 7 mmt marine finfish production system. It is also proposed to identify 1% of the country's EEZ (i.e. 20,200 km^2) and earmark for mariculture operations, which may enable the country to achieve 7 mmt of marine finfish and equal quantities of shellfish and aquatic plants. Considering the huge availability of suitable sea space in the 20–100 m depth zone compared to the near shore environments, it is proposed to utilize the 20 to 50 m depth zone, primarily to produce 85% of the target and the remaining 15% from the near shore using the 20 and 9 m Ø cages, respectively. The proposal with an inbuilt motto of "Farm the Ocean to Feed the Nation" when implemented, will result in the employment generation of 1.6 million direct jobs i.e. 40% of the country's fishermen population which can be converted from wild hunting to fish farming, thereby reducing the fishing pressure on the wild stock.

6 Ocean Observation and Hazard Mitigation

Hazards for the ocean were present from time immemorial. Storms, high wave actions, tidal waves, tsunamis etc. are some of them. With the advent of technology, there are ways and means to predict many of these hazards. Mathematical models have been developed which more or less provide solution to complex non-linear equations. However, they require data from ocean and atmospheric observations above the ocean surface. The data buoys precisely provide these in real time. The enhancement of measurement techniques such as Autonomous Ocean Observation Systems (AOOS) and establishment of state–of-the-art National Ocean Observation Control Centre are the future needs to meet the increasing demand of advanced monitoring techniques and the systematic maintenance of the ocean data collection requirements. To enhance the measurement techniques and to implement the state-of-the-art facilities for testing/calibration and systematic data collection, new systems are needed. Prototype development of moored data buoy and tsunami buoy require further enhancement. The present tsunami buoy system detects the Rayleigh waves generated by an earth quake/land slide/volcanic eruption which lead to a slight delay in detecting an impending tsunami. The introduction of a seismometer will reduce the latency in detecting the tsunami events. The moored buoy network provides time series met-ocean observations at selected locations in Indian Seas. One major lacuna is the availability of data at limited locations. The development of newer observational platforms such as Robofish, Glider, AOOS etc. can contribute to the spatial coverage and vertical profiling at areas of interest, which includes establishment of Marine Bionic laboratory, Design & Development of Autonomous Ocean Observation Systems (AOOS) such as Surface Sampling System, underwater vertical profiling system, underwater surveillance system and underwater internet. Further, a state-of-the-art National Ocean Observation Control Centre is the need of the hour to meet the increasing demand to diversified ocean data. The ocean observation system would enable quick access to get data from the sea bed to Desktop and projects such as Lab on Chip for ocean data collection. Real time ocean observation network establishment in the Polar region for Long-term monitoring of Arctic region for climate change studies is also planned. Standardization of ocean data collection is another major initiative proposed.

Testing and demonstration of equipment at various depths and under varying meteorological conditions call for extensive participation of vessels and are dependent on the sea and weather conditions. Hence, it is proposed to develop a multi-purpose ocean observation platform not only for the measurements of the surface and subsurface oceanographic parameters but also to serve the needs of deployment and retrieval operations. These observational platforms can be semi-permanent in nature and shall be able to serve as coastal laboratories with working accommodation for the scientists and are accessible by boats and coastal research vessels. These observation platforms can be linked to the mainland by satellite as well as RF communications for data transmission and other communications.

7 Ocean Acoustics and Systems Development

The future needs are autonomous systems for deep water noise measurements, polar region measurements and underwater acoustic communication systems for long range and high rate of data transfer. It is aimed to bring the state-of-the-art underwater Acoustic Test Facility as Nationally Designated Laboratory for underwater acoustics in India.

Sensors are heart of the Ocean Observation Systems like Buoys, ROVs and AUVs and they are currently imported for oceanographic measurements. Frequent replacements of sensors due to biofouling necessitated development of sensors indigenously for continuous and quality data. The cost and lead time for the procurement of sensors is very high. The experience gained in the development of the NTC thermistor-based SST sensor for drifting buoys can be further enhanced to develop sensors like wind speed and direction, conductivity, depth and turbidity, chlorophyll and oxygen that are mainly suitable for ocean observation systems which will make us self-reliant.

There is a specific need to have data transfer through Indian satellites as foreign satellites do not have hubs in India. This raises issues related to security. The communication requirement of Oceans and Polar Region platforms can be achieved by dedicated payloads on Geostationary and Polar satellites with support from ISRO. Underwater oceanographic observation calls for many types of new devices. Development of expendable CTD profilers, C profilers and satellite communication will be realized in the future. Development & establishment of high altitude coastal & ocean observation system for understanding the ocean & atmospheric boundary layer is another long-term strategy.

Significant progress made on Drifter buoy with INSAT communication will be further enhanced. Development of deep sea autonomous underwater profiling drifters (DAUPD) using the experience gained in glider operations and profiling systems is proposed to augment the deep-sea observations.

8 Conclusion

The material given is only tip of the ice berg due to space limitations. There are much more technological details in the area of ocean technology, especially as India is still in the nascent stage of development. The need of the hour is having more youngsters taking up their study in ocean technology. Having programs like student competitions in building autonomous underwater vehicles and such other systems will enable to encourage students to take up their career in ocean technology. Collaboration with other International Institutes like Woodshole Oceanographic Institution, USA, Scripps Institute of Oceanography, USA, Institut Français de Recherche pour l'Exploitation de la Mer (IFREMER); French for 'French Research Institute for Exploitation of the Sea', France, Japan Agency for Marine-Earth Science and Technology (JAMSTEC), Japan and student exchange programs will have to be taken up on a war footing.

References

http://oceanservice.noaa.gov/facts/ninonina.html

http://www.skymetweather.com/content/weather-faqs/what-is-indian-ocean-dipole-iod

http://www.imo.org/en/OurWork/Environment/BallastWaterManagement

Burgholzer P, Hofer C, Reitinger B, Mohammed A, Degischer HP, Loidl D, Schulz P (2005) Non-contact determination of elastic moduli of continuous fiber reinforced metals. Compos Sci Technol 65:301–306

Chartering the Course for Ocean science for the United States for the next decade, an ocean Research Plan and Implementation Strategy (2007) NSTC joint subcommittee on ocean science and technology

Creating A nation Strategy for Increasing Science knowledge, Chapter 25 of Report (2004) An Ocean Blue print for the 21st Century, from the U.S. Commission on Ocean Policy, September 20, 2004.

Navigating the Future (2013) Position Paper 20, European marine board
Sankar U (2007) The Econmomics of India's space Programme: an exploratory analysis. Oxford University Press
Wang X, Xu X (2001) Thermoelastic wave induced by pulsed laser heating. Appl Phys A 73:107–114
"Wealth in the ocean: deep sea mining on the horizon?" (2014) UNEP global environmental alert service

Deep-Sea Mineral Resources and the Indian Perspective

P. John Kurian and Parijat Roy

Abstract Depleting terrestrial mineral deposits coupled with the ever-increasing demand for metals and minerals led to the resurgence of interest in the exploration of marine mineral resources. The majority of the marine resources, such as Polymetallic Nodules (PMN), Polymetallic Sulfides (PMS), and Cobalt-rich Fe–Mn Crusts (CC) being largely found in the "Area" beyond the national jurisdiction of maritime countries, the role of International Seabed Authority (ISA) in managing the regulatory challenges in a socially responsible and innovative way is important. These deposits are distinct in their geological settings, metal content, abundances, global distribution, processes of formation, associated ecosystem, etc., and also, the exploration and exploitation approaches are different in terms of technology, scientific, economic, and regulatory aspects. India, a maritime country with a rich maritime heritage, is actively engaged in exploring the resourceful ocean bed for meeting the future metal demands of the country. Having entered two exploration contracts with ISA, one each for PMN and PMS, Indian activities are spearheaded by the Ministry of Earth Sciences with the active participation of many research institutions and universities in the country. Continuous and dedicated efforts in up-scaling exploration activities as well as the development of efficient exploitation tools and metal extraction technologies by India shall strengthen the long-term development of the country on the social, economic, and societal fronts.

1 Introduction

The fast-depleting terrestrial mineral resources and the increasing demand for metals and strategic minerals forced mankind to explore alternative resources in the oceanic domain. The successful recovery of potential seabed mineral resources in turn imparted great impetus to the widespread search and greater interest for such

---

P. John Kurian (✉) · P. Roy
National Centre for Polar and Ocean Research (NCPOR), Ministry of Earth Sciences, Headland Sada, Vasco-da-Gama 403804, Goa, India
e-mail: john@ncpor.res.in; kurianjohn@gmail.com

V. K. Gahalaut and M. Rajeevan (eds.), *Social and Economic Impact of Earth Sciences*,
https://doi.org/10.1007/978-981-19-6929-4_17

resources in the deep-sea domains, which led to the rapid development of our knowledge too. Advancements in research on emerging techniques of seabed exploration, exploitation, and mineral prospecting have paved the way for the greater involvement of scientists and engineers worldwide in this endeavor. Many of these deposits being in the international waters have necessitated the formulation of appropriate regulations under the UN Convention on the Law of the Sea (UNCLOS) through the establishment of the International Seabed Authority (ISA).

The sustainable development goals, increasing population, transition to green technologies, etc. are bound to increase the demand for metals, and hence, the development of marine mineral resources can be considered as complementary to replenish the requirements for future industrial and economic progress. The present-day demands for metals are met through mining on land, which represents less than one-third of the earth's surface. The over-exploitation of these resources is eventually leading to scarcity of high-grade land deposits. This scenario is posing a serious question in front of mankind whether the oceans can feed, strengthen, and offer the mineral resources for the generations to come.

Deep-Sea mineral resources such as polymetallic nodules (PMN), polymetallic hydrothermal sulfides (PMS), and cobalt-rich ferromanganese crusts (CC) have attained immense attraction as being considered as the future alternative of land resources and source for global metal supply. Due to the rise in commercial interest and increased marine scientific activities, the current focus for exploration and future exploitation is primarily on these three types of deep ocean metal-rich mineral resources. Each of these deposits is distinct in its geological settings, metal content, abundance, global distribution, processes of formation, associated ecosystem, etc., and also, the exploration and exploitation attributes are different in terms of technology, scientific, economic, and regulatory aspects (Fig. 1, Table 1). Although, exploration of all these deep-sea minerals is considered to be challenging, searching for PMS deposits is a challenge, as they are confined to smaller areas and the exploration for active sites is by trailing the tracers in the water column and the seafloor. However, the common factor for these three mineral deposits is their occurrence in the deep waters far beyond the shelf break.

A review of the various deep-sea mineral resources, which India is presently focusing on their exploration, assessment, and future exploitation, is addressed in this article, with emphasis on the global distribution and presenting a status of India's exploration activity. A brief history of marine mineral exploration and also the present global scenario on the various environmental issues and exploitation concerns are also discussed.

2 History of Seabed Mineral Exploration

The presence of mineral deposits in deep-sea domain was totally unknown to the humans before 1870s. The discovery of marine manganese nodules commences with the voyage of the HMS Challenger, 1872–1876, and for the first time on 18 February

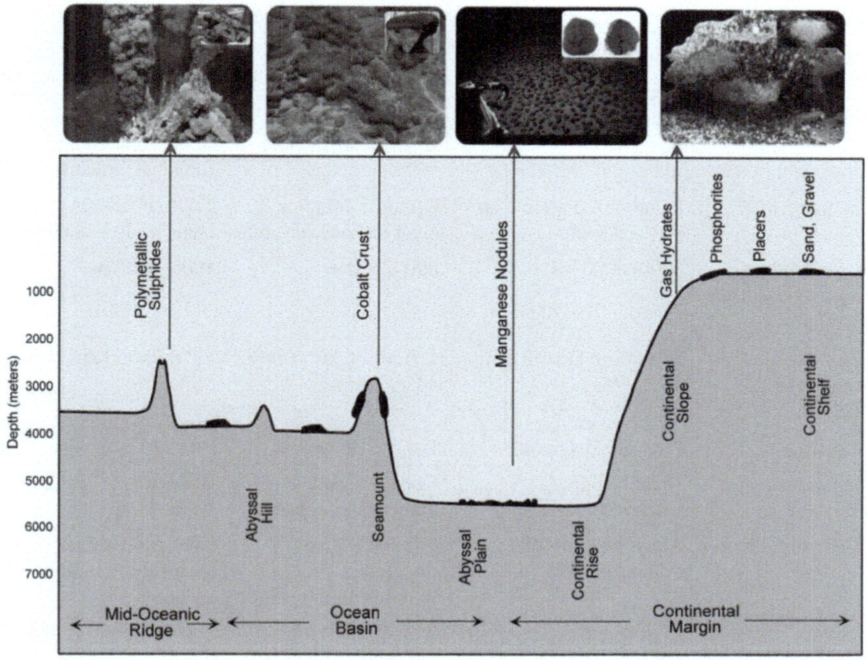

Fig. 1 Schematic diagram showing the occurrence of various mineral deposits on the seafloor in relation to sea-bottom topography

1873 approximately 300 km southwest of the island of Ferro in the Canary Group deep-sea nodules were recovered (Murray and Renard 1891). Subsequent to the Challenger expedition more extensive collections of nodules were made during the Albatross expeditions of 1899–1900 and 1904–1905 from the Pacific and their distribution in the equatorial Pacific was mapped (Agassiz 1903, 1906; Murray and Lee 1909, Glasby 1977). Sediments characterized by high iron and low aluminum content were dredged near the East Pacific Rise (EPR) and the peculiar black potato-shaped samples recovered from the seabed were revealed by chemists as rich manganese oxides. Such unusual deposits were collected again in the 1940s on the US Ship "Carnegie" (Revelle 1944). These discoveries could not attain any attraction probably due to the lack of convincing scientific explanation. Later, Skornyakova et al. (1964), Boström and Peterson (1966), Bonatti and Joensuu (1966) reported deep-sea metalliferous sediments and crusts enriched in iron and manganese hydroxides and inferred their origin to the hydrothermal activity. Later, it was the American mining engineer John L. Mero, in his book "The Mineral Resources of the Sea", (1965) sparked considerable interest in the scientific world when he estimated huge ferromanganese (Fe–Mn) nodule resources in the Pacific Ocean and inferred considerable amount of metals such as Mn, Cu, Ni, and Co in these resources. This prompted the aspirations of the world scientific community in developing such deep-sea deposits

Table 1 Characteristics of deep-sea mineral resources (modified after Peterson et al. 2016)

	Manganese nodules	Co-rich ferromanganese crusts	Seafloor massive sulfides
Geological settings	Sedimented abyssal plains	Upper flanks of old volcanic seamounts	Oceanic spreading centers and young island arc volcanoes
Characteristics	Potato-sized nodule on soft sediments	Up to 25 cm thick crust on hard substrate	Ten to hundreds of meter wide mounds
Water depth of greatest economic potential	3000–6000 m	800–2500 m	1000–5000 m
Favorable area ("Area", EEZ, ECS)	38 million km^2 (81%, 14%, 5%)	1.7million km^2 (46%, 44%, 10%)	3.2 million Km^2 (58%, 36%, 6%)
Dimensions	Large2-Ddeposits	Large2-Ddeposits	Small3-Ddeposits
Main metals of interest	Nickel, copper, manganese, cobalt	Cobalt, nickel, manganese, copper	Copper, zinc, gold, silver
Other commodities	Molybdenum, lithium, titanium	Titanium, REEs, platinum, molybdenum, bismuth	Cadmium, gallium, germanium, indium, antimony
Resource estimate	21,100 million tonnes in the Clarion-Clipperton-Zone 1335[@] million tones Indian Application Area in Central Indian Basin	7533 million tons in the Prime Crust Zone	600 million tons in the neovolcanic zone of mid-ocean ridges
Grades	(Clarion-Clipperton) 2.4 wt% Cu + Ni 0.2 wt% Co 28 wt% Mn (India's Retained area in Central Indian Basin[$]) Mn 22–30% Ni 1–1.7% Cu 0.9–1.5% Co 0.08–0.12% Ni + Cu + Co 2.2–3.3%	(Prime Crust Zone) 0.5wt%Cu + Ni 0.7 wt% Co 23 wt% Mn (Afanasiy Nikitin Seamount, Indian Ocean[a]) 0.26 to 0.44 wt% Cu + Ni 0.36 to 0.86 wt% Co 15.5 to 24.9 wt% Mn	(Occurrence median) 3wt%Cu 9 wt% Zn 2 ppm Au 100 ppm Ag
Grade distribution	Homogeneous on regional scale	Homogeneous on regional scale	Very heterogeneous on regional and local
Footprint of 2 mio ton mining activity on the seafloor	150 km^2	25 km^2	<0.2 km^2
Knowledge base for resource estimate	Good in the CCZ[a] CIB	Poor	Poor
Resource potential	High	High	High

(continued)

Table 1 (continued)

	Manganese nodules	Co-rich ferromanganese crusts	Seafloor massive sulfides
Global impact on mining on metal markets	High	High	Low

[@]Nair, R.R.; Jauhari, P. (1993) Polymetallic nodule resources of the Indian Ocean. Science and quality of life, Ed. by: Qasim, S.Z. 393–405p
[$]ShyamPrasad (2005)
[a]Rajani et al. (2005)

and economic mining thus became the starting point for deep-sea mining projects worldwide. Following this, comprehensive studies focusing on the exploration and economic potential of deep-sea minerals were undertaken by many aspirant nations as well as international scientific groups. Many workers have provided considerable scientific insights into the understanding of these deep-sea mineral deposits, especially for PMN deposits (Glasby 1977; Siddiquie et al. 1978; Hein et al.1979; Cronan 1980; Frazer and Wilson 1980; Cronan and Moorby 1981; Frazer and Fisk 1981; Thijssen et al. 1981; Glasby et al. 1982; Rao and Nath 1988; Martin-Barajas et al. 1991; Usui and Someya 1997). Initially, volcanism was considered as one of the potential sources of metals in oceans (e.g., Murray and Renard 1891). However, the discovery of metal-bearing "submarine thermal springs" or "hydrothermal vents" on the seafloor of the Galapagos Rift in 1977 (Corliss et al. 1979) gave a direct evidence for the same. Actively forming sulfide deposits, known as seafloor massive sulfide or polymetallic sulfide deposits, were subsequently discovered in the East Pacific Rise (Francheteau et al. 1979). During the 1980s and 1990s, similar interest has also grown in the exploration of Fe–Mn-rich crusts occurring on topographic highs such as seamounts (Halbach et al. 1989; Hein et al. 1997).

3 India and the International Seabed Authority

India played a pivotal role during the various deliberations that took place in the formulation of the United Nations Convention on Law of the Sea (UNCLOS) and as well in protecting the special interest of the country. Indian marine scientists from the 1980s were actively involved in exploration and research activities pertaining to polymetallic nodules in the deep ocean realms of the Indian Ocean. As a great success to the Indian researchers, the first set of nodules was collected onboard by R.V. Gaveshani from the flanks of Carlsberg Ridge in the Arabian Sea during January 1981. Considering India's long history of marine scientific research and also considering the commendable work carried out in the exploration and research of PMN in the deep ocean basin, United Nations recognized India as a Pioneer Investor in April 1982. Subsequently, in August 1987, India was allocated the Pioneer area of 150,000 km^2

in the Central Indian Ocean Basin (CIOB) for exploration of PMN. India was the first country to achieve this unique status. The Preparatory Commission had registered seven pioneer investors: China, France, India, Japan, the Republic of Korea, and the Russian Federation, as well as a consortium known as the Interoceanmetal Joint Organization (IMO) including the Czech Republic, Slovakia, and Poland. With the Convention in force and the International Seabed Authority being functioning, these pioneer investors became contractors along the terms contained in the Convention and the Agreement, as well as regulations established by the International Seabed Authority. Six entities got rights in the Pacific Ocean (France, Russia, Japan, China, IMO, Korea, and Germany) and one in the Indian Ocean (India) for exploration.

The enhanced interest shown by many entities in undertaking comprehensive marine mineral exploration over the large deep-sea domains having potential resources resulted in laying their claims under the provisions of UNCLOS for exclusive rights for the development of mining sites in the region. This ultimately led to the establishment of "The International Seabed Authority (ISA)" with its headquarters in Kingston, Jamaica in 1994. Under As per UNCLOS, exploration and exploitation of any seabed mineral resources in the International waters outside the national jurisdiction of any country (referred to as "Area") can only be carried out under a contract with the ISA and subject to its rules, regulations, and procedures. For this purpose, ISA has developed internationally legally binding regulations with respect to exploration as well as exploitation. Regulations governing exploration for nodules (PMN), cobalt crusts (CC), and seafloor massive sulfide deposits(PMS) are already in place and accordingly, as of 2019, a total of 30 exploration contracts, spanning 15 years each, have already been signed by ISA. Of these, eighteen contracts are for the exploration of polymetallic nodules (PMN), seven contracts for polymetallic sulfides (PMS), and the remaining five contracts for cobalt-rich crusts (CC) (Fig. 2). The regulations pertaining to the exploitation of deep-sea mineral resources are being developed, and consultations and discussions are in progress, which are in a very advanced stage now in the case of PMN. The mineral resources of the "Area" are being considered as the common heritage of mankind, and ISA is using an innovative international consultation process with the world nations on how these resources are to be exploited and how the proceeds are to be allocated, thus to assist it in the development of these regulations (Lodge and Verlaan 2018). Assessment of the geographical distribution of the three types of deep-sea mineral resources under consideration (PMN, PMS, and CC) reveals their distribution to fall in all the three different areas of jurisdiction, namely the Exclusive Economic Zone (EEZ), the Extended Continental Shelf (ECS), and the "Area" (the international seabed). The "Area" constitutes the largest share which will be managed by the ISA.

Pursuant to the approval of the ISA adopting regulations on prospecting and exploration for polymetallic nodules in the "Area" in 2000, India signed the first 15-year contract for exploration of PMN in the Central Indian Ocean Basin on 25-March-2002, which was later extended for another 5-year term in 2017. India also signed the 15-year contract for polymetallic sulfides with ISA on 26 September 2016 for undertaking exploration and other developmental activities in the allotted area of 10,000 km^2 in parts of the Central Indian Ridge (CIR) and South west Indian Ridge

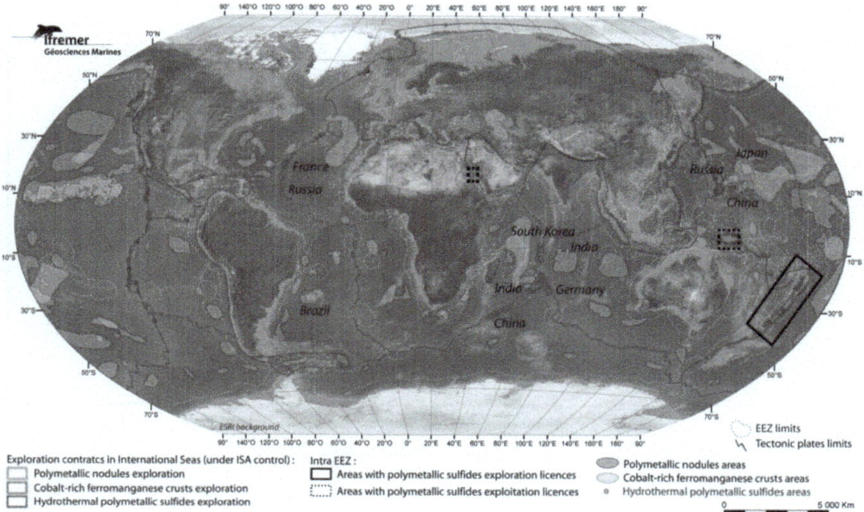

Fig. 2 Map showing the locations of various exploration licenses issued by ISA (*Source* IFREMER 2017)

(SWIR) in the Indian Ocean. India is also undertaking preparatory work toward the filing of the application to ISA for a grant of license for the exploration of cobalt crusts.

4 Polymetallic Nodules (PMN)

Polymetallic nodules (also called manganese nodules) are mineral concretions, composed of Fe oxyhydroxide and Mn oxide, and vary in shape and size between 1 and 12 cm in diameter. They widely occur in oceanic abyssal plains, at water depths of 3,000–6,500 m, and are found on or directly below the sediment-covered seafloor where sedimentation rates are lower than 20 mm per thousand years (Petersen et al. 2016). Polymetallic nodules can form by a combination of hydrogenetic (precipitate from ambient seawater) and diagenetic (precipitate from sediment pore waters) processes. Hydrogenetic nodules grow at a very slow rate at 1–10 mm per million year, whereas diagenetic nodules grow several hundred mm per million year. The physicochemical properties of the Fe and Mn colloids under oxic conditions make these nodules excellent for scavenging dissolved metals from seawater and get strongly enriched in Ni, Cu, Co, Mo, Zr, Li, Y, and rare-earth elements (REEs) relative to the Earth's crust (Hein et al. 2013; Lusty and Murton 2018).

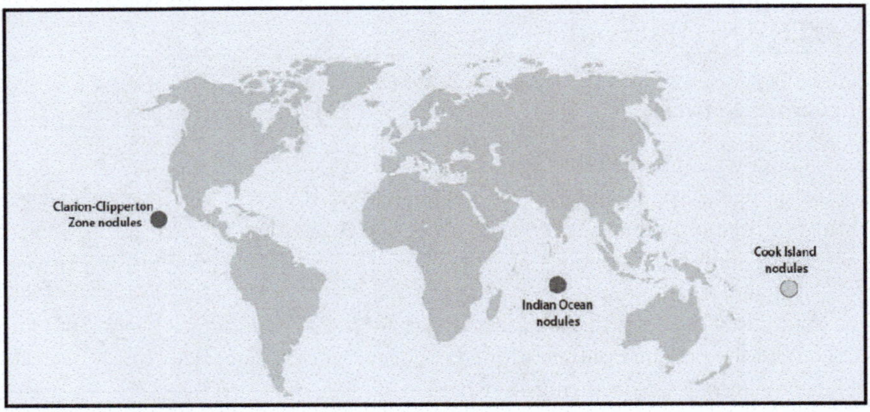

Fig. 3 Global picture of PMN occurrence (Hein and Peterson 2014)

4.1 Global Occurrence

Polymetallic nodule coverage is known in more than 50% of areas of the abyssal Pacific seabed and Central Indian Ocean Basin (Rona 2008). They also occur in the Atlantic Ocean. The most promising regions of polymetallic nodules rich in Ni and Cu are found in the Clarion–Clipperton Zone (CCZ) which extends from off the west coast of Mexico to as far west as Hawaii in the equatorial Pacific, the Peru Basin off South America near the Cook Islands, and the Central Indian Ocean Basin (Hein and Koschinsky 2014) (Fig. 3).

4.2 Indian Endeavor

India's endeavors in the survey and exploration for polymetallic nodules from the deep seafloor cherished on 26th January, 1981 when the first polymetallic nodule from the deep seafloor was picked up onboard by the Indian research vessel, RV Gaveshani. Since then the country has had a sustained R&D program for the survey and exploration of polymetallic nodules involving several national institutions, agencies, and Universities. Spearheaded by the Ministry of Earth Sciences (erstwhile Department of Ocean Development), extensive exploration work had been carried out in the Central Indian Ocean Basin (CIOB) by the nodal agency CSIR-National Institute of Oceanography in identifying potential areas of nodule occurrence in the Pioneer Areas and Allocated Area. A massive exercise of sampling at 11,000 stations in an area of 300,000 km^2 in ~72 expeditions (Sharma 2010) was undertaken in CIOB. As per the regulatory system of UNCLOS, phase-wise relinquishment of 50% was carried out in the initial allocated area of 150,000 sq. km to finally have a retained area of 75,000 km^2 in the CIOB (ShyamPrasad 2005) containing estimated metal resources of >6 mt (Cu + Ni + Co) (e.g., Sharma 2010). Subsequently, a First Generation Mine Site (FGM) for nodules was identified in this Area. The final phase of relinquishment was carried out in 2002. The first 15-year contract for the exploration of PMN was signed on 25 March 2002 between ISA and the Government of India, which was later renewed for another 5-year term in 2017. The present focus is to identify the most potential zone in the Retained Area which would form the nucleus of the First Generation Mine Site (FGM) for nodules in the Central Indian Basin. Detailed fine-scale surveys are in progress in the area, and environmental impact studies are also been planned now in strict adherence to ISA guidelines for undertaking the test mining in the area. The major milestones achieved by India in the exploration of PMN are listed in Table 2.

India started undertaking environmental studies in the CIB even before ISA formulated any formal guidelines. A multidisciplinary scientific team from the nodal agency, CSIR-NIO has undertaken the EIA studies, as a part of the MoES project, since the year 1994 and Indian scientists have contributed to the formulation of initial guidelines/protocols of ISA for collecting the environmental studies in the Area (ISA Report 2001). The first two phases of the EIA study, comprising of baseline studies (Sharma and Nath 2000; Sharma et al. 2000) and the benthic disturbance experiment (Sharma and Nath 1997; Sharma et al. 2000), were carried out by NIO, Goa, in technical cooperation with State Geological Enterprise, Gelendzhik, Russia, under an MOU between the Department of Ocean Development, the Government of India, and the Ministry of Natural Resources, the Government of the Russian Federation (Muthunayagam and Das 1999). Later environmental studies of monitoring

Table 2 Milestones achieved by India in polymetallic nodule exploration (modified from Shyam-Prasad 2005; Sharma 2010)

Period	Milestones
April, 1982	United Nations recognized India as the Pioneer Investor
December, 1982	India Signed UNCLOS III
August 1983	First metal from nodules extracted and 2 mil. km^2 area explored (the exploration and the metallurgy teams were working in tandem)
August 1987	Allocated the Pioneer area of 150,000 km^2 in Central Indian Ocean Basin to India
July 1994	20% area relinquished (block size 25 km)
October 1996	10% area relinquished (block size 12.5 km)
Year 2002	20% final relinquishment
March 2002	India signed a contract with the International Seabed Authority for exploration for nodules in the area for a period of 15 years
Year 2008	First Generation Mine Site Identified
Year 2013	Test Mine Site identified
Present status	Process of developing a technology for exploration and utilization of nodules from an area of 75,000 km^2 retained in the Central Indian Ocean Basin and allotted to it by the Authority

and restoration of benthic impact experiment (Sharma et al. 2005; Nath et al. 2012) and the regional environmental database creation were entirely carried out by Indian researchers.

4.3 Exploitation and Mining

The major factors favoring toward the mining of the polymetallic nodules are as follows: (i) they occur in millions of tons in all the oceans of the world; (ii) they are sources of several metals (Mn, Fe, Ni, Cu, Co, Pb, Zn, etc. aggregating up to 40%); and (iii) they are loosely strewn on the seafloor and so are easy to mine (Sharma 2010) and can be alternative potential ore reserves of strategic importance to India. However, the exploitation and mining of nodules from the deep ocean floor at a depth of 5000–6000 m is a major technological challenge and the development of appropriate technology has been taken up by the Ministry of Earth Sciences in a phase-wise manner. It is pertinent to mention here that many of the advanced countries like the USA, France, Russia, Norway, Japan, China, etc. have developed a number of capabilities including underwater vehicles (ROV, AUV, manned submersibles, etc.) which are rated for water depths of even up to 6000 m. However, in the field of deep-sea mining, it is to be noted that most of the players in the field are still in the experimental stage only, except for the initial experiments and successful demonstration of mining of polymetallic nodules by a USA consortium and detailed review on the R&D efforts of the mining technologies from the 1970s until the 2000s by international consortia and national programs are summarized in Yamazaki and Brockett (2017).

Under the aegis of MoES, ESSO-NIOT is entrusted with the responsibility for the design and development of deep-sea mining system. The various underwater systems developed include the following:

- Underwater mining system for long-term operation using Dynamic Positioning System.
- Underwater collection and crushing system for manganese nodule mining.
- Soil tester for in-situ measurement of soil properties in the Central Indian Ocean Basin.

Based on the successful tests at 500 m depth and measuring the soil strength at the polymetallic nodule site, a deep water mining machine for 6000 m depth has been developed and tested up to 3000 m depth. The challenge of developing a flexible riser for pumping the nodules to the ship has been taken up and its work is in progress. The first exploratory mining at the site in the Indian Ocean is expected by the year 2024 (G A Ramadass, personal communication).

5 Hydrothermal Sulfides

Submarine hydrothermal activity is a common feature that occurs along the rift valleys and subduction zones (Fig. 1). The discovery of submarine hydrothermal vents at the Galápagos Rift in 1977 initiated the exploration of hydrothermal systems in global oceans which got intensified subsequently and is still continuing (Hannington et al. 2011). Modern seafloor hydrothermal systems are excellent natural laboratories for understanding the genesis of volcanic hosted massive sulfide deposits(VMS) and thus can be considered as modern analogs of ancient Volcanogenic Massive Sulfide (VMS) ore deposits, where evidence of their origin and nature are often obscured by millions of years of geological history.

The discovery of hydrothermal systems in the deep oceanic realm has kindled a lot of interest primarily on account of the high concentration of base metals (Cu, Pb, Zn) and many noble metals (Au, Ag, Pd, Pt) in them and thus known as polymetallic massive sulfides. Apart from their economic potential, active hydrothermal vent fields are also significant because of the presence of diverse types of living organisms around the vent sites, and their biological evolution in a sulfidic environment can be indicative of conditions during the evolution of early life on Earth. Seafloor hydrothermal activity associated with concomitant sulfide minerals and biological resources is a major research topic with vital scientific significance and economic considerations.

The basic paradigm of hydrothermal activity involves the sub-seafloor convective circulation of seawater through permeable rocks mainly driven by the upper mantle heat sources. The fractures, fissures, and cracks on the seafloor serve as conduits for the bottom cold seawater to penetrate which reaches depths of several kilometers below the seafloor surface and is heated up to temperatures of more than 400 °C during water–rock interaction above a magma chamber. In spreading centers, prevailing kinematics favor rigorous churning of water and downward propagation supplements the circulation of the hydrothermal fluid manifold. This heated seawater during its transit leaches out metals from the surrounding rock and gets modified to a hydrothermal fluid with low pH, low Eh, high temperature, and dissolved metals and sulfur (e.g., Von Damm 1990). The high–temperature water being less dense escapes through the vents on the seafloor as enriched plumes and rises in the water column until it attains buoyancy. Copper, zinc, and Iron sulfides are the first to be precipitated on the vent orifices thus forming hydrothermal vents on the seafloor. After attaining buoyancy, these plume fluids take a horizontal course depending on the hydrodynamic conditions prevailing in the region and in the transit process precipitate many of the metals into the surrounding bottom thus forming metal-rich sulfide deposits on the seafloor. The rest which remain dissolved in the water column are carried away by the currents over tens to hundreds of kilometers into the deep ocean (Resing et al. 2015).

The hot metal-rich fluids emanating out of hydrothermal vents support the sustenance of bacteria that use chemicals in the vent fluids to generate cellular energy. These chemosynthetic bacteria feed plentiful communities of strange invertebrates around the vent system thus making it a unique seascape in the sea-bottom which is otherwise devoid of any other living organisms. The survival of such strange communities on the seafloor has also gained significant insights into the extremes at which life can exist on Earth and potentially elsewhere in the Universe. Species inhabiting the deep-sea hydrothermal vents are strongly influenced by the geological settings in terms of the host rock and the deep-crustal/ mantle composition since the chemical/mineral-rich fluids are basically controlled by the geological condition (Mullineauex et al. 2018). Considering the unique invertebrate communities in the sea-bottom, some national governments, such as those of Canada, Portugal, Mexico, and the United States, have introduced marine parks to protect vent fields of particular scientific interest within their EEZ (Van Dover 2011).

Plume surveys are considered as the main tool for exploring active hydrothermal systems. However, they fail to locate older and hydrothermally extinct seafloor massive sulfide deposits due to their lack of distal water column signals. Hence, very less is known about the occurrence and distribution of extinct seafloor massive sulfide deposits, which would be buried under sediments or lava and located far from zones of active venting. These extinct seafloor massive sulfide deposits have gone through the complete hydrothermal cycle and reached their maximum size and hence likely to be more abundant and larger than the hydrothermally active massive sulfide deposit sites that are still forming. Currently, very little is known about the deposits that are moved away from the ridge axis by seafloor spreading.

5.1 Global Distribution

Polymetallic sulfide deposits are found on fast, intermediate, slow-spreading mid-ocean ridges, on axial and off-axis volcanoes, seamounts, in sedimented rifts associated to continental margins, and in subduction-related arc and back-arc environments (Fig. 4). Occurrences of seafloor massive sulfide deposits can be found at the plate boundaries, where there is a strong correlation between magmatism, seismicity, and high-temperature hydrothermal venting (Hannington et al. 2011).

The present-day quest for deep-sea exploration being on an increasing trend, the number of hydrothermal vent discoveries has also been steadily increasing, since the first discovery at the Galapagos Rift in 1977 (Corliss et al. 1979). The total number of vent sites that exist along the ridges and volcanic arcs is still not known and the inferences on their abundances are mostly based on the heat flow measured from plume sources (Peterson et al. 2016). An earlier estimate of the number of high-temperature active vent sites globally suggests that there are ~1300 sites of which about 1000 remain to be explored (Beaulieu et al. 2015). Most of the findings are mainly at active high-temperature hydrothermal vent fields that are restricted to the young and volcanically active parts of the ocean floor (Fig. 5).

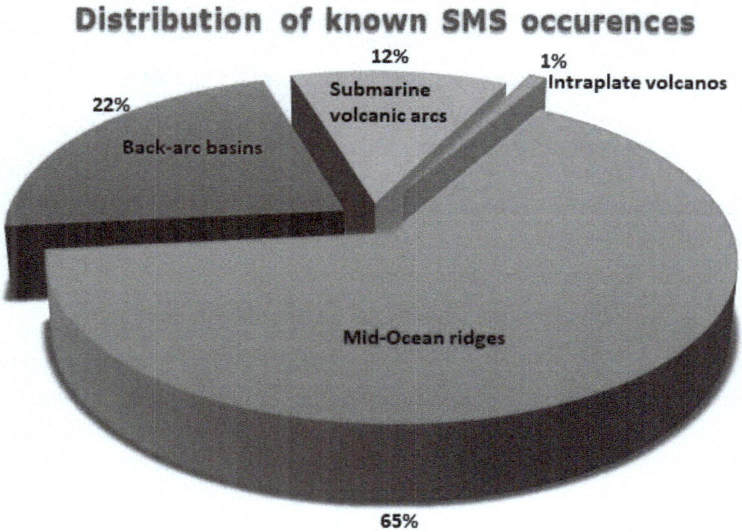

Fig. 4 Distribution of known seafloor massive sulfide occurrences in different geological domains (Hannigton et al. 2011)

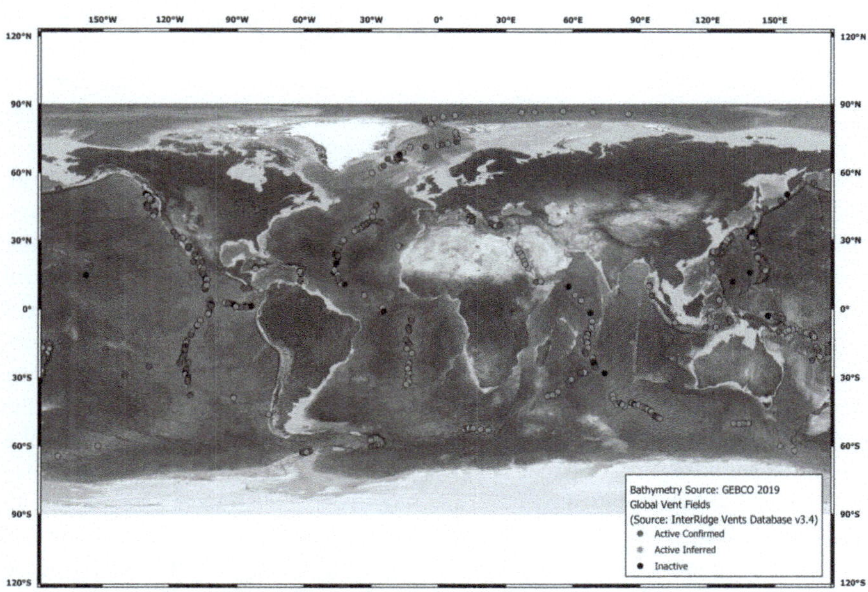

Fig. 5 Distribution of global vent fields (*Source* Interridge database v 3.4)

5.2 Indian Endeavor

Exploration for seafloor hydrothermal activity along the Indian Ocean ridges was initiated with the GEMINO (Geothermal Metallogenesis Indian Ocean), an Indo-German collaborative program way back in 1983 undertaken onboard by RV SONNE for the purpose of locating hydrothermal activity in the region in which Indian scientists have taken part. Investigations revealed the presence of hydrothermally influenced sediment cores, altered basalts, anomalous concentrations of total dissolved Mn and CH_4 (Herzig and Pluger 1988), and mixed hydrothermal–hydrogenous ferromanganese crusts (Nath et al. 1997). Followed by that, discoveries of more fossil and active hydrothermal systems along the Central Indian Ridge (CIR) and the South West Indian Ridge (SWIR) as well as further evidence of hydrothermal plume and mineralization signatures in this region suggest the possibility of many hydrothermal sulfide mineralization zones in the region. This prompted India's aspirations to undertake a comprehensive research and exploratory program, which not only aims at the identification of new locals of hydrothermal activity, but also can address the complex physical–chemical–biological–geological scenario of the hydrothermal systems and thus entailing the genesis, environment, and distribution of hydrothermal plumes and mineralization in the region.

Indian endeavors started on a small scale in years 1985 and 1986, but the major efforts were initiated in the year 2002 with the high-resolution mapping of the Carlsberg Ridge (CR) and the northern Central Indian Ridge (CIR) segments by CSIR-NIO. Autonomous Underwater Vehicle (AUV) investigations carried out over a segment of the CIR provided high-resolution bathymetry and near seabed water column characteristics. Detailed multidisciplinary investigations carried out eventually led to the discovery of few hydrothermal plume signatures in the region (Ray et al. 2012, 2020). Further, in 2011, the Ministry of Earth Sciences entrusted the responsibility of leading the hydrothermal exploration program to the ESSO-NCPOR. After conducting preliminary surveys and exploratory works in 2012–13 in parts of Central and South West Indian ridges, close to the Rodriguez Triple Junction, and followed by the analysis and interpretation of the data/samples acquired, probable and potential zones of hydrothermal activity have been identified in the region. The promising outcome of the studies paved the way for the submission of India's application to the International Seabed Authority (ISA) in May 2013 for a grant of license for initiating exploration activities for an area of 10,000 km^2. In July 2014., ISA granted approval for the plan of work submitted by India and further India entered an agreement with ISA for undertaking 15 years of exploration work in the region for identifying and developing mining sites in the region. The major milestones achieved so far in the PMS program are summarized in Table 3.

Pursuant to the grant of exploration license, NCPOR undertook an extensive exploration campaign in the region engaging the latest techniques and technologies. Systematic water column studies using CTD and MAPR (Miniature Autonomous Plume Recorder) were undertaken in the region, thus locating many plume signatures. The hydrothermal origin of these plumes was further corroborated from the

Table 3 Major Milestones achieved by India in polymetallic sulfide exploration

Period	Milestones
July 2014	India's application for grant of license for exploration of polymetallic sulphides in parts of CIR and SWIR approved in General Assembly of ISA
September 2016	India Signed 15 year of contract with ISA for exploration of polymetallic sulphides in an area of 10,000 km^2 in parts of CIR and SWIR
January 2017	Initiated the exploration activity in the contract area by undertaking comprehensive exploration cruises in the area
May 2020	India discovered first hydrothermal mineralisation site in the contract area in CIR, apart from identifying more than a dozen number of potential zones
Present status	Further exploration activity together with generation of baseline environmental data is under progress

anomalies of Fe, Mn, methane, and helium isotopes in the water column. Morpho-tectonic studies in the region could infer many potential zones, wherein dredging and sediment sampling were carried out, the analysis of which provided further evidence of mineralization at a few places. Integrated bathymetric and geological studies could infer the identification of new Oceanic Core Complexes (OCCs), which are considered as favorable sites for hydrothermal activity.

Having located many potential zones of activity in the form of plumes, sediment mineralization, etc. the main focus is now to locate the active and inactive zones of mineralization. Modern surveys primarily utilizing autonomous Underwater Vehicles (AUV) and remotely Operated Vehicles (ROV) will be undertaken further for the purpose. The efficiency of the modern approach utilizing AUV has been demonstrated well in the case of Okinawa Trough (off Japan) where AUV surveying over the past four years has doubled the number of known active vent sites from 11 (found between 1988 and 2013) to 23 in 2017 (Peterson et al. 2018). As per the present plans, India is planning to deploy a combination of AUV and ROV mounted with multibeam bathymetry, SBP, SAS, SP/iSP, magnetometer, video/photography, CTD, and sampling gadgets.

6 Cobalt-Rich Crusts

Cobalt-Rich Crusts (CC) are important metal resources formed on the seafloor by the precipitation of dissolved and colloidal components from ambient seawater onto the indurated substrate in the deep ocean. The crusts are also commonly called Fe–Mn crusts, because their major constituents are iron (Fe) and manganese (Mn), although a host of other minerals occur in them in variable amounts, including cobalt, and hence also often called cobalt-rich crusts or cobalt-rich Fe–Mn crusts. The cobalt-rich crusts may be the slowest growing of any earth material, accumulating one molecular layer every 1 to 3 months or one to seven millimeters per million years (Hein et al. 2000). The slow growth rates together with the enormous high surface area in these Fe–Mn crusts (mean 325 m^2 per gram of crust) and high porosity (mean 60%) stimulate the adsorption of many rare metals, which cannot be found in such high concentrations in any other geological–oceanographic setting (Hein et al. 2010). The thickness of the Fe–Mn crusts varies usually from <1 mm to about 260 mm.

Fe–Mn crusts in the open oceans are more enriched in cobalt than any other widely distributed sediments or rocks. The key requirements for Fe–Mn crusts formation are (a) availability of sediment-free hard substrate, (b) oxic ambient water, and (c) sufficient supply of oxidizable Mn. Such conditions are generally found on seamounts rising into the Oxygen Minimum Zone (OMZ), where a large supply of dissolved Mn can be expected due to intense oxidation of the organic matter.

The mean cobalt concentrations in Fe–Mn crusts are three to ten-fold higher than those in land-based deposits. Tellurium also has significant enrichment compared to both seawater and Earth's crust and has a mean global concentration of about 50 ppm in Fe–Mn crusts and a maximum value of 206 ppm (Hein et al. 2010). The concentration of metals other than Fe and Mn in the crusts depends on the concentration of metals in seawater (the source), the Fe/Mn ratio of colloids in seawater, and the surface charge of the Fe–Mn colloids (Koschinsky and Hein 2003).

Cobalt-rich Fe–Mn crusts have gained recognition as potential future resources for a wide variety of elements such as Co, Ti, Mn, Ni, Pt, Zr, Nb, Te, Bi, Mo, W, Th, and rare-earth elements (REEs) essential for emerging high- and green-technology applications and having strategic importance. About one-fourth to one-half of cobalt consumption is used by the aerospace industry in making superalloys. These metals are also employed in chemical and high-technology industries, for products such as photovoltaic and solar cells, superconductors, advanced laser systems, catalysts, fuel cells, and powerful magnets, as well as for cutting tools. Hence, cobalt-rich Fe–Mn crusts constitute significant and important future mineral resources for the mankind.

6.1 Global Distribution

Oxidized deposits of cobalt-rich Fe–Mn crusts are found throughout the global oceans on the flanks and summits of seamounts (submarine mountains), ridges, and plateaus, where seafloor currents sweep the ocean floor and clear off the sediment for millions of years. During volcanically inactive periods, seamount surfaces come in contact with enormous quantities of ocean water that flows past and forms Fe–Mn crusts (Hein 2008; Staudigel and Clague 2010) Global estimates indicate there are more than 33,400 seamounts rising 1000 m or more from the seafloor, and more than 138,000 smaller features known as knolls (rising 500–1,000 m) and hills (rising < 500 m) (Pitcher 2007).

Fe–Mn crusts vary in metal concentrations on global, regional, local, and intracrust scales. The region where Fe–Mn crusts show the highest metal content is the central and western equatorial Pacific (Fig. 6), hence called the Pacific Prime Crust Zone (PPCZ). However, thorium is higher in the Atlantic and Indian Ocean crusts. Yesson et al. (2011) estimated that seamounts occupy an area of ~17.2 million km^2 or 4.7%

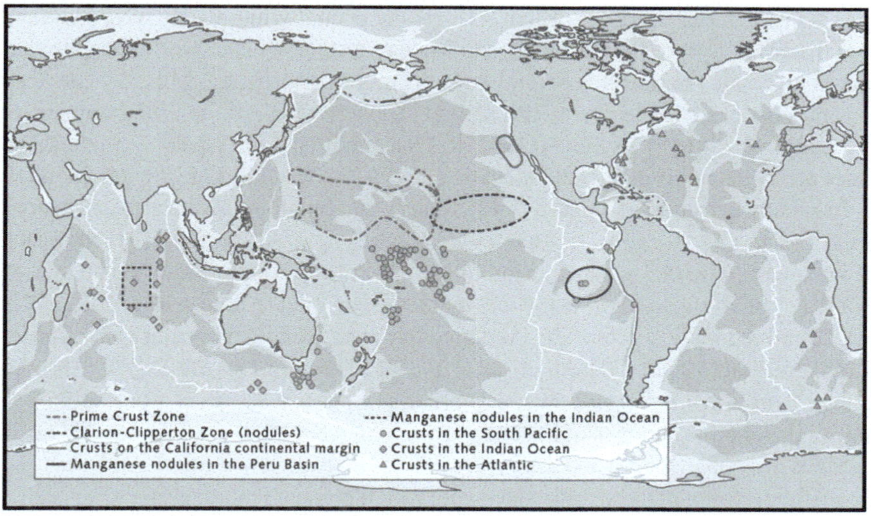

Fig. 6 Cobalt crust occurrence in different ocean regions (after Hein et al. 2010)

of the global ocean floor. Only a few of the estimated seamounts that occur in the Pacific have been mapped and sampled in detail. The Atlantic and Indian Oceans contain fewer seamounts but have been sampled far less.

6.2 Indian Endeavor

Presently, considerable progress has been made on the cobalt-rich Fe–Mn crusts exploration in India. However, quantitative resource evaluation for Fe–Mn crusts is not yet available due to limited technology. The first recovery of cobalt-enriched Fe–Mn Crusts was made in 1994 from the Afanasiy-Nikitin Seamount (ANS) onboard by RV Sidorenko (Banakar et al. 1997). A detailed multibeam bathymetric survey in the ANS region revealed clusters of seamounts at a water depth ranging between 1.7–3 km, with Fe–Mn crusts, the recovery, and analysis of which reveal Co metal up to 0.9%, Pt up to 1 ppm, and Ce up to 0.22% (Banakar et al. 2007; Rajani et al. 2005). The National Institute of Oceanography, under the aegis of MoES, launched a major project to explore seamounts in the equatorial and northern Indian Ocean for Co-enriched Fe–Mn crusts. The National Geophysical Research Institute (NGRI) and National Centre for Polar and Ocean Research (NCPOR) were other partners in this program. The high Ce concentrations in the ANS appear to be the highest Ce-enrichment known so far in Fe–Mn crusts. Subsequently the seamounts in Laxmi Basin (western EEZ of India) viz. Raman Seamount, Panikker Seamount, and Wadia guyot were mapped by the multibeam (Bhattacharrya et al. 1994), and the occurrence of Fe–Mn crusts was expected because of very strong Oxygen Minimum

Zone (OMZ) in the overlying water. Also, some encouraging results were obtained from the seamounts of the Lakshadweep Sea and further detailed work is required in order to estimate the resource potential for the metal deposits. Fe–Mn crusts are also known to occur in the deeper areas (>3,000 m) on the flank of the Carlsberg Ridge. The Geological Survey of India (GSI 1992) has reported the recovery of manganese crusts off Tillanchard in the Andaman Sea, and from the slopes of Chetam Island in the Lakshadweep group (Chandra 1992). Many other seamounts were also discovered in the Indian Ocean region (Pranab et al. 2007; Iyer et al. 2012, 2018) and in the Andaman Sea viz. the cratered seamount (CSM) in the Nicobar earthquake swarm area (Kamesh Raju et al. 2008) and the Southern seamount (Kamesh Raju et al. 2004) located immediately south of the Andaman back-arc spreading center (ABSC) and in Indian western EEZ (identified under the EEZ program of MoES), which may be interesting sites for cobalt-crust exploration. However, the work done so far has indicated workable deposits of Fe–Mn crusts in the ANS region. As the price of Co is ten times more than the Cu and the Pt is two times more expensive than gold (Au), the cobalt-rich ferromanganese crusts deposits which contain tentatively ~0.6% Co and ~0.5 ppm of Pt would yield considerable revenues and will be alternative potential ore reserves of strategic importance to India, making the country more independent in strategic mineral resources. Further, the enrichment of Pt (Banakar et al. 2007) and Te (Hein et al. 2003) in seamount crusts may add additional value to these deposits. Since India has no known resources of cobalt and tellurium, the discovery of potential deposits of these metals in the Indian EEZ and in the International waters would be significant.

7 Environmental and Ecological Concerns

The environmental and ecological risks associated with the exploration and exploitation of deep-sea mineral resources are to be addressed to mitigate any potential impact on the fragile ecosystem associated with many of these deposits. It is important that before any exploration or extraction of deep-sea minerals, a clear defining environmental objective that will guide the management of any exploration or mining operation should be prepared. The main objectives of the environmental guidelines are to maintain the overall biodiversity and ecosystem health and function as mandated by international law and to reduce, mitigate, and, wherever possible, prevent the impacts of mining and pollution that can affect wider habitats and ecosystems.

Depending on the deposit type and the extraction methodology, the impacts could potentially be severe and long-lasting, and therefore, an environmental risk assessment based on in-situ experiments at different scales is essential (Ahnert and Borowski 2000). The deposits differ in mineralogy, metal composition, surface expression, morphology, and spatial extent, resulting in different ecosystem structures and functions and different disturbance risks (Jones Daniel et al. 2018). Also, the area and volume of mining activity would also be different since mining for manganese nodules would cover several hundreds of sq.km, while for cobalt crusts

Table 4 Principal environmental impacts and advantages of seabed mining (Lodge and Verlaan 2018)

Environmental Impacts of Seabed Mining	Environmental Advantages of Seabed Mining (Contrasted with Terrestrial Mining)
• Permanent removal of hard substrate required by certain organisms/faunal communities • Effects of sediments, wastes, and other effluents (at bottom, mid-water and surface) • Noise, Vibration, Light • Leaks, spills, the effects of infrastructure corrosion • Operational discharges from the surface vessels[a] • Slow and different biological regeneration (especially of sessile communities) • Uncertain remediation potential • Vessel traffic for ore transport to shore[a] for land-based processing • Vessel-source air pollution[a] • Surface and mid-water marine community disturbance, especially if mining vessels remain on location for many months	• Little or no overburden to remove (e.g. overlying rock, soil, vegetation cover) • Ore grades can be significantly higher than on land, meaning that less ore is required to provide the same amount of metal • Multiple metals can be obtained from a single site due to polymetallic nature of deposits • No local human populations to be disrupted • No permanent infrastructure

[a]*Governed by International Maritime Organisation treaties and regulations*

would be several tens of sq. km and while for sulfides would be a few or even less than a sq.km. Nevertheless strong international regulations must be carefully considered to ensure fair use and minimal environmental impact.

Since the viability of any deep-sea mining in the future largely depends on its impact on the environmental and ecological factors, the potential stakeholders are actively engaged in fully assessing the situation. An assessment of various potential environmental impacts of seabed mining and the environmental advantages while compared to land mining is presented in Table 4.

8 Exploitation and Mining

Deep-sea mining presents complex regulatory challenges due to its multi-faceted political, economic, technological, scientific, environmental, social, industrial, and legal aspects. While examining the progress trend in the exploitation and mining front for the three mineral resources, it is noticed that sulfides might be the first seabed mineral which is likely to be mined in the near future, considering the significant progress made due to the commercial interests in Papua New Guinea and by strong domestic interests in Japan (Peterson et al. 2016). However, Nautilus Minerals, the company which is supposed to undertake the Papua New Guinea sulfide mining, run into troubled waters due to multiple issues including economic terms (Sparenberg 2019). Japan state agencies also conducted the sulfide mining test within the Japanese EEZ in 2017, which was rated as successful. However, the report on the assessment mentioned a list of the challenges, including finding highly valuable deposits, establishing efficient production technologies, and a higher level of metal prices (Sparenberg 2019).

Considering the huge volume of resources available, nodule mining is expected to have a more substantial impact on the seabed mining industry and global metal markets. Even though exploration, environmental impact studies, and resource assessment are in a very advanced stage for nodules, the development of an advanced mining system is yet to take place probably due to many challenges viz. the technology constraints, the limiting depth ratings of deep-sea equipments/sensors, economic beneficial aspects, etc. The Belgian company Global Sea Mineral Resources (GSR) is planning to deploy a tracked collector vehicle in the Pacific to validate the technology and assess the environmental impact. GSR is pushing for a mining code by 2020 and says that it needs these regulations in place to launch full-scale mining by 2026 (Heffernan 2019). As mentioned in the earlier section, India is also in an advanced stage of development for undertaking pilot mining and actual mining operations in the future in the deep seabed once the mining exploitation code is approved by the ISA.

In comparison to polymetallic sulfide and polymetallic nodules, the exploration for Co-rich ferromanganese crusts is still in its infancy and the technologies to assess and recover crusts have not been built or tested (Peterson et al. 2016). Mining of the crusts is considered as most technically challenging among the deep ocean mineral deposits because they are firmly attached to often steep and uneven rock surfaces and hence become difficult to separate them from the hard substrata.

9 Discussion and Conclusion

Depleting land resources are forcing the mankind to look into the immeasurable treasures of seabed for fulfilling its insatiable desires for various kinds of natural resources, including metals and minerals. The growing demand for advanced and green technologies and standard of living in daily life is fueling a steady increase in the need for many metals and minerals. Most of these metals have strategic and many irreplaceable industrial applications. Realizing the need to be cognizant for the optimum utilization of marine mineral resources, its effective and economic mining, and giving due considerations in protecting the associated fragile marine environment, a large group of intellectuals under the International Seabed Authority is deliberating and designing appropriate regulatory aspects thus to achieve commercially viable and socially responsible deep-sea mining.

Apart from providing inputs for industrial sectors, mining to end-product and also its allied trade are also vital for sustaining population well-being, which eventually forms a key factor for the country's social and economic development. Contribution of minerals and metals are fuelling the manufacturing sector and also creating jobs and value-added products along the supply chains of material goods.

India, a fast-developing country having ambitious, sustainable, and development goals dwells on mineral resources to a large extent. The long coastline of India and its proximity to the vast oceanic domains, as well as the long marine scientific history of the country, demand venturing into the oceanic domain for exploring and utilizing the huge treasures of mineral resources in the deep-sea domains.

Exploring new mineral deposits generally depends on their utility, cost, and availability on the land. India does not possess much land resources for a few of the key metals such as nickel and cobalt, and resources for a few other metals like copper are diminishing. Also, in recent years, certain metals have been designated as "critical", mainly due to their economic importance and the possibility of a supply shortage due to uneven resource distribution and geopolitics (Lusty and Murton 2018). These factors together with the increasing challenges of land-based mining (Calas 2017) are encouraging the search for alternative sources of minerals and metals. The scenario compiles the country to advance its efforts in exploring the alternative available resources in the marine domain.

MoES is spearheading the deep-sea mineral exploration program of India, with the active participation of many national institutions, organizations, and universities, including ESSO-NCPOR, ESSO-NIOT, CSIR-NIO, CSIR-IMMT, CSIR-NGRI, Delhi University, etc. So far India entered into exploration contracts with ISA for PMN in the Central Indian Ocean basin and PMS in CIR/SWIR. Extensive activities of exploration, identification, and development of suitable mining sites in the contract area are in progress. MoES is also in the process of applying for another exploration license for CC. The technology development for effective and efficient mining of these deep-sea mineral resources and the development of ore processing and extraction technology are also in progress. All the activities are being undertaken following strict compliance with the environmental regulation protocols specified by ISA. Apart from any geopolitical interests, the continuous and dedicated efforts in the exploration and exploitation of marine mineral resources by India will definitely strengthen the long-term development of the country on social, economic, and societal fronts.

Acknowledgements We gratefully acknowledge the encouragement and support of Director, NCPOR for the compilation of this article. Also would like to thank the Ministry of Earth Science (MoES) and the Hydrothermal team at NCPOR for providing all requisite help in the compilation of the information. This is NCPOR contribution #.

References

Agassiz A (1903) Three cruises of the United States Coast and Geodetic Survey steamer "Blake" in the Gulf of Mexico, in the Caribbean Sea, and along the Atlantic coast of the United States, from 1877 to 1880. Vol II. Bull Museum Comp Zoology 15:1–220

Ahnert A, Borowski C (2000) Environmental risk assessment of anthropogenic activity in the deep-sea. J Aquat Ecosyst Stress Recover 7:299–315

Agassiz Alexander (1906) General report of the expedition. Memoirs of the museum of comparative zoology, vol. XXXIII. Cambridge, USA

Banakar VK, Pattan JN, Mudholkar AV (1997) Palaeoceanographic conditions during the formation of ferromanganese crusts on the Afanasiy-Nikitin seamount, North central Indian Ocean: geochemical evidences. Mar Geol 136:299–315

Banakar VK, Hein JR, Rajani RP, Chodankar AR (2007) Platinum group elements and gold in ferromanganese crusts from Afanasiy-Nikitin seamount, equatorial Indian Ocean: Sources and fractionation. J Earth Syst Sci 116:3–13

Beaulieu SE, Baker ET, German CR (2015) Where are the undiscovered hydrothermal vents on oceanic spreading ridges? Deep Sea Res II Topical Stud Oceanogr 121:202–212

Bhattacharya GC, Chaubey AK, Murty GPS, Srinivas K, Sarma KVLNS, Subrahmanyam V, Krishna KS (1994) Evidence for seafloor spreading in the LaxmiBasin, northeastern Arabian Sea. Earth Planet Sci Lett 125:211–220

Bonatti E, Joensuu O (1966) Deep Sea Iron Deposits from the South Pacific. Science 154:643–645

Bostrom K, Peterson MNA (1966) Precipitates from hydrothermal exhalations on the East Pacific Rise. Econ Geol 61:1258–1265

Calas G (2017) Mineral resources and sustainable development. Elements 13:301–306

Chandra PR (1992) Records, geological survey of India, p 125

Corliss JB, Dymond J, Gordon LI, Edmond JM, von Herzen RP, Ballard RD, Green K, Williams D, Bainbridge A, Crane K, van Andel TH (1979) Submarine thermal springs on the Galápagos Rift. Science 203:1073–1083

Cronan DS (1980) Underwater minerals. Academic Press, London

Cronan DS, Moorby SA (1981) Manganese nodules and other ferromanganese oxide deposits from the Indian Ocean. J Geol Soc London 138:527–539

Francheteau J, 14 coauthors (1979) Massive deep-sea sulphide ore deposits discovered on the East Pacific Rise. Nature 277:523–528

Frazer JZ, Fisk MB (1981) Geological factors related to characteristics of sea-floor manganese nodule deposits. Deep-Sea Res 28(A):1533–1551

Frazer JZ, Wilson LL (1980) Manganese nodule resources in the Indian Ocean. Marine Mineralogy 2:257–292

Glasby GP (1977) Marine manganese deposits. Elsevier Oceanogr Ser 15:1–9. https://doi.org/10.1016/S0422-9894(08)71015-8

Glasby GP, Stoffers P, Sioulas A, Thijssen T, Friedrich G (1982) Manganese nodule formation in the Pacific Ocean: a general theory. Geo-Mar Lett 2:47–53

GSI (1992) Records, geological survey of India, . 125

Halbach P, Nakamura K, Wahsner M, Lange J, Sakai H, Kaselitz L, Hansen RD, Yamano M, Post J, Prause B, Seifert R, Michaelis W, Teichmann F, Kinoshita M, Marten A, Ishibashi J, Czerwinski S, Blum N (1989) Probable modem analogue of Kuroko-type massive sulphide deposits in the Okinawa Trough back-arc basin. Nature 338:496–499

Hannington M, Jamieson J, Monecke T, Petersen S, Beaulieu S (2011) The abundance of seafloor massive sulfi de deposits. Geol Soc Amer 39:1155–1158

Heffernan O (2019) Scientists track damage from controversial deep-sea mining method. Nature 567:294

Hein JR, Yeh HW, Alexander E (1979) Origin of iron-rich montmorillonite from the manganese nodule belt of the north equatorial Pacific. Clay Mineralogy 27:185194

Hein JR, Koschinsky A, Halliday AN (2003) Global occurrence of tellurium-rich ferromanganese oxyhydroxide crusts and a model for the enrichment of tellurium. Geochimica Cosmochimica Acta 67(117–1):127

Hein JR, Mizell K, Koschinsky A, Conrad TA (2013) Deep-ocean mineral deposits as a source of critical metals for high- and green-technology applications: comparison with land-based resources ore geology review 51:1–14

Hein JR, Koschinsky A (2014) Deep-ocean ferromanganese crusts and nodules (Ed Scott S.), 13, chapter 11, Treatise on Geochemistry

Hein JR, Peterson S (2014) The geology of manganese nodules (Eds: Elaine Baker and Yannick Beaudoin) 1 B pp 7–13 Secretariat of the Pacific Community

Hein JR, Trace Y, Conrad A, Hubert S (2010) Seamount mineral deposits a source of rare metals for high-technology industries. Oceanography 23:184–189

Hein JR, Koschinsky A, Halbach P, Manheim FT, Bau M, Jung-Keuk K, Lubick N (1997) Iron and manganese oxide mineralization in the Pacific. (Eds: Nicholson K, Hein JR, Bfihn B, Dasgupta S) pp 123–138. Manganese mineralization: geochemistry and mineralogy of terrestrial and marine deposits, Geological Society Special Publication No. 119

Hein JR, Koschinsky A, Bau M, Manheim FT, Kang J-K, Roberts L (2000) Cobalt-rich ferromanganese crusts in the Pacific (Eds: Cronan DS) Handbook of marine mineral deposits, pp 239–279. Boca Raton, FL: CRC Press

Hein JR (2008) Geologic characteristics and geographic distribution of potential cobalt-rich ferromanganese crusts deposits in the area, pp 59–90. Mining Cobalt-rich ferromanganese crusts and polymetallic Sulphide Deposits: Technological and economic considerations. Proceedings of the International Seabed Authority

Herzig PM, Plüger WL (1988) Exploration for hydrothermal activity near the Rodriguez Triple Junction. Indian Ocean. Canad Mineralogist 26:721–736

IFREMER (2017). https://wwz.ifremer.fr/en/Public-policy-support/Raw-materials-and-resources/Contracts-metallic-mineral-resources-in-international-waters

Interridge database v 3.4. https://vents-data.interridge.org/

Iyer SD, Das P, Kalangutkar NG (2012) Seamounts—windows of opportunities and the Indian scenario. Curr Sci 102(10):1382–1391

Iyer SD, Amonkar AA, Das P (2018) Genesis of Central Indian Ocean basin seamounts: morphological, petrological, and geochemical evidence. Int J Earth Sci 107:2517–2538. https://doi.org/10.1007/s00531-018-1612-z

Jones Daniel OB, Ardron, Jeff A, Colaço Ana, Durden Jennifer M (2018) Environmental considerations for impact and preservation reference zones for deep-sea polymetallic nodule mining Marine Pollution. https://doi.org/10.1016/j.marpol.2018.10.025 (In Press)

ISA report (2001) standardization of environmental data and information: development of guidelines. Proceedings of the international seabed authority's workshop, Kingston, Jamaica, June 25–29, 2001

Kamesh Raju, Chaubey AK, Amarnath D, Mudholkar A (2008) Morphotectonics of the Carlsberg ridge between 62 20′ and 66 20′ E, northwest Indian Ocean. Marine Geol 252:120–128

Koschinsky A, Hein JR (2003) Uptake of elements from seawater by ferromanganese crusts: solid-phase associations and seawater speciation. Mar Geol 198(3–4):331–351

Lodge MW, Verlaan PA (2018) Deep-Sea mining: international regulatory challenges and responses. Elements 14:331–336

Lusty PAJ, Murton BJ (2018) Deep-ocean mineral deposits: metal resources and windows into earth processes. Elements 14:301–306

Martin-Barajas A, Lallier-Verges E, Leclaire L (1991) Characteristics of manganese nodules from the Central Indian Basin: Relationship with the sedimentary environment. Marine Geol 101:249–265

Mero JL (1965) The mineral resources of the sea. Elsevier, Amsterdam, p 312

Mullineaux LS, Metaxas A, Beaulieu SE, Bright M, Gollner S, Grupe BM, Herrera S, Kellner JB, Levin LA, Mitarai S, Neubert MG, Thurnherr AM, Tunnicliffe V, Watanabe HK and Won Y-J

(2018) Exploring the ecology of deep-sea hydrothermal vents in a Metacommunity framework. Front Marine Sci.https://doi.org/10.3389/fmars.2018.00049

Murray J, Renard AF (1891) Deep-sea deposits (based on the specimens collected during the voyage of HMS Challenger in the years 1872 to 1876). Report on the scientific results of the voyage of H.M.S. Challenger during the years 1873–76; John Menzies and Co., Endinburgh, United Kingdom. http://www.vliz.be/nl/open-marien-archief?module=ref&refid=41584, 688 pp hdl:10013/epic.45942.d002

Murrey J, Lee JV (1909)The depth and marine deposits of pacific. Memoirs of the museum of comparative zoology, vol. XXXVIII. Cambridge, USA

Muthunayagam AE, Das SK (1999) Proceedings of the third ISOPE Ocean mining symposium, Goa, India, November 8–10

Nath BN, Bau M, Ramalingeswara Rao B, Rao ChM (1997) Trace and rare earth elemental variation in Arabian Sea sediments through a transect across the oxygen minimum zone. Geochim Cosmochim Acta 61(12):2375–2388

Nath BN, Khadge NH, Nabar S, RaghuKumar C, Ingole BS, Valsangkar AB, Sharma R, Srinivas K (2012) Monitoring the sedimentary carbon in an artificially disturbed deep-sea sedimentary environment 184:2829–2844

Petersen S, Krätschell A, Augustin N, Jamieson J, Hein JR, Hannington MD (2016) News from the seabed—Geological characteristics and resource potential of deep-sea mineral resources. Marine Pollut 70:175–187

Petersen S, Lehrmann B, Murton BJ (2018) Modern seafloor hydrothermal systems: new perspectives on ancient ore-forming processes. Elements 14:307–312

Pitcher TJ (2007) Preface. In: Pitcher TJ, Morato T, Hart PJB, Clark M, Haggan M, Santos RS (eds) Seamounts: ecology, fisheries, and conservation, vol 17 Oxford: Blackwell Publishing

Pranab D, Iyer SD, Kodagali VN (2007) Morphological characteristics and emplacement mechanism of the seamounts in the Central Indian Ocean Basin. Tectonophysics 443(1–2):1–18

Rajani RP, Banakar VK, Parthiban G, Mudholkar AV (2005) Compositional variation and genesis of ferromanganese crusts of the Afanasiy-Nikitin Seamount, Equatorial Indian Ocean. J Earth Syst Sci 114:51–61

Raju K, Ramprasad T, Rao PS, Rao BR, Varghese J (2004) New insights into the tectonic evolution of the Andaman basin, northeast Indian Ocean. Earth Planet Sci Lett 221:145–162

Rao VP, Nath BN (1988) Nature, distribution and origin of clay minerals in grain size fractions of sediments from manganese nodule field, Central Indian Ocean Basin. Indian J Marine Sci 17:202–207

Ray D, KameshRaju KA, Baker ET, Srinivas Rao A et al (2012) Hydrothermal plumes over the Carlsberg Ridge. Indian Ocean. Geochem Geophys Geosyst 13:Q01009. https://doi.org/10.1029/2011GC003888

Ray D, KameshRaju KA, Srinivas Rao A, SuryaPrakash L, Mudholkar AV, Yatheesh V, Samudrala K, Kota D (2020) Elevated turbidity and dissolved manganese in deep water column near 10°47’S Central Indian Ridge: studies on hydrothermal activities. Geo-Marine Lett 40:619–628

Resing J, Sedwick P, German C, Jenkins WJ, Moffett JW, Sohst BM, Tagliabue A (2015) Basin-scale transport of hydrothermal dissolved metals across the South Pacific Ocean. Nature 523:200–203

Revelle R (1944) Marine bottom samples collected in the Pacific Ocean by the Carnegie on its seventh cruise. Carnegie Inst. Washington Publ., 556, pp 1–180

Rona PA (2008) The changing vision of marine minerals. Ore Geol Rev 33:618–666

Sharma R (2010) First nodule to first mine-site: development of deep-sea mineral resources from the Indian Ocean. Curr Sci 99(6):750–759

Sharma R, Nath BN (2000) Selection of test and reference areas for the Indian: Deep sea Environment Experiment (INDEX). Mar Georesour Geotechnol 18:177–187

Sharma R, Nath BN, Jaisankar S (2005) Monitoring the impact of simulated deep-sea mining in central Indian Basin. Mar Georesour Geotechnol 23(4):339–356

Sharma R, Nath BN (1997) Benthic disturbance and monitoring experiment in the Central Indian Ocean Basin Paper presented at the second ISOPE ocean mining symposium, Seoul, Korea, November 1997.Paper Number: ISOPE-M-97–025. Published, November 24, 1997

Sharma, Nath RBN, Valsangkar AB, Parthiban G, Sivakolundu KM, Walker G (2000) Benthic disturbance and impact experiments in the central Indian Ocean Basin. Marine Georesources Geotechnol 18 (3):209–221

ShyamPrasad M (2005) Exploration for nodules in the Central Indian Ocean Basin: Past present and the future. In: Anand S, Sanjay K (eds) Proceedings of national seminar on polymetallic nodules (PMN 2005), RRL, Bhubaneswar, pp 22–25

Siddiquie HN, Das Gupta DR, Sen Gupta NR, Shrivastava PC, Mallik TK (1978) Manganese-Iron nodules from the Indian Ocean. Ind J Mar Sci 7:239–253

Skornyakova NS, Andrushchenko PF, Fomina LS (1964) Chemical composition of the Pacific ocean's iron-manganese concretions. Deep Sea Res Ocean. Abstracts 11:93–104. https://doi.org/10.1016/0011-7471(64)91086-1

Sparenberg O (2019) A historical perspective on deep-sea mining for manganese nodules, 1965–2019. Extractive Ind Soc 6:842–854

Staudigel H, Clague DA (2010) The geological history of deep-sea volcanoes: biosphere, hydrosphere, and lithosphere interactions. Oceanography 23:58–71

Thijssen T, Glasby GP, SchmitzWiechowski A, Friedrich G, Kunzendorf H, Müller D, Richter H (1981) Reconnaissance survey of manganese nodules from the northern sector of the Peru Basin. Mar Minealogy 2:385–428

Usui A, Someya M (1997) Distribution and composition of marine hydrogenetic and hydrothermal manganese deposits in the northwest Pacific. (Eds: Nicholson K, Hein JR, Bu¨hn B, Dasgupta S). Manganese mineralization: geochemistry and mineralogy of terrestrial and marine deposits, Geological Society London, p 119

Van Dover CL (2011) Tighten regulations on deep-sea mining. Nature 470:31–33

Von Damm KL (1990) Seafloor hydrothermal activity: black smoker chemistry and chimneys. Annu Rev Earth Planet Sci 18(1):173–204

Yamazaki T, Brockett FH (2017) History of deep-ocean mining, encyclopedia of maritime and offshore engineering, 1–9. https://doi.org/10.1002/9781118476406.emoe067

Yesson C, Clark MR, Taylor ML, Rogers AD (2011) The global distribution of seamounts based on 30 arc seconds bathymetry data. Deep Sea Res. Part I Oceanogr Res 58:442–453

Tsunami Early Warning Services

T. Srinivasa Kumar, E. Pattabhi Rama Rao, Ch. Patanjali Kumar, Sunanda Manneela, B. Ajay Kumar, Dipankar Saikia, R. S. Mahendra, P. L. N. Murty, and J. Padmanabham

Abstract India's coastline stretches about 7516 km and about 15% of the Indian population lives around the coast. Coastal population, ecosystem and infrastructure in the coastal zone are becoming increasingly vulnerable to natural disasters such as tsunamis and storm surges. These events not only destroy the life and ecosystem along the coast, but they also affect many economic sectors like agriculture, housing, tourism, nuclear and conventional power plants and coastal transportation network including ports and harbours. Major Tsunamis are rare events, but they cause substantial loss of lives and property when they happen. They are transborder hazards, and the communities especially those living in low-lying coastal regions across the globe are the most vulnerable to tsunamis. Large tsunamis occur with relatively low frequency but have potentially high impact. The tsunami on 26 December 2004 has been one of the worst in the world and the deadliest of all time by an order of magnitude. It emphasized the need for setting up of an early warning system for tsunamis in India. The Indian Tsunami Early Warning Centre (ITEWC) was established at the Indian National Centre for Ocean Information Services (INCOIS), Hyderabad, Ministry of Earth Sciences, in October 2007 to provide early warnings for tsunamis caused by earthquakes in the Indian Ocean. In this paper, we describe various components of Indian Tsunami Early Warning System (ITEWS), recent developments, including capacity-building initiatives for community awareness and preparedness to mitigate the tsunami hazard. We also discuss issues and challenges in the tsunami early warning system and briefly the economic benefits.

Keywords Oceanic hazards · Tsunami · Early warning · Mitigation · Disaster preparedness

---

T. Srinivasa Kumar · E. Pattabhi Rama Rao (✉) · Ch. Patanjali Kumar · S. Manneela ·
B. Ajay Kumar · D. Saikia · R. S. Mahendra · P. L. N. Murty · J. Padmanabham
Indian National Centre for Ocean Information and Services (INCOIS), Ministry of Earth Sciences (MoES), Hyderabad 500090, India
e-mail: pattabhi@incois.gov.in

V. K. Gahalaut and M. Rajeevan (eds.), *Social and Economic Impact of Earth Sciences*,
https://doi.org/10.1007/978-981-19-6929-4_18

1 Introduction

Tsunamis in the Indian Ocean are not as frequent as in the Pacific Ocean, but they pose a great threat to all the countries of the region. In the last 300 years, Indian Ocean region recorded thirteen tsunamis and three of them occurred in the Andaman and Nicobar region for which the details of the location of the epicentre, death/damage caused, etc. are not known (NDMA Guidelines 2010). The three tsunamis which affected Andaman and Nicobar Islands occurred on 19 August 1868 (Newcomb and McCann 1987), 31 December 1881 and 26 June 1941 (Alam and Dominey-Howes 2016). The 1945 tsunami following an earthquake of magnitude 8.2 M in the Arabian Sea is reported to have a maximum run-up of 13 m in Pakistan and resulted in the death of 4000 people (NDMA Guidelines 2010). Overall the run-up levels in the pre-2004 Indian Ocean tsunamis varied from 1 to 13 m. The Indian Ocean Tsunami of 26 December 2004 that triggered by the third-largest (long lasting about 10 to 12 min rupture time) earthquake of magnitude Mw 9.2 occurred near off coast of Banda Aceh Indonesia is one of the most destructive Tsunamis known to have hit India and 13 other countries in the Indian Ocean region. With a combined toll of 238,000 causalities and roughly more than 1.6 million people displaced in 14 countries, this tsunami resulted in damage and destruction of property, assets and infrastructure in the coastal areas of over USD 14 billion. The tsunami caused enormous loss of life and damage in the coastal villages of Kerala, Tami Nadu, Andhra Pradesh, Puducherry and Andaman & Nicobar Islands (NDMA Guidelines 2010). In India, a population of 26.63 lakhs in 1396 villages in five states and Union Territories were affected by this tsunami. 10,749 people lost their lives due to the tsunami and 5640 people were missing in the tsunami-affected areas. Maximum damage occurred in low-lying coastal areas and high casualties were found in thickly populated areas. In addition to the fatalities and direct economic damage, the occurrence of such disasters hinders the process of growth and development.

The tsunami risk and vulnerability which the coastal communities in India are exposed to even by a distant high-intensity earthquake in Indonesia came as a shock and surprise to the unsuspecting public. The tsunami had badly affected the fishermen community who not only lost their near and dear ones but also lost their means of livelihood. The absence of an effective TEWS and the last mile connectivity to disseminate alert and early warning messages to the coastal communities as well as the lack of public awareness and emergency response preparedness along the various stakeholder groups caused the widespread damage and enormous loss of lives.

Aftermath of the boxing day tsunami in the year 2004, the Indian government established Indian Tsunami Early Warning Center at Indian National Center for Ocean Information Services (INCOIS), Ministry of Earth Sciences (MoES), based out of Hyderabad. ITEWC became operational on 15 October 2007 and has since been providing tsunami early warning services on 24 X 7 basis (Nayak and Kumar 2008a, b, c). The Indian Tsunami Early Warning System (ITEWS) has mainly four components to address tsunami risk in Indian Ocean (i). Detection of large earthquakes occurring in two subduction zones of the Indian Ocean (Andaman-Nicobar-Sumatra Island Arc in the Bay of Bengal and the Makran subduction zone in Northern Arabian Sea) with land-based seismic stations, (ii). Monitoring sea-level changes for confirmation of tsunami generation through deep ocean Tsunami Bottom Pressure Recorders near epicentral region and coastal Tide gauges, (iii). Identifying coastal areas under risk using numerical model outputs and (iv). Disseminating bulletins to all coastal state disaster management offices as well as other stakeholders.

ITEWS is a typical multi-institutional framework of one of its kind and was established in a record time of less than 2 years. The observation networks of ITEWS comprise land-based seismic stations, tsunami buoys and tide gauges. Dedicated advanced modelling computational and communication facilities were established to receive the data in near-real-time as well as to disseminate timely tsunami advisories to different stakeholders with the necessary level of redundancy at each stage. The warning centre operates round the clock to monitor the tsunami activities in the Indian Ocean by processing the data received in real-time from observation networks, assesses the tsunami forecast model results, and provides timely advisories, also, equal importance is given to various community awareness and preparedness activities (Fig. 1). Since inception, ITEWC monitored more than 630 earthquakes of more than 6.5 M out of which 101 tsunamigenic events in the Indian Ocean region. Out of them, 7 large events occurred and only on one occasion ITEWC issued tsunami warning (Kumar et al. 2012), that too only for three Nicobar Islands.

Table: List of ITEWC issued threat (Warning/Alert/Watch) bulletins.

S. No	Date & Time (UTC)	M	Region name	ITEWC evaluation	Tsunami observation
1	12-Sep-2007 11:10:26	8.5	Southern Sumatra, Indonesia	Tsunami WATCH for A&N Islands, Orissa, Andhra Pradesh, Tamilnadu, Kerala	1 m at Padang, Indonesia and 15 cm at Cocos Islands
2	12-Sep-2007 23:49:05	7.8	Kepulauan Mentawai Region, Indonesia	Tsunami WATCH for few regions in A&N Islands	2 cm at Christmas Island, Sibolga, Sabang
3	30-Mar-2010 16:54:50	6.9	A&N Islands, India	Tsunami WATCH for West & land fall islands, Flat islands, North Sentinel Islands, Port Blair	No tsunami
4	12-Jun-2010 19:26:47	7.5	Nicobar Island, India	Tsunami WATCH for Nicobar, Komatra & Katchal Island	3 cm at Trincomalee, Sri Lanka
5	10-Jan-2012 18:37:00	7.1	Off West Coast of Northern Sumatra	Tsunami WATCH for Nicobar Islands	No tsunami

(continued)

(continued)

S. No	Date & Time (UTC)	M	Region name	ITEWC evaluation	Tsunami observation
6	11-Apr-2012 08:38:36	8.5	Off west Coast of Northern Sumatra	Tsunami WARNING for Indira Point, Car Nicobar, Komatra & Katchal Islands of A&N Islands ALERT for rest of A&N Islands, Tamilnadu, Andhra Pradesh. WATCH for few areas in mainland	1 m at Meulaboh and 0.35 m at Sabang Indonesia and 0.30 m at Campbellbay
7	11-Apr-2012 10:43:10	8.2	Off West Coast of Northern Sumatra	Tsunami ALERT for Nicobar Islands and WATCH for Andaman Islands and east coast of India	20 cm at Meulaboh, Indonesia

ITEWC successfully monitored and issued subsequent bulletins for the earthquake of magnitude 8.5 that occurred off the west Coast of Northern Sumatra on 11 April 2012. ITEWC detected the earthquake within 3 min 52 s and issued six bulletins. As per Standard Operating Procedure and pre-run model simulations, only 3 zones in Nicobar Islands were placed under warning that called for evacuation of public to higher grounds. Andaman Islands as well as east coast of India were placed under alert status that implicated a marine threat and hence only clearing the beaches. Thus, the timely advisories generated for this event avoided false alarms and unnecessary public evacuations in the Indian mainland. ITEWC issued six bulletins for the event revising the alert levels based on revised earthquake parameters and water level observations from tsunami buoy and tide gauge. The earthquake generated a small ocean-wide tsunami and it was observed maximum sea level height of 1.06 m at Meulboh tide gauge, Indonesia, and 0.30 m at Campbell bay tide gauge, India. Subsequently, after confirmation from sea level gauges that no significant tsunami was generated, ITEWC issued a Threat Passed bulletin. Thus, ITEWC has provided qualitative and quantitative tsunami advisories based on the model results and observations.

The warning centre is designated as one of the Tsunami Service Provider (TSP) for the Indian Ocean region by the Intergovernmental Oceanographic Commission (IOC)-UNESCO in October 2011. Since then, ITEWC has been providing tsunami services to 25 Indian Ocean countries, along with similar centres in Indonesia and Australia.

Tsunamis are triggered not only by submarine earthquakes, but also by landslides which often accompany large earthquakes, and submarine volcanic eruptions. The recent 2018 tsunamis in Palu (Heidarzadeh et al. 2019) and Krakatau (Lingling Ye 2020) in Indonesia were great examples which generated destructive local tsunamis due to submarine landslides and volcanic flank collapses. Current early warning

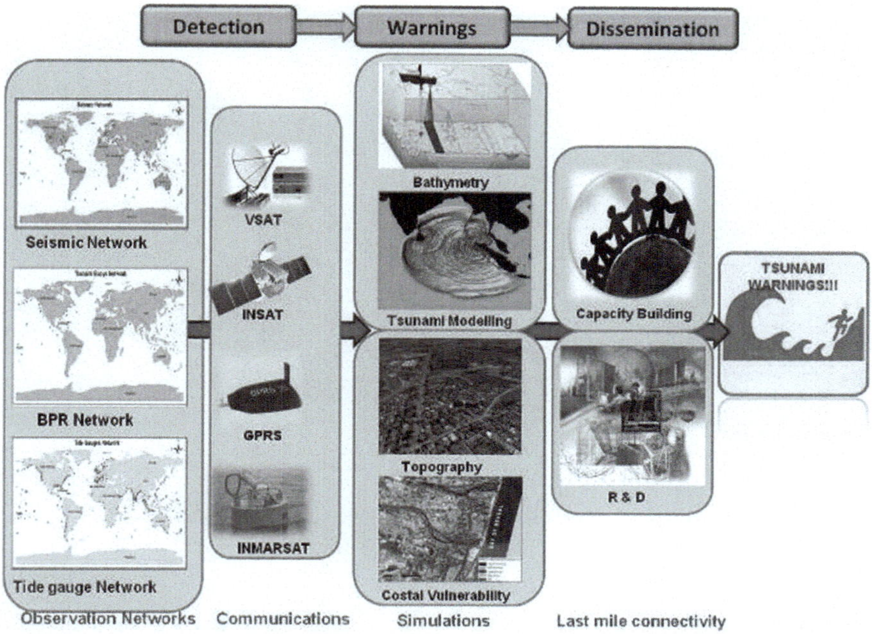

Fig. 1 Components of Indian Tsunami Early Warning System (ITEWS)

systems are most effective for tsunami generated by subduction zone earthquakes, but have limitations to handle atypical (landslide and volcanic) and/or near-field tsunamis.

Trans-ocean tsunamis generated by giant earthquakes and volcanic eruptions have been recorded in the past (Satake et al 2020). The 2004 Indian Ocean tsunami was one of the examples of trans-ocean tsunami which was also recorded in Pacific Ocean.

2 Components of Indian Tsunami Early Warning System (ITEWS) Observation Networks

(i) Seismic Network

The tsunami warning process starts with real-time detection of a submarine or near-shore earthquakes of magnitude ≥ 6.5 with the help of a network of seismic stations. A network of approximately 500 broadband seismic stations (Fig. 2) comprising of ~430 global stations, 17 stations of Indian Real-Time Seismic Monitoring Network (RTSMN) and ~50 stations from Indian Seismic and GNSS Network (ISGN) have been configured in the earthquake autolocation software at ITEWC (Manneela et al. 2016). With the present configuration, the system is capable of detecting any earthquake of magnitude >5 occurring anywhere in the world within 10 min. However, the

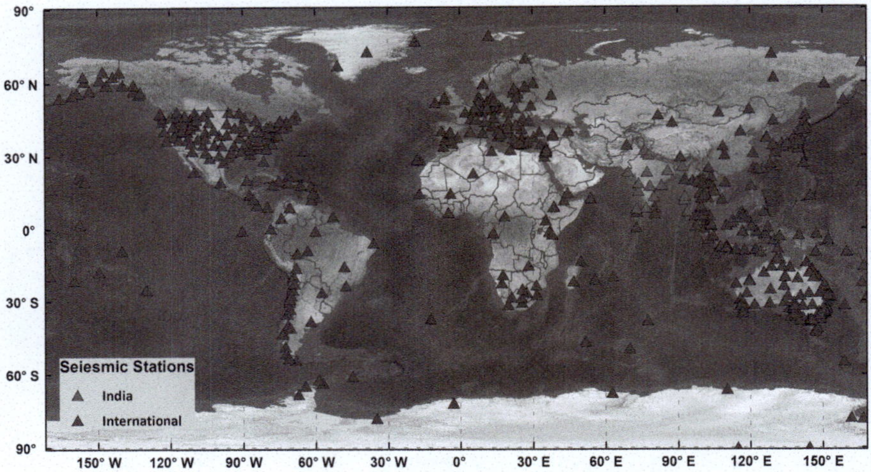

Fig. 2 The global network of seismic broadband stations including Indian stations (RTSMN) used for real-time earthquake monitoring

first detection of an earthquake depends strongly on the density of seismic stations near the epicentre of the earthquake. As the seismic stations are mostly land-based and their global distribution is not uniform, the first detection times vary significantly based on the earthquake locations.

In case of a significant earthquake (an earthquake of magnitude $\geq$6.5 and location is under the sea or near the coast), the earthquake parameters are further refined by manual processing and seismic moment tensor is calculated to access the tsunamigenic the potential of the earthquake.

(ii) **Sea Level Network**

Sea level network is a very crucial component of tsunami warning system to confirm the generation of a tsunami and to predict the tsunami hazard for locations where the waves are yet to strike. Mainly two types of sea level sensors, tsunami buoys and tide gauges are used to detect and monitor tsunami generation and propagation. Tsunami buoys are deployed in the open ocean near the tsunamigenic source zones and tide gauges are installed along the coast.

(a) **Tsunami Buoys**

To detect the propagation of tsunami waves in the open ocean, 7 Tsunami buoys are deployed close to tsunamigenic source regions in the Bay of Bengal and the Arabian Sea (Fig. 3). The deployment and maintenance of these buoys are being done by INCOIS and National Institute of Ocean Technology (NIOT), Chennai. The data are transmitted in real-time through satellite communication for processing and interpretation. Tsunami buoy system consists of two parts, Bottom Pressure Recorder (BPR) placed on the seafloor which measures the pressure change in water level and derives the difference in vertical water level, and the Surface Buoy placed on the sea surface is used for communication purpose between BPR and satellite. The system has two data reporting modes: normal and event mode. The system operates routinely in normal mode, in which four values are recorded at 15-min intervals. When the internal detection software detects a change in water level above the threshold value, the system automatically switches into event/tsunami response mode transmission, where values are transmitted every one minute. The buoys remain in the tsunami response mode for 3 h and automatically return to the normal mode. The tsunami buoys are capable of detecting minor water level changes of even 1 cm at water depths up to 6 km (Meinig et al. 2005). In addition to national tsunami buoys, real-time data from about 60 tsunami buoys operated by other countries in Indian and Pacific Oceans are also received at ITEWS. ITEWC is sharing real-time data of all Tsunami Buoys (STB01, STB02, STB03,STB05, ITB05, ITB09 and ITB12) with the international community.

The de-tided data is being computed in 'real-time' using Principal Component Analysis of Tsunami Buoy Record (Tolkova 2009, 2010) for analysis of time series observations.

(b) **Tide Gauges**

Tide gauges are very critical to monitor the tsunami progress and the coastal sea-level changes. A network of 36 tidal gauge stations has been established along strategic locations of the Indian coastline to monitor the progress of tsunami waves (Fig. 4). There are many types of tide gauge sensors to measure sea-level changes. INCOIS installed 3 types of sensors, Radar gauges, Pressure gauges and Shaft encoder gauges. Each tide gauge measures the sea level data sampling rate of one minute and transmits

Fig. 3 Indian Tsunami Buoy Network

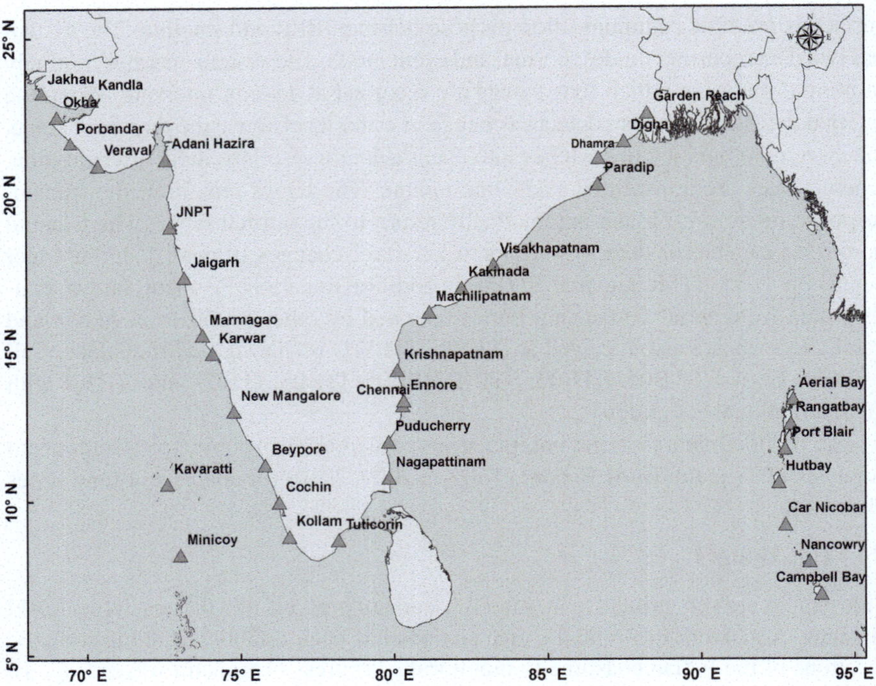

Fig. 4 Indian Tide Gauge Network

it after every 3 min (Islands stations every 1 min). The data are transmitted in real-time through different modes of communication like INSAT, GPRS simultaneously for processing and interpretation.

The installation and maintenance of these tide gauges are being done by INCOIS in collaboration with Survey of India (SoI), Dehradun. In addition to national tide gauges, real-time data from around 400 tide gauges operated by other countries in Indian and Pacific Oceans are also received at INCOIS. ITEWC is sharing real-time data of 8 tide gauge stations (Chennai, Cochin, Marmagao, Minicoy, Nancowry, Port Blair, Veraval and Visakhapatnam) with the international community. The de-tide data has been computed using predicted software of SLPR2 (Caldwell 1998) for all sea-level stations to distinguish tsunami wave from normal tidal waves.

ITEWC developed techniques for inversion of sea level data from the tsunami buoys and tide gauges to quickly estimate and forecast the tsunami heights and arrival times accurately to coastal regions following large, potentially tsunamigenic earthquakes that occur in the Indian Ocean region. This sea level inversion technique is very much useful to detect the tsunamis caused by atypical sources—volcanic eruptions, land based—strike slip earthquakes. In this context of atypical tsunamis real-time tide gauge data plays a crucial role in sea level inversion in order to characterise the tsunami source.

3 Numerical Modelling and Quantitative Forecasting of Tsunamis

The forecast of Indian Tsunami Early warning system is quantitative in nature with back-bone support of Tsunami Open Ocean Propagation Scenario Database for the Indian Ocean. The TUNAMIFF (Tohoku University's Numerical Analysis Model for Investigation of Far-Field tsunamis) Version 2011 model (TUNAMI FF—CUDA Version 2011 May 21) used in the operational context consists of a set of pre-computed model runs. The Standard TUNAMIN2 (Tohoku University's Numerical Analysis Model for Investigation of Near-field tsunamis, No. 2) model by Dr Fumihiko Imamura 2006 under IOC TIME Project (IOC Manuals and Guides No. 35), is also adopted and customised for inundation studies (linear theory in the deep sea, shallow-water theory in shallow sea and run-up on land with constant grids) of Indian coastal region. The TUNAMI N2 model validation studies (Ramana Murty et al. 2005; Tune Usha et al. 2009; Rao et al. 2011; Ramana Murty et al. 2011) were carried out for the 26 December 2004 Sumatra earthquake field observations by NCCR. The model results are matching 80% with actual inundation of 26 December 2004 tsunami measured from inundation field survey.

On 11 March 2011, an earthquake of magnitude Mw 9.0 on the coast of Tohoku, Japan, triggered a great tsunami in the Pacific Ocean. Though this event was outside the Indian Ocean region, ITEWC succeeded in providing timely national bulletins by executing global tsunami model in real-time and assessing the possible impact of the tsunami on the Indian Ocean coastal regions. Successful handling of this event by executing real-time model simulations opened up possibilities of ITEWC expanding its services to 'real-time forecast' of global ocean tsunamis, i.e. those occurring in other than the Indian Ocean.

ITEWC is working on enhancing its capabilities to extend the services with real-time tsunami modelling using in-house High-performance computing facilities, in addition to existing Open Ocean Propagation Scenario Database covering Andaman-Sumatra and Makran subduction zones in the Indian Ocean.

(i) **Open Ocean Propagation Scenario Database (OOPSDB)**

At the time of an earthquake, only the hypocentral parameters and magnitude are available in near-real-time. The information about fault geometry arrives much later, which cannot be used for real-time tsunami prediction for near-source regions. In addition, tsunami modelling, especially the coastal inundation, cannot be run in real-time for generating operational tsunami advisories as the model run takes large computing time. For operational quantitative tsunami forecast, there needs to be a method to quickly estimate the expected tsunami arrival (ETA) times and expected tsunami wave amplitudes (EWA) based on the quickly available earthquake parameters (based on magnitude and hypocenter of earthquake only) without running numerical simulations right after the occurrence of a large tsunamigenic earthquake. This issue can be dealt with by creating a database of pre-run scenarios.

A ‘scenario’ is a single tsunami model simulation that is calculated from the required initial seismic deformation condition with pre-defined earthquake source parameters (i.e. fault location, depth, length, width, displacement, strike angle, dip angle and slip angle). Each Scenario output contains the expected tsunami wave travel times, run-up heights and directivity maps. The fault geometry parameters are carefully selected so that they are likely to represent an actual Tsunamigenic earthquake based on knowledge of the historical earthquakes in two Tsunamigenic source regions of the Indian Ocean.

The scenario database consists of approximately 975 simulation points, each with separation of half a degree, located all along the subduction zones in two Tsunamigenic source regions of Indian Ocean (Fig. 5). From the sensitivity analysis and based on historical earthquakes, it is concluded that each of these locations should have multiple scenarios associated with it for a combination of 6 different depths (10, 20, 40, 60, 80 and 100 km) and 7 moment magnitudes (Mw) (6.5, 7.0, 7.5, 8.0, 8.5, 9.0 and 9.5), providing a total number of approximately 50,000 scenarios. Each simulation covers the entire Indian Ocean domain with 15 h of simulation time and a time step of 5 s. The tsunami profiles of 15 h for every 15 s are saved at coastal forecast points for each scenario. The coastal forecast points are selected at 30 m bathymetry assuming that till such depth, the computation is linear. About 1800 Coastal Forecast Points (CFPs) are selected for the tsunami domain separated by ~50 km apart

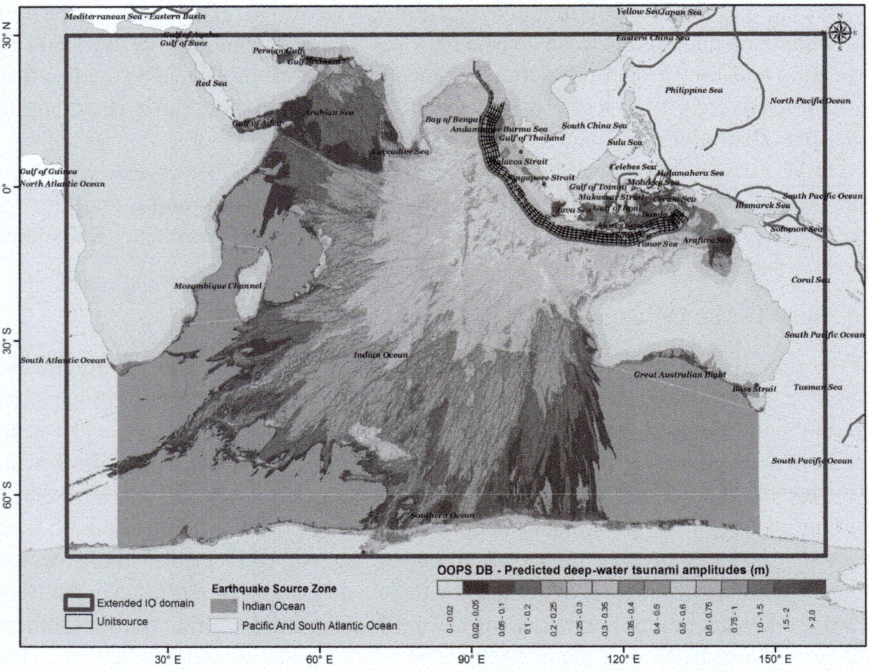

Fig. 5 Open Ocean Propagation Scenario DataBase (OOPSDB) for the Indian Ocean

covering the entire Indian Ocean rim countries. Arrival times and wave heights at specific coastal locations for each scenario are stored in a database. Travel times and Surge heights on 30 m bathymetries are interpolated to get the values at Coast. Travel Times to coast are calculated by considering the speed of the wave at different depths (30, 20, 15 and 10 M). The distance to the coast is divided by the average speed to get the travel times at the coast. Tsunami Wave height (H) at the coast is calculated by Green's function which is the fourth root of bathymetry depth (h1 = 30 m) multiplied by the wave height (H1) at 30 m depth. Whenever an earthquake occurs, the closest scenario to the event is extracted from the scenario database based on magnitude and hypocenter location to identify the regions at risk.

The model performance has been validated against the 12 September 2007 Tsunami (Kumar et al. 2010) which was generated from the earthquake of magnitude 8.4 that occurred in Southern Sumatra, Indonesia. The pre-run scenario for the 12 September 2007 event was used to calculate the estimated travel time and run-up heights at various coastal locations and water level sensors (Tide gauges and Tsunami Buoys). The directivity map generated from the picked scenarios showed that south-east and south-west Indian coast were likely to be affected by a minor tsunami (~ 20 cm) and Andaman and Nicobar Islands (~10 cm) which was evident from the observations of tidal stations at Chennai and Port Blair. The estimates from the model scenario matched well with the observations from Tsunami Buoys and tide gauges.

4 Standard Operating Procedure (SOP) and Decision Support System (DSS)

Considering the near-filed and far-field tsunamis, ITEWC has designed a unique Standard Operating Procedure (SOP) for generating and issuing tsunami information to national and international warning centres.

ITEWC services for an earthquake event commence whenever earthquakes are recorded with magnitudes $\geq$6.5 within the Indian Ocean and magnitudes $\geq$8.0 outside the Indian Ocean. Upon detecting a tsunamigenic earthquake, the duty officers respond immediately and begin to analyse the event. The analysis includes manual processing of seismic data for fine-tuning the earthquake's epicentre, depth and origin time, as well as its magnitude. The criteria for generation of advisories for a specific region are (WARNING/ALERT/WATCH) based on available warning time (i.e. time taken by the tsunami wave to reach the particular coast). The ITEWC SOP and DSS are described Nayak et al. (2008a, b, c), Kumar et al. (2010) and Pattabhi Rama Rao et al. (2020) (Fig. 6).

The 'codified' version of SOP is Decision Support System (DSS) that enables and supports the 24 $\times$ 7 operations at tsunami warning centre during major tsunamigenic event (Fig. 7). DSS at early warning centre is intended to monitor the real-time data from individual sensors, generate automatic alarms based on preset decision rules

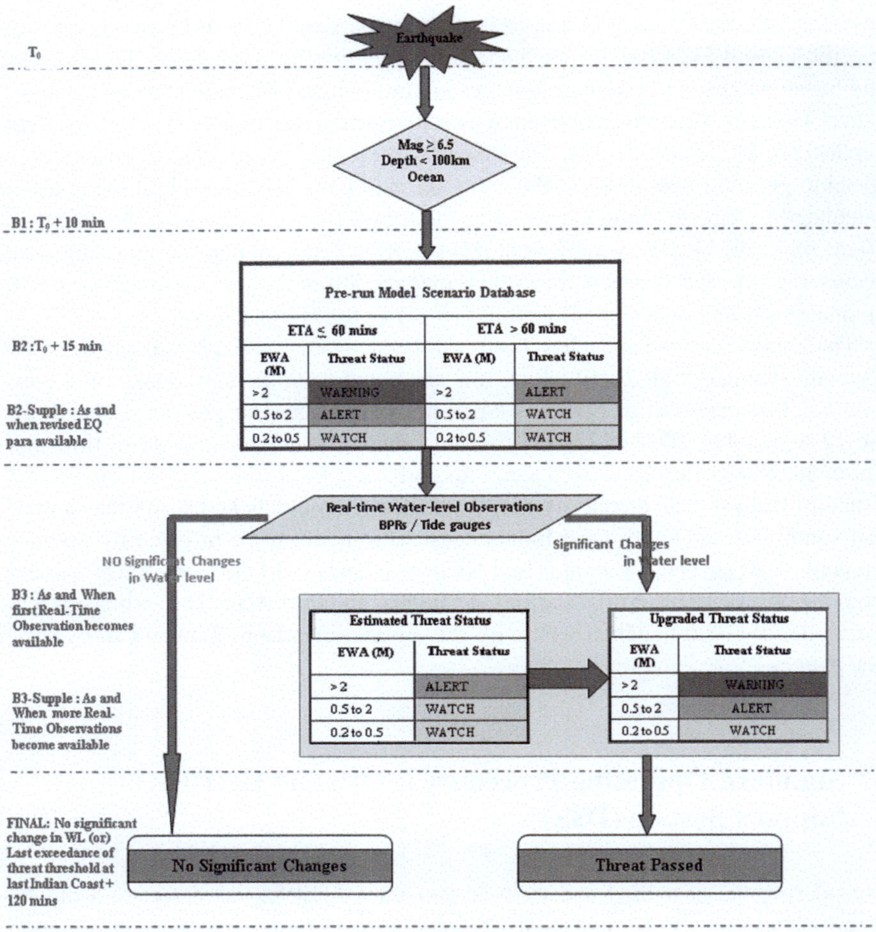

Fig. 6 Standard Operating Procedure and Bulletin Time Lines of ITEWC

for one or many of the input parameters and carry out criteria-based analysis for one or many of the above-mentioned input parameters to generate online advisories. The DSS software aid shift operator at tsunami warning centre to confirm/change the decision made or put out by software for 'timely' dissemination of tsunami information with 'single' mouse click. The tsunami early warning centre is equipped with the latest dissemination facilities such as SMS, E-mail, Fax, Phone, Website, FTP, VOIP, Hotline, Electronic Display Boards (EDB) and a satellite-based Virtual Private Network for Disaster Management Support (VPN-DMS) to disseminate the tsunami advisories to various stakeholders.

Considering that A&N islands are close to the tsunamigenic zones, special priority is given to disseminate tsunami bulletins to the concerned authorities during an event through multiple dissemination modes. For instance, tsunami bulletins are also

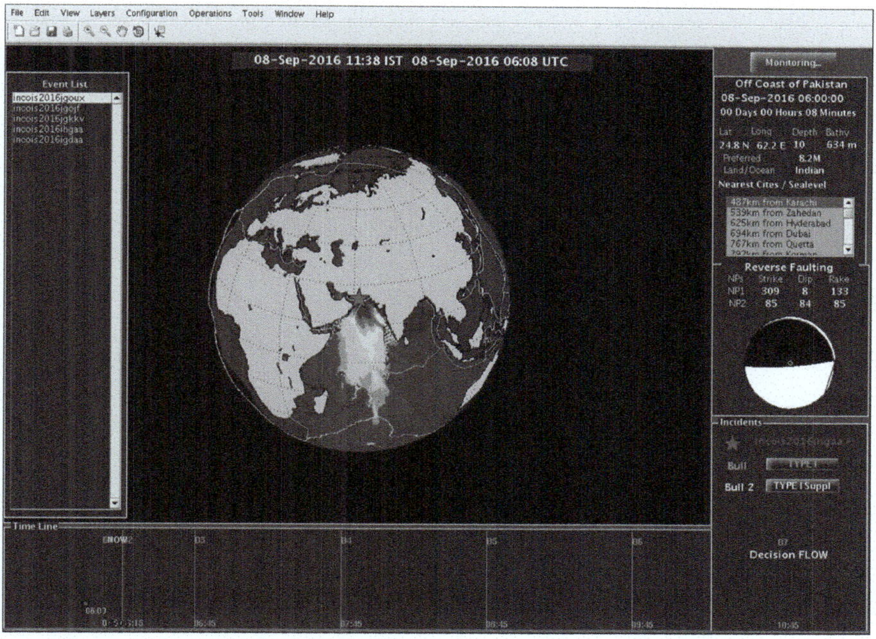

Fig. 7 Dashboard of DSS 2016 version in 'operation' during IOWAVE16

disseminated through VSAT-aided Emergency Communication System, Earthquake Alert System, Electronic Display Boards installed in A&N Islands. Earthquake information, tsunami bulletins as well as real-time sea-level observations are also made available on a dedicated website for officials, public and media.

5 Geospatial Applications: Coastal Vulnerability and Hazard Assessment

As geospatial technology plays a vital role in disaster management, the national tsunami and storm surge warning centre extensively utilized geospatial technology for the operational warning system as well as the coastal vulnerability assessment. Very high-resolution topographic data representing coastal Digital Elevation Model (DEM) generated using Airborne Lidar Terrain Mapping (ALTM). ALTM survey was carried out for the Indian mainland coast with 5 m horizontal posting and vertical accuracy of 25–30 cm. These data are being merged with bathymetry data to generate the high-resolution modelling grids for the coastal inundation forcing for tsunami and storm surges (Fig. 8). These data are critical in producing more accurate estimation of coastal inundation by tsunami and storm surges.

Fig. 8 3D view of Visakhapatnam coast overlaid with elevation dontours generated using ALTM topography draped with aerial images

Multi-hazard Vulnerability Mapping (MHVM): The purpose of multi-hazard vulnerability mapping (MHVM) is holistic approach to identify the most vulnerable and high-risk coastal areas which are exposed to inundation from oceanogenic disasters. MHVM was carried out using the parameters: sea-level change; shoreline change rate; elevation contours; extreme water level from tide gauges and the return periods (Mahendra et al. 2010, 2011). All these parameters were synthesized in Geographic Information System (GIS) to derive the composite hazard zones. These maps depict the coastal zones that are exposed to coastal inundation from oceanogenic hazards including tsunami and storm surges. An atlas comprising MHVMs on 1:25,000 scale was prepared, and sample map is provided as shown in Fig. 9.

3D GIS Mapping: The high vulnerable coastal zones were identified based on the MHVM to setup detailed 3D GIS database. This 3D GIS database comprises 3D buildings associated with attributed details of the owner, address and other socio-demographic details for the locations Cuddalore, Pondicherry, Chennai, Tuticorin, Rameshwaram, Machilipatnam, Nizampatnam, Kakinada, Puri, Kochi and Alleppey areas. A desktop application 3D/2D data Visualization and Analysis System (3DVAS) is developed and integrated with 2D GIS database and other relevant 2D geospatial data as shown in Fig. 10. 3DVAS can overlay tsunami inundation details and generate the building level risk maps on-the-fly and these maps can be disseminated to the disaster management authority along with tsunami/storm surge advisories.

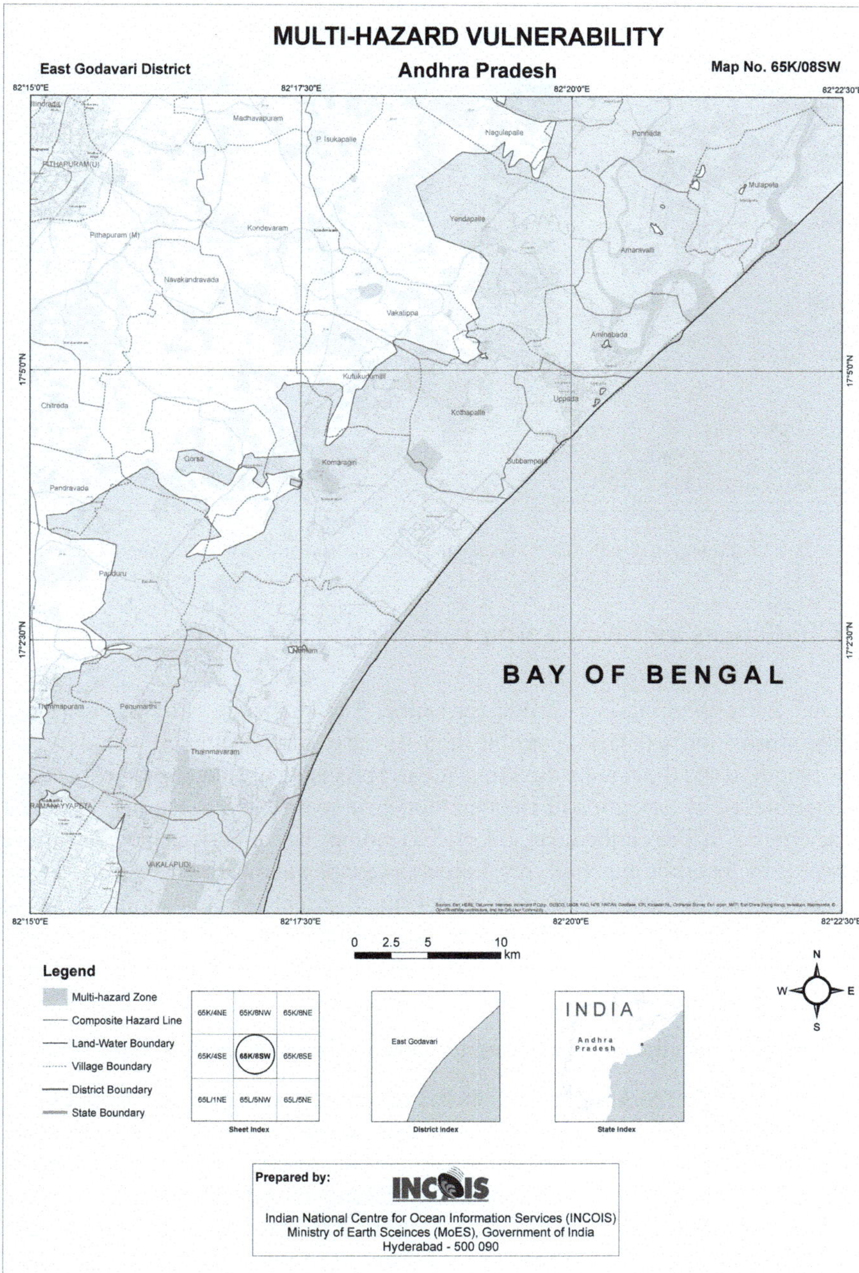

Fig. 9 Multi-hazard vulnerability map showing the part of Andhra Pradesh coast

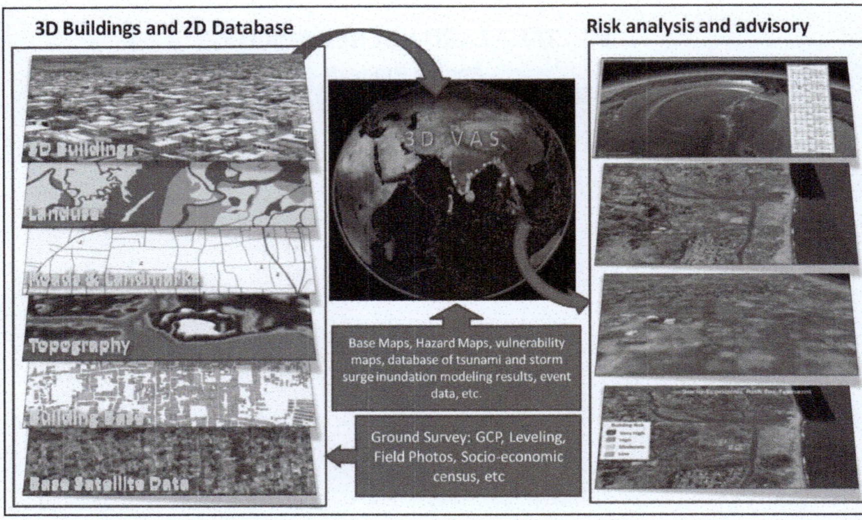

Fig. 10 3D VAS integrated with 3D/2D geospatial data and tsunami risk assessment

6 Challenges and New Initiatives

Tsunami Early Warning is a race against time. The challenge is to provide high-quality information in the least possible time. As the resulting tsunami depends on the type of faulting and the rupture direction, greater precision of tsunami early warnings can be achieved by execution of tsunami models in real-time considering the actual characteristics of the earthquake source. Extending the numerical models further landwards by incorporating high-resolution topography and bathymetry data is very important for accurate estimation of the level and extent of inundation due to the incursion of the tsunami wave. With increasing computing power and availability of high-resolution topographic data, INCOIS is working towards execution of real-time coastal inundation models for tsunami and storm surges.

(i) **Real-time Inundation Modelling for Indian Coast using ADCIRC**

The execution of the real-time tsunami model will avoid the errors occurring due to the interpolation of tsunami wave characteristics at the coast. A finite element model with the capability to deal with wet and dry grids that can preserve the volume of water in each grid is necessary to achieve the capability of estimating the inundation over land. The finite element-based Advanced CIRCulation (ADCIRC) model, mostly used to estimate the surge height and inundation during the landfall of cyclone/hurricane can be used as tsunami model and also to estimate the inundation during a tsunami. ADCIRC is a finite element shallow-water model that solves the water level and currents at different ranges of scales (Luettich et al. 1992, 2004). ITEWC successfully configured the ADCIRC model for the entire Indian Ocean region to compute tsunami propagation and extent of inundation along the east coast of India. The model

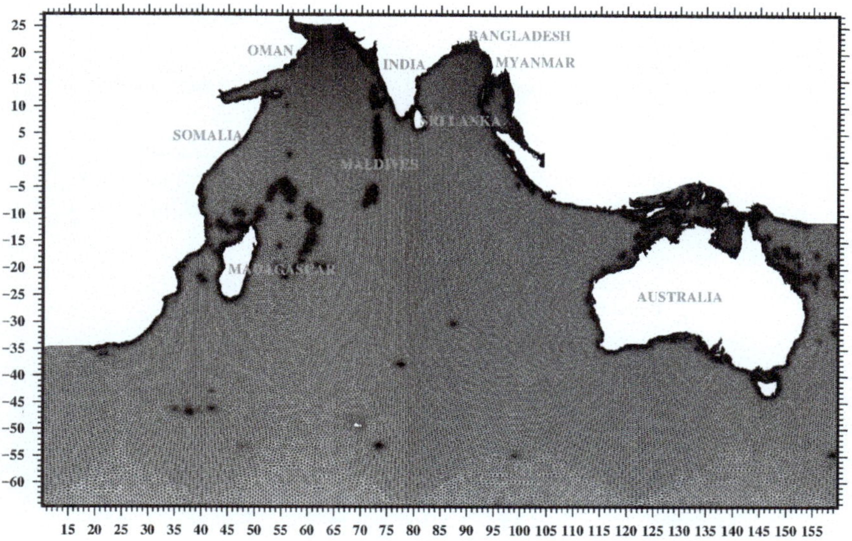

Fig. 11 Model domain and mesh

is successfully evaluated with the benchmark case 26 December 2004 tsunami observations. Computed water levels and associated inundation extents along the Tamil Nadu coast clearly show that the wide coastal stretch is experiencing the inland flooding due to tsunami (Satake et al. 2020) (Figs. 11, 12 and 13).

Results showed that the model could capture far-field wave arrival time, amplitude, as well as the inundation with good accuracy. Model agrees well in the crucial quantities such as arrival time of the first wave crest, the magnitude of the first crest and inundation extents. The run time of the model simulation is quite optimal to be able to launch in real-time using a high-performance computing facility.

ADCIRC model will be made operational soon to run in real-time to estimate the tsunami run-heights and inundation extents for Indian coasts, which is vital information for local coast administrators to make right and confirmative decisions for necessary coastal evacuation measures and mitigation efforts.

(ii) **GNSS and SMA Network in Andaman and Nicobar Islands**

Another major challenge in tsunami early warning system is the accurate estimation of magnitude and characterization of the source (i.e. earthquake fault geometry rupture extent and direction, etc.) at the earliest possible time. To estimate the source parameters, we need long-duration seismic waves in the conventional method, as for large magnitude earthquakes using only body waves may lead to underestimation of the magnitude. The other way is to derive it from sea-level data inversion. However, in the former method, for large magnitude earthquakes we need to wait for >20 min to get the accurate parameters and in case of the latter method, we need tide gauges or bottom pressure recorder data. The data from the sea-level sensors is the indication that the tsunami wave has already reached coast or on its way from the open ocean to reach the coast with short time left for warning the public.

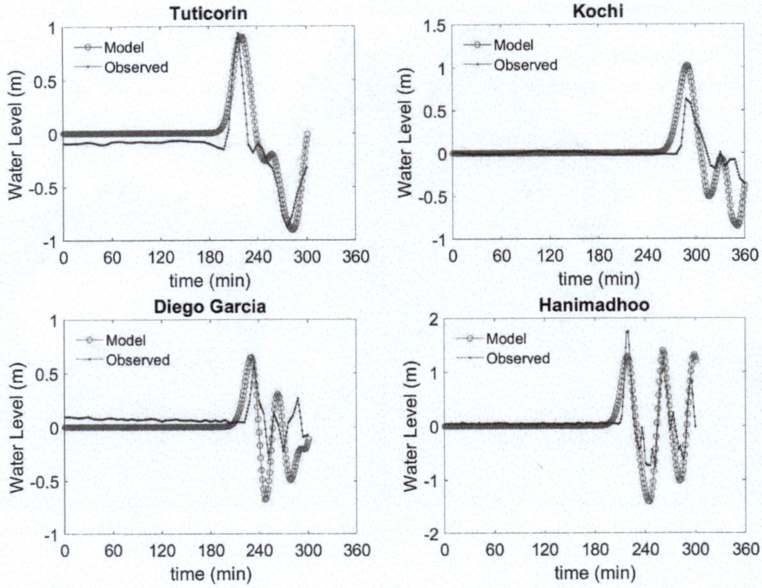

Fig. 12 Validation of computed tsunami levels against the tide gauge records

To measure coseismic displacements associated with an earthquake in near real-time which are the direct indication of the amount of rupture and direction, INCOIS has established a network of 35 GNSS and Strong Motion Accelerometers (SMA) at Andaman & Nicobar Islands (Fig. 14). The Andaman & Nicobar Islands lie right on the top of a subduction zone and at the least proximity zone for measuring the coseismic displacements immediately from the source region. Unlike the broadband sensors, strong motion sensors do not saturate for higher magnitude earthquakes and play a crucial role in calculating magnitudes and coseismic displacement. However, strong motion sensors are plagued with an issue called baseline offset, caused by small distortion in the reference level of motion. Although small in the acceleration time series, it may produce unrealistic displacement value after double integration. On the other hand, the GNSS/GPS provide a direct measurement of ground displacement and do not have such issues. Also compared with seismic sensors, GNSS receivers are comparatively less costly, independent of temperature and pressure changes and they are generally easy to maintain. The displacement obtained from real-time GNSS/GPS data are used for estimation of Moment Magnitude which is a critical indicator of the tsunamigenic potential of an earthquake. Also, the directly estimated rupture area and vertical and horizontal displacements of the seafloor from coseismic offsets give us the information to force the tsunami propagation and inundation models. The estimated run-up heights will be more accurate with real source parameters compared with worst-case scenario slip distribution, which is the current operational strategy for fast tsunami warnings.

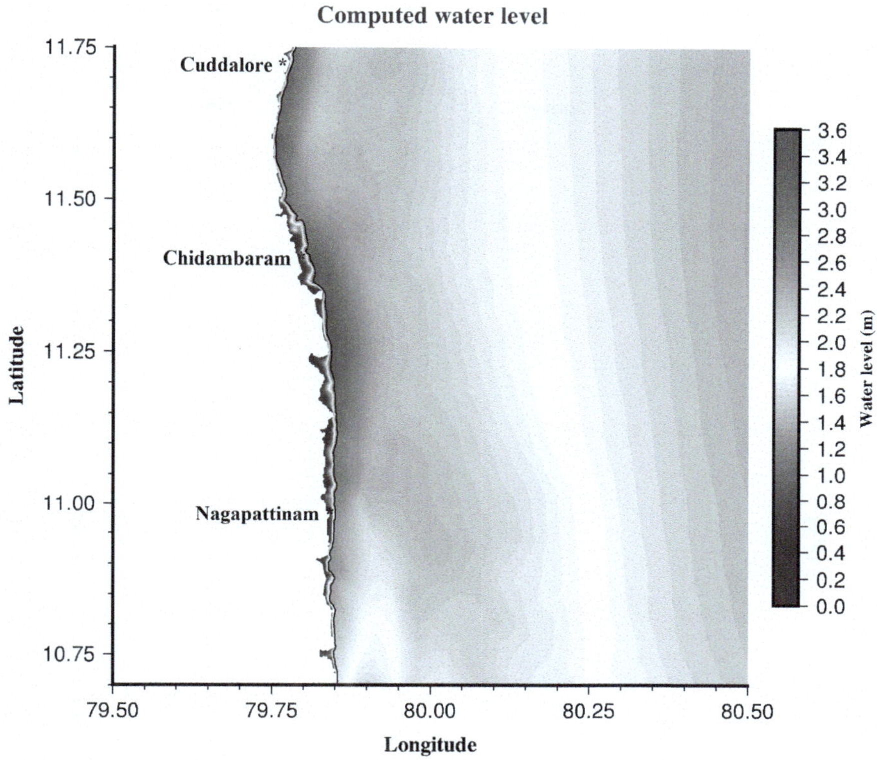

Fig. 13 Computed water level and inundation extent along the Tamil Nadu coast

To complement the direct displacement measures obtained from GNSS/GPS, the data from strong-motion accelerometers (SMA)will also be used.

The GNSS and SMA sensors locations are planned in such a way that they could be co-located with Emergency Operations Centres (EOCs) of Andaman & Nicobar Islands so that VSAT communication links used for transmission of data can also be used for enabling emergency communication for voice/fax as well as activation of the sirens.

7 Tsunami Capacity Building for Community Awareness and Preparedness

Disasters cannot be prevented, but their impact can be mitigated through community and emergency preparedness, timely warnings, effective response and public education. Through the continuous effort of scientists and disaster management officers, there has been a substantial improvement in tsunami warnings and timelines by

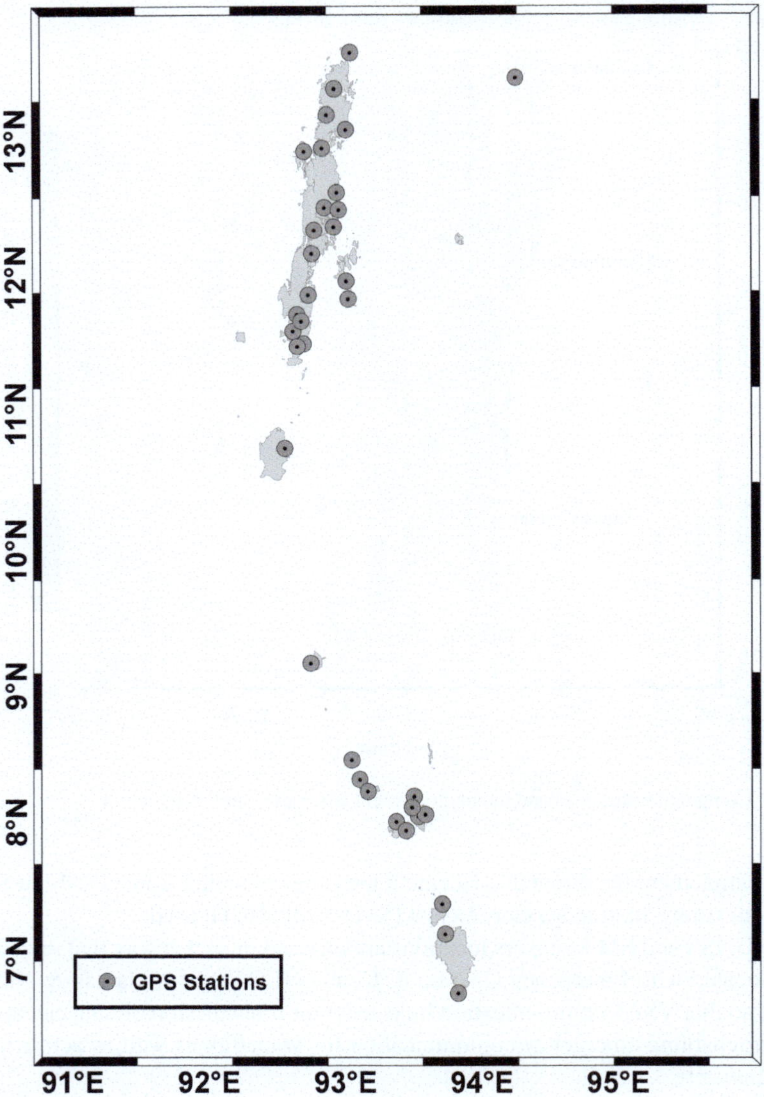

Fig. 14 GNSS and SMA Network in Andaman and Nicobar Islands

deploying better sensors, accurate models and concurrent multiple modes of dissemination. However, the success of a warning is measured by what actions people take once the respective authorities issue a public warning. Surviving a tsunami depends on the ability of an individual in the hazard zone to recognize warning signals, make correct decisions and act quickly.

Tsunami impact can be mitigated through public education, community awareness and preparedness. ITEWC has been organizing regular workshops, training and seminars to create awareness on tsunamis to Disaster Management Officials (DMOs) and other stakeholders. Tsunami exercises are a very effective tool for creating tsunami awareness and preparedness among the stakeholders. The purpose of Tsunami exercise is to evaluate the readiness of the stakeholders and to identify the changes that can improve its effectiveness. The exercises not only emphasize the testing of communications from warning centre to its stakeholders but also provides an opportunity for testing national/state/local chains of command and decision-making, including the alerting and evacuation of people from selected coastal communities. ITEWC regularly conducts communication tests and mock exercises to test the efficiency of communication links to evaluate the readiness to handle emergency situations. The communication (COMMs) tests are conducted twice a year in coordination with Intergovernmental Coordination Group for the Indian Ocean Tsunami Warning and Mitigation System (ICG/IOTWMS) of the Intergovernmental Oceanographic Commission (IOC) of UNESCO ensuring the participation of all countries of the Indian Ocean rim. Since March 2011, 21 COMMs tests have been conducted involving 25 countries. Since 2009, ITEWC coordinated 6 Indian Ocean wide tsunami mock exercises called IOWAVE (IOWAVE-2009, IOWAVE-2011, IOWAVE-2014, IOWAVE-2016, IOWAVE-2018 and IOWAVE-2020) in coordination with ICG/IOTWMS. In addition, ITEWC also conducted 4 mock exercises for the Indian coastal states (Andaman & Nicobar Islands 2013, mock exercise to East Coast of India 2015, mock exercise to Kerala coast 2016 and Mega mock exercise for East Coast of India 2017) in collaboration with National Disaster Management Authority (NDMA), Ministry of Home Affairs (MHA) and State/UTs Disaster Management Organizations where more than 1,00,000 people participated actively. Though there have been significant scientific and technological developments in the upstream to improve the tsunami early system, mitigation of disasters depends on the response and actions by the community which are vulnerable. The recent tsunami events in 2018 in Palu and Krakatau in Indonesia triggered by atypical sources such as submarine landslides and volcanic flank collapses pose a challenge to the existing tsunami early warning systems. There is an immediate need to strengthen the downstream, i.e. at the coastal community level for better preparedness to face the tsunami threat.

The Tsunami Ready (TR) programme by IOC-UNESCO addresses such gaps in enhancing the awareness of vulnerable communities on how to respond appropriately to these warnings and preparedness. ITEWC being the nodal agency for Tsunami is playing a lead role in implementing the TR concept in India guided by the National Board established by the Ministry of Earth Sciences with the support from NDMA and ICG/IOTWMS. ITEWC initiated the implementation of IOC-UNESCO's TR programme at the community level in the country on a pilot basis. TR is a voluntary community-based program that facilitates tsunami preparedness as an active collaboration of the public, community leaders, local and national emergency management agencies (Guidelines for Tsunami Ready by IOC-UNESCO 2017). The main objective of TR is to improve coastal community preparedness for tsunami emergencies and to minimize the loss of life and property. Odisha State Disaster Management

Authority (OSDMA), Odisha implemented the program on a pilot basis in Venkatraipur and Noliasahi villages. Based on the recommendations of the National Board after the evaluation process, the IOC-UNESCO has recognized Venkatraipur and Noliasahi villages of Odisha as TR communities. With the IOC-UNESCO recognition of these two communities from Odisha, India became the first country in the Indian Ocean Region and Odisha became the first state to achieve the TR recognition. The TR Programme is planned to extend to all the coastal states and Union Territories of India.

8 Economic and Social Impact of Tsunami Early Warning Services

Quantifying the economic and social benefits of the early warning systems is a difficult task. There is no doubt that effective early warning systems have substantially reduced the causalities and damages. The Ministry of Earth Sciences, Government of India engaged the National Council of Applied Economic Research (NCAER) to conduct a survey on 'Economic Benefits of Dynamic Weather and Ocean Information Services' provided by NCMRWF and INCOIS. NCAER used an innovative approach that departs from convention. The economic benefits of providing ocean state and 'No Tsunami Threat' forecasts were estimated by analysing the expenditure saved on activities that were cancelled due to advance availability of these forecasts. The economic benefits of the Tsunami Early Warning Centre can be assessed by the list of undersea earthquakes in the Indian Ocean Region for which a 'No Tsunami Threat' advisory issued by ITEWC, avoids relocation and rehabilitation expenditures of human settlement in the affected regions. The report considered 23 cases from 2007 to 2014, where 'No Tsunami Threat' was issued and assumed that an expenditure saving of around Rs 3,500 crore for one relocation and rehabilitation (based on the Phailin case study), would result in cumulative savings that would amount to Rs 80,500 crore. This would translate into an annuity (savings due to 'No Tsunami Threat') of Rs 11,500 crore. The compounded investment in the Tsunami Early Warning Centre since its inception was a mere Rs 133 crore in 2014 prices at the social discount rate of 12%. The gross economic benefit far exceeds such gross investments.

9 Conclusions

After the devastating Indian Ocean Tsunami on 26 December 2004, India established the Indian Tsunami Early Warning Centre at INCOIS and fully operational since 15 October 2007. The ITEWC has completed 13 years of services to the nation, by providing timely and accurate tsunami early warnings thus helping to reduce damage to life and property. To date, no false alarms have been issued and no casualties were reported in India. Further, ITEWC was designated as a Tsunami Service Provider (TSP) for the ICG/IOTWMS in 2011 along with TSP Australia and TSP Indonesia and providing Services to 25 member states in the Indian Ocean region. Since its inception, ITEWC has introduced several innovative concepts in tsunami modelling, inundation mapping, Decision Support System, SOPs to address the emerging challenges to provide timely and accurate tsunami early warnings.

Natural disasters cannot be prevented, but their impact can be minimized and mitigated through preparedness, timely warnings, effective response and public education. Several workshops, training and seminars have been organized to create awareness about tsunamis and to build confidence in the stakeholders of tsunami disaster management. ITEWC has been conducting regular communication tests, tsunami mock exercises to test the efficiency of communication links and to strengthen the readiness to handle the emergency situations with stakeholders. The recent tsunami mock exercises IOWave18 conducted in September 2018 and Multi-State mega mock exercise conducted in November 2017 were great examples of building a tsunami resilient community.

Though tsunamis are most often caused by major earthquakes, Palu (28 September 2018) and Sunda strait (23 December 2018) tsunamis are examples of atypical tsunamis triggered by landslides (caused by the earthquake) and volcanic eruptions and generated near-field tsunamis causing damage and loss of lives. These tsunamis showed that significant challenges remain, despite the advanced Tsunami Early Warning Systems are in place. There is an immediate need to strengthen the downstream, i.e. at the coastal community level for better preparedness to face the tsunami threat. Lessons learnt from these events reiterate the importance of tsunami awareness and preparedness and also the need for society-based/people-centric tsunami early warning systems to enhance the capabilities and resilience of the communities to protect themselves. ITEWC is leading the implementation of Tsunami Ready programme of IOC-UNESCO in India under the guidance of the National Board to promote tsunami preparedness through the active collaboration of public, community leaders, and national and local emergency management agencies. It provides a structural and systematic approach to build community preparedness for tsunami emergencies and to minimize loss of life and property.

References

Alam E, Dominey-Howes D (2016) (2016) A catalogue of earthquakes between 810BC and 2012 for the Bay of Bengal. Nat Hazards 81:2031–2102

Caldwell P (1998) Sea level data processing on IBM-PC compatible computers, Version 3.0, JIMAR Contribution No. 98–319, Univ. of Hawaii at Manoa

Economic Benefits of Dynamic Weather and Ocean Information and Advisory Services in India' and 'Coast and Pricing of Customized Products and Services of ESSO-NCMRWF & ESSO-INCOIS' (2015) National council of applied economic research

Guidelines for Indian Ocean Tsunami Ready (IOTR) Programme by IOC-UNESCO Indian Ocean Tsunami Information Centre (2017) version 7. www.ioc-tsunami.org/IOTRguidelines

Heidarzadeh M, Muhari A, Wijanarto AB (2019) Insights on the source of the 28 September 2018 Sulawesi Tsunami, Indonesia based on spectral analyses and numerical simulations. Pure Appl Geophys 176:25–43. https://doi.org/10.1007/s00024-018-2065-9

IOC Manuals and Guides No.35 "IUGG/IOC TIME PROJECT: numerical method of Tsunami simulation with the Leap-Frog Scheme". http://www.jodc.go.jp/info/ioc_doc/Manual/122367eb.pdf

Kumar TS, Nayak S, Kumar CP, Yadav RB, Kumar BA, Sunanda MV, Devi EU, Kumar NK, Kishore SA, Shenoi SS (2012) Successful monitoring of 11 April 2012off coast of Sumatra tsunami by Indian Tsunami Early WarningCenter (ITEWC). Curr Sci 102(11):1519–1526

Luettich Jr, RA, Westerink JJ (2004) Formulation and numerical implementation of the2D/3D ADCIRC finite element model version 44.XX, 74

Luettich RA Jr, Westerink JJ, Scheffner NW (1992) ADCIRC: an advanced three dimensional circulation model for shelves coasts and estuaries, report 1: theory and methodology of ADCIRC-2DDI and ADCIRC-3DL. Dredging Research Program Technical Report DRP-92–6, U.S. Army Engineers Waterways Experiment Station,Vicksburg, MS, USA, p 137

Mahendra RS, Mohanty PC, Bisoyi H, Srinivasa Kumar T, Nayak S (2011) Assessment and management of coastal multi-hazard vulnerability along the Cuddalore-Villupuram, East Coast of india using geospatial techniques. Ocean Coast Manag 54(4):302–311. https://doi.org/10.1016/j.ocecoaman.2010.12.008

Mahendra RS, Mohanty PC, Srinivasa Kumar T, Shenoi SSC, Nayak S (2010) Coastal multi-hazard vulnerability mapping: a case study along the coast of the Nellore District, Andhra Pradesh, East Coast of India. Italian J Remote Sens 42(3):67–76.https://doi.org/10.5721/ItJRS20104235

Manneela S, Devi U, Saikia S, Srinivasa Kumar T, Shenoi SSC (2016) Recent Advances in the Indian Tsunami Early Warning System

Meinig C, Stalin SE, Nakamura AI, González F, Milburn HG (2005) Technology developments in real-time tsunami measuring, monitoring and forecasting. In: Oceans 2005 MTS/IEEE, 19–23 September 2005, Washington, D.C

National Disaster Management Guidelines: Management of Tsunamis (2010) A publication of the national disaster management authority, Government of India. 978-93-80440-06-4, August 2010, New Delhi

Nayak S, Kumar TS (2008a) The First Tsunami early warning Centre in the Indian Ocean, Risk Wise.Tudor Rose Publications, pp 175–177

Nayak S, Kumar TS (2008b) Addressing the risk of the Tsunami in the Indian Ocean. J South Asia Disaster Stud 1(1):45–57

Nayak S, Kumar TS (2008c) Indian tsunami warning system. In: Proceedings of the ISPRS technical commission IV, Vol XXXVII, 2008c, pp 1501–1506

Newcomb KR, McCann WR (1987) Seismic history and seismotectonics of the Sunda Arc. J Geophys Res 92(B1):421–439

Ramana Murthy MV, Reddy NT, Pari Y, Usha T, Mishra P (2011) Mapping of seawater inundation along Nagapattinam based on field observations. Natural Hazards.https://doi.org/10.1007/s11069-011-9950-1

RamanaMurty MV, Subrmanyam BR et al (2005) Preliminary assessment of impact of Tsunami in selected coastal areas of India (2005), Department of Ocean Development, Integrated Coastal and Marine Area Management, Project Directorate, Chennai, February 2005

Rama Rao P, Patanjali Kumar ECh, Ajay Kumar B, Sunanda MV, Mahendra RS, Murty PLN, Padmanabham J, Saikia D, Shenoi SSC (2020) Indian Tsunami early warning system: future developments, Geography and You (GnY) magazine, INCOIS special issue on Sentinels of the Seas, vol 20, Issue 6–4, No.144–145, 2020, pp 40–47

Rao VR, Reddy NT, Sriganesh J, Murthy MR, Murty TS (2011) Tsunami hazard evaluation at selected locations along the south Andhra coast: numerical modeling and field observations. Marine Geodesy 34(1):29–47. https://doi.org/10.1080/01490419.2011.547816

Satake K, Heidarzadeh M, Quiroz M, Cienfuegos R (2020) History and features of trans-oceanic tsunamis and implications for paleo-tsunami studies, Earth-Science Reviews, Volume 202, 2020. ISSN 103112:0012–8252. https://doi.org/10.1016/j.earscirev.2020.103112
Srinivasa Kumar T, Kumar P, Nayak S (2010) Performance of the Indian Tsunami Early warning system, international archives of the photogrammetry, remote sensing and spatial information science, Volume XXXVIII, Part 8, Kyoto Japan 2010
Tolkova E (2009) Principal component analysis of Tsunami buoy record: tide prediction and removal. Dyn Atmos Oceans 46/1–4:62–82. https://doi.org/10.1016/j.dynatmoce.2008.03.001
Tolkova E (2010) EOF analysis of a time series with application to tsunami detection. Dyn Atmos Oceans 50/1:35–54. https://doi.org/10.1016/j.dynatmoce.2009.09.001
TUNAMI FF—CUDA Version 2011 May 21 (tunamiff20110521). https://github.com/tunamiff2011cuda/tunamiff2011. Accessed Dec 2014
Usha T, Murthy MR, Reddy NT, Murty TS (2009) Vulnerability assessment of car Nicobar to tsunami hazard using numerical model. Sci Tsunami Hazards 28(1):15
Ye L, Kanamori H, Rivera L, Lay T, Zhou Y, Sianipar D, Satake K (2020) The 22 December, 2018 tsunami from flank collapse of Anak Krakatau volcano during eruption. Sci Adv 6:eaaz1377

Landslide Hazard and Monitoring

A. Jayakumar, T. Arulalan, Robert Neal, and A. K. Mitra

Abstract Landslides are a widespread geological event with nearly 12% of India's land area being prone to this natural hazard. The development of a reliable regional landslide early warning system is an immediate requirement for the country in the perspective of proper disaster management. This chapter begins by describing the most common morphological, geological, meteorological, and anthropogenic factors that may influence slope stability. It then explores the potential future use of Numerical Weather Prediction (NWP) forecast products for the assessment of landslide risk and early warning issuance. Related to this, a major discussion is about an objectively derived set of 30 daily weather patterns for India, which provide a historical classification allowing us to relate large-scale circulation at the country level to observed rainfall and landslide occurrence at a particular location—therefore providing us with a set of high-risk weather patterns which are susceptible to landslide occurrence. From a medium-range forecast perspective (out to two weeks), this chapter then describes how ensemble members from ensemble prediction systems (such as those run by the Indian National Centre for Medium-Range Weather Forecasting (NCMRWF)) can be objectively assigned to the closest matching weather pattern in order to produce probabilistic forecasts highlighting when the large-scale circulation may be susceptible to landslide occurrences. In addition, this chapter conducts a review of the local rainfall thresholds liable for triggering landslides, covering both intensity and antecedent thresholds. Nilgiri is used as a case study location, for determining the quantitative temporal probability of landslide initiation, which is essential for utilizing short-range forecasts from NWP models in a more efficient way.

A. Jayakumar (✉) · T. Arulalan · A. K. Mitra
National Centre for Medium Range Weather Forecasting, Noida, India
e-mail: jkumar@ncmrwf.gov.in

R. Neal
Met Office, Exeter, Devon EX1 3PB, UK

T. Arulalan
Centre for Atmospheric Sciences, Indian Institute of Technology Delhi, New Delhi, India

V. K. Gahalaut and M. Rajeevan (eds.), *Social and Economic Impact of Earth Sciences*,
https://doi.org/10.1007/978-981-19-6929-4_19

1 General Introduction

Landslides are the downslope movement of earth's material under the influence of gravity. Landslides are classified generally based on the type of movements (slide, fall, topple, spread, and flow) and the material in the slope (rock, earth, and debris) as well as other factors contributing the slope failures. The major classification of the landslides in terms of the material type and type of movement is originally devised by Varnes (1978), for example, rock fall, rock slide, earth flow, etc. (Fig. 1). Slide materials are moved along a roughly planar surface in the translational slide whereas a coherent movement in a single unit is displayed in the block slide. Falls and topples are the abrupt movements of slope materials which are influenced by gravity, mechanical weathering, pore water pressure, etc. Lateral spreads are even occurring in flat terrain mainly caused by sand liquefaction (transformation from a solid into a liquefied state). Debris flows indicate the mobilization of soil or rock caused by intense surface-water flow, due to heavy precipitation or rapid snowmelt. A detailed review of the landslide classification is by Cruden and Varnes (1996) and further modifications to this categorization could be found in Hungr et al.(2013).

Among all the natural hazards, landslides affect nearly 12% of India's land mass hence the development of an advanced landslide forecasting system is a thrust area needing focus from the perspective of natural disaster management in India.

Morphological, geological, meteorological, and anthropogenic factors all influence slope stability and the resulting landslide risk for a particular area. Morphological factors included the tectonic uplift, glacial rebound, and erosion of the hill slope. Causes of landslides from a geological perspective include factors such as slope, soil thickness, relative relief, drainage, material contrasts, permeability contrasts, and sheared, jointed, or fissured materials. Meteorological factors, such as the weather conditions over the previous days as well as on the day of any landslide also significantly influence the stability of slopes, particularly when significant rainfall or rapid thawing of lying snow is involved. Soil texture and structure greatly influence water infiltration, permeability, and water-holding capacity. The water-holding capacity of a landmass could be reduced through rapid infiltration (volume flux of water entering into the soil by rainfall/snowfall) through the pore space within rocks/soils. The infiltration rate is higher for coarse-textured soils than for fine-textured soils. The factor of safety, F_s, the ratio of shear strength to shear stress in a slope is also dependent on the cohesive nature of the soil, parent material (rock type), soil texture, etc. Moreover, the soil moisture retention capacity is dependent on the adhesive nature of the soil, which in turn relates to the stability of the hill slope (Kannan 2014). Even low soil moisture presence in non-cohesive soil would make it vulnerable to landslides. On the other hand, the anthropogenic factors such as the construction of roads, infrastructure, blockage of drainage systems deforestation are the major cause of landslides.

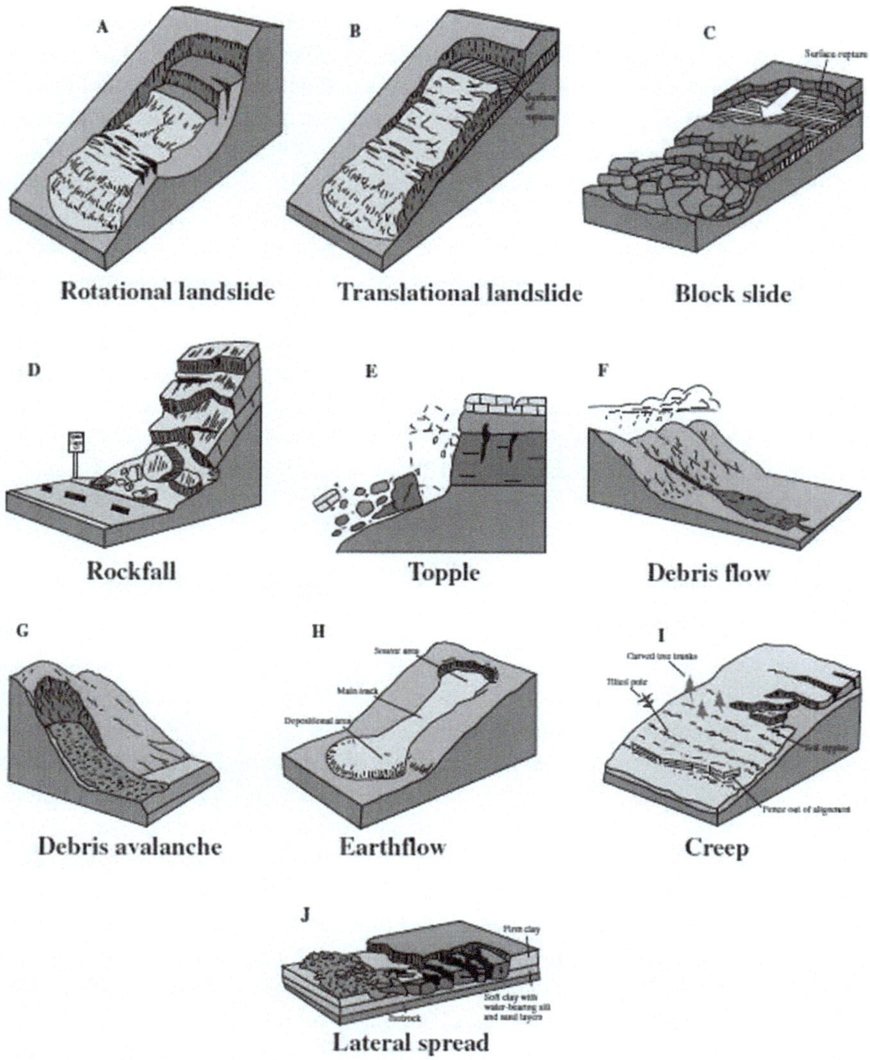

Fig. 1 Schematic illustration of major types of landslide movements. Image courtesy is USGS fact sheet 2004–3072(https://pubs.usgs.gov/fs/2004/3072/)

Though the landslide susceptible areas have been identified by the Geological Survey of India (GSI), the factors deforestation, urbanization as well as encroachments often change the susceptibility conditions. With advancements in computational and numerical modeling capabilities, the Indian Meteorological Department (IMD) is able to provide reliable meteorological input to localized regions from a high-resolution numerical weather prediction (NWP) model. One use of this forecast output is to provide guidance for forecaster-issued warnings of severe weather, such as thunderstorms, strong winds, heavy snowfall, and heat waves. However, opportunities still exist to exploit this forecast information further in applications of weather impacts. In this context, a recent international research collaboration between Indian, Italian, and UK research institutes has been working on designing a new framework for the better use of forecast products within forecasts of rainfall-induced landslides. This monitoring system may help in the Early warning of landslides especially debris flows, shallow slides, etc.

The major discussion of this chapter is about the potential future use of meteorological data or forecast products for landslide forecasts. The chapter is arranged with the following major sections: A discussion of how large-scale weather patterns over India can be related to the occurrence of meteorological-induced hazards such as landslides; Review of the local rainfall thresholds for triggering landslides, specifically focusing on Nilgiri, where the Western and Eastern Ghat meet. The final section discusses a methodological way of using NWP models from the National Centre for Medium-Range Weather Forecasts (NCMRWF) within future landslide warnings.

2 Relating Weather Patterns to an Increased Likelihood of Landslide Occurrences

A weather pattern can be described as one of many circulation types over a defined region (e.g. the Indian subcontinent), which differs in its characteristics from other weather patterns over the same region and varies on a daily basis. The term 'weather regimes' can also be used to describe a defined circulation type. Weather regimes are typically larger in scale, fewer in number, and persist for more days than weather patterns. However, to a certain extent, the two terms can be used interchangeably. This section describes recent research led by the UK Met Office and with support from the NCMWRF among others (as part of the international collaboration described in Sect. 1) to objectively define large-scale circulation patterns over India, which can be used in forecasting applications for weather impacts. Examples from the UK include relating weather patterns to coastal flooding (Neal et al. 2018), meteorological drought (Richardson et al. 2018), flooding (Richardson et al. 2020), temperature-related excess mortality (Huang et al. 2020) and landslide occurrences (Robbins et al. 2018). The weather pattern for landslides will also be one of the first forecasting applications is being used in India, where recent research investigated the ability of weather pattern forecasts to identify periods more susceptible to landslide occurrences.

The weather over the Indian subcontinent is highly modulated by the transition of the prevailing easterly trade winds in winter to the westerly trade winds during the summer monsoon. It is possible to identify circulation pattern variability within this seasonality as the weather in India is highly influenced by the Himalayas in the north, the Bay of Bengal to the East, and the Arabian Sea to the west. In addition, along with the variability of the low-level jet of oceanic origin, a mid-latitude origin western disturbance is a major factor controlling both the summer and winter seasons over India.

Research led by the UK Met Office has developed an objectively derived weather pattern classification for India using long historical meteorological data, providing a set of plausible weather pattern definitions throughout the year, with a catalog of observed patterns available over the last 40+ years (Neal et al. 2020). These pattern definitions were generated using cluster analysis. Specifically, a non-hierarchical k-means clustering algorithm was applied to daily gridded reanalysis fields from ERA-Interim (Dee et al. 2011) covering the period from 1979 to 2016. Other studies to generate circulation patterns using cluster analysis include Fereday et al. (2008) who generated a set of 10 weather regimes for each two-month period throughout the year over Europe, Ferranti et al. (2015) who generated a set of four large-scale weather regimes over Europe and Neal et al. (2016) who generated a set of 30 weather patterns over the UK and surrounding European area. Few similar studies exist over India—at least for weather patterns relevant for the whole of India and representative of circulation variability over the full year.

Large-scale weather regimes over India follow a relatively predictable evolution throughout the course of the year, driven predominantly by the onset and retreat of the Asian summer monsoon (Rao et al. 2005). However, the objectively derived weather pattern definitions derived by the UK Met Office open up opportunities for future research and forecasting applications over India. Firstly, the new weather pattern definitions may help understand the synoptic-scale driving forces behind the occurrence of high-impact weather (e.g. intense thunderstorms) or the occurrence of specific meteorologically driven hazards (e.g. landslides). Secondly, they could be used within the probabilistic medium- to long-range weather pattern forecasting tools, where daily weather pattern forecast probabilities are derived by assigning multiple forecast scenarios (ensemble members) from global ensemble prediction systems (such as the one run by the NCMWRF) to the closest matching weather pattern definition. This application is similar to what is already done operationally in Europe; e.g., Ferranti and Corti (2011) and Neal et al. (2016).

Results from the recent research led by the UK Met Office suggest that 30 weather patterns are enough to sufficiently represent variability within different phases of the Indian climate (Neal et al. 2020), where weather patterns are generated by clustering daily wind fields around 1.5 km above sea level. Each of the 30 weather patterns can be categorized into one of seven broad-scale regimes, called (1) winter dry period, (2) western disturbances, (3) pre/post-summer monsoon, (4) monsoon onset, (5) active monsoon (Fig. 2 left image), (6) break monsoon and (7) retreating monsoon (Fig. 2 right image). The dominant precipitation patterns can be shown to vary between these seven regimes as well as between the sub-patterns within each regime. This is important for identifying the most prone landslide periods within a large-scale regime.

Once the weather patterns have been defined, the daily historical weather pattern catalog introduced earlier in this section can then be used to relate observed weather patterns to past occurrences of landslides. This helps build up a picture of the

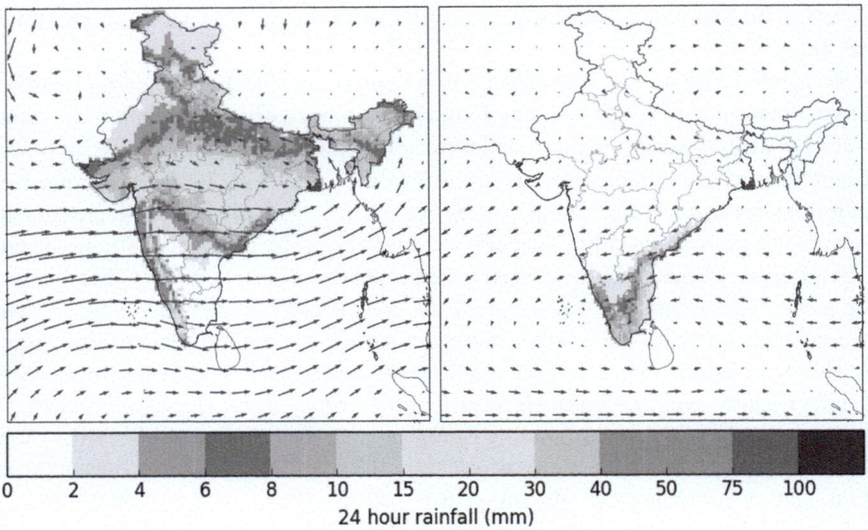

Fig. 2 An example of two contrasting Indian weather patterns. Left: An example of an active monsoon weather pattern (under south-west monsoon period) showing the prevailing westerly winds and heavy rain across western, central and north-eastern parts of India. Right: An example of a retreating monsoon (under north-east monsoon period) weather pattern, showing the prevailing north-easterly winds and rainfall affecting the southern tip on India

overall synoptic-scale driving mechanisms behind meteorologically induced landslides. Then, a high confidence forecast for a particular weather pattern allows climatological assumptions to be made about where in India may see an increased likelihood of landslides. This approach should work best for medium to long-range forecast periods (one week onwards) when high-resolution NWP forecasts of rainfall have less skill.

3 Rainfall Thresholds for Landslide Occurrences

A landslide rainfall threshold is defined as the minimum intensity or duration of rainfall needed to trigger a landslide (Varnes 1978) and a similar threshold value is estimated for other meteorological and hydrological parameters such as soil moisture, pore pressure, etc. The paper by Kanungo and Sharma (2014) documents the thresholds analyzed for landslide occurrences. Both empirical and physical models are used as a methodology for landslide occurrence using antecedent and triggering rainfall. Xiaohui et al. (2018) used both empirical and physical models, in that the former is based on the type of rainfall measurements (e.g., intensity and duration of the rainfall event), whereas the latter one is making a relationship between safety factors, F_S, of the slope with the rainfall depth (in terms of the daily or hourly rainfall amount). A detailed description of one of the physical models can be found in Montrasio and Valentino (2008). A process-based model is a model combining both meteorological and geotechnical parameters by incorporating a link between the slope stability condition to the rainfall pattern and infiltration (e.g. Iverson 2000). Empirical rainfall thresholds are broadly classified based on the following rainfall measurements; (a) rainfall events, (b) antecedent conditions, and (c) hydrological conditions (Reichenbach et al. 1998). Furthermore, threshold analysis based on the rainfall events is segregated into intensity–duration thresholds, thresholds based on the total event rainfall, rainfall event–duration thresholds, and rainfall event–intensity thresholds (Guzzetti et al. 2007). Area-specific threshold analysis is broadly subdivided into global, regional, or local thresholds depending on the geographical extent and scale effect of meteorological and hydrological conditions.

Antecedent thresholds are based on the amount of the antecedent rainfall prior to the slope failure or landslides. Accumulated rainfall controls soil saturation level and infiltration through the rock and increases the pore-water pressure. The different criteria under the period of consideration for the antecedent threshold analysis in the Himalayan region are reviewed in the paper by Kaungo and Sharma (2014). The result from their study shows a minimum 10-day antecedent rainfall of 55 mm and a 20-day antecedent rainfall of 185 mm are required for the initiation of landslides over the Chamoli-Joshimath region of the Garhwal Himalayas. Dikshit and Satyam (2018) introduced a two-dimensional probabilistic thresholds by applying a statistical method of Bayesian analysis using a combination of intensity, duration, and event rainfall. The use of a probabilistic approach over deterministic methods is a better approach to employing an early warning system for landslide-affected areas and can be used as the first line of action.

There are many studies documented about the rainfall triggering analysis over the Himalayan region, but a systematic study is lagging at the rainfall threshold analysis over Nilgiri and Tamil Nadu, which are locations prone to massive landslides during the north-east monsoon period. Sujatha and Suribabu (2017) discussed that the landslides at Coonoor may be triggered by high pore-water pressures generated during the prolonged and intense rainfall. They report that the landslides over this location may happen when a heavy rainfall spell is followed by a period of continuous rainfall.

Under the FSP projects on the development of a landslide early warning system in Nilgiris, GSI provided data on landslide dates, latitude, and longitude and observed daily rainfall from 24 rain gauges in the Nilgiri district of Tamil Nadu State. Nilgiri receives maximum rainfall during retreating monsoon periods and withdrawal periods of the summer monsoon (Sujatha and Suribabu 2017). Hence this section is more focused on establishing the relationship between the intensity of rainfall versus antecedent rainfall prior to landslide events in Nilgiri, whereas the seasonality in the rainfall measured among two monsoon periods is not considered here. Figure 3 shows the location map of observed rain gauges and landslide incidents reported by GSI. Both landslide catalog and rainfall measurement data are available from 1987 to 2017. The rain gauges and observed landslides are located between

11°14′ and 11°26′ North latitude and between 76°40′ and 76°53′ East longitude. The 24 rain gauges are located at places like valleys, tea estates, and villages in the Nilgiri district, namely Adderley, Benhope, Burliyar, Carolina, Coonoor, Glendale, Hillgrove, Indco, Kairabetta, Kallar, Kattabettu, Katteri, Ketty, Singara (lower, upper divisions), Moutere, Mullur, Ooty, Runnymede, Selvaganpathy, Simspark, Tigerhill, Tuttapallam, and Upasi, and their locations are marked in Fig. 3. There are three hill stations in the Nilgiri district, namely Ooty, Coonoor, and Kotagiri.

This study has chosen the nearest rain gauge (within 1.5 km distance) from the landslide locations that gives a maximum rainfall intensity (filtered-out rain gauge with a lower rainfall intensity than 5 mm) on the day of each landslide event. Figure 4 is representing the relationship between antecedent rainfall prior to the landslide (cumulative rainfall calculated from the past 1, 3, 7, 10, 15, 20, and 30 days before the landslide event date) and the intensity of rainfall observed on the day of the landslide event reported at Nilgiri. GSI has identified 509 landslide events which

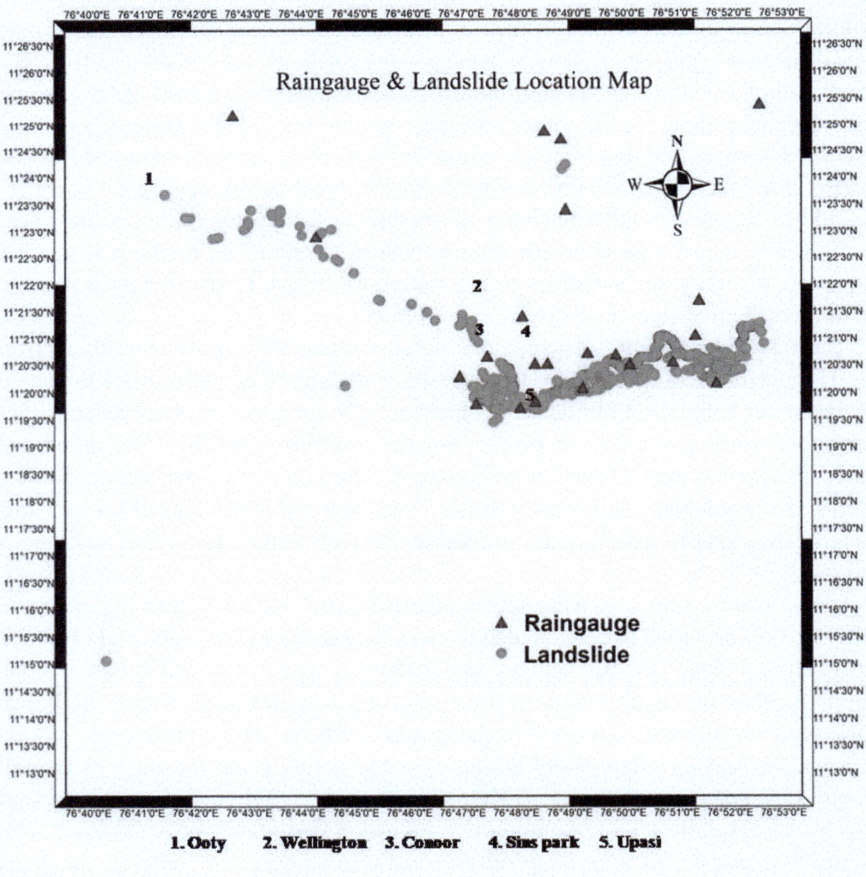

Fig. 3 Location map of observed rain gauges and landslide incidents at Nilgiri district in the Tamil Nadu State, reported by GSI

were caused by 116 rainfall events. In Fig. 4, all panels (a–f) are divided into two halves by the diagonal line in order to distinguish between the scattering bias of rainfall intensity (y-axis) and antecedent rainfall (x-axis). The diagonal divider itself indicates that the daily rainfall on the day of the landslide and the antecedent rainfall prior to the landslide is the same.

As observed from Fig. 4 a-f, the majority of landslide events are biased towards the antecedent rainfall prior to the landslide in comparison to the daily rainfall on landslides of different time durations (3, 7, 10, 15, 20, and 30 days). Figure 4a is illustrating the 3-day antecedent rainfall prior to a landslide event (x-axis) against the daily rainfall intensity on a landslide event (y-axis). From the scattered diagram, 48.1% of the total landslide events (i.e. 245 landslides out of a total of 509 were caused by 68 rainfall events) are biased towards the daily rainfall intensity and the rest of 51.9% of landslide events (i.e. rest of 264 landslides were caused by 48 rainfall events) are biased towards 3-day antecedent rainfall prior to landslide events. When it is analyzed for another cumulative period, the ratio of biases towards daily rainfall intensity and antecedent rainfall is estimated as 35.8:64.2, 16.9:83.1, 14.3:85.7, 12.4:87.6 and 4.7:95.3 for the case of 7-day, 10-day, 15-day, 20-day and 30-day antecedent rainfall respectively (Fig. 4b–f). Hence, it may be stated that a maximum of 95.3% of the landslide events (i.e. 485 landslides caused by 95 rainfall events) occurred under the influence of 30-day antecedent rainfall prior to landslide events, and the effect of increasing chances of landslide is dominated by cumulative rainfall over the intensity of rainfall on the day of landslide incidents when duration increases prior to the event.

The general conclusion on the basis of 3-day antecedent rainfall and intensity rainfall on the day of landslide events occurred at Nilgiri district is that both might be contributing equally, and Fig. 4a gives only the bias difference of 3.7% towards antecedent rainfall, whereas 7-day to 30-day antecedent rainfall are having more biased towards duration increments compared to the intensity of rainfall on the day of landslide events (Fig. 4b–f).

4 Meteorological Forecast Data Used for Early Warning

NCMRWF operates state-of-the-art numerical weather prediction (NWP) models–with both global and regional configurations with forecasts covering time scales from a few days to a few weeks in advance over India. Recent advances in this capability are in the use of the Unified Model, which is a collaborative model developed with the UK Met Office. The NCMWRF version is called the NCMRWF Unified model (NCUM). Different configurations of the NCUM are given in Table 2 by Rajagopal et al. (2019). This section of the chapter mainly focuses on the potential use of seamless forecast products in generating a framework for landslide predictions.

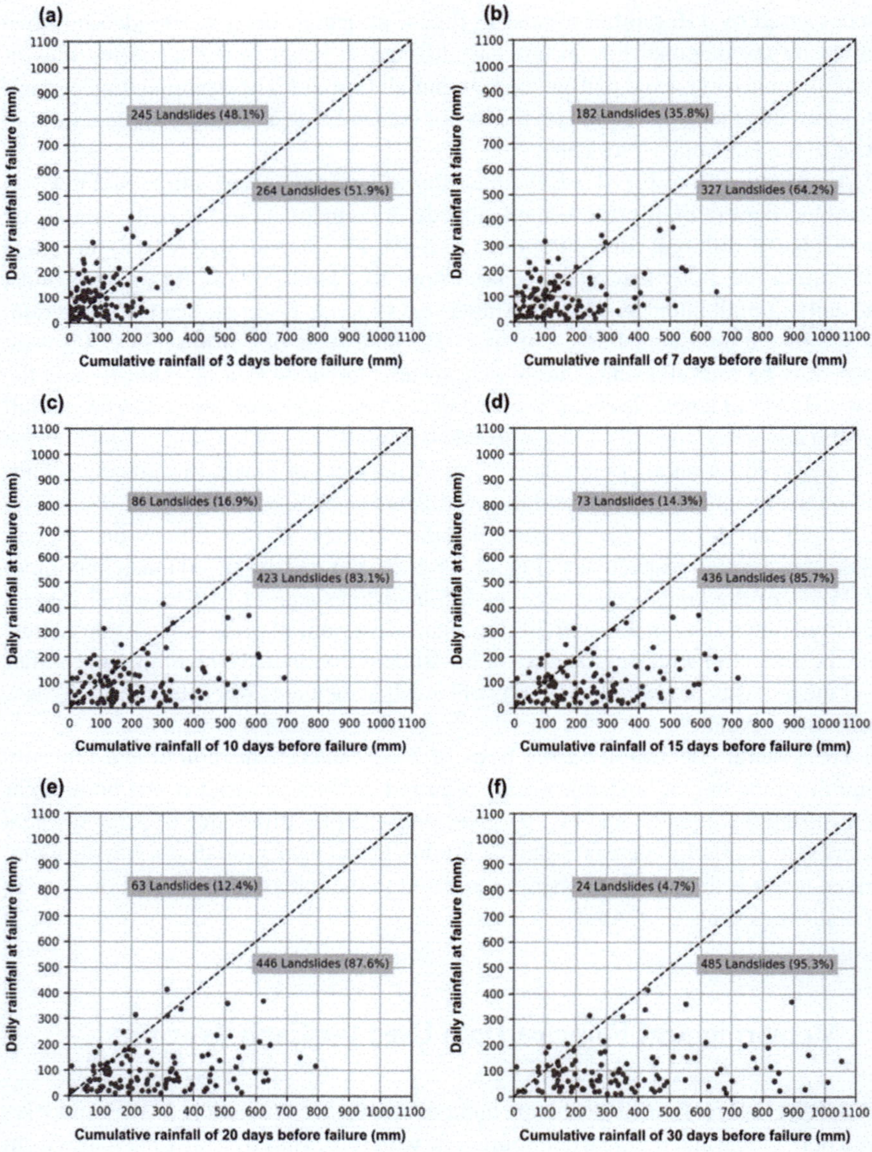

Fig. 4 Relationship between antecedent rainfall prior to failure (3, 7, 10, 15, 20 and 30 days shown at panels from a-f respectively) and daily rainfall at failure for 509 landslide occurrences caused by 116 rainfall events at Nilgiri district in Tamil Nadu

Fig. 5 A debris flow occurred on 14th June, 2018 at Chettiyampara colony in Kurumbilangod Village in Malappuram District of Kerala and is triggered by three days of antecedent rainfall. Image courtesy is GSI Field Season: 2018–19 (Item No.: M4SI/NC/SR/SU-KRL/2018/21108)

The current NCMRWF global ensemble prediction system has among the highest horizontal resolution (~12 km) in the world. There are 23 ensemble members (22 perturbed and 1 control), among that 11 members and control members are initialized at 00UTC of the present day and the remaining 11 members are from 12 UTC of the previous day (Mamgain et al. 2018). This ensemble system provides up to 10-day global forecasts, whereby forecast scenarios (ensemble members) can be objectively assigned to the closest matching weather pattern definition, such as those introduced in Sect. 2. The probabilities for each weather pattern occurring at each forecast lead time from the ensemble members are evaluated based on the observed training parameters used in the clustered analysis. Statistics referred to in Neal et al. (2016), such as the mean area weighted distance between members and their assigned weather patterns and the mean spatial correlation between members and their assigned weather patterns, can be used to match ensemble members to weather patterns. These probabilistic weather pattern forecasts provide useful guidance within the medium-to-long-range forecast periods (week one onwards).

For short-range forecasts and warnings of landslides, threshold analysis from the high-resolution model forecast based on the intensity and antecedent rainfall is used. The previous section discussed multiple rainfall parameters which can be used as a landslide-triggering threshold for the Nilgiri location in the Western Ghats. Power law distribution of the rainfall conditions such as rainfall duration and intensity resulting in landslides is written as the equation as $I = \alpha D^{\beta}$, where I is the rainfall intensity, D is the duration, α and β are the empirically derived parameters (e.g., Kanungo and Sharma 2014). Thresholds for the landslide occurrence with respect to rainfall intensity, and duration generated from power-law will be tested in the real-time for the selected pilot region from the model forecast. These combinations of parameters from the observed training sets will be compared with the regression values for the aforementioned parameters from the forecasted fields. We suggest that the power–law relationship between rainfall level and landslide occurrence can be used to estimate the probability of various levels of landslide activity during a rainfall event as a guide for making decisions related to emergency preparedness. This combination of parameters may be appropriate for shallow landslides but for complex failure types we may need a modified algorithm. For example, a Bayesian approach may be employed by combining observed rainfall data with the rainfall amount from the forecasted string to obtain antecedent rainfall.

For example, a landslide in Malappuram, Idukki, and Wayanad districts during June 14, 2018, lead to debris flow over a number of houses associated with a heavy spell of rain. Figure 5 is an image of one of those sites located in Malappuram where the landslide caused severe impacts. The site was marked with heavy rainfall that lasted for three days preceding the event leading to the build-up of an excess pore water pressure. Figure 6 (panel 3) shows the skill of the NCMRWF ensemble model in predicting antecedent high rainfall during the date of 9–13 June prior to slope failure in the Malappuram district on 14th June 2018 (Fig. 5). Other panels are the surface atmospheric parameters modulating the safety factor of the slope location through soil saturation level and pore water pressure. Hence predicting most of these variables from the models in 10-day advance (that is current maximum lead time forecast available) will show the robustness of the model in picking the closest matching weather regimes and threshold analysis for initiating the landslide occurrence (e.g., Neal et al. 2022).

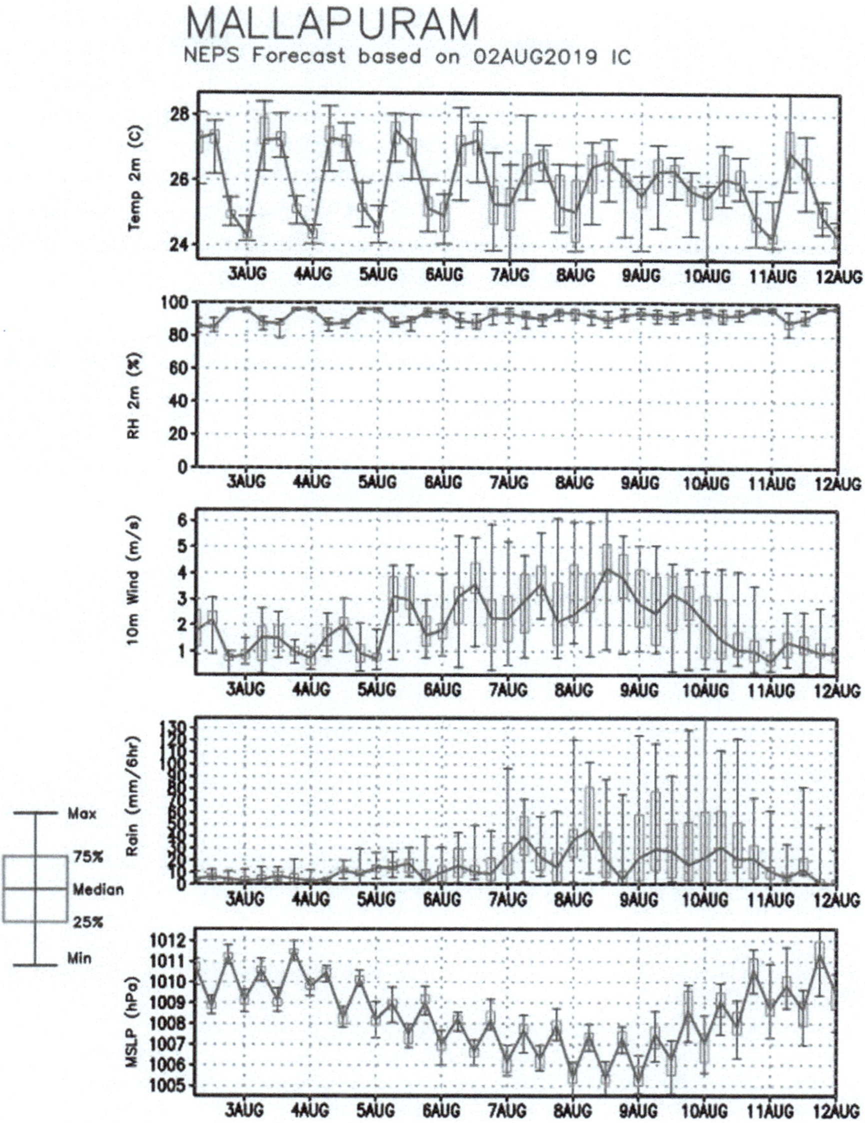

Fig. 6 A 10 day forecast of surface temperature, surface humidity, 10-m wind, rain, and MSLP from NCMRWF ensemble forecast based on an initial condition of 00UTC 04 June 2018

Under the international collaboration project introduced in Sect. 1, an experimental prototype of a regional landslide early warning system (EWS) is being set up for two pilot regions:- by GSI over Darjeeling in West Bengal and Nilgiri in Tamil Nadu. Following this project, EWS for other landslide-prone areas is being taken up beginning with the Rudraprayag district, in Uttarakhand with a joint effort of GSI, National Remote Sensing Centre, Indian Space Research Organisation

(ISRO), Disaster Mitigation and Management Center (DMMC) and NCMRWF. Effective use of the new parameters such as pore pressure, soil type, and moisture holding capacity will be considered for the development of the EWS system. In future, the NWP model used in the landslide EWS will be upgraded to employ a 30 m resolution ISRO land use land cover (LuLc) over the Indian region along with Cartosat 30 m digital elevation map orography. ISRO LuLc provides recent and updated information about the vegetative and non-vegetative model tiles over the Indian region, where soil moisture/temperature forecasts play a critical role in the slope failure predictability within the landslide EW system.

Acknowledgements The data generated by GSI for the FSP projects on the development of Landslide EWS in Nilgiris. We would like to warmly thank Dr. Pankaj Jaiswal, Director of GSI Eastern Region, Jharkhand, and Shri Eldhose K, Geosciences Division National Centre for Earth Science Studies, Kerala for the scientific discussion.

Author Contributions A. J initiated the chapter. R.N. and T.A. contributed the Sects. 2 and 3, respectively. A.J and A.K.M contributed to the writing of the manuscript.

References

Cruden DM, Varnes DJ (1996) Landslide types and processes. In: Turner AK, Schuster RL (eds) Landslides investigation and mitigation. Transportation research board, US National Research Council. Special Report 247, Washington, DC, Chapter 3, pp 36–75

Dee DP, Uppala S, Simmons A, Berrisford P, Poli P, Kobayashi S et al (2011) The ERA-Interim reanalysis: configuration and performance of the data assimilation system. Q J R Meteorol Soc 137:553–597

Dikshit A, Satyam N (2018) Estimation of rainfall thresholds for landslide occurrences in Kalimpong, India, technical note. Innov Infrastr Solut 3:24. https://doi.org/10.1007/s41062-018-0132-9

Fereday DR, Knight JR, Scaife AA, Folland CK, Philipp A (2008) Cluster analysis of North Atlantic-European circulation types and links with tropical Pacific sea surface temperatures. J Clim 21:3687–3703

Ferranti L, Corti S (2011) New clustering products. ECMWF Newsl 127:6–11

Ferranti L, Corti S, Janousek M (2015) Flow-dependent verification of the ECMWF ensemble over the Euro-Atlantic sector. Q J R Meteorol Soc 141:916–924. https://doi.org/10.1002/qj.2411

Guzzetti F, Peruccacci S, Rossi M, Stark CP (2007) Rainfall thresholds for the initiation of landslides in central and southern Europe. Meteorol Atmospheric Phys 98:239–267

Huang WTK, Charlton-Perez A, Lee RW, Neal R, Sarran C, Sun T (2020) Weather regimes and patterns associated with temperature-related excess mortality in the UK: a pathway to sub-seasonal risk forecasting. Environ Res Lett 15:124052

Hungr O, Leroueil S, Picarelli L (2013) The Varnes classification of landslide types, an update Landslides. Landslides. https://doi.org/10.1007/s10346-013-0436-y

Iverson RM (2000) Landslide triggering by rain infiltration. Water Resour Res 36:1897–1910

Kannan (2014) Landslide hazard zonation analysis and management in Bodi—Bodimettu Ghat section, Theni district, Tamil Nadu, PhD thesis Sastra University. http://shodhganga.inflibnet.ac.in/handle/10603/76444

Kanungo DP, Sharma S (2014) Rainfall thresholds for prediction of shallow landslides around Chamoli-Joshimath region, Garhwal Himalayas, India, Landslides 11:629:638. https://doi.org/10.1007/s10346-013-0438-9

Mamgain A, Sarkar A, Dube A, Arulalan T, Chakraborty P, George JP, Rajagopal EN (2018) Implementation of very high resolution (12 km) global ensemble prediction system at NCMRWF and its initial validation, Technical report, NMRF/TR/02/2018

Montrasio L, Valentino R (2008) A model for triggering mechanisms of shallow landslides. Nat Hazards Earth Syst Sci 8:1149–1159. www.nat-hazards-earth-syst-sci.net/8/1149/2008/

Neal R, Fereday D, Crocker R, Comer R (2016) A flexible approach to defining weather patterns and their application in weather forecasting over Europe. Meteorol Appl 23:389–400

Neal R, Dankers R, Saulter A, Lane A, Millard J, Robbins G, Price D (2018) Use of probabilistic medium- to long-range weather pattern forecasts for identifying periods with an increased likelihood of coastal flooding around the UK. Meteorol Appl 25:534–547

Neal R, Robbins J, Dankers R, Mitra A, Jayakumar A, Rajagopal EN, Adamson G (2020) Deriving optimal weather pattern definitions for the representation of precipitation variability over India. Int J Climatol 40:342–360

Neal R, Guentchev G, Arulalan T, Robbins J, Crocker R, Mitra A, Jayakumar A (2022) The application of predefined weather patterns over India within probabilistic medium-range forecasting tools for high-impact weather. Meteorol App 29. https://doi.org/10.1002/met.2083

Rajagopal EN, Mitra AK, Gupta MD, George JP, Ashrit R, Sarkar A, Jayakumar A (2019) Current and future activities in unified modelling and data assimilation at NCMRWF in the book 'current trends in the representation of physical processes in weather and climate models' edited by Randall DA, Srinivasan Ravi J, Nanjundiah S, Mukhopadhyay P, Springer Atmospheric Sciences. https://link.springer.com/book/10.1007%2F978-981-13-3396-5

Rao PLS, Mohanty UC, Ramesh KJ (2005) The evolution and retreat features of the summer monsoon over India. Meteorol Appl 12:241–255

Reichenbach P, Cardinali M, De Vita P, Guzzetti F (1998) Regional hydrological thresholds for landslides and floods in the Tiber River Basin (Central Italy). Environ Geol 35(2–3):146–159

Richardson D, Fowler H, Kilsby C, Neal R (2018) A new precipitation and drought climatology based on weather patterns. Int J Climatol 38:630–648

Richardson D, Neal R, Dankers R, Mylne K, Cowling R, Clements H, Millard J (2020) Linking weather patterns to regional extreme precipitation for highlighting potential flood events in medium- to long-range forecasts. Meteorol Appl 27:e1931

Robbins J, Dankers R, Dashwood C, Lee K, Neal R, Reeves H (2018) Early warning of landslides in Scotland using probabilistic weather pattern forecasts. EGU, Vienna, 8th to 13th April 2018

Sujatha R, Suribabu CR (2017) Rainfall Analyses of Coonoor Hill Station of Nilgiris District for Landslide Studies. IOP Conf Ser Earth Environ Sci 80(1):012066. https://doi.org/10.1088/1755-1315/80/1/012066

Varnes DJ (1978) Slope movement types and processes. In: Schuster RL, Krizek RJ (eds) Landslides, analysis and control, special report 176: Transportation research board National Academy of Sciences, Washington, DC., 11–33

Xiaohui C, Ma T, Li C, Liu H, Ding B, Peng W (2018) The catastrophic 13 November 2015 rock-debris slide in Lidong, south-western Zhejiang (China): a landslide triggered by a combination of antecedent rainfall and triggering rainfall. Geomat Nat Hazards Risk 9(1):608–623https://doi.org/10.1080/19475705.2018.1455750

Seismic Microzonation of Indian Cities and Strategy for Safer Design of Structures

O. P. Mishra, H. S. Mandal, Priya Singh, Ravikant Mahato, Sasi Kiran Gera, Vikas Kumar, Vandana, Babita Sharma, Shashank Shekhar, Poorti Gusain, Sanjay K. Prajapati, Anurag Tiwari, and Sireesha Jaladi

Abstract Resiliency to structures during earthquake shaking is extremely important for cities that are associated with earthquake source zones or located in the vicinity of earthquake prone belts. Present state of knowledge does not allow us to predict the occurrence of earthquakes with respect to location, time, and size of the earthquake. Here, we made an attempt to address the issues involved with earthquake risk resiliency to structures in the light of availability of GIS-based comprehensive seismic microzonation maps, comprising of multi-thematic maps pertaining to se-smological, geophysical, and geotechnical characteristics of the sub-surface formation on 1:10,000 and 1:25,000 scales for many Indian cities. This becomes very important since diagnostic earthquake precursors and earthquake early warning system for the earthquake prone-regions are not yet feasible till date. Successful case studies for some of Indian cities (Delhi-NCR, Bengaluru, Guwahati, Kolkata, Jabalpur, and entire Sikkim state) aptly demonstrated the efficacy of different thematic maps that are generated through seismic microzoning areas under investigations, which can be used by civil engineers and city planners during construction of safer dwelling and infrastructures. Here, we present a detailed discussion about seismic macro-hazard zonation vis-a-vis the seismic microzonation with reference to the evaluation of several geotechnical parameters to highlight the importance of adoption of seismic microzonation program as an earthquake risk mitigation strategy which can help in promoting the construction of earthquake risk resilient structures. Extended form of the methodology of seismic microzonation of a region with finer details along with a way forward for its diverse applications has been discussed here. We have also provided a cyclic seismic risk resilient strategy for sustainable development of infrastructures for safer urban agglomerates through land use planning that may help in taking seismic risk mitigation decisions.

Keywords Risk resilient structures · Seismic microzonation · Site response · Site vulnerability · Geotechnical parameters · Seismic risk mitigation decision

---

O. P. Mishra (✉) · H. S. Mandal · P. Singh · R. Mahato · S. K. Gera · V. Kumar · Vandana · B. Sharma · S. Shekhar · P. Gusain · S. K. Prajapati · A. Tiwari · S. Jaladi
National Center for Seismology, Ministry of Earth Sciences, Lodhi Road, New Delhi 110003, India
e-mail: omp.mishra@nic.in

V. K. Gahalaut and M. Rajeevan (eds.), *Social and Economic Impact of Earth Sciences*,
https://doi.org/10.1007/978-981-19-6929-4_20

1 Introduction

Structures that are designed well to protect them from damage during earthquake shaking are referred as Earthquake Risk Resilient Structures (ERRS) and such structures make the urban system earthquake risk resilient (Srivastava et al. 2017; Gupta 2020). It is practically impossible to make the buildings and structures completely risk proof due to ambiguity in the precise prediction of the earthquake loads, however, these can be built as risk resilient which substantially reduces the damage as they resist the severe impact of earthquake shaking on them. The seismic macro-hazard zone map of India, in general indicates the hazard level of different zones. It can be divided into several classes based on the empirical attenuation law for a known maximum credible earthquake for a possible epicentre distance of the earthquake source without any consideration of the in-situ material heterogeneities and geotechnical properties of the sub-surface sites (Gupta 2000; Mishra 2012; Vandana and Mishra 2019; Sharma et al. 2016; Gupta 2020; Mishra 2020; Mishra et al. 2020a, b; Sharma and Mishra 2020). Different researchers determined different parameters for different areas (Arya 1992; Bolt 1987; Champatiray et al. 2001; Dutta et al. 2012; Immè and Morelli 2012; Mahajan et al. 2007; Mishra 2011; Molnar 1990; NDMA 2007; Oldham 1882; Pande et al. 2006; Paudiyal and Singh 2005; Riggio and Santulin 2012, 2014; SAARC 2009; Yadav et al. 2019; Zhang et al. 1999). On the other hand, ERRS takes into the account various parameters, such as, geology, geomorphology, seismicity, past earthquakes, and liquefaction history of the area through detailed seismic zoning and seeks fine scale information of the zone by dividing into several smaller zones of the seismic macro-hazard zone within the study area. Resistant building codes can be generated for the probability of the highest magnitude that can occur in that area. This approach may help prevent damage to infrastructures caused by the earthquakes, which can help saving lives of people and property. Numerous studies exist elsewhere in the world which stressed up on the need for risk resilient structures due to rapid urbanization since such an endeavour has paramount importance for the construction of safer structures, particularly those under the threat of natural disasters falling under a rapid and unplanned urbanization (Rai 2000; Bilham et al. 2001; Bruneau et al. 2003; Sahabi et al. 2018; Gupta 2020; Mishra et al. 2020). This is especially relevant in the earthquake prone areas where no earthquake early warning system exists and no diagnostic precursors which may precede earthquakes exist (Mishra 2020). It is reported that about 57% of total area of India falls under the seismically vulnerable zones, of which more than 12 Crore buildings fall under seismically hazard prone zones. Most of these buildings are not earthquake-resistant and are potentially vulnerable to collapse in the event of a large magnitude earthquake. As it is rightly said that "it is not the earthquake that kills people, but the structures that kill people". It is necessary to find a solution for safer structures and it is observed that many professionals are engaged in the area of Earthquake Resistant Design (EQRD) to create various cost-effective design solutions that make structures less vulnerable to earthquakes. An Earthquake Resistant Design (EQRD) must satisfy several criteria (Rai 2000), such as, the effect of earthquake on infrastructure; infrastructure's ability to resist the impact of shaking without failure, feasible economic cost of earthquake risk resiliency, etc. The constructional design techniques have been developed primarily for the last five decades, mostly in the developed countries which are under seismically active regions such as the United States, Japan, New Zealand, and China. It has been observed that traditional structures in earthquake-prone areas include special construction features, which made them less vulnerable to earthquakes (Sahabi et al. 2018). Even then sustainability of such structures should be ensured based on scientific and engineering solutions and this is where seismic microzonation may serve the purpose of constructing earthquake risk resilient structures.

This paper presents a conceptual framework to define seismic risk as a function of seismic activity and vulnerability of the built environment in a given area. Since the earthquake engineer has no control over the earthquake itself, mitigation of seismic risk (Rai 2000), which needs to be addressed to find very apt and logical solution of such unresolved issue. Therefore, engineers must construct structures by taking care of the design of foundation of the structures, which can withstand and resist the severe impacts of earthquake ground motions, preferably with the use of new seismic risk resilient building design codes that are derived from the comprehensive seismic microzonation studies of the areas under study. Such step, in turn, not only reduces the recurrent loss of economic and human life but also help to adopt the plan for safer and sustainable structures with minimum cost of expenditure for the construction of the said structures.

2 Seismic Microzonation and Risk Resiliency of Infrastructure

Seismic Hazard Microzonation is one of the most powerful methodologies to make a comprehensive quantification of earthquake risks for a specific area of interest, and requires detailed information of several pertinent factors (e.g., geological, geophysical, seismological, topological, geomorphological, and geotechnical) that characterize the subsurface formations to understand more about soil properties/depositional history, site specific effects along with the typology and design of the constructions.

A comprehensive work plan is executed to estimate various parameters by applying geophysical, seismological, and geotechnical tools (Fig. 1b) to understand strong bearing of these factors on dictating the nature and extent of the damage pattern (Gupta 2020; Mishra 2020).

Seismic Hazard Microzonation is based on detailed study by dividing the existing seismic hazard macro-zones into several smaller zones to investigate variability of the physical properties of the sub-surface materials related to geo-mechanical strengths of sub-surface formations at sites at a local-scale for ascertaining the seismic safety of

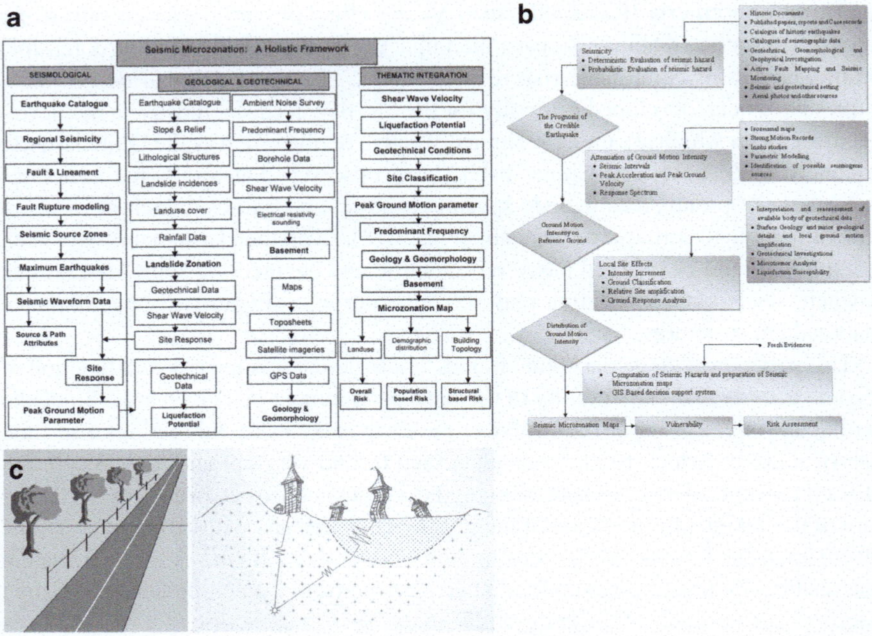

Fig. 1 **a** A comprehensive work-plan for systematic study of Seismic Microzonation of cities in India (adapted from NCS-MoES (2016), Mishra (2020)). **b** A comprehensive Seismic Microzonation Framework from seismicity to Risk Assessment (after Mishra (2020)). **c** A snap showing varying impact of earthquake shaking on closely spaced objects that advocated the deployment of Seismic microzonation (adapted from Mishra (2020))

the foundational design of built infrastructures under the given framework (Fig. 1a, b). The site-specific-based information (e.g., spectral amplification; peak frequency, site specific risk index; peak ground acceleration; safety factor of the site, liquefaction, etc.) can be used by the city/town planner for the development of risk resilient infrastructures for the city by mapping/highlighting the most and least seismically vulnerable sites. The impact of earthquake shaking is astonishingly different on the same objects even if these are closely located due to varying geo-mechanical strength of the sub-surface formations or foundational material beneath the objects that get impacted as shown in Fig. 1c.

Seismic microzonation of a city is helpful for new settlements as well as for the development of sustainable smart cities in India whilst the investigations further furnish significant information about the existing structures that may require engineering solutions of retrofitting and some structures may be demolished for new constructions as per new building design codes (Mishra 2020; Mishra and Ghatak 2020). Most of the existing seismic design codes are based on the empirical knowledge accumulated through systematic earthquake damage data collection and analysis by characterizing earthquake source and seismic quality factors of the media through which seismic wave propagates (Sharma et al. 2016; Vandana and Mishra

2019). Adequate levels of protection and seismic design requirements have gradually increased from one catastrophe to other due to increased information on the effect of different earthquakes of varying magnitudes under the plan of revision of seismic codes, especially for earthquake prone countries, which are economically developed (e.g., U.S.A, Japan, China, Italy, and Spain). Significant efforts were made by structural engineers to use the knowledge acquired through experimental and analytical studies for detailed information on structural parameters to bring out various strategies to increase inelastic capacity of buildings and structures to resist earthquake ground motions with acceptable damage levels. In recent years, the general concept on seismic design codes describes the performance objectives, which practically are unable to quantify the losses in terms of safety of structures. This concept requires further improvement and is only possible by the estimation of comprehensive geophysical, seismological, and geotechnical parameters through extensive seismo-geophysical surveys for better understanding of earthquake genesis, hazards, and mitigation (Gupta 2000; Gahalaut 2010) which will constitute an important ingredient for the seismic hazard microzonation studies that help in bringing out new earthquake risk resilient building codes.

3 Seismic Macrozonation vis-à-vis Seismic Microzonation

India plate is one of the seismically most active plate with variable types of earthquakes that occur in different regions along the periphery of Indian landmass or in different parts of India such as the earthquakes in the Himalayan region, Andaman–Nicobar region plate interior (referred as stable continental region), earthquakes triggered by surface reservoirs, rock burst, and mining activities (Fig. 2a). These are associated with very complex and intricate seismotectonic settings with varying structural heterogeneities of different types (Bilham 1995; Bilham et al. 2001; Zhao et al. 2002; Ambraseys and Jackson 2003; Bilham 2004; Zhao et al. 2004; Valdiya 2010; Mishra 2014; Panda et al. 2020) as shown in Fig. 2b. As a result of the varying earthquake activity, many Indian cities fall under seismic hazard zones III; IV; V (Fig. 3). According to the latest seismic zoning map of India (Bhatia et al. 1999: BIS 2002) about 59% of India's land area falls under vulnerable to moderate or severe seismic hazard, and is very much prone to shaking of MSK intensity VII and above. In the recent past, most Indian cities have witnessed the phenomenal growth of multi-storied buildings, super malls, luxury apartments and social infrastructure as a part of the development. Rapid expansion of the built environment in moderate or high-risk cities makes it imperative to incorporate seismic risk reduction strategies in various aspects of urban planning and construction of new structures.

During the period 1990–2015, India has experienced the impact of at least nine damaging earthquakes and the most devastating tsunamigenic earthquake, which resulted in over 40,000 deaths and caused enormous damages to property, assets, and infrastructure in India. The entire Himalayan region is considered to be vulnerable to great earthquakes of exceeding 8.0, and in a relatively short span of about 50—years,

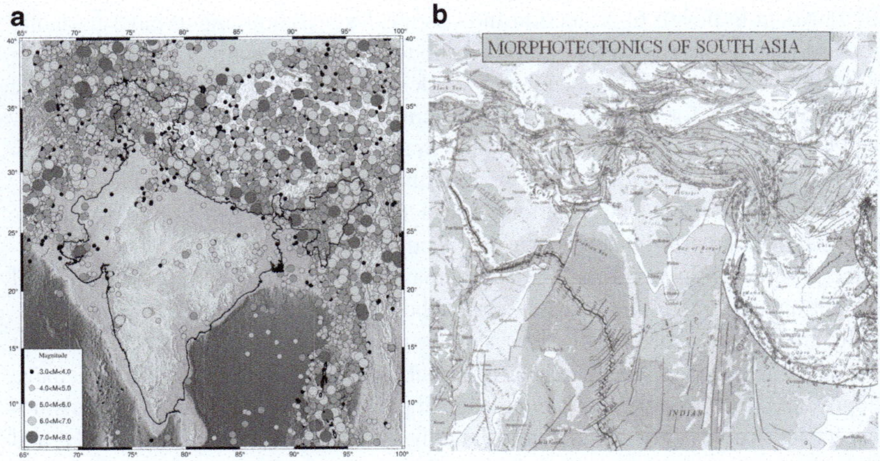

Fig. 2 **a** Earthquake distribution in India and its adjoining region of various strengths. **b** Complex geo-morphotectonic set up of South Asia associated with faults, lineaments, rift and ridges denoted by broken and thick lines along with past seismicity (in circles) (adapted from online sources and Valdiya (2010), Mishra (2020))

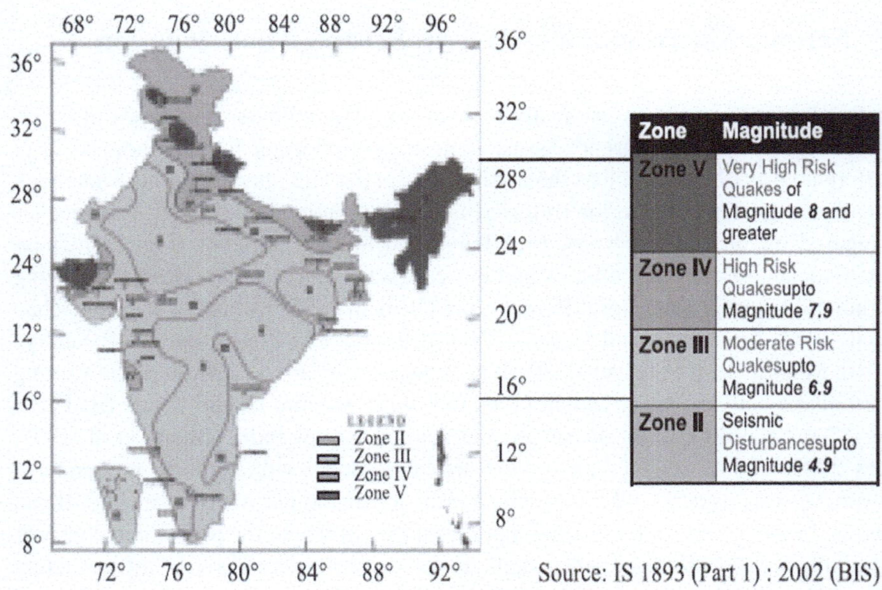

Fig. 3 A map showing Seismic Hazard Zones of India (adapted from, BIS (2002))

four such earthquakes have occurred, these are 1897 Shillong (M 8.1) 1905 Kangra (M.7.8) 1934 Bihar–Nepal (M 8.3), and 1950 Assam–Tibet (M 8.6). Several scientific publications suggest for very severe earthquakes that may likely to occur anytime in the Himalayan Region, which could adversely affect the lives of several million people in India (Bilham et al. 2001; Bilham 2019). Similarly the plate boundary in the east, the Sumatra–Andaman subduction zone, has witnessed giant earthquake in 2004 which caused catastrophic tsunami. In the plate interior regions, extensive research suggest several regions are active which have witnessed earthquakes in past, such as the Koyna earthquake of 1967; the Latur earthquake of 1993; the 1997 Jabalpur earthquake, and the 2001 Bhuj earthquake.

It is worth to mention that Seismic Hazard Zonation map of India is a macro zonation map derived from the extrinsic attenuative behaviour of seismic energy generated from the earthquake by analysing the regional earthquake catalogue data with reference to some of the known maximum credible earthquakes that occurred in past (Mandal et al. 2013, 2014; Mishra 2020). Thus, the seismic hazard zonation map is qualitative in nature and does not appropriately provide comprehensive site-specific information for the zone on a local scale. It is evident that earthquake hazards have strong bearing on in-situ material heterogeneities and geotectonic settings. Secondly, geological-geophysical vulnerability index of the area constitutes an important ingredient for earthquake risk assessment since earthquake vulnerability for the area depends on the size (magnitude) and depth of the earthquakes along with the geotectonic settings of the area. The comprehensive assessment of damage and loss, however, can be estimated once the socio-economic-demographic strata of the area under investigations would be known or be made available. Most importantly, earthquakes neither be predicted nor be stopped but one can develop a safer earthquake risk resilient structures to mitigate the extent of loss to people and properties (Mishra 2020).

Under such scenario, it is envisaged that there is a need to evolve a suitable plan for earthquake risk mitigation. Accordingly, a comprehensive plan for seismic microzonation for cities in India for suitable mitigation measures can be adopted for making structures earthquake risk resilient for city planning to be sustainable (RFP-MoES 2017; NCS-MoES 2016) as per the international guideline of earthquake mitigating design of structures (FEMA 2010).

It is important to note that each of the above-mentioned zones can reasonably expect to have earthquake shaking of more or less same maximum intensity in future. It is, however, the maximum seismic ground acceleration in each zone which cannot be presently predicted with accuracy either on a deterministic or on a probabilistic basis. The basic zone factors included in the code are reasonable estimate of effective PGA for the design of various structures. In accordance to GSHAP programme, PGA shows good correspondence with the existing Seismic Hazards of the region (Gupta 2000) as shown in Fig. 3.

The degree of damage to structures, loss of lives, flora and fauna, and the environmental degradation during earthquake depends mainly on two factors, the first one, Earthquake inherent parameters such as the location (Latitude & Longitude);

Earthquake origin time; focal depth or hypocenter depth; magnitude of the earthquake; type of faulting; maximum credible earthquakes triggered by past known and exposed faults or even by the hidden causative fault. The second one is the **vulnerability index** of the area where earthquake occurs. The vulnerability index is influenced by the demographic set up of the earthquake prone area, status of buildings and infrastructures; social, economic and environmental set up of the people who live in the earthquake prone region.

4 Paradigm Shift in Disaster Risk Assessment

Microzonation is an important methodology to generate parameters for site-specific structural designs, land use plans, and disaster mitigation, and therefore, **Seismic Hazard and Risk Microzonation (SHRM)** forms a significant tool in risk evaluation and mitigating the disastrous impact of earthquakes. All over the world, there is paradigm shift for the management of earthquake disaster from the **response-centric** regime consisted of 'Rescue', 'Relief', and 'Rehabilitation (3R)', to mitigation and preparedness-centric regime as pre-disaster management includes: Prevention, Mitigation and Preparedness (PMP) (Mishra 2020), where efforts and funds are to be invested to address the underlying causes of vulnerability for the design of safer structures under the preparedness as the pre-disaster risk response (Singh et al. 2017; Srivastava et al. 2017). There is a paradigm shift of disaster risk resilient assessment based on the Hyogo Frame for Reference (**2005–2015**) and post Hyogo Frame of Reference referred to as the Sendai Disaster Risk Management reference (Sendai 2015) that stresses upon effective steps to be taken prior to expected disastrous events that struck the area under the pre-disaster responses measures. Based on several factors that impacted structures during earthquake shaking, a holistic cyclic seismic risk resilient strategy has been that can shed light on sustainable development of earthquake prone belts of the region. The strategy also demonstrates the significance of seismic risk resilient infrastructures by addressing several issues related to safety of school, hospitals, prisons, vital installations, besides mitigating social chaos, and earthquake traumas for providing safety of flora and faunas to ensure economic prosperity of the area. Seismic hazard microzonation is an important ingredient of this strategy.

On the basis of the above discussion in accordance to the model (Fig. 1a), it has been decided to conduct seismic microzonation study for cities in India in a scale bringing down to as precise as 1: 25,000/1:10,000 scale, based on decision by a group of Indian experts. The work components and applied methodology of estimating several parameters related to "Ground Characterization"; "Site Response"; and "Source Characterization" and outcomes of investigations are shown under Tables 1 and 2.

Estimates of average shear wave velocity (Vs 30); site amplification; PGA; liquefaction potential; and hazard index and with corresponding predominant frequency are made.

Table 1 Different seismo-geophysical-geotechnical investigations adopted for seismic microzonation study

Methodology	In-situ and experimental techniques
Geophysical and seismological	• Microtremor Survey • Down-Hole Shear Wave Velocity Test (DHT) • Multi-Channel Analysis of Surface Waves (MASW)
Geotechnical (in-situ)	• Boreholes of each 30 m depth for SPT/SCPT/DCPT • Standard Penetration Test (SPT) • Seismic Cone Penetration Test (SCPT) • Dynamic Cone Penetration Test (DCPT)
Geotechnical (laboratory)	• Cyclic Tri-axial test • Resonant Column Test • Atterberg limits • Bulk Density • Specific Gravity • Natural Water Content • Coefficient of Consolidation • Complete Grain Size Analysis • Direct Shear

5 Seismic Microzonation of Indian Cities

MoES has developed guidelines for seismic microzonation which agree with the international practices as described under Seismic Microzonation Manual and Seismic Microzonation Handbook developed and published by the Ministry of Earth Sciences (MoES) in the year 2011 (MoES 2011a, b). Detailed investigations have been carried out to generate information through seismic microzonation. Accordingly, different institutes of India have completed comprehensive microzonation for several cities (e.g., Bangalore, Delhi-NCR, Dehradun, Jabalpur, Gandhinagar, Kolkata, and Sikkim state). National Centre for Seismology, Ministry of Earth Sciences, has taken up this project as its one of the flagship missions, entitled "Geotechnical and Geophysical Investigation under the project of Seismic Microzonation of 30- cities in India", which are under execution since October 2018. Currently, four cities (e.g., Bhubaneshwar, Chennai, Coimbatore, and Mangalore) are going to be completed and are under full swing investigation, while cities, namely, Agra, Amritsar, Dhanbad, Kanpur, Lucknow, Meerut, Patna, Varanasi (Fig. 5) are also being investigated since 15 February 2021. We attempted to highlight that the intensity and attenuation-based seismic hazard zonation map is not sufficient to provide quantitative and qualitative information on damage scenario of infrastructures during earthquake shaking. We demonstrated that the in-situ material heterogeneity and the geo-mechanical strength of soil play an important role in estimating risks to infrastructures by using comprehensive seismic microzonation in different cities in India (see Fig. 4).

Table 2 Seismo-geophysical-geotechnical parameters generated by comprehensive seismic micro-zonation studies for earthquake risk resilient structures (RRS)

Parameters	Test	Purpose	Result/outcome
Geophysical and seismological	Microtremor survey (In a grid pattern of 500 m × 500 m)	To know the peak frequency and amplification of a site	• Peak Frequency (F_0) • Peak Amplification (A_0)
	Down-hole shear wave velocity test (DHT)	Down-hole techniques provide accurate shear wave velocity measurements to depths of a few hundred meters	• Shear wave velocity (V_s)
	Multi-channel analysis of surface waves (MASW)	To determine the shear wave velocity (VS) and evaluate the dynamic properties of soil in the shallow sub-surface	• Generation of dispersion curve from captured Rayleigh wave
Geotechnical (in-situ)	Boreholes of each 30 m depth for SPT/SCPT/DCPT	To get the layer wise information of soil	• SPT, SCPT, DCPT, etc.
	Standard penetration test (SPT)	To identify the relative densities of granular deposit and evaluate the engineering strength of the soil material	• SPT-N values • The SPT is used to provide results for empirical determination of a sand layer's susceptibility to soil liquefaction based on research performed by Harry Seed and T. Leslie Youd • Collection SPT disturbed samples for the laboratory analysis (Grain size analysis, Atterberg limits, • Specific gravity, Natural water content, Direct Shear, Tri-axial shear consolidated un-drained and drained)
	Seismic cone penetration test (SCPT)	To determine the shear wave velocity (VS) and Poisson's ratio and soil profile	Shear Modulus, Poisson's ratio and soil profile

(continued)

Table 2 (continued)

Parameters	Test	Purpose	Result/outcome
	Dynamic cone penetration test (DCPT)	To determine the effort required to force a point through the soil and so obtain resistance value	• N-values • Identification of variability of subsoil profile and can locate the soft pockets such as filled up ponds
Geotechnical (laboratory)	Cyclic tri-axial test	The material properties of the soil like shear resistance, cohesion and the dilatancy stress is determined from this test	Shear resistance of the sample
	Resonant column test	The resonant column test is used to determine the shear or elastic modulus and damping characteristics of soils based on the theory of wave propagation in prismatic rods	• Shear modulus • Damping ratio
	Atterberg limits	The Atterberg limits can be used to distinguish between silt and clay, and to distinguish between different types of silts and clays	• **Plasticity index** (PI = LL-PL) • **Liquidity index** LI = (W-PL)/(LL-PL) where W is the natural water content • **Consistency index** CI = (LL-W)/(LL-PL), where W is the existing water content
	Natural water content	Determination of bearing capacity and settlement in the soils. The natural water content will give an idea of the state of soil in the field	The natural water content of the soil sample
	Bulk density	To express soil physical, chemical and biological measurements on a volumetric basis for soil quality	Bulk Density Density = mass/volume

(continued)

Table 2 (continued)

Parameters	Test	Purpose	Result/outcome
	Specific gravity	Determination of specific gravity of soil will help in the calculation of void ratio, degree of saturation and other different soil properties	$G = \frac{G_L(m2-m1)}{(m4-m1)-(m3-m1)}$ **CL** = specific gravity of the liquid used, at the constant temperature; **m1** = mass of density bottle in g; **m2** = mass of bottle and dry soil in g; **m3** = mass of bottle, soil and liquid in g; and **m4** = mass of bottle when full of liquid only in g
	Coefficient of consolidation	It is used to describe the rate at which saturated clay or other soil undergoes consolidation, or compaction, when subjected to an increase in pressure	Shear resistance of the sample
	Complete grain size analysis	This test is performed to determine the percentage of different grain sizes contained within a soil	Percentage of distribution of grain size and grain size maturity
	Direct shear	To determine the shear strength of the soil by direct shear	The shear strength (s) = c1 + $\bar{\sigma}$ tan Ø1 Where C1 = Effective cohesion $\bar{\sigma}$ = Effective stress Ø1 = Effective angle of shearing resistance

In assessing the seismic hazard of any urban centre, ambient noise measurements are quite popular in estimating the amplification and the dominant frequencies of the soil column using experimental microtremor method. The method requires a seismic broad band station with three components for estimating the site response using Nakamura technique where the resonance frequency is obtained by evaluating the horizontal to vertical spectral ratios (Nakamura 1989). The main consideration of this technique is the micro tremors which are primarily composed of Rayleigh waves, produced by local sources. These waves propagate in a surface layer over a half space, considering the motion at the interface of the surface layer and half space is not affected by the source effect and the horizontal and vertical motion at this interface are approximately equal. Site response studies mainly deal with the determination of peak frequency (PF) of the soft soil, amplification and the nature of response

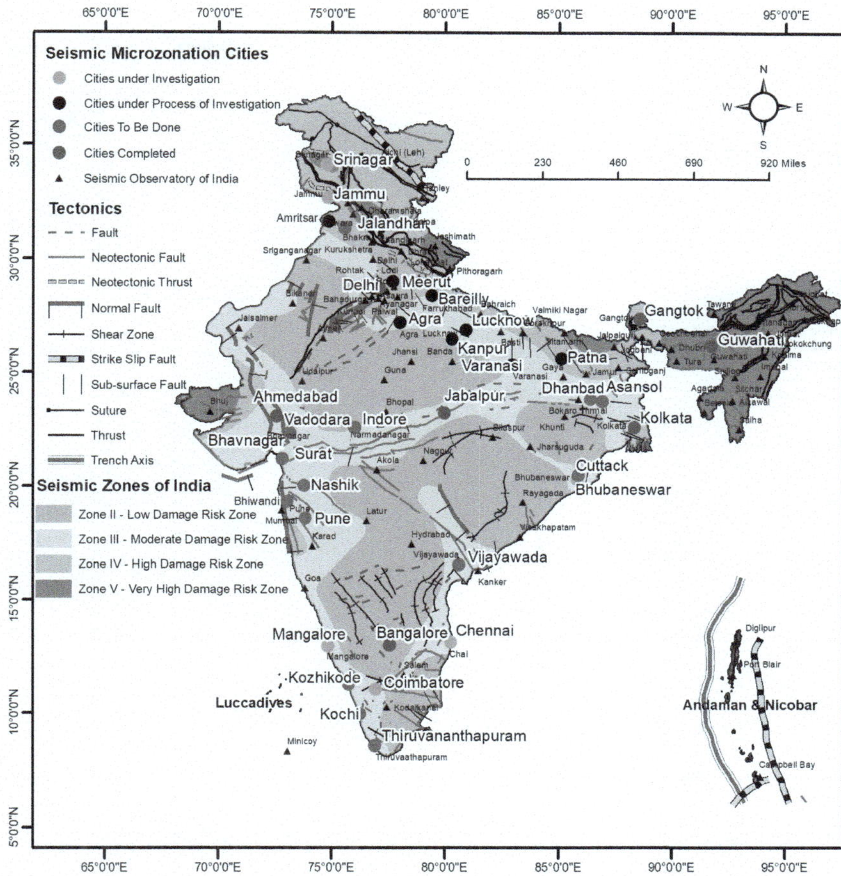

Fig. 4 Distribution of cities in India under seismic microzonation by NCS, MoES

curve that defines the transfer function at the site which forms an important input for evaluating and characterizing the ground motion for seismic hazard quantification. Here, we discuss the efforts of Ministry of Earth Sciences to generate comprehensive reports on seismic microzonation of a few cities in India under its flagship project of earthquake disaster risk management.

5.1 *Seismic Microzonation of Delhi-NCR*

Delhi-NCR has been studied under seismic microzonation plan to estimate detailed parameters based on seismo-geophysical and geotechnical investigations for generating various thematic maps, such as Peak Amplification factor; predominant frequency, Peak Ground Acceleration (PGA) at the surface as well in the engineering

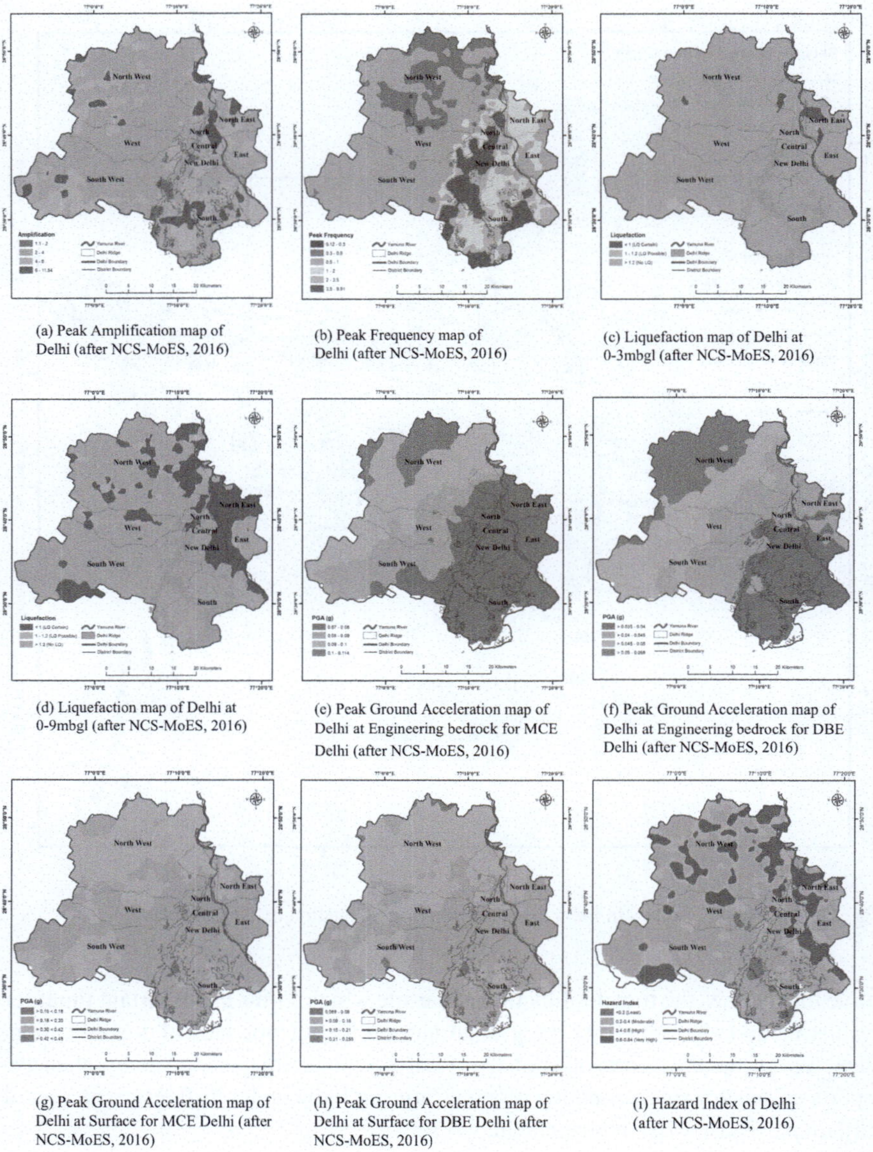

(a) Peak Amplification map of Delhi (after NCS-MoES, 2016)

(b) Peak Frequency map of Delhi (after NCS-MoES, 2016)

(c) Liquefaction map of Delhi at 0-3mbgl (after NCS-MoES, 2016)

(d) Liquefaction map of Delhi at 0-9mbgl (after NCS-MoES, 2016)

(e) Peak Ground Acceleration map of Delhi at Engineering bedrock for MCE Delhi (after NCS-MoES, 2016)

(f) Peak Ground Acceleration map of Delhi at Engineering bedrock for DBE Delhi (after NCS-MoES, 2016)

(g) Peak Ground Acceleration map of Delhi at Surface for MCE Delhi (after NCS-MoES, 2016)

(h) Peak Ground Acceleration map of Delhi at Surface for DBE Delhi (after NCS-MoES, 2016)

(i) Hazard Index of Delhi (after NCS-MoES, 2016)

Fig. 5 **a** Peak Amplification; **b** Peak Frequency; **c** Liquefaction map at 0–3 mbgl; **d** Liquefaction map at 0–9 mbgl; **e** Peak Ground Acceleration at Engineering bedrock for MCE; **f** Peak Ground Acceleration map at Engineering bedrock for DBE; **g** Peak Ground Acceleration at Surface for MCE; **h** Peak Ground Acceleration at Surface for DBE; **i** Hazard Index map for the city of Delhi-NCR (after NCS-MoES (2016)) bed rocks; liquefaction at various sub-surface levels, and, finally hazard index in scale of 1:10,000 as shown in Fig. 6a–i, which are self-explanatory that can be used by the construction engineers, policy and planners, disaster practitioner for development of the risk resilient structures for the city. We found that NCR-Delhi has heterogeneous distribution of soil thicknesses with comparatively thicker alluvial soils in the vicinity of the river Yamuna in the east Delhi and in the region around the Delhi ridge area it is found to be associated with loose sediments. Peak ground acceleration and liquefaction map at varying depths corroborate with the corresponding peak frequency and site amplification distribution map made for NCR Delhi. We infer from uneven distribution of hazard Index map that NCR-Delhi is conspicuously associated with high (0.4–0.6) to very high hazard (<0.6) index with some zones in pocket associated with the least (<0.2) to moderate (0.2–0.4) hazard index having good corroboration with site-specific geological formations and geo-tectonics. It is imperative to adopt site-specific-based new structural design codes derived from seismic microzonation to mitigate risks to structures for the city that help evolve effective strategy for safer and sustainable infrastructures for NCR-Delhi.

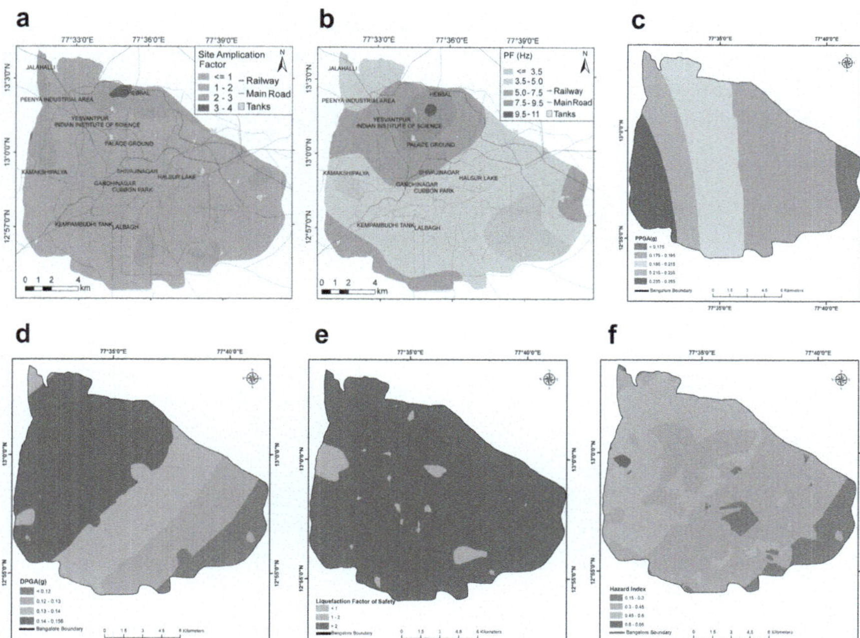

Fig. 6 **a** Amplification Factor; **b** Peak Frequency; **c** Distribution of PGA by PSHA; **d** Distribution of PGA by DSHA; **e** Liquefaction Factor; **f** Hazard Index map for the city of Bengaluru (adopted from Anbazhaganet al. (2010))

5.2 Seismic Microzonation of Bengaluru

Seismic microzonation study of Bengaluru provided a deep insight into the subsurface layers, which controls the extent of peak ground acceleration that plays an important role to categorize which type of structures will collapse or survive for what ground acceleration during shaking. It is, therefore, survival of structures that can be ensured by estimating peak ground acceleration at sites estimated from the seismic hazard zonation map for the city using the concept of both probabilistic and deterministic approaches as shown in Fig. 7a–f. The estimated peak amplification factor (PA) is found to vary in the range $4.0 \leq PA \leq 1.0$ with corresponding variability of Predominant frequency (PF) being $11.0 \leq PF \leq 3.5$.

Expected PGA at rock level using Deterministic Seismic Hazard Assessment for Bengaluru is about 0.15 g, which is found to corroborate with the 3-D geotechnical model based on analyses of large amount of borehole data for making estimates of the corrected SPT ‘N’ values and soil shear wave velocity using Multichannel Analysis of Surface Wave (MASW). It is reported that Bengaluru corresponds to site class D associated with the amplification factor varies from 1 to 4.7 for the corresponding predominant frequency that varies from 3 to 12 Hz. The results of the liquefaction hazard map using factor of safety demonstrated that Bangalore is safe against liquefaction except at few locations where the overburden is sandy silt with presence of shallow water table (Anbazhagan 2010).

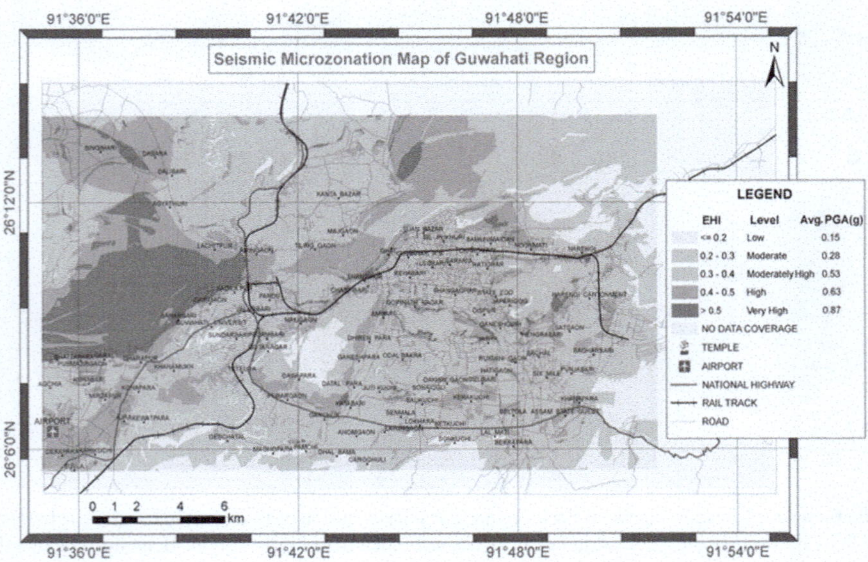

Fig. 7 Hazard Index map of Guwahati, India (adopted from Nath and Thingbaijam (2008))

5.3 Seismic Microzonation of Guwahati

Guwahati city has been microzoned with local and regional site conditions incorporated through GIS to find the seismic ground motion hazard map and the corresponding Hazard Index map for the city (Fig. 8a, b), comprehensively reported by Nath and Thingbaijam (2008).

The microzonation map (Fig. 8) of the city of Guwahati divided into five hazard sub-zones, namely very low with average hazard level of 0.16; low with average hazard level of 0.26; moderate with average hazard level of 0.35; high with hazard level of 0.44; and very high having hazard level greater than 0.55. It is interesting to note that very high hazard zone falls under the site class IIIA where average shear wave velocity (Vs 30) is 228 m/s with site amplification is larger than 5.5 and average predominant frequency is 1.15 Hz with corresponding very high PGA value greater than 0.81 g, distributed to various sites of the residential areas ascribed to the western part of Guwahati. The high hazard zone under site class IIIB shows average shear wave velocity (Vs 30) of 260 m/s with average predominant frequency of 2.93 Hz for the corresponding average site amplification of 4.81.It is observed that low and very low hazard zones are under site class D which is almost covered by Hillocks, demonstrating high average shear wave velocity of 340 m/s while average predominant frequency of 7.1 Hz with corresponding very low PGA value of 0.11 and the lowest site amplification less than 1.5 that falls in the periphery of hills.

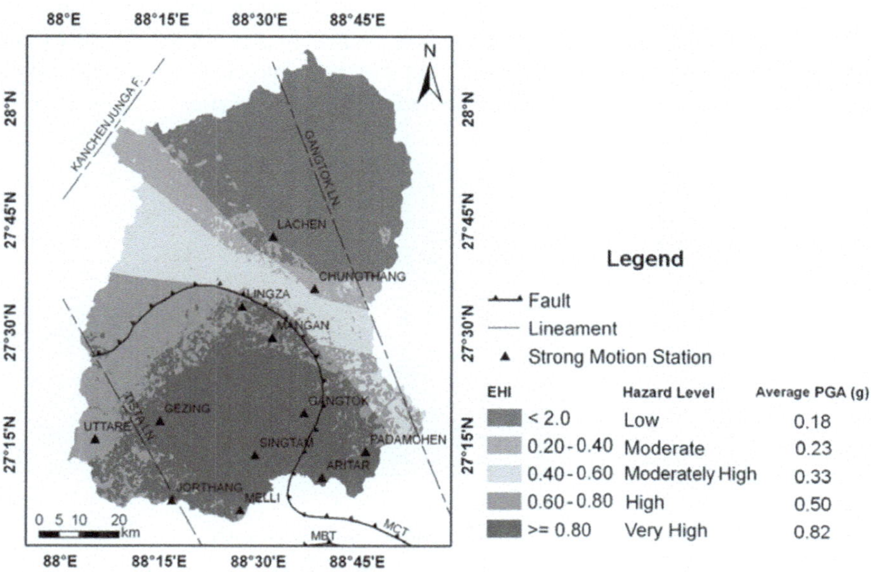

Fig. 8 Hazard Index map of Sikkim, India (adopted from Nath and Thingbaijam (2008))

5.4 Seismic Microzonation of Sikkim

Sikkim has unique location among the tectonically active northeastern states of India associated with seismically active Darjeeling—Sikkim Himalaya. Multi-disciplinary effort resulted in a GIS based seismic hazard map through the integration of litho-unit, soil site class and slope convergence provided the site condition of the Sikkim region on which seismic microzonation map for the region was prepared by taking a series of factors including geology, topology, sub-soil condition, building morphology, earthquake ground motion amplification into account (Fig. 9). The study unravelled the fact that the southern Sikkim has higher seismic risk values in comparison to that of the northern Sikkim, which is mostly associated with competent bedrock while southern part of Sikkim is covered by coarse textured skeletal soils susceptible to water erosion and landslide hazards, representing high site amplification values due to weak geological formation, and equally influenced the seismic risk factor and peak ground acceleration of the area (Nath and Thingbaijam 2008).

Average PGA and resonance or predominant frequency in the Lesser Himalaya is higher than those in the Higher Himalayan crystalline rocks as the northern Sikkim being seismically more stable. Different seismic hazard sub- zones are identified with different hazard index values in which sites in Singtam are found to be associated with maximum seismic hazard probability of experiencing earthquake hazard classified into low, fair, moderate, high and very high hazard potentials as shown in Fig. 9. The maximum risk probability is found to be greater than 78% in the Singtam and adjoining area. It is suggested that the microzonation map of the Sikkim region is very much useful for land use planning for development of seismic hazard mitigation decisions as the areas demonstrates different shaking characteristics and potential during future earthquakes beneath Sikkim and its adjoining region (see Fig. 10).

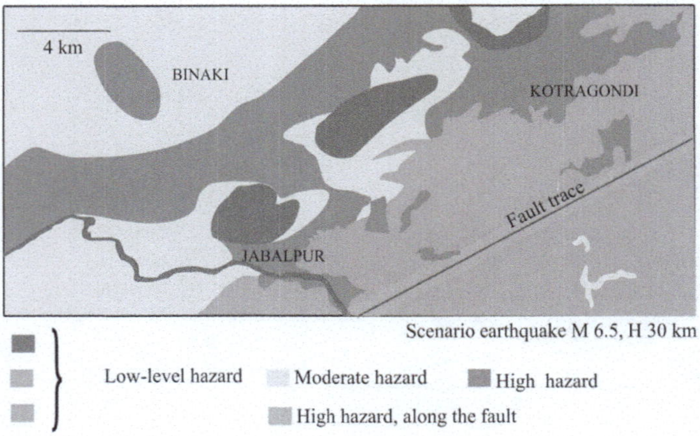

Fig. 9 Seismic Microzonation map for the city of Jabalpur (after PCRSMJUA (2005))

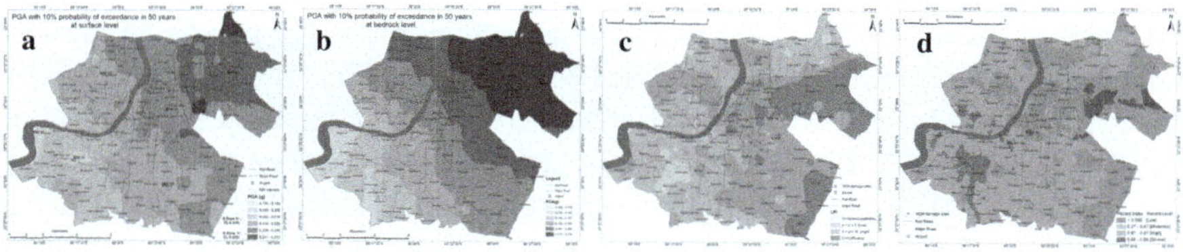

Fig. 10 Showing **a** Distribution of PGA at surface; **b** Distribution PGA at Bedrock level; **c** Liquefaction Potential map; **d** Hazard Index map for the city of Kolkata (adopted from Nath (2016))

5.5 Seismic Microzonation of the City of Jabalpur

Seismic Microzonation of the city of Jabalpur involves detailed characterization of the ground amplifications into 11 types of response curves that help delineating the basement by estimating velocities and thicknesses of the subsurface layers reported by PCRSMJUA (2005). Shear wave velocity (Vs) estimated using MASW shows their variation of 250 m/s for top soil, less than 750 m/s for the weathered rock/saturated soil and Vs > 750 m/s for the engineering bedrock.

Results of numerical modelling and shear wave velocity estimates are found well correlated with the estimated values of site amplification factor and the corresponding predominant frequency for the Jabalpur area. Seismic Microzonation of the area demonstrates that western part of Jabalpur with thick alluvial cover of about 50 m thickness is more vulnerable to earthquake shaking of even moderate sized earthquakes whilst, the Gondwana formations exhibit much lower amplification of less than 3 despite of thick sedimentary thickness of about 30 m. Most interesting observation is found for Mahakoshals, the oldest, metamorphic rocks in the region having the least or negligible amplification and so seismically very safe with minimal loss of property and lives in this sub- zone. The estimated predominant frequencies of amplification in Jabalpur are found to vary in range of 4–5 Hz as the very important information for civil engineering applications.

5.6 Seismic Microzonation of the City of Kolkata

The metro-city of Kolkata is regarded as the socio-political and economic nerve centre of Eastern India and hence, its extensive seismic microzonation provided various thematic maps consisted of PGA at surface; PGA at Bedrock level; Liquefaction Potential map; Hazard Index map for the city of Kolkata (adopted from Nath 2016) as shown in Fig. 11. It is pertinent to note that the extent of seismic hazard is found to be at par of high damage scenario of different cities of India and abroad, including the Ahmedabad city during 2001 Bhuj earthquake; Maxico city during 1985 Mexico earthquake; and of the Northridge city during 1994 Northridge earthquake (Nath et al. 2014; Nath 2016). Kolkata showed probabilistically predicted ground motion at the bedrock level with 50% probability of exceedence as 0.118 g, which is almost similar to the codal value (0.12 g) for zone IV-IS 1893 (I)-2002. The variation of PGA at the surface level is 0.096–0.206 g. Larger PGA has been reported for DumDum; Mall Road; Belgachia, Bhowanipore; Tollyganj and Bangur. The study of seismic microzonation of Kolkata unravelled the fact that Southeast Kolkata is safer during earthquake shaking because of lower estimates of PGA; PGD; PGV along with the lower value of response spectral acceleration. That corresponds to a thick low velocity soil layer at high depth with appreciable impedance contrast of its upper and lower interface. Some locations are found to be associated with high response spectral acceleration in a low time period zone that provides safety to low rise building whilst some locations show lower PGA due to larger amplification in the low frequency zone, which signifies that those locations' response spectral accelerations at larger time period is more.

Hence, in these locations, high rise buildings will be affected so it warrants that high-rise buildings in these locations must be designed taking high spectral acceleration values into account. Thus, seismic microzonation study of the city of Kolkata provides detailed information on the fundamental frequency and amplification factor; PGA; PGV; PGD; maximum response spectral acceleration at different time period bands can be used by different stakeholders that consisted of structural engineers,

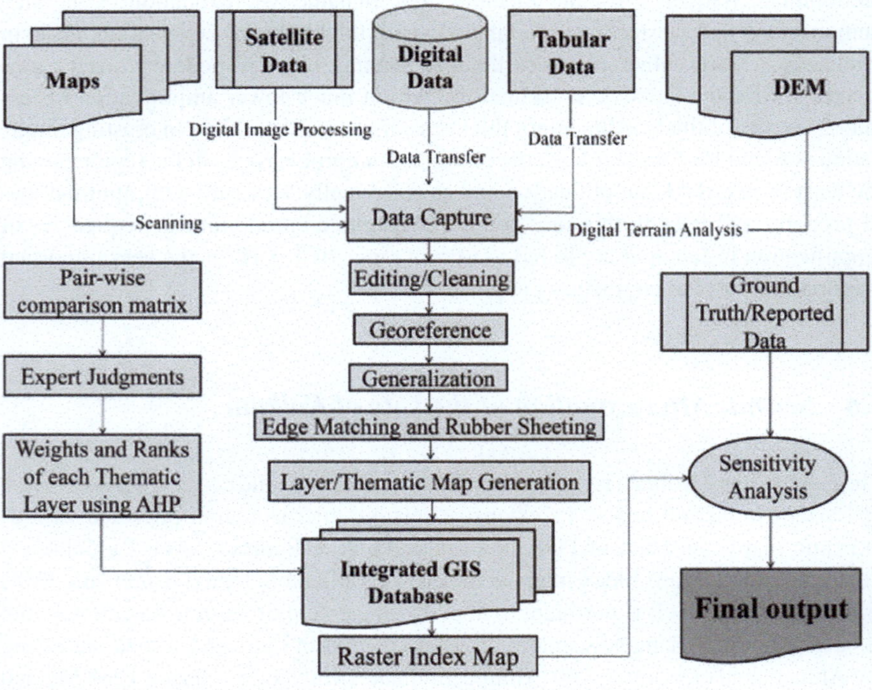

Fig. 11 A schematic process flow in GIS based Thematic Mapping and Spatial analysis (adapted from Nath 2016)

land use planners, private business establishments; emergency managers as well as by general public.

6 Seismic Microzonation: A Way Forward

A detailed and comprehensive assessment of different physical properties of soil and basement rock with their reliability at the finest grid of observational points is very much necessary and the entire data set is required to be processed and interpreted on a GIS-based platform as suggested by Nath (2016) for easy reference and their utility to engineers for constructing earthquake risk resilient structures, which may help develop new building design codes based on seismic microzonation inputs of pertinent information as shown in Figs. 1 and 11. Mishra (2020) advocated that the assessment of the impacts of earthquake shaking at closely spaced observational points is of the utmost importance by conducting geophysical and geotechnical investigations, which has ample scope of extending the mode of investigations of different parameters of seismo-geophysical-geotechnical domains at fine grid spacing under seismic microzonation of the city under the novel concept of seismic nano-zonation, seismic pico-zonation, and seismic femto-zonation (Fig. 12). Such detailed scale of investigation can provide deep insight into the smallest scale anomalies in the subsurface, which are one of the important responsible factors of influencing the extent of earthquake shaking. Use of such detailed information may help to construct the structures not only of earthquake risk resilient but earthquake risk proof using the newest dynamic design codes (Fig. 13), which enhance our capability of detailed investigations to reduce earthquake risks of strategic cities as well as of the smart cities in India.

All the above discussion demonstrates that the impact of earthquake shaking even within the same Seismic Hazard zone, is different. It also signifies that the basic rules of intensity attenuation may not work in the region where the damage pattern does not follow anticipated damage pattern as per the earthquake hazard macro-zones (II–V zones in Fig. 3) and thus the earthquake risk becomes ambiguous. Earthquake Risk is defined as:

$$\begin{aligned}\text{Earthquake Risk} &= \text{Earthquake Hazards} \times \text{Risk Value} \\ &\quad \times \text{Vulnerability} \times 1/\text{Capacity}\end{aligned}$$

In order to assess the above parameters, we have to undertake rigorous seismic microzonation studies so that a comprehensive Earthquake Risk Mitigation model can be evolved and applied for earthquake risk mitigation as shown in Fig. 12.

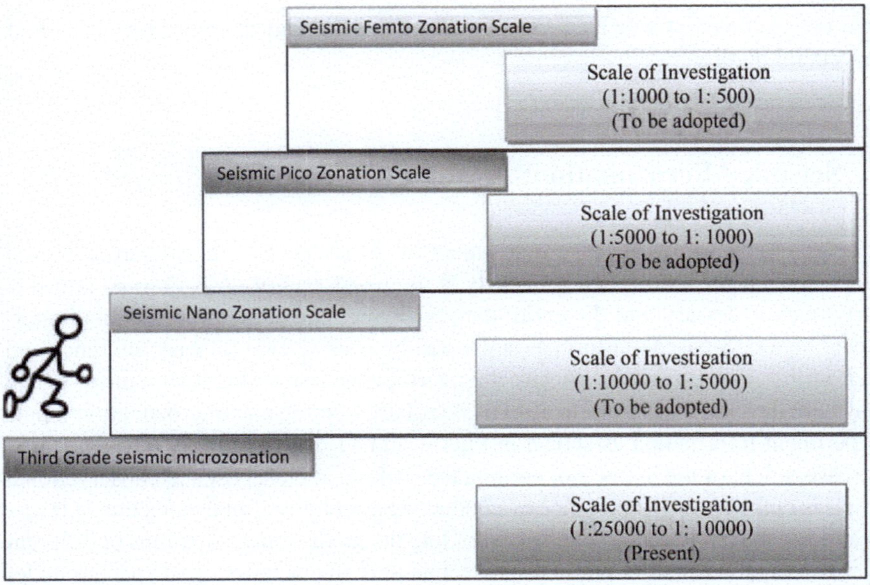

Fig. 12 Advanced Grades of Seismic Zonation for safer and Earthquake Risk Resilient Urban Agglomerates proposed in this study (Modified after SAARC (2011) & adopted from Mishra (2020))

7 Conclusions

Our study emphasizes the efficacy of seismic microzonation that demonstrates how the variability of site characteristics are associated with peak amplification and corresponding predominant frequency, which control the degree of damage to infrastructures. The geo-scientific constitution of the area provides a highly variable domain with a complex scenario which need closer evaluation of Seismic Hazard. Realizing that earthquakes cannot be predicted with the present state of knowledge of science, seismic hazard prediction and seismic hazard analysis are very much required to provide adequate safety measures. The areas located away from any riverine and lacustrine alluvial deposits are relatively safer where losses to structures are comparatively minimal. For several Indian cities and regions, multi-thematic maps pertaining to geo-scientific, geotechnical, and site-response characterizations have been generated and interpreted on thematic maps in the scale of 1:10,000 on GIS platform. The derived comprehensive seismic microzones consisted of peak amplification map with corresponding peak or predominant frequency variation; site vulnerability index; liquefaction potential at varying depth layers to detect and delineate seismically high hazard potential layers with an estimate of the site-specific risk index map. Peak ground acceleration and liquefaction map at varying depths corroborate with the corresponding peak frequency and site amplification distribution map made for few cities of India. The task needs to be completed under the flagship project of Ministry of Earth Sciences, Government of India, to accomplish seismic microzonation of the

Fig. 13 A holistic Cycle model based on this study showing Seismic Risk Resilient strategy for Sustainable development of infrastructures

already proposed 30-Cities as well as of all the smart cities in India in future. The concept of Mishra (2020) for conducting the comprehensive study for the safety of vital infrastructures of Indian cities with finer grid spacing at a varying scale for generating information under the name of seismic nano-zonation; seismic pico-zonation, and seismic femto-zonation that has become very necessary, and which would be supported by the cyclic seismic risk resilient strategy for sustainable development of infrastructures in cities as well as in rural areas.

The role of the National Disaster Management Authority (NDMA), Govt. of India (http://www.ndma.gov.in/images/guidelines/earthquakes.pdf) is paramount, which can steer different institutions for deployment of detailed seismic zonation (on micro-; nano-; pico-; femeto- scale) using the parameters derived from the studies in the structural design code for the city for the safety of structures/infrastructures/flora/faunas with strict implementation of methodologies for sustainable earthquake risk resiliency plan in India.

Acknowledgements This piece of comprehensive interpretation and thorough review of seismic microzonation for few Indian cities has become possible because of acquisition of high-quality of data for those cities on which the final reports are available in the public domain. The authors thank the editors: Dr. M. Rajeevan, Secretary, MoES; and Dr. Vineet K. Gahalaut, Chief Scientist, CSIR-NGRI, Hyderabad, India for providing us opportunity to contribute one of the articles on this pertinent issue to this volume. The authors are grateful to all experts of the subject and contributors for NCT-Delhi (Dr. A. K. Shukla; Dr. H.P. Shukla; Dr. Rajesh Prakash; Shri Dal Singh; Dr. A. P. Pandey; Shri R.K. Singh; Dr. H. S. Sisodia, Prof. M.L. Sharma; and all supportive staffs of IMD, NCS-MoES); for Sikkim and Kolkata (Prof. S. K. Nath and his associates); for Bengaluru (Prof. T.G. Sitharam & Prof. P. Anbazhagan and associates); for Guwahati (Prof. S.K. Nath; Dr. Thingbaijam; Dr. B.K. Bansal); Jabalpur (Dr. P.S. Mishra and other Scientists of GSI; Dr. A.K. Shukla; Dr. M. Ravi Kumar ; Dr. Purnachandra Rao; and other associated Scientists from Government of India) who contributed towards the high quality of report for different Indian cities on sponsorship of Ministry of Earth Sciences & Department of Science and Technology (DST), Government of India. The authors also acknowledge all colleagues and former senior Scientists and the entire team of NCS-MoES for their support and active involvement in stimulating discussions. New and revised Figures are prepared using ArcGIS at the Data Processing and Computational Laboratory of Seismic Hazards Risk Assessment (SHRA), NCS-MoES, New Delhi.

References

Anbazhagan P, Thingbaijam KKS, Nath SK, Narendara Kumar JN, Sitharam TG (2010) Multi-criteria seismic hazard evaluation for Bangalore city, India. J Asian Earth Sci 38:186–198

Ambraseys N, Jackson D (2003) A note on early earthquakes in northern India and southern Tibet. Curr Sci 84(4):571–582

Arya AS (1992) Possible Effects of a Major Earthquake in Kangra Region of Himachal Pradesh. Curr Sci 62:251–256

Bhatia SC, Ravikumar M, Gupta HK (1999) A probabilistic seismic hazard map of India and adjoining regions. Ann Geofis 42(6):1153–1164

Bilham R (1995) Location and Magnitude of the 1833 Nepal Earthquake and its relation to the rupture zones of the contiguous great Himalayan earthquakes. Curr Sci 6992:155–187

Bilham R, Gaur VK, Molnar P (2001) Himalayan seismic hazard. Science 293:1442–4

Bilham R (2004) Earthquakes in India and the Himalaya: Tectonics, Geodesy and History. Ann Geophys 47(2):839–858

Bilham R (2019) Himalayan earthquakes: a review of historical seismicity and early 21st century slip potential. Geol Soc Lond Specl Publ 483(1):423–482

BIS (2002) IS:1893-2002. Criteria for earthquake resistant design of Structures, part 1-general provisions and buildings, bureau of Indian standards, New Delhi

Bolt BA (1987) Earthquakes. W.H. Freeman, New York, p 282

Bruneau M, Stephanie E, Chang RT, Eguchi GC, Lee TD, O'Rourke AM, Reinhorn MS, Tierney K, Wallace WA, von Winterfeldt D (2003) A framework to quantitatively assess and enhance the seismic resilience of communities. Earthq Spectra 19(4):733–752

Champatiray PK, Foreste K, Roy PS (2001). Bhuj earthquake, vol V, no 8. GIS Development

Champatiray PK, Perumal RJG, Thakur VC, Bhat MI, Mallik MA, Singh VK, Lakhera RC (2005) A quick appraisal of ground deformation in Indian region due to the October 8, 2005 earthquake, Muzaffarabad, Pakistan. J ISRS 33(4):465–473

Choudhuri SN, Singh OP, Mishra OP, Kayal JR (2008) Microzonation study from ambient noise measurement for assessing site effects in Krishnagar area and its significance with the damage pattern of Ms 4.3 of the 24th September, 1996 earthquake. Spl Issue Indian Minerals 61(3–4):183–92

Dutta PK, Naskar MK, Mishra OP (2012) Test of strain behavior model with radon anomaly in earthquake prone zones. Himalayan Geol 33(1):23–28
FEMA (2010) Earthquake-resistant design concepts: an introduction to the NEHRP recommended seismic provisions for new buildings and other structures. https://cdn.ymaws.com/www.nibs.org/resource/resmgr/BSSC/FEMA_P-749.pdf
Gahalaut VK (2010) Earthquakes in India: hazards, gensesis and mitigation measures. In: Jha MK (eds) Natural and anthropogenic disasters. Springer, Dordrecht. https://doi.org/10.1007/978-90-481-2498-5_2
Gupta HK (2000) Earthquake Hazard in developing countries and GSHAP (Global Seismic Hazard Assessment Programme). In: Proceedings of the second international workshop on earthquakes and megacities held during December 1–3, 1999 at Makati city, Philippines, pp 1–8
Gupta HK (2020) Recent earthquakes in Delhi and developing an earthquake resilient society. J Geol Soc India 96:107–110. https://doi.org/10.1007/s12594-020-1520-2
Immè G, Morelli D (2012) Radon as earthquake precursor, earthquake research and analysis-statistical studies, observations and planning. In: D'Amico S (ed) Tech Europe, Rijeka, Croatia, 470 pp. https://doi.org/10.5772/2461
Mahajan AK, Slob S, Ranjan R, Sporry R, Champatiray PK, van Westen CV (2007). Seismic microzonation of Dehradun city using geophysical and geotechnical characteristics in the upper 30-meters of soil column. J Seismol 4:355–370
Mandal HS, Shukla AK, Khan PK, Mishra OP (Dec 2013) A new insight into probabilistic seismic hazard analysis of central India 170(12):2139–2161
Mandal HS, Khan PK, Shukla AK (2014) Soil responses near Delhi Ridge and adjacent region in greater Delhi during incidence of a local earthquake. Nat Hazards 70(1):93–118
Mishra OP (2011) Three-dimensional tomography of Northeast India and Indo-Burma region and its implications for earthquake risks. National workshop on earthquake risk mitigation: strategy in NE, pp 40–54
Mishra OP (2012) Seismological research in India. Proc. Indian Natl Sci Acad 78(3):361–371
Mishra OP (2014) Intricacies of the Himalayan seismotectonics and seismogenesis: need for integrated research. Special section: science of the Himalaya, current science, vol 106, no 2
Mishra OP, Ghatak M (2020) SAARC: a regional perspective. Geogr You 20(1–2):80–87
Mishra OP (2020) Seismic microzonation study of South Asian cities and its implications to Urban risk resiliency under Climate Change Scenario. Inter J Geosci (IJG) 11(4). https://doi.org/10.4236/ijg.2020.114012
Mishra OP, Priya Singh B, Ram SK, Gera OP, Singh KK, Mukherjee GK, Chakrabortty SVN, Chandrasekhar A, Selinraj SKS (2020a) Seismic site specific study for seismic microzonation: a way forward for risk resiliency of vital infrastructure in Sikkim, India. Inter J Geosci (IJG) 11(3):125–144. https://doi.org/10.4236/ijg.2020.113008
Mishra OP, Vandana, Kumar V (2020b). A new insight into Seismic Attenuation characteristics of Northwest Himalaya and its surrounding region: implications to structural heterogeneities and earthquake hazards. Phys Earth Plant Inters. https://doi.org/10.1016/j.pepi.2020.106500
Molnar P (1990) A Review of the Seismicity and the rates of active under thrusting and deformation at the Himalaya. J Himal Geol 1:131–154
MoES (2011a) Seismic microzonation manual, 228p. https://moes.gov.in/writereaddata/files/seismic_mcrozonation_manual.pdf
MoES (2011b) Seismic microzonation handbook, 507p. https://moes.gov.in/writereaddata/files/seismic_microzonation_handbook.pdf
Nakamura Y (1989) A method for dynamic characteristics estimation of subsurface using micro tremors on the ground surface. Q Rep Railw Tech Res Inst Tokyo 30(1):25–33
Nath SK (2016) Seismic Hazard, vulnerability and risk microzonation atlas of Kolkata. Geoscience Division, Ministry of Earth Sciences, Government of India, New Delhi.
Nath SK, Das Adhikari M, Maiti SK (2014). Earthquake scenario in West Bengal with emphasis on seismic hazard microzonation of the city of Kolkata, India. Natl Hazards Earth Syst Sci 14(9)

Nath SK, Thingbaijam KKS (2008) Earthquake hazard in Northeast India—a seismic microzonation approach with typical case studies from Sikkim Himalaya and Guwahati city. J Earth Syst Sci

NCS-MoES (2016) A report on seismic microzonation of NCT-Delhi on 1:10,000 scale. National Centre for Seismology, Ministry of Earth Sciences (NCS-MoES), 2016, 387p. https://www.moes.gov.in/writereaddata/files/Delhi_microzonation_report.pdf

NDMA (2007) National disaster management guidelines- management of earthquakes. https://reliefweb.int/report/india/india-national-disaster-management-guidelines-management-earthquakes; http://www.ndma.gov.in/images/guidelines/earthquakes.pdf

Oldham T (1882) A catalogue of Indian earthquakes from the earlies time to the end of A.D. 1869, Memoirs. Geol Surv India 19(3):163–215

Panda D, Kundu B, Gahalaut VK, Rangin C (2020). India-Sunda plate motion, crustal deformation, and seismic hazard in the Indo-Burmese Arc, Tectonics 39:e2019TC006034. https://doi.org/10.1029/2019TC006034

Pande P, Joshi DD, Kandpal GC, Joshi KC, Singh RJ, Singh BK, Singh J, Johri B (2006) Macroseismic studies of 8th October 2005 Kashmir earthquake, 13 see 06, Roorkee

PCRSMJUA (2005) Project completion report of seismic microzonation of Jabalpur urban area, vol 2. Department of Science and Technology, Government of India, India

Paudiyal H, Singh VP (2005) Estimation for the occurrence of moderate size earthquakes in central Himalaya, Nepal. In: Proceedings of symposium on seismic hazard analysis and microzonation, vol I, pp187, IIT Roorkee (UP) India

Rai DC (10 Nov 2000) Future trends in earthquake-resistant design of structures. In: Current science, vol 79, no 9

RFP-MoES (2017) Geotechnical and geophysical investigation of 30- cities under the project "Seismic Microzonation of 30- cities in India. https://www.moes.gov.in/writereaddata/files/RFP_NCS-SHRA_Final_15052017.pdf

Riggio A, Santulin M (2012) Precursors: analysis of the periods preceding the recent earthquakes and problems related to interpretation, Atti 31° Convegno Gruppo Nazionale di Geofisica Terra Solida, Potenza, Italy, pp 356–363

Riggio A, Santulin M (2014) Earthquake forecasting: a review of radon as seismic precursor. Istituto Nazionale di Oceanografia e di GeofisicaSperimentale (OGS), Trieste, Italy, December 22, 2014

SAARC (2009) Workshop on earthquake, risk management in south Asia, 8–9 October 2009, Islamabad, Pakistan

SAARC (2011) Seismic microzonation: methodology for vulnerable cities of South Asian countries. SAARC Disaster Management Centre, New Delhi, India

Sahabi A, Reis E, Khorram D (2018) The case for earthquake resilience: why safer structures protect and promote social and economic vitality. UNISDR, prevention web: the knowledge platform for disaster risk reduction, Feb 23, 2018

Sendai (2015) Sendai framework for disaster risk reduction 2015–2030. https://www.preventionweb.net/files/43291_sendaiframeworkfordrren.pdf

Sharma B, Mishra OP (2020) Simulation of strong ground motion for earthquake (Mw 7.0) beneath the Bhutan Himalaya and its implications to trans-boundary impacts for North-East India. Phys Earth Plant Inters 106603. https://doi.org/10.1016/j.pepi.2020.106603

Sharma B, Sharma V, Kumar V, Prashanta, Mishra OP (2016) Characteristic ground motions of the 25th April 2015 Nepal earthquake (Mw 7.9) and its implications for the structural design codes for the Border Areas of India to Nepal. J Asian Earth Sci. https://doi.org/10.1016/j.jseaes.2016.07.021

Singh P, Srivastava RK, Mall RK, Mishra OP (2017) A lesson from 2004 Tsunamigenic Sumatra-Andaman earthquake and its damage scenario of Andaman-Nicobar Islands: future strategy for sustainable Coastal Risk Management for safer cities under climate change scenario. Int J Adv Res (IJAR) 6(4):822–830

Srivastava RK, Mall RK, Mishra OP, Singh N, Banerjee T (2017) Integrating Smart city concept in planning and management of heritage city for Disaster Risk Management: a case study of Varanasi. Int J Manag Res (IJMR) 5(1):45–55

Valdiya KS (2010) The making of india geodynamic evolution. Macmillan Publ, 816p

Vandana, Mishra OP (2019) Source characteristics of the NW Himalaya and its adjoining region: geodynamical implications. Phys Earth Planet Inter 294:106277. https://doi.org/10.1016/j.pepi.2019.106277

Yadav RK, Gahlaut VK, Bansal AK, Sati SP, Catherine J, Gautam P, Kumar K, Rana N (2019) Strong seismic coupling underneath Garhwal-Kumaun region, NW Himalaya, India. Earth Planet Sci Lett 506:8–14. https://doi.org/10.1016/j.epsl.2018.10.023

Zhang P, Yang ZX, Gupta HK, Bhatia SC, Shedlock KM (1999) Global Seismic Hazard Assessment Program (GSHAP) in continental Asia. Annali di Geofisica

Zhao D, Mishra OP, Sanda R (2002) Influence of fluid and magma on earthquakes: seismological evidence. Phys Earth Planet Inter 132(249–267):2002

Zhao D, Tani H, Mishra OP (2004) Crustal heterogeneity in the 2000 western Tottori earthquake region: effect of fluids from slab dehydration. Phys Earth Planet Inter 145:161–177

Earthquake Monitoring in India by National Center for Seismology, India

Vineet K. Gahalaut

Abstract The Indian national seismological network currently has 150 observatories which are equipped with real-time data connectivity feeding data to an auto-location software at the headquarters of the National Center for Seismology (NCS), New Delhi. The earthquake information is disseminated to the public and all stakeholders in real-time, and the data are supplied to all scientists for their research purposes. The past five years witnessed two strong/moderate magnitude earthquakes, namely, the 2016 Tamenglong and 2017 Tripura earthquakes, hydroseismicity in the Amravati district, and an earthquake swarm in the Palghar district of Maharashtra. Besides these, several small-magnitude earthquakes occurred in the Delhi region in 2020. In all these cases, NCS personnel analyzed the seismicity data and established additional temporary stations to closely monitor earthquake swarms. Quick dissemination of earthquake information, understanding of earthquake processes in various tectonic regions, and assessment of hazard potential of strong earthquakes, etc., helped in allaying the panic in the general public and enabled the government authorities to take quick decisions on relief and rescue operations.

Keywords Earthquakes · Seismology · Earthquake monitoring

1 Introduction

Among all the geological hazards, earthquakes cause the most damage and are responsible for the maximum loss of lives, particularly in Asian countries. Earthquakes in the Indian subcontinent occur due to the northeastward motion of the Indian plate at a rate of ~5 cm/year, and its interaction with the Eurasian plate along the Himalayan arc and with the Sunda plate along the Indo-Burmese and Andaman Sumatra arcs. Besides these two major plate boundary regions, earthquakes occur within the Indian plate as a result of intraplate deformation. The ground shaking due

V. K. Gahalaut (✉)
CSIR-National Geophysical Research Institute, Hyderabad, India
e-mail: vkgahalaut@ngri.res.in

Earlier at National Center for Seismology, Ministry of Earth Sciences, New Delhi, India

V. K. Gahalaut and M. Rajeevan (eds.), *Social and Economic Impact of Earth Sciences*,
https://doi.org/10.1007/978-981-19-6929-4_21

to an earthquake poses a hazard. A building will survive if it can withstand ground shaking. The impact of ground shaking may cause partial/total collapse of built environments, cracks on the ground, soil liquefaction and landslides, etc. depending on the ground conditions and soil characteristics. So, the collapse of buildings is the primary cause of injuries and deaths of human and animal lives, though other processes may also result in loss of lives. Thus it is not the earthquakes which kill people, it is the structures, particularly those that are poorly built. A list of damaging earthquakes in India is given in Table 1.

Since the prediction of earthquakes is not possible and they cannot be stopped or prevented from the occurrence, our improved understanding through earthquake monitoring, seismological research, and hazard assessment has brought us to a stage where we can live relatively safely with earthquakes. Earthquake monitoring is the most important component of such studies since it provides the basic data for research toward the mitigation of losses.

Understanding earthquake processes, the structure of the Earth's interior, and related studies require earthquake data from all over the globe. This was understood

Table 1 Earthquakes in India and adjoining regions with large magnitude and/or large fatalities

Date	Location	Mag	Deaths
2016-01-04	Tamenglong, Manipur	6.7 M_W	8
2015-04-25	Gorkha, Nepal	7.8 M_W	8,964
2005-10-08	Kashmir	7.6 M_W	86,000–87,351
2004-12-26	Off northern Sumatra and Andaman	9.1–9.3 M_W	227,898
2001-01-26	Bhuj, Gujarat	7.7 M_W	13,805–20,023
1999-03-29	Chamoli, Uttarakhand	6.8 M_W	~103
1993-09-30	Latur, Maharashtra	6.2 M_W	9,748
1991-10-20	Uttarkashi, Uttarakhand	6.8 M_W	768–2,000
1988-08-21	Udayapur, Nepal	6.9 M_W	709–1,450
1967-12-11	Koyna, Maharashtra	6.6 M_W	177–180
1950-08-15	Assam, Tibet	8.6 M_W	1,500–3,300
1941-06-26	Andaman Islands	7.7–8.1 M_W	8,000
1935-05-31	Quetta, Baluchistan	7.7 M_W	30,000–60,000
1934-01-15	Nepal-Bihar	8.0 M_W	6,000–10,700
1905-04-04	Kangra	7.8 M_S	>20,000
1897-06-12	Shillong, India	8.0 M_W	1,542
1881-12-31	Nicobar Islands	7.9 M_W	
1869-01-10	Assam, Cachar	7.4 M	2
1833-08-26	Kathmandu, Nepal	8.0 M_S	
1819-06-16	Allah Bund, Kachchh, Gujarat	7.7–8.2 M_W	>1,543
1505-06-06	Saldang, Karnali zone (S Tibet, Nepal)	8.2–8.8	6,000

by the world community and several collective programs for global earthquake monitoring were initiated. It all started with the setting up of a World Wide Standardized Seismic Network (WWSSN) in 1960 by the U. S. Geological Survey. Today, we have two Global Seismic Networks (GSN). One, the GSN itself which is established and maintained by the United States Geological Survey (USGS), and the other, GEOFON, operated by the GeoForschungsZentrum (GFZ), Potsdam, Germany. A more detailed description may be found in Prakash et al. (2020).

2 Earthquake Monitoring in India

The National Centre for Seismology (NCS) as an attached office of the Ministry of Earth Sciences, is the nodal agency of the Government of India for earthquake monitoring and related matters in the country. It was established in 2014 by reorganizing the seismological activities of the India Meteorological Department (IMD), the then nodal agency of the Government of India and the Ministry of Earth Sciences.

Prior to 1990, the national/Indian seismological network consisted of the conventional type of seismograph systems viz., analog type (photographic/smoke/ink) recording systems with a short period and long period sensors to record the near and distant events separately. The development of earthquake monitoring network/system in India may be categorized into four phases: The pre-instrumental era (prior to 1900), Early-instrumental era (1900–1963), Modern-instrumental era (1964–1995), and Digital-instrumental era (1996 onwards). The first observatory was set up in 1898 at Alipore in Kolkata and by 2011, the National Seismological Network (NSN) consisted of 82 seismological observatories. The present network (as of 2020) comprises 115 state-of-art seismic observatories with communication and auto-location facilities. The ground motion data are processed in real time with set parameters for triggering/assessment of the occurrence of an earthquake followed by the determination of earthquake parameters (epicenter, focal depth, magnitude, focal mechanism), which are displayed on a computer screen. The earthquake information is disseminated to the designated destinations. The earthquake parameters are assessed by a scientist on operation duty and the final earthquake information is disseminated for follow-up actions. The response time which was about an hour before 1996, which got reduced to less than 10 min now and the earthquake location capability has increased significantly.

The event database is compiled on a monthly basis, as ‘Monthly Seismological Bulletins’. These bulletins contain information about the arrival times of different phases at the stations, root mean square error in the earthquake parameters. These bulletins are archived in the NCS database and are also sent to International Seismological Center (ISC), UK, on a regular basis as per a fixed schedule for their incorporation in the ISC’s Seismological Bulletins, containing data of all global stations. These data may be accessed at www.isc.ac.uk.

Realizing the necessity to further upscale this network, another 30 stations are being added and it is expected that by the end of 2021, the strength of the national network will be 145. However, even with this number, it is not possible to estimate all the hypocentral parameters (particularly the focal depth) reliably. At some places, particularly in the Himalayas, the network appears to be sparse and is not sufficient for reliably estimating the hypocentral parameters. It is also desirable to integrate seismic networks operated by other institutes in India into the national network for operational purposes.

Although it is imperative to strengthen the network in a phased manner, it is not very straightforward to pinpoint the ideal number of stations network. However, it is intuitive to add at least 40–50 stations every year to be able to locate earthquakes of magnitude more than 2.5 in all regions of India, irrespective of their tectonic setting. Such strengthening of the network will not only meet the operational requirements of NCS, but will also serve the research community in understanding the earthquake processes, deciphering earth structure, and assessing seismic hazards in a more objective manner (see Fig. 1).

3 Other Significant Activities of NCS in the Past years

3.1 Shaking in Delhi During the October 26, 2015, Hindukush Earthquake

It has been observed in several cases that the shaking due to the earthquakes in the Hindukush Pamir region is felt quite strongly in the Delhi region. The October 26, 2015, Hindukush earthquake with an Mw of 7.5 was one such earthquake which caused panic in the Delhi region due to shaking. Delhi is located about 1100 km southeast of the epicenter. Surprisingly, the 2015 Mw 7.8 Gorkha earthquake which is ~800 km east of Delhi, did not cause that much shaking in the same region. The researchers of NCS analyzed the data from the two earthquakes recorded at the same stations in Delhi. It was found that the shaking due to earthquakes in Hindukush is felt after ~300 s of their origin time when the surface waves arrive at Delhi. The shaking due to these earthquakes from the Hindukush region is attributed to (i) deeper depth, and (ii) azimuth of Delhi with respect to the fault/rupture plane of the Hindukush earthquakes. In case of the Gorkha earthquake, the Delhi region is located close to the nodal plane, implying relatively low amplitude body waves. Also, the rupture directivity towards the east in the Gorkha earthquake caused lower shaking in the Delhi region. Because of these two reasons, even the higher magnitude Himalayan earthquakes, e.g., the 2015 Gorkha earthquake, are not felt that strongly.

It is interesting to note here that auto-location software being used by NCS detects strong magnitude earthquakes from the Hindukush Pamir region before the shaking in Delhi is experienced. Also, the shaking (long period/low frequency) experienced during these earthquakes is distinct from that experienced in local earthquakes. This information dissemination helps in reducing panic among the public.

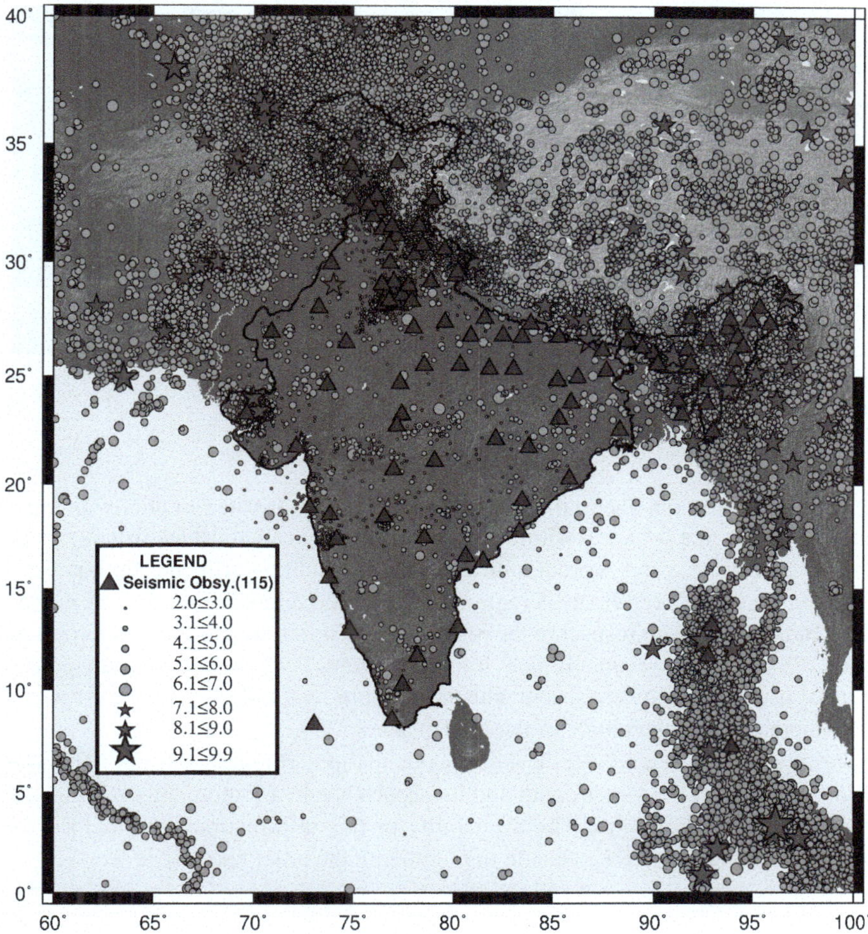

Fig. 1 Earthquakes in and around India since 1505 until 2018, along with the stations of the national seismological network (blue traingles)

3.2 *Seismological Investigations on the 2016 Mw 6.7 Tamenglong (Manipur) Earthquake*

The Mw 6.7 Tamenglong earthquake which occurred in the state of Manipur in north-eastern India on 4 January 2016 with a focal depth of 59 km, was the largest earthquake in the region in the past few decades (Gahalaut et al. 2016). Although the large magnitude earthquake caused anxiety amongst the people, due to the very low population density in the epicentral region, it did not cause any serious damage there (Fig. 2). However, in the state capital, Imphal, which is ~30 km east of the earthquake epicentre, it caused some damage to buildings and to the community markets. The

Fig. 2 A tilted hut at Kabuikhullen village in the epicentral region of the 2016 Tamenglong earthquake. Right panel shows a meeting and counselling of the villagers

total lives lost in India due to this earthquake was eight. It was essentially an intra-slab earthquake that occurred within the Indian plate beneath the thick frontal wedge of the Indo-Burmese arc. A predominant strike-slip motion during the earthquake is consistent with the overall stress regime and with the dominant northward motion of the Indian plate with respect to the Sunda plate. The strong motion accelerograph data, GPS-derived coseismic offsets from the region, field surveys on damage and intensity assessment, are consistent with these results, particularly with the estimated focal depth and the magnitude of the earthquake.

Despite its high magnitude, the damage due to this earthquake was less. This was mainly because of the larger depth and its occurrence in a remote hilly region with very light weight buildings. The availability of this information again helped the government to take a quick decision in mobilizing the relief and rescue work.

3.3 Seismological Investigations of the 2017 Mw 5.6 Manu, Tripura Earthquake

On 3 January 2017, an Mw 5.6 earthquake occurred in the Dhalai district of Tripura (India) with its epicentre at 24.018°N and 91.964°E, and a focal depth of 31 $\pm$ 6 km (Debbarma et al. 2017). The focal mechanism solution determined using data from seismological observatories in India indicated a predominantly strike-slip motion on a steeply dipping plane. The estimated focal depth and focal mechanism solution place this earthquake within the Indian plate that lies beneath the overlying Indo-Burmese wedge. As in the 2016 Manipur earthquake, a strong motion record from Shillong, India, appears to suggest site amplification possibly due to topographic effects. In the epicentral region in Tripura, the damage assessed from a field survey and from media reports indicated that the macroseismic intensity approached 6–7 on European Macroseismic Scale (EMS) with damage also reported in the adjacent

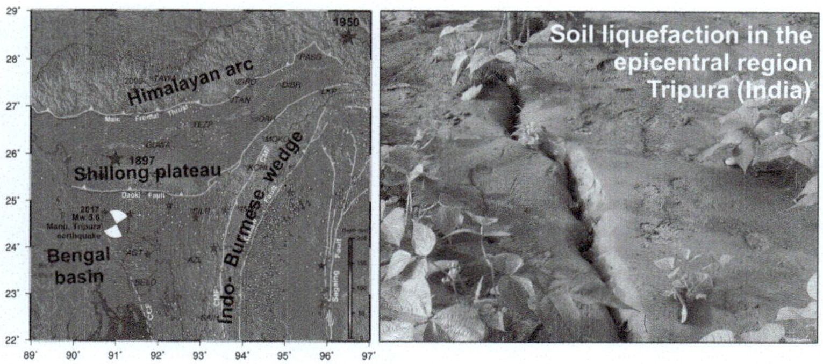

Fig. 3 The epicentre and focal mechanism of the 2017 Manu earthquake are shown in the left panel. Soil liquefaction and sand blows (24.1202°N and 91.9926°E) in the epicentral region (right panel) of the earthquake

parts of Bangladesh. A striking feature of this earthquake was the numerous reports of liquefaction that were forthcoming from fluvial locales in the epicentral region and at anomalous distances farther north in Bangladesh. The occurrence of the 2017 Manu earthquake emphasizes the hazard posed by intraplate earthquakes in Tripura and in the neighboring Bengal basin region where records of past earthquakes are scanty or vague, and where the presence of unconsolidated deltaic sediments and poor implementation of building codes pose a significant societal and economic threat during larger earthquakes in the future.

Quick information about the depth of the earthquake and its moderate magnitude helped the central as well as the local government to take quick decisions in mobilizing the relief and rescue work (see Fig. 3).

3.4 Palghar Earthquake Swarm Modulated by Rainfall

A low-magnitude earthquake swarm started in the western coastal region of the stable continental region of India in November 2018 (and continuing even now, i.e., December 2020, though with low earthquake frequency) (Sharma et al. 2020). The swarm started much after the monsoon season and was on a decline in May–June 2019. However, the monsoonal rainfall in June 2019 again led to an increase in the earthquake frequency. A network of 4 broadband seismometers was installed in January 2019. Until July 2019 it had recorded more than 16,000 earthquakes in the magnitude range of −0.5 to 3.8. All the earthquakes are tightly clustered in a region of 10×6 km^2 and occur through normal slip on a steep north–south oriented east dipping fault which extends down to a depth of 6–7 km only. The occurrence of tightly clustered shallow focus earthquakes causing subsidence due to normal slip in an overall compressive regime of the stable part of the Indian plate interior, implies

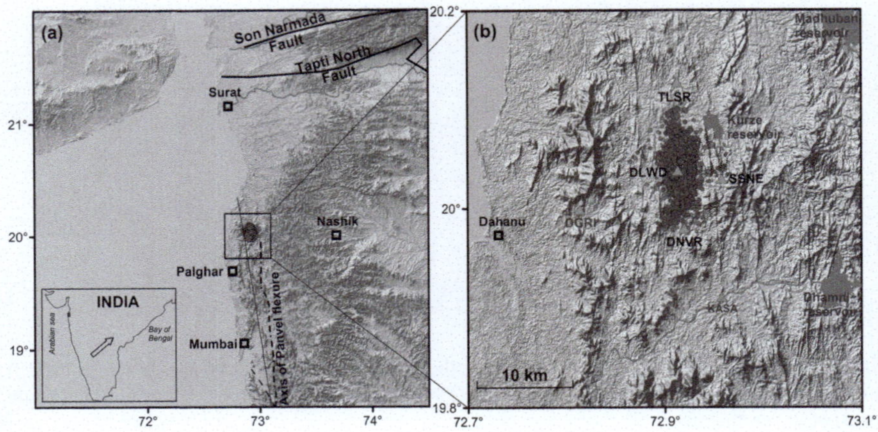

Fig. 4. **a** Location of the Palghar swarm with its location on India map (shown in the inset). **b** Earthquakes of Palghar swarm (filled circles) that occurred during January 26, 2019 to May 30, 2019, shown along with the seismic network (green-filled triangles). Nearby reservoirs are also shown. Several north–south oriented lineaments can be seen on the SRTM topographic image

some surficial process of water infiltration causing the collapse of cavities and may not be linked with the tectonics of the region.

This swarm created lots of panic amongst the local residents and also alerted the local authorities. However, once it was understood that this earthquake sequence is mainly of swarm type and may not lead to a strong or even a moderate earthquake, the local public and administration were relieved. In fact, the low-magnitude swarm activity is still continuing, but the local residents have become used to it and do not panic anymore (see Fig. 4).

3.5 Hydroseismicity in the Amravati District, Maharashtra, Central India

The highly clustered shallow (<0.4 km) earthquake activity of low magnitude with accompanying rumbling sound in Sadrabadi and Zilphi villages in Dharni Taluka of Amravati district, Maharashtra during the monsoon period of 2018 is reported through the installation of a small aperture network of four stations (Wadhawan et al., 2020). The earthquake sequence was found to have characteristics of a swarm. Its close association with the monsoon implies that the activity was probably triggered by the rainfall and may be categorized as hydro-seismicity, resulting from hydro-fracturing of the soil/weathered basalt and collapsing and caving of the rocks. In the past, no such activity has been reported from the region during or after the monsoon, despite the fact that there was more rainfall in 2019. We suggest that the low-magnitude earthquake swarm at a very shallow depth might have been induced by the percolation of monsoonal rainwater through the weathered and fractured rock mass associated with the fault system of the Narmada Son failed rift region. The feature of the swarm and its association with the monsoonal rainfall allowed us to propose that it may not lead to a large-magnitude earthquake. This inference contributed significantly in reducing panic among the residents of the local villages. At the same time, it enhanced our understanding about the processes of hydro-fracturing and added a new monsoonal rainfall-induced earthquake swarm in the list of fluid-induced earthquakes, especially in a stable continental region of the world.

This short-duration low magnitude swarm created some panic amongst the local residents and also alerted the local authorities. However, once it was understood that this earthquake sequence is mainly of swarm type, rainfall-induced and extremely shallow, and thus may not lead to a strong or even a moderate earthquake, the local public and administration were relieved (see Fig. 5).

3.6 Low-Magnitude Earthquakes in and Around Delhi Region in 2020

A low-magnitude earthquake sequence started in and around the Delhi region in April 2020 with reduced activity after August 2020. Over 30 earthquakes ($ML > 1.8$) were recorded between April and August by the network of NCS. Many of these shocks were also felt in the national capital and added to an already elevated sense of public angst during a strict nationwide lockdown in the early months of the Covid-19 pandemic in 2020. Although the precise mechanism of earthquakes in the Aravalli Delhi fold belt (ADFB) is not known, they appear to occur due to the interaction of the Marwar and Bundelkhand cratons along the ADFB. The frequency of their occurrence appears to be modulated by the seasonal rainfall. The mechanism that governs earthquake genesis in this intraplate region is a matter of debate warranting future investigations (Yadav et al. 2022).

The earthquake sequence in Delhi created panic amongst the local residents, media and also alerted the government authorities. Good knowledge about the tectonics of the region, occurrence pattern, and history of earthquakes in the Delhi-Aravalli region helped in convincing the public and authorities that such small magnitude earthquakes do occur in the Delhi region. Also, it has been communicated that there have been several occasions in the past when several earthquakes occurred, closely clustered in space and time, and none of them have led to the occurrence of a strong earthquake. Hence, even this sequence may not lead to any strong earthquake and the sequence may subside with time after the monsoon. Indeed, that was the case for the Delhi earthquake sequence in 2020 (see Fig. 6).

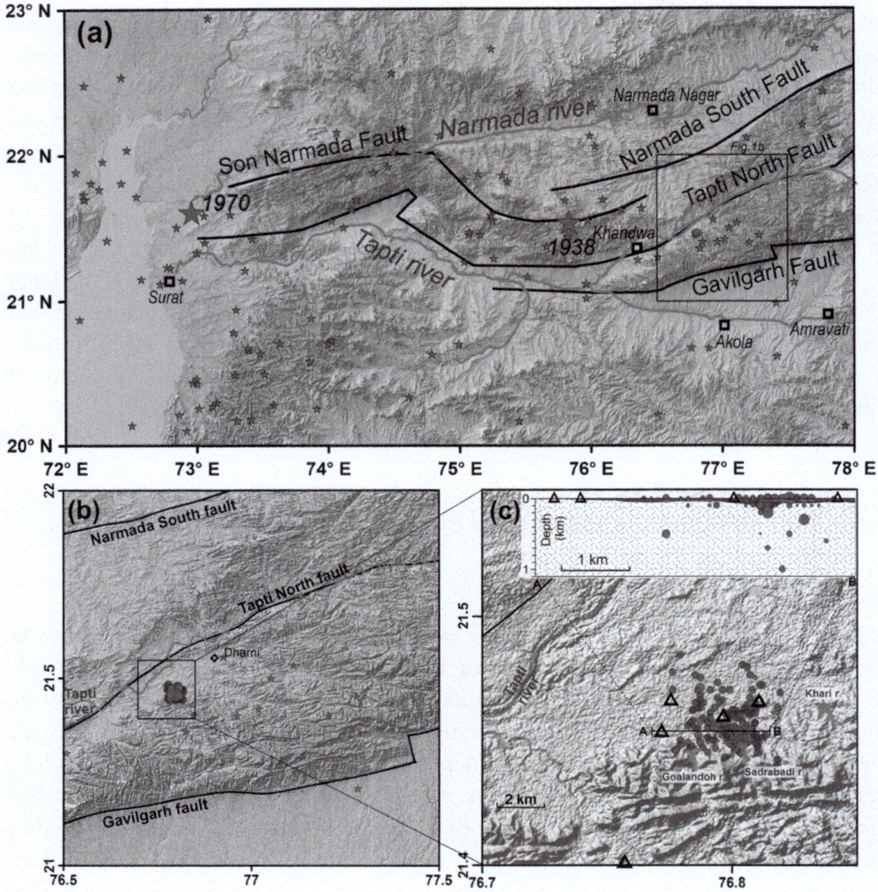

Fig. 5 **a** Broad features of the western and central parts of the Narmada Son failed rift region with earthquakes during 1967–2018 from the ISC catalogue (which also includes events located by IMD/NCS). The epicentres of 1970 Bharuch and 1938 Khandwa earthquakes are also shown. **b** Enlarged view of the region within the square in **a** along with the earthquakes recorded locally. **c** Further enlarged view of the region shown in **b**. Earthquakes in the Sadrabadi Zilphi villages of Dharni Taluka of Amravati district, Maharashtra during September 1, 2018 to December 9, 2018 located using data from local seismological stations. Solid red colour circles are the best locations (error not more than 0.5 km, local magnitude ranges between 0.1 and 2.3) estimated from five stations. Other locations (magenta circles) are derived from three stations and their locations may not be as accurate (error of up to 1 km) as those of the events shown as red circles. Five black triangles are the seismic stations. Small reservoirs (r) in the region are also shown. 1 arc second SRTM Topography data are used in all these maps. The inset on top of Figure c shows a west to east vertical depth section, depicting the focal depths of the earthquakes. Schematically, it also shows the soil cover (a few meter thick) and weathered basalt of a few tens of m thickness, which could be deeper at the center of the earthquake cluster (after Wadhawan et al. 2020)

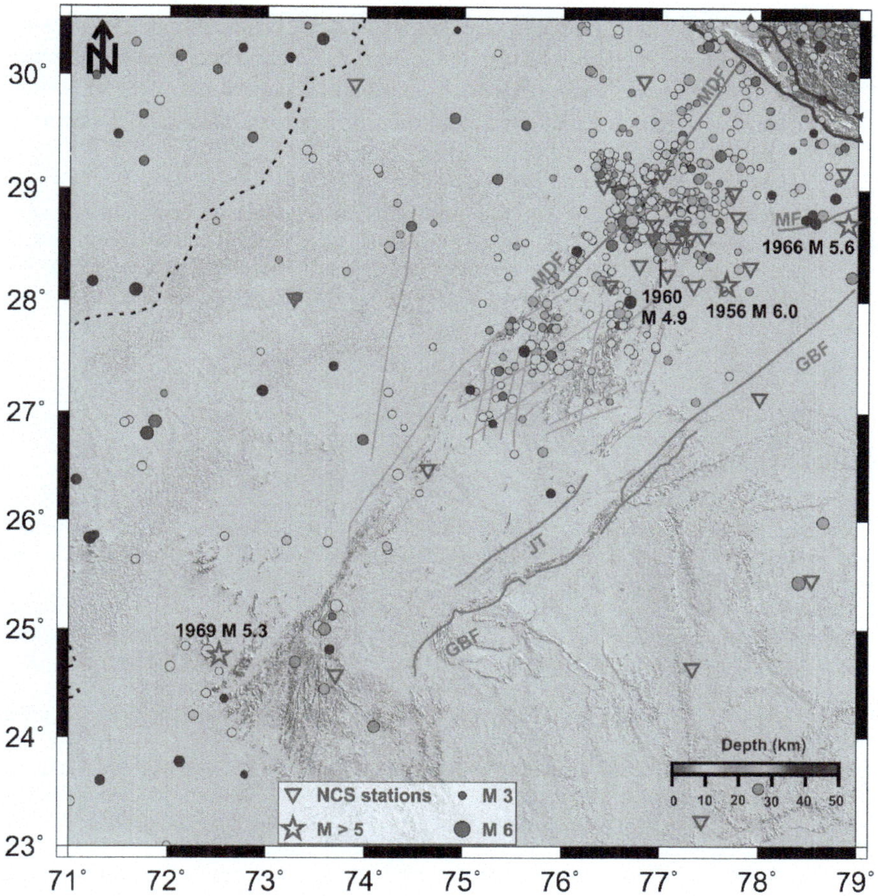

Fig. 6 Epicentral distribution of earthquakes (M ≥ 2.5) in the ADFB since 1966 (*Source* NCS/IMD). Current locations of the national seismological stations of the NCS are also shown (inverted red triangles). MDF-Mahendragarh-Dehradun Fault; MF- Moradabad Fault; GBF-Great Boundary Fault; JT- Jahazpur Thrust

4 Conclusions

The upgradation of broadband digital seismograph stations with VSAT communication system, strengthening of the National Seismic Network, and state-of-art real-time earthquake monitoring in and around the country have improved the operational capabilities of the NCS. Currently, there are 150 stations of the NSN. The NSN is being strengthened with about 50 more stations to further improve the location and uniform detection capability down to magnitude 3.0.

A dense seismic network with quick dissemination of earthquake information, good knowledge of earthquake occurrence in various tectonic domains, and quick assessment of hazard potential due to strong earthquakes (which basically depends upon the earthquake parameters, local soil properties, population density, and building typology) help in assessing the seismic hazard and deciding the quantum of rescue and relief work in case of a strong earthquake occurrence. The NCS used this information and demonstrated its capabilities during earthquakes which occurred in the past five years.

Acknowledgements I am thankful to the Ministry of Earth Sciences where i served as the Director of the new institute, NCS, during 2015–2019. The data collected and reported here is the collective effort of all the employees of NCS. I also benefitted greatly from the paper of Prakash et al. (2020). I am thankful to Drs M Ravikumar and Sumer Chopra for reviewing the ms and providing constructive comments.

References

Debbarma J, Martin SS, Suresh G, Ahsan A, Gahalaut VK (2017) Preliminary observations from the 3 January 2017, MW 5.6 Manu, Tripura, (India) earthquake. J Asian Earth Sci 148:173–180. https://doi.org/10.1016/j.jseaes.2017.08.030

Gahalaut VK, Stacey Martin D, Srinagesh SL, Kapil GS, Saikia S, Kumar V, Dadhich H, Prajapati SK, Gautam JL, Baidya P, Mandal S, Jain A (2016) Constraints from seismological, geodetic and macroseismic observations on the 4 January 2016 Tamenglong, Manipur earthquake. Tectonophysics 688:36–48

Prakash R, Suresh G, Gahalaut VK (2020) Earthquake monitoring in India. Proc Indian Natn Sci Acad 86(1):631–642

Sharma V, Wadhawan M, Naresh Rana KM, Sreejith RA, Charu Kamra KS, Hosalikar KV, Narkhede GS, Gahalaut VK (2020) A long duration non-volcanic earthquake sequence in the stable continental region of India: the Palghar swarm. Tectonophysics. https://doi.org/10.1016/j.tecto.2020.228376

Wadhawan M, Rana N, Gahalaut V, Singh M, Singh K, Suresh G, Mishra OP, Joshi AK, Kulkarni AV, Singh M, Das AK (2020) Monsoonal rainfall induced shallow earthquake Swarm in the Amravati district of the central India. J Earth Syst Sci 130(1):1–7https://doi.org/10.1007/s12040-020-01511-z

Yadav RK, Martin SS, Gahalaut VK (2022) Intraplate seismicity and earthquake hazard in the Aravalli–Delhi fold belt, India. J Earth Syst Sci 131:204. https://doi.org/10.1007/s12040-022-01957-3

The manufacturer's authorised representative in the EU is Springer
Nature Customer Service Centre GmbH, Europaplatz 3, 69115 Heidelberg,
Germany. If you have any concerns regarding our products, please
contact ProductSafety@springernature.com

Printed and bound by CPI Group (UK) Ltd, Croydon, CR0 4YY
30/09/2026
02225249-0003